AF609334

Infosys Science Foundation Series

Infosys Science Foundation Series in Mathematical Sciences

Series Editors

Gopal Prasad, University of Michigan, Ann Arbor, USA

Irene Fonseca, Carnegie Mellon University, Pittsburgh, PA, USA

Editorial Board

Chandrashekhar Khare, University of California, Los Angeles, USA

Mahan Mj, Tata Institute of Fundamental Research, Mumbai, India

Manindra Agrawal, Indian Institute of Technology Kanpur, Kanpur, India

Ritabrata Munshi, Tata Institute of Fundamental Research, Mumbai, India

S. R. S. Varadhan, New York University, New York, USA

Weinan E, Princeton University, Princeton, USA

The **Infosys Science Foundation Series in Mathematical Sciences,** a Scopus-indexed book series, is a sub-series of the *Infosys Science Foundation Series.* This sub-series focuses on high-quality content in the domain of mathematical sciences and various disciplines of mathematics, statistics, bio-mathematics, financial mathematics, applied mathematics, operations research, applied statistics and computer science. With this series, Springer and the Infosys Science Foundation hope to provide readers with monographs, handbooks, professional books, edited volume, and textbooks of the highest academic quality on current topics in relevant disciplines. Literature in this sub-series will appeal to a wide audience of researchers, students, educators, and professionals across mathematics, applied mathematics, statistics and computer science disciplines.

More information about this subseries at https://link.springer.com/bookseries/13817

Debargha Banerjee · Kiran S. Kedlaya ·
Ehud de Shalit · Chitrabhanu Chaudhuri
Editors

Perfectoid Spaces

Editors
Debargha Banerjee
Department of Mathematics
Indian Institute of Science Education and Research
Pune, Maharashtra, India

Ehud de Shalit
Institute of Mathematics
Hebrew University
Jerusalem, Israel

Kiran S. Kedlaya
Department of Mathematics
University of California San Diego
La Jolla, CA, USA

Chitrabhanu Chaudhuri
Department of Mathematics
National Institute of Science Education
Jatni, Odisha, India

ISSN 2363-6149 ISSN 2363-6157 (electronic)
Infosys Science Foundation Series
ISSN 2364-4036 ISSN 2364-4044 (electronic)
Infosys Science Foundation Series in Mathematical Sciences
ISBN 978-981-16-7123-4 ISBN 978-981-16-7121-0 (eBook)
https://doi.org/10.1007/978-981-16-7121-0

Mathematics Subject Classification: 11F72, 14G40, 11F37, 11F03, 11G50, 11F70, 11F80

© The Editor(s) (if applicable) and The Author(s), under exclusive license to Springer Nature Singapore Pte Ltd. 2022
This work is subject to copyright. All rights are solely and exclusively licensed by the Publisher, whether the whole or part of the material is concerned, specifically the rights of translation, reprinting, reuse of illustrations, recitation, broadcasting, reproduction on microfilms or in any other physical way, and transmission or information storage and retrieval, electronic adaptation, computer software, or by similar or dissimilar methodology now known or hereafter developed.
The use of general descriptive names, registered names, trademarks, service marks, etc. in this publication does not imply, even in the absence of a specific statement, that such names are exempt from the relevant protective laws and regulations and therefore free for general use.
The publisher, the authors and the editors are safe to assume that the advice and information in this book are believed to be true and accurate at the date of publication. Neither the publisher nor the authors or the editors give a warranty, expressed or implied, with respect to the material contained herein or for any errors or omissions that may have been made. The publisher remains neutral with regard to jurisdictional claims in published maps and institutional affiliations.

This Springer imprint is published by the registered company Springer Nature Singapore Pte Ltd.
The registered company address is: 152 Beach Road, #21-01/04 Gateway East, Singapore 189721, Singapore

Preface

This book contains selected chapters and the introduction as well as several applications of perfectoid spaces, as invented by Peter Scholze in his Fields medal winning work. These papers are presented at the conference on "Perfectoid Spaces" held at the International Centre for Theoretical Sciences, Bengaluru, India, from 9 to 20 September 2019.

p-adic methods play a key role in the study of arithmetic properties of modular forms. This theme takes its origins in Ramanujan congruences between the Fourier coefficients of the unique eigenform of weight 12 and the Eisenstein series of the same weight modulo the numerator of the Bernoulli number B12. After the work of Deligne on Ramanujan's conjecture, it became clear that congruences between modular forms reflect deep properties of corresponding p-adic representations. The general framework for the study of congruences between modular forms is provided by the theory of p-adic modular forms developed in fundamental papers of Serre, Katz, Hida, and Coleman (1970's–1990's).

p-adic Hodge theory was developed in pioneering papers of Fontaine in 1980s as a theory classifying p-adic representations arising from algebraic varieties over local fields. It culminated with the proofs of Fontaine's de Rham, crystalline and semistable conjectures (Faltings, Fontaine–Messing, Kato, Tsuji, Niziol, ...). In order to classify all p-adic representations of Galois groups of local fields, Fontaine (1990) initiated the theory of (φ, Γ)-modules. This gave an alternative approach to classical constructions of the p-adic Hodge theory (Cherbonnier, Colmez, Berger). The theory of (φ, Γ)-modules plays a fundamental role in Colmez's construction of the p-adic local Langlands correspondence for GL2. On the other hand, in their famous paper on L-functions and Tamagawa numbers, Bloch and Kato (1990) discovered a conjectural relation between p-adic Hodge theory and special values of L-functions. Later Kato discovered that p-adic Hodge theory is a bridge relating Beilinson–Kato Euler systems to special values of L-functions of modular forms and used it in his work on Iwasawa–Greenberg main conjecture. One expects that Kato's result is a particular case of a very general phenomenon.

The above work of Scholze represents the main conceptual progress in p-adic Hodge theory after Fontaine and Faltings. Roughly speaking, it can be seen as a wide generalization, in the geometrical context, of the relationship between p-adic representations in characteristic 0 and characteristic p provided by the theory of (φ, Γ)-modules. As an application of his theory, Scholze proved the monodromy weight conjecture for toric varieties in the mixed characteristic case. On the other hand, in a series of papers, Scholze applied his theory to the study of the cohomology of Shimura varieties. In particular, the construction of mod p Galois representations is predicted by the conjectures of Ash (see, P. Scholze. "On torsion in the cohomology of locally symmetric space". *Ann. of Math.* 182: 2015). Another striking application of this theory is the geometrization of the local Langlands correspondence in the mixed characteristic case. Here, the theory of Fontaine–Fargues plays a fundamental role.

Pune, India — Debargha Banerjee
La Jolla, USA — Kiran S. Kedlaya
Jerusalem, Israel — Ehud de Shalit
Jatni, India — Chitrabhanu Chaudhuri

Contents

About the Editors

Debargha Banerjee is Associate Professor of mathematics at the Indian Institute of Science Education and Research (IISER), Pune, India. He earned his Ph.D. from the Tata Institute of Fundamental Research, Mumbai, in 2010, under the guidance of Prof. Eknath Ghate. He worked at the Australian National University, Canberra, and the Max Planck Institute for Mathematics, Germany, before joining the IISER, Pune. He works in the theory of modular forms. He published several articles in reputed international journals and supervised several students for their Ph.D. and master's degree at the IISER, Pune.

Kiran S. Kedlaya is Professor and Stefan E. Warschawski Chair in Mathematics at the University of California San Diego, USA. He earned his Ph.D. from Massachusetts Institute of Technology (MIT), USA. He is an Indian--American mathematician, he has previously held a faculty appointment at MIT as well as visiting appointments at the Institute for Advanced Study, Princeton and the University of California, Berkeley. He is an expert in *p*-adic Hodge theory, *p*-adic/non-Archimedean analytic geometry, *p*-adic differential equations, and algorithms in arithmetic geometry. He gave an invited talk at the ICM 2010.

Ehud de Shalit is Professor of Mathematics, The Einstein Institute of Mathematics, Hebrew University, Giv'at-Ram, Jerusalem, Israel. A number theorist, Prof. Shalit has worked on topics related to class field theory, Iwasawa theory of elliptic curves, modular forms, *p*-adic L-functions, and *p*-adic analytic geometry. Current projects involve studying the cohomology of *p*-adic symmetric domains and the varieties uniformized by them.

Chitrabhanu Chaudhuri is Assistant Professor at the National Institute of Science Education and Research, Bhubaneswar, Odisha, India. His research revolves around the topology and geometry of the moduli of curves. The moduli of curves parametrize algebraic curves or Riemann surfaces up to isomorphisms. He earned his Ph.D. from Northwestern University, USA.

On ψ-Lattices in Modular (φ, Γ)-Modules

Elmar Grosse-Klönne

Introduction

Let $F/\mathbb{Q}_p$ be a finite field extension. It was a fundamental insight of Fontaine [1] that (p-adically continuous) representations of the absolute Galois group $\mathrm{Gal}(\overline{F}/F)$ of F on finite free modules over $\mathbb{Q}_p$ or $\mathbb{Z}_p$ or $\mathbb{F}_p$—for short: p-adic Galois representations—can equivalently be described by linear algebra objects which he called étale (φ, Γ)-modules. These are modules over certain Laurent series rings, in one variable, endowed with commuting semilinear actions by a Frobenius operator φ and the group $\Gamma = \mathcal{O}_F^{\times}$.[1] Approaching Galois representations through their associated étale (φ, Γ)-modules has proven to be an extremely powerful method in numerous contexts. We mention here only the important role which it plays in Colmez' work [2] on the p-adic local Langlands program. Also, the theory of (φ, Γ)-modules has been vastly generalized since into numerous directions. Among these generalizations is the work by Zábrádi [3] who showed that (for $F = \mathbb{Q}_p$), the representations of the d-fold self product $\mathrm{Gal}(\overline{F}/F) \times \cdots \times \mathrm{Gal}(\overline{F}/F)$ are in category equivalence with étale (φ, Γ)-modules over certain Laurent series rings in d variables (multivariable étale (φ, Γ)-modules).

In this note we restrict attention to p-modular coefficients only, i.e., our Galois representations (which however remain entirely in the background) and étale (φ, Γ)-modules are (in particular) $\mathbb{F}_p$-vector spaces. In this context, an étale (φ, Γ)-module is a finite dimensional $k((t))$-vector space $\mathbf{D}$, for a finite extension k of the residue field of F, endowed with said actions by φ and Γ. A critical ingredient in the aforementioned work of Colmez was the detection and study of finite $k[[t]]$-lattices spanning $\mathbf{D}$, stable under Γ and a certain operator ψ left inverse to φ, on which ψ in fact acts

E. Grosse-Klönne (✉)
Humboldt University, Berlin 12489, Germany
e-mail: gkloenne@math.hu-berlin.de

[1] In fact, Γ was taken to be a slightly different group in [1], but there is no substantial difference to the point of view taken here.

© The Author(s), under exclusive license to Springer Nature Singapore Pte Ltd. 2022
D. Banerjee et al. (eds.), *Perfectoid Spaces*, Infosys Science Foundation Series in Mathematical Sciences https://doi.org/10.1007/978-981-16-7121-0_1

surjectively. Among these lattices he identified a minimal one, denoted by $\mathbf{D}^{\natural}$, and a maximal one, denoted by $\mathbf{D}^{\sharp}$.

The purposes of this note are the following. Firstly, we want to explain that the study of $\mathbf{D}^{\natural}$ and $\mathbf{D}^{\sharp}$ makes sense similarly in the context of (p-modular) *multivariable* étale (φ, Γ)-modules. Secondly, we want to advocate an approach toward these lattices $\mathbf{D}^{\natural}$ and $\mathbf{D}^{\sharp}$ in which we rather construct and study their k-linear duals. This approach goes hand in hand with a method for constructing étale (φ, Γ)-modules. In this latter function, i.e., as a tool for the explicit construction and description of étale (φ, Γ)-modules (which typically is quite delicate, since, e.g., testing if candidates for φ- and Γ-actions, given by explicit power series, really commute with each other can be quite challenging), the method was introduced by Colmez in [2] and then used later in [4], both in the one-variable case. We intend to use the constructions presented here (the construction and analysis of $\mathbf{D}^{\natural}$ and $\mathbf{D}^{\sharp}$) to generalize the work [4] to a multivariable setting in the future. Thirdly, by discussing several examples we try to shed some more light on the behavior of $\mathbf{D}^{\natural}$ and $\mathbf{D}^{\sharp}$. Actually, most of these examples pertain to the one-variable case, yet we think that they demonstrate some features of $\mathbf{D}^{\natural}$ and $\mathbf{D}^{\sharp}$ which at least have not yet been documented in the literature (although undoubtedly known to the experts).

1 Multivariable Modular étale $(\varphi_\bullet, \Gamma_\bullet)$-Modules

Notations: Let $F/\mathbb{Q}_p$ be a finite field extension. Denote by q the number of elements of the residue field $\mathbb{F}_q$ of F. Let π be a uniformizer in the ring of integers $\mathcal{O}_F$ of F. Let k be a finite extension field of $\mathbb{F}_q$. Put $\Gamma = \mathcal{O}_F^{\times}$.

There is a unique (up to isomorphism) Lubin-Tate group for F with respect to π. Fixing a coordinate t we write $\Phi(t)$ for the corresponding Lubin-Tate formal power series describing multiplication by π. Equivalently, to any power series $\Phi(t) \in \mathcal{O}_F[[t]]$ with $\Phi(t) \equiv \pi t$ modulo $t^2\mathcal{O}_F[[t]]$ and $\Phi(t) \equiv t^q$ modulo $\pi\mathcal{O}_F[[t]]$ is associated a Lubin-Tate formal group law, with $\Phi(t)$ representing multiplication by π, and the resulting formal group (with multiplication by $\mathcal{O}_F$) is independent (up to isomorphism) on the specific $\Phi(t)$. For $\gamma \in \Gamma$ let $[\gamma]_\Phi(t) \in \mathcal{O}_F[[t]]$ denote the power series describing the action of γ in the Lubin-Tate group. Let D be a finite set, and for each $d \in D$ let t_d be a free variable. Put

$$k[[t_\bullet]] = k[[t_d]]_{d\in D}, \qquad k((t_\bullet)) = k[[t_\bullet]][t_D^{-1}] \quad \text{with } t_D = \prod_{d\in D} t_d.$$

For each $d \in D$ let Γ_d be a copy of Γ. For $\gamma \in \Gamma$ let γ_d denote the element in $\Gamma_\bullet = \prod_{d\in D} \Gamma_d$ whose d-component is γ and whose other components are trivial. The formulae

$$\gamma_d(t_d) = [\gamma]_\Phi(t_d), \qquad \gamma_{d_1}(t_{d_2}) = t_{d_2}$$

with $\gamma \in \Gamma$ and $d, d_1, d_2 \in D$ such that $d_1 \neq d_2$ define an action of $\Gamma_\bullet$ on $k((t_\bullet))$. Consider the $k[[t_\bullet]]$-algebra

$$k[[t_\bullet]][\varphi_\bullet, \Gamma_\bullet] = k[[t_d]]_{d\in D}[\varphi_d, \Gamma_d]_{d\in D}$$

with commutation rules given by

$$x_{d_1} \cdot y_{d_2} = y_{d_2} \cdot x_{d_1},$$

$$\gamma_d \cdot \varphi_d = \varphi_d \cdot \gamma_d, \qquad \gamma_d \cdot t_d = (\gamma_d(t_d)) \cdot \gamma_d, \qquad \varphi_d \cdot t_d = \Phi(t_d) \cdot \varphi_d = t_d^q \cdot \varphi_d$$

for $\gamma \in \Gamma$ and $x, y \in \Gamma \cup \{\varphi, t\}$ and $d, d_1, d_2 \in D$ with $d_1 \neq d_2$. Similarly we define the $k((t_\bullet))$-algebra $k((t_\bullet))[\varphi_\bullet, \Gamma_\bullet]$ and its subalgebra $k((t_\bullet))[\Gamma_\bullet]$.

Definition An étale $\varphi_\bullet$-module over $k((t_\bullet))$ is a $k((t_\bullet))[\varphi_\bullet]$-module $\mathbf{D}$ which is finitely generated over $k((t_\bullet))$ such that for each $d \in D$ the linearized structure map

$$\mathrm{id} \otimes \varphi_d : k((t_\bullet)) \otimes_{\varphi_d, k((t_\bullet))} \mathbf{D} \longrightarrow \mathbf{D}$$

is bijective. An étale $(\varphi_\bullet, \Gamma_\bullet)$-module over $k((t_\bullet))$ is a $k((t_\bullet))[\varphi_\bullet, \Gamma_\bullet]$-module whose underlying $\varphi_\bullet$-module is étale. In the case $|D| = 1$ we drop the indices $(.)_d$ resp. $(.)_\bullet$ and simply talk about étale (φ, Γ)-modules over $k((t))$.

Remark The action of $\Gamma_\bullet$ on an étale $(\varphi_\bullet, \Gamma_\bullet)$-module is automatically continuous for the weak topology.

Lemma 1.1 *([5] Lemma 4, Proposition 6)*
The category of étale $(\varphi_\bullet, \Gamma_\bullet)$-modules over $k((t_\bullet))$ is abelian.
Regard $k((t))$ as a $k((t_\bullet))$-module by means of $t_d \cdot x = tx$ for $d \in D$ and $x \in k((t))$. The functor

$$\mathbf{D} \mapsto k((t)) \otimes_{k((t_\bullet))} \mathbf{D}$$

is an exact and faithful functor from the category of étale $(\varphi_\bullet, \Gamma_\bullet)$-modules over $k((t_\bullet))$ to the category of étale (φ, Γ)-modules over $k((t))$.[2]

Theorem 1.2 *(a) (Fontaine [1], Kisin-Ren [6], Schneider [7]) There is an equivalence between the category of étale (φ, Γ)-modules over $k((t))$ and the category of continuous representations of* $\mathrm{Gal}(\overline{F}/F)$ *on finite dimensional k-vector spaces.*
(b) (Zábrádi [3]) Assume $F = \mathbb{Q}_p$. There is an equivalence between the category of étale $(\varphi_\bullet, \Gamma_\bullet)$-modules over $k((t_\bullet))$ and the category of continuous representations of $\mathrm{Gal}(\overline{\mathbb{Q}_p}/\mathbb{Q}_p) \times \cdots \times \mathrm{Gal}(\overline{\mathbb{Q}_p}/\mathbb{Q}_p)$ *(with $|D|$ many factors indexed by D) on finite dimensional k-vector spaces.*

---

[2] We will not make use of the second statement of Lemma 1.1 in the following.

(c) *Statements (a) and (b) merge into a common generalization (Pupazan [8]): For any F, there is an equivalence between the category of étale* $(\varphi_\bullet, \Gamma_\bullet)$*-modules over* $k((t_\bullet))$ *and the category of continuous representations of* $\mathrm{Gal}(\overline{F}/F)\times\cdots\times\mathrm{Gal}(\overline{F}/F)$ *on finite dimensional k-vector spaces.*

Definition A ψ-operator on $k[[t_\bullet]]$ is a system $\psi_\bullet=(\psi_d)_{d\in D}$ of additive maps

$$\psi_d : k[[t_\bullet]] \longrightarrow k[[t_\bullet]]$$

for $d\in D$ such that $\psi_{d_1}(\gamma_{d_2}(t_{d_1}))=\gamma_{d_2}(\psi_{d_1}(t_{d_1}))$ for all $\gamma\in\Gamma$ and $d_1,d_2\in D$, such that $\psi_{d_1}(t_{d_2})=t_{d_2}$ for $d_1\neq d_2$ and such that the following holds true: If we view the φ_d as acting on $k[[t_\bullet]]$, then $\psi_{d_1}\circ\varphi_{d_2}=\varphi_{d_2}\circ\psi_{d_1}$ for $d_1\neq d_2$, but

$$\psi_d(\varphi_d(a)x)=a\psi_d(x)$$

for $a,x\in k[[t_\bullet]]$.[3]

Lemma 1.3 *([5] Lemma 3) There is a* ψ*-operator* $\psi_\bullet$ *on* $k[[t_\bullet]]$ *such that each* ψ_d *is surjective.*

To be explicit, in the case where $|D|=1$ (the general case is handled factor by factor) and $\Phi(t)=\pi t+t^q$, we may choose $\psi_{k((t))}$ on $k((t))$ such that for $m\in\mathbb{Z}$ and $0\leq i\leq q-1$ we have[4]

$$\psi_{k((t))}(t^{mq+i})=\begin{cases}\frac{q}{\pi}t^m & : \quad i=0\\ 0 & : \quad 1\leq i\leq q-2\\ t^m & : \quad i=q-1\end{cases}. \tag{1}$$

In the following, we fix $\psi_\bullet$ as in Lemma 1.3.

Let $\mathbf{D}$ be an étale $(\varphi_\bullet,\Gamma_\bullet)$-module over $k((t_\bullet))$. For $d\in D$ we define the composed map

$$\psi_d : \mathbf{D} \longrightarrow k((t_\bullet))\otimes_{\varphi_d,k((t_\bullet))}\mathbf{D}\longrightarrow\mathbf{D}$$

where the first arrow is the inverse of the structure isomorphism $\mathrm{id}\otimes\varphi_d$, and where the second arrow is given by $a\otimes x\mapsto\psi_d(a)x$.

Lemma 1.4 *For all* $x\in\mathbf{D}$, $\gamma\in\Gamma$, $a\in k((t_\bullet))$ *and* $d\in D$ *we have*

$$\psi_d(a\varphi_d(x))=\psi_d(a)x,\qquad \psi_d(\varphi_d(a)x)=a\psi_d(x),\qquad \gamma_d(\psi_d(x))=\psi_d(\gamma_d(x)).$$

For all $x\in\mathbf{D}$ *and* $d_1\neq d_2\in D$ *we have*

$$\psi_{d_1}(\psi_{d_2}(x))=\psi_{d_2}(\psi_{d_1}(x)),\qquad \psi_{d_1}(\varphi_{d_2}(x))=\varphi_{d_2}(\psi_{d_1}(x)).$$

---

[3] Notice that we do not require $\psi_d(1)=1$.

[4] Notice that $\frac{q}{\pi}=0$ (in k) if $F\neq\mathbb{Q}_p$.

Proof The formula $\psi_d(a\varphi_d(x)) = \psi_d(a)x$ is immediate from the construction. To see the formula $\psi_d(\varphi_d(a)x) = a\psi_d(x)$, write $x = \sum_i a_i\varphi_d(e_i)$ with $e_i \in \mathbf{D}$ and $a_i \in k((t_\bullet))$ (this is possible as $\mathbf{D}$ is étale). We then compute

$$\begin{aligned}\psi_d(\varphi_d(a)x) &= \sum_i \psi_d(\varphi_d(a)a_i\varphi_d(e_i)) = \sum_i \psi_d(\varphi_d(a)a_i)e_i \\ &= a\sum_i \psi_d(a_i)e_i = a\sum_i \psi_d(a_i\varphi_d(e_i)) = a\psi_d(x).\end{aligned}$$

To see the formula $\gamma_d(\psi_d(x)) = \psi_d(\gamma_d(x))$ observe that, since the actions of γ_d and φ_d on $k[[t_\bullet]]$ commute, and since $\Gamma_\bullet$ acts semilinear on $\mathbf{D}$, the additive map

$$k((t_\bullet)) \otimes_{\varphi_d, k((t_\bullet))} \mathbf{D} \to k((t_\bullet)) \otimes_{\varphi_d, k((t_\bullet))} \mathbf{D},$$

$$a \otimes b \mapsto \gamma_d(a) \otimes \gamma_d(b)$$

is the map corresponding to the action of γ_d on $\mathbf{D}$ under the isomorphism $\mathrm{id} \otimes \varphi_d$, and under $a \otimes x \mapsto \psi_d(a)x$ it commutes with γ_d acting on $\mathbf{D}$ since the actions of γ_d and ψ_d on $k((t_\bullet))$ commute. The remaining commutation formulae are clear. □

Definition For a k-vector space Δ we write $\Delta^* = \mathrm{Hom}_k(\Delta, k)$. We say that a $k[[t_\bullet]]$-module Δ is admissible if it is a torsion module[5] over $k[[t_d]]$ for each $d \in D$ and if

$$\Delta[t_\bullet] = \{x \in \Delta \mid t_d x = 0 \text{ for each } d \in D\}$$

is a finite dimensional k-vector space.

Proposition 1.5 *([5] Proposition 5) Let Δ be a finitely generated $k[[t_\bullet]][\varphi_\bullet, \Gamma_\bullet]$-module which is admissible as a $k[[t_\bullet]]$-module and satisfies $\Delta = k[[t_\bullet]]\varphi_d(\Delta)$ for each $d \in D$. Then $\Delta^* \otimes_{k[[t_\bullet]]} k((t_\bullet))$ is in a natural way an étale $(\varphi_\bullet, \Gamma_\bullet)$-module over $k((t_\bullet))$. The functor $\Delta \mapsto \Delta^* \otimes_{k[[t_\bullet]]} k((t_\bullet))$ is exact.*

We remark that the $k[[t_\bullet]][\Gamma_\bullet]$-action on $\Delta^* \otimes_{k[[t_\bullet]]} k((t_\bullet))$ results from the $k[[t_\bullet]][\Gamma_\bullet]$-action on Δ^* given by the formulae

$$(a \cdot \ell)(\delta) = \ell(a\delta),$$

$$(\gamma \cdot \ell)(\delta) = \ell(\gamma^{-1}\delta)$$

for $a \in k[[t_\bullet]]$, $\ell \in \Delta^*$, $\delta \in \Delta$ and $\gamma \in \Gamma_\bullet$. The $\varphi_\bullet$-action on $\Delta^* \otimes_{k[[t_\bullet]]} k((t_\bullet))$ is a certain right inverse to the dual of the $\varphi_\bullet$-action on Δ; its construction involves the ψ-operator $\psi_\bullet$ on $k[[t_\bullet]]$.

---

[5] In [5] we had not included this torsion condition into the definition of admissibility; however, for staying consistent with established terminology we should have done so. (Yet, the omission of this condition in the paper [5] does not invalidate its results.)

Lemma 1.6 *Let* Δ *and* $\mathbf{D} = \Delta^* \otimes_{k[[t_\bullet]]} k((t_\bullet))$ *be as in Proposition 1.5 and suppose that each* t_d *acts surjectively on* Δ.

(a) *The natural map* $\Delta^* \to \mathbf{D} = \Delta^* \otimes_{k[[t_\bullet]]} k((t_\bullet))$ *is injective. Each* ψ_d *respects* Δ^* *and acts on it by the rule*

$$[\ell : \Delta \to k] \quad \mapsto \quad [\Delta \to k, x \mapsto \ell(\varphi_d(x))].$$

(b) *If each* φ_d *acts injectively on* Δ *then each* ψ_d *acts surjectively on* Δ^*.

Proof This follows immediately from the construction (given in [5] Proposition 5) referred to in Proposition 1.5. □

2 The lattices $\mathbf{D}^\natural$ and $\mathbf{D}^\sharp$ inside $\mathbf{D}$

We write $\psi_D = \prod_{d\in D} \psi_d$ and $\varphi_D = \prod_{d\in D} \varphi_d$ (as k-linear operators on $k((t_\bullet))$).

Let $\mathbf{D}$ be an étale $(\varphi_\bullet, \Gamma_\bullet)$-module over $k((t_\bullet))$. We call a finitely generated $k[[t_\bullet]]$-submodule of $\mathbf{D}$ a lattice in $\mathbf{D}$ if it generates $\mathbf{D}$ (as a $k((t_\bullet))$-module).

Lemma 2.1 *Let* E *be a lattice in* $\mathbf{D}$, *let* $d \in D$.

(a) $\psi_d(E)$ *is a* $k[[t_\bullet]]$*-module.*
(b) *If* $\varphi_d(E) \subset E$ *then* $E \subset \psi_d(E)$.
(c) *If* $E \subset k[[t_\bullet]] \cdot \varphi_d(E)$ *then* $\psi_d(E) \subset E$.
(d) *If* $\psi_d(E) \subset E$ *then* $\psi_d(t_d^{-1}E) \subset t_d^{-1}E$. *For each* $x \in \mathbf{D}$ *there is some* $n(x) \in \mathbb{N}$ *such that for all* $n \geq n(x)$ *we have* $\psi_D^n(x) \in t_D^{-1}E$.

Proof (a) Use $\psi_d(\varphi_d(a)x) = a\psi_d(x)$ for $a \in k((t_\bullet))$ and $x \in \mathbf{D}$.
(b) Choose $a \in k[[t_\bullet]]$ with $\psi_d(a) = 1$. For $e \in E$ we have $e = \psi_d(a\varphi_d(e))$ which belongs to $\psi_d(E)$ since $\varphi_d(E) \subset E$.
(c) Let $e \in E$. By assumption there are $e_i \in E$ and $a_i \in k[[t_\bullet]]$ with $e = \sum_i a_i\varphi_d(e_i)$, hence $\psi_d(e) = \sum_i \psi_d(a_i)e_i \in E$.
(d) For $i \geq 1$ we have

$$\psi_d(\varphi_d^i(t_d^{-1})E) \subset \varphi_d^{i-1}(t_d^{-1})\psi_d(E) \subset \varphi_d^{i-1}(t_d^{-1})E$$

where the second inclusion uses the assumption. Taking the product over all d this implies

$$\psi_D(\varphi_D^i(t_D^{-1})E) \subset \varphi_D^{i-1}(t_D^{-1})\psi_D(E) \subset \varphi_D^{i-1}(t_D^{-1})E. \tag{2}$$

From $\varphi_d(t_d^{-1}) = t_d^{-q}$ we get

$$\psi_d(t_d^{-1}E) \subset \psi(\varphi_d(t_d^{-1})E) \subset t_d^{-1}E$$

and taking the product over all d we thus get

$$\psi_D(t_D^{-1}E) \subset \psi(\varphi_D(t_D^{-1})E) \subset t_D^{-1}E.$$

Moreover, if $n(x) \in \mathbb{N}$ is such that $x \in \varphi_D^n(t_D^{-1})E$ for $n \geq n(x)$, then iterated application of formula (2) shows

$$\psi_D^n(x) \in \psi_D^n(\varphi_D^n(t_D^{-1})E) \subset \psi_D^{n-1}(\varphi_D^{n-1}(t_D^{-1})E) \subset \ldots \subset t_D^{-1}E$$

for $n \geq n(x)$. □

Lemma 2.2 *(a) There are lattices E_0, E_1 in* **D** *with*

$$\varphi_D(E_0) \subset t_D E_0 \subset E_0 \subset E_1 \subset k[[t_\bullet]]\varphi_D(E_1).$$

(b) For any $n \geq 0$ we have $\psi_D^n(E_0) \subset \psi_D^{n+1}(E_0) \subset E_1$.

Proof (a) Let $e_1, \ldots, e_r$ be a generating system of **D** as a $k((t_\bullet))$-module. Then also $\varphi_d(e_1), \ldots, \varphi_d(e_r)$ is such a generating system of **D**, for each $d \in D$. We therefore find elements f_{ij}, g_{ij} in $k((t_\bullet))$ such that $\varphi_D(e_j) = \sum_{i=1}^r f_{ij}e_i$ and $e_j = \sum_{i=1}^r g_{ij}\varphi_D(e_i)$ for all $1 \leq j \leq r$. Choose $n \in \mathbb{N}$ with $t_D^{n(q-1)} f_{ij} \in t_D k[[t_\bullet]]$ and $t_D^{n(q-1)} g_{ij} \in t_d k[[t_\bullet]]$ for all i, j. Then

$$E_0 = \sum_{i=1}^{r} t_D^n k[[t_\bullet]]e_i, \qquad E_1 = \sum_{i=1}^{r} t_D^{-n} k[[t_\bullet]]e_i$$

work as desired.

(b) Choose $a \in k[[t_\bullet]]$ with $\psi_D(a) = 1$. For $x \in E_0$ we have $\psi_D^n(x) = \psi_D^{n+1}(a\varphi_D(x)) \in \psi_D^{n+1}(E_0)$ since $\varphi_D(E_0) \subset t_D(E_0)$ implies $\varphi_D(x) \in E_0$ and hence $a\varphi_D(x) \in E_0$. This shows $\psi_D^n(E_0) \subset \psi_D^{n+1}(E_0)$. As $E_0 \subset E_1 \subset k[[t_\bullet]]\varphi_d(E_1)$, an induction using Lemma 2.1 shows $\psi_D^{n+1}(E_0) \subset E_1$. □

Proposition 2.3 *(a) There exists a unique lattice* $\mathbf{D}^\sharp$ *in* **D** *with* $\psi_D(\mathbf{D}^\sharp) = \mathbf{D}^\sharp$ *and such that for each* $x \in \mathbf{D}$ *there is some* $n \in \mathbb{N}$ *with* $\psi_D^n(x) \in \mathbf{D}^\sharp$.
(b) For any lattice E in **D** *we have* $\psi_D^n(E) \subset \mathbf{D}^\sharp$ *for all* $n >> 0$.
(c) For any lattice E in **D** *with* $\psi_D(E) = E$ *we have*

$$t_D\mathbf{D}^\sharp \subset E \subset \mathbf{D}^\sharp.$$

Proof Let E_0, E_1 be as in Lemma 2.2. For $n \in \mathbb{N}$ put $F_n = \psi_D^n(E_0)$. For $x \in E_0$ we have $\psi_D^n(x) = \psi_D^{n+1}(\psi_D(1)\varphi_D(x))$; since $\psi_D(1)\varphi_D(x) \in E_0$ this shows that $(F_n)_n$ is an increasing sequence of lattices in **D**. As $E_1 \subset k[[t_\bullet]]\varphi_D(E_1)$, Lemma 2.1 (c) shows $F_n \subset E_1$. As $k[[t_\bullet]]$ and hence E_1 is noetherian, there is some n_0 with $F_n = F_{n_0}$ for all $n \geq n_0$, and hence with $\psi_D(F_{n_0}) = F_{n_0}$. For $m \in \mathbb{N}$ put

$$G_m = \psi_D^m t_D^{-1} F_{n_0}.$$

Lemma 2.1 (d) shows that $(G_m)_m$ is a descending sequence of lattices in $\mathbf{D}$, containing F_{n_0} since $\psi_D(F_{n_0}) = F_{n_0}$. As $t_D^{-1}F_{n_0}/F_{n_0}$ is artinian we therefore find some m_0 with $G_m = G_{m_0}$ for all $m \geq m_0$, and hence with $\psi_D(G_{m_0}) = G_{m_0}$. Moreover, Lemma 2.1 (d) shows that for each $x \in G_{m_0}$ there is some $i(x) \in \mathbb{N}$ with $\psi_D^i(x) \in t_D^{-1}F_{n_0}$ for all $i \geq i(x)$. We then have $\psi_D^{m_0+i}(x) \in G_{m_0}$ for all $i \geq i(x)$. Thus, $\mathbf{D}^\sharp = G_{m_0}$ works as desired.

To see the uniqueness of $\mathbf{D}^\sharp$, assume that there is another candidate $\tilde{\mathbf{D}}^\sharp$ satisfying the same properties. Then so does $\mathbf{D}^\sharp + \tilde{\mathbf{D}}^\sharp$, hence we may assume $\mathbf{D}^\sharp \subset \tilde{\mathbf{D}}^\sharp$. But ψ_d for any $d \in D$ acts both surjectively and nilpotently on the finite dimensional k-vector space $\tilde{\mathbf{D}}^\sharp/\mathbf{D}^\sharp$, hence $\mathbf{D}^\sharp = \tilde{\mathbf{D}}^\sharp$. □

Proposition 2.4 *(a) For any lattice E in $\mathbf{D}$ contained in $\mathbf{D}^\sharp$ and stable under ψ_d for $d \in D$ we have $\psi_d(E) = E$.*
(b) The intersection $\mathbf{D}^\natural$ of all lattices in $\mathbf{D}$ contained in $\mathbf{D}^\sharp$ and stable under ψ_d for all $d \in D$ is itself a lattice, and it satisfies $\psi_d(\mathbf{D}^\natural) = \mathbf{D}^\natural$ for all $d \in D$.

Proof (a) Since $\mathbf{D}^\sharp$ as well as E and $\psi_d(E)$ are lattices in $\mathbf{D}^\sharp$, both $\mathbf{D}^\sharp/E$ and $\mathbf{D}^\sharp/\psi_d(E)$ are finite dimensional k-vector spaces. ψ_d induces an isomorphism $\mathbf{D}^\sharp/E = \mathbf{D}^\sharp/\psi_d(E)$ (as $\psi_d(E) \subset E$), hence $\psi_d(E) = E$.
(b) For any lattice E in $\mathbf{D}$ contained in $\mathbf{D}^\sharp$ and stable under ψ_d for all $d \in D$ we have $t_D\mathbf{D}^\sharp \subset E$ by what we saw in (a) together with proposition 2.3. This shows $t_D\mathbf{D}^\sharp \subset \mathbf{D}^\natural$, hence $\mathbf{D}^\natural$ is indeed a lattice, and $\psi_d(\mathbf{D}^\natural) = \mathbf{D}^\natural$ follows by applying (a) once more. □

Lemma 2.5 *Let Δ be as in Lemma 1.6, with each φ_d acting injectively on Δ. If $\Delta[t_\bullet]$ generates Δ as a $k[[t_\bullet]][\varphi_\bullet]$-module then $\Delta^* = \mathbf{D}^\natural$.*

Proof For $i = (i_d)_{d\in D} \in \mathbb{Z}_{\geq 0}^D$ let

$$F^i\Delta^* = \{\ell \in \Delta^* \mid \ell((\prod_{d\in D} t_d^{n_d}\varphi_d^{i_d})(x)) = 0 \text{ for all } n_d > 0, x \in \Delta[t_\bullet]\}.$$

This is a $k[[t_\bullet]]$-submodule of Δ^*. Let E be a lattice in $\mathbf{D}$ contained in $\mathbf{D}^\sharp$ with $\psi_d(E) \subset E$ for all $d \in D$. We have $\cap_i F^i\Delta^* = 0$ since $\Delta[t_\bullet]$ generates Δ as a $k[[t_\bullet]][\varphi_\bullet]$-module. As E generates $\mathbf{D}$ we therefore find $F^i\Delta^* \subset E$ for some i. But

$$(\prod_{d\in D}\psi_d^{i_d})F^i\Delta^* = \{\ell((\prod_{d\in D}\varphi_d^{i_d})(.)) \mid \ell \in \Delta^*\} = \Delta^*$$

where the second equality follows from the injectivity of the φ_d. We thus obtain $\Delta^* \subset E$ as the ψ_d respect E. □

Proposition 2.6 (Colmez) *Suppose* $|D| = 1$. *If* $\mathbf{D}$ *is an irreducible étale* (φ, Γ)*-module with* $\dim_{k((t))}(\mathbf{D}) \geq 2$, *then* $\mathbf{D}^\natural = \mathbf{D}^\sharp$. *If* $\dim_{k((t))}(\mathbf{D}) = 1$ *then* $\dim_k(\mathbf{D}^\sharp/\mathbf{D}^\natural) = 1$.

Proof See [9] Corollaire II.5.21 for the first statement. The second one follows, e.g., from I.3.2. Exemple in [2], but also from the discussion of example (a) below. □

Remark It is easy to see that both $\mathbf{D} \mapsto \mathbf{D}^\natural$ and $\mathbf{D} \mapsto \mathbf{D}^\sharp$ are functors from the category of étale $(\varphi_\bullet, \Gamma_\bullet)$-modules to the category of $\psi_\bullet$-modules (obvious definition). Moreover, if

$$0 \longrightarrow \mathbf{D}_1 \longrightarrow \mathbf{D} \longrightarrow \mathbf{D}_2 \longrightarrow 0 \tag{3}$$

is an exact sequence of étale $(\varphi_\bullet, \Gamma_\bullet)$-modules, then the sequences

$$0 \longrightarrow \mathbf{D}_1^\natural \longrightarrow \mathbf{D}^\natural \longrightarrow \mathbf{D}_2^\natural \longrightarrow 0, \tag{4}$$

$$0 \longrightarrow \mathbf{D}_1^\sharp \longrightarrow \mathbf{D}^\sharp \longrightarrow \mathbf{D}_2^\sharp \longrightarrow 0 \tag{5}$$

are both exact on the left and on the right (see [9] Proposition II 4.6 and Proposition II 5.19 for the case $|D| = 1$). However, in general they need not be exact in the middle. We are going to exemplify this below.

3 Examples

(a) By Proposition 2.6, if $|D| = 1$ then a rank one étale (φ, Γ)-module contains precisely two (ψ, Γ)-stable lattices with surjective ψ-operator. If $|D| > 1$ we find more.

Fix some $c_d \in k^\times$ and some $m_d \in \mathbb{Z}/(q-1)\mathbb{Z}$ for each $d \in D$. Put

$$B = \bigoplus_{C \subset D} k.e_C,$$

the k-vector space with basis $\{e_C\}_{C \subset D}$ indexed by the subsets C of D. Let $k[[t_\bullet]][\Gamma_\bullet]$ act on B by requiring

$$t_d \cdot e_C = \begin{cases} 0 & : \ d \in C \\ e_{C \cup \{d\}} & : \ d \in D - C \end{cases},$$

$$\gamma_d \cdot e_C = \begin{cases} \gamma_d^{m_d+1} e_C & : \ d \in C \\ \gamma_d^{m_d} e_C & : \ d \in D - C \end{cases}.$$

(On the right-hand side of the defining formula for $\gamma_d \cdot e_C$ we refer to multiplication with the scalar in $k^\times$ to which $\gamma \in \Gamma = \mathcal{O}_F^\times$ is mapped.)

Let $\mathcal{D}$ be a set of subsets of D such that for any $C \in \mathcal{D}$ and $d \in D - C$ we also have $C \cup \{d\} \in \mathcal{D}$. It is clear that $\sum_{C \in \mathcal{D}} k.e_C$ is a k-sub vector space of B stable under $k[[t_\bullet]][\Gamma_\bullet]$, hence $k[[t_\bullet]][\Gamma_\bullet]$ acts on

$$B_{\mathcal{D}} = \frac{B}{\sum_{C\in\mathcal{D}} k.e_C}.$$

Define

$$\Delta_{\mathcal{D}} = \Delta_{\mathcal{D}}(c_\bullet, m_\bullet) = \frac{k[[t_\bullet]][\varphi_\bullet] \otimes_{k[[t_\bullet]]} B_{\mathcal{D}}}{\langle t_d^{q-1}\varphi_d \otimes e_\emptyset - 1 \otimes c_d e_\emptyset\rangle_{d\in D}}$$

where $\langle ?\rangle$ indicates the $k((t_\bullet))[\varphi_\bullet]$-sub module generated by all expressions within the brackets (and $e_\emptyset$ actually means the class of $e_\emptyset \in B$ in $B_{\mathcal{D}}$). One checks that this submodule $\langle t_d^{q-1}\varphi_d \otimes e_\emptyset - 1 \otimes c_d e_\emptyset\rangle_{d\in D}$ is in fact also stable under the action of $\Gamma_\bullet$; indeed, the $t_d^{q-1}\varphi_d \otimes e_\emptyset - 1 \otimes c_d e_\emptyset$ are eigenvectors under the action of $\Gamma_\bullet$. It follows that $\Delta_{\mathcal{D}}$ becomes a $k[[t_\bullet]][\varphi_\bullet, \Gamma_\bullet]$-module. It is finitely generated over $k[[t_\bullet]][\varphi_\bullet]$, admissible over $k[[t_\bullet]]$ and each ψ_d acts surjectively on $\Delta_{\mathcal{D}}^* = (\Delta_{\mathcal{D}})^*$. Thus $\Delta_{\mathcal{D}}^*$ is a lattice inside

$$\mathbf{D} = \mathbf{D}(c_\bullet, m_\bullet) = \Delta_{\mathcal{D}}^* \otimes_{k[[t_\bullet]]} k((t_\bullet)).$$

The natural projections $B_{\mathcal{D}} \to B_{\mathcal{D}'}$ for $\mathcal{D} \subset \mathcal{D}'$ induce $k[[t_\bullet]]$-linear inclusions $\Delta_{\mathcal{D}'}^* \to \Delta_{\mathcal{D}}^*$ which, when tensored with $k((t_\bullet))$, become isomorphisms. In particular, the rank one étale $(\varphi_\bullet, \Gamma_\bullet)$-module $\mathbf{D}$ is in a natural way independent of $\mathcal{D}$, and (inside $\mathbf{D}$) we have

$$\Delta_{\mathcal{D}}^* \neq \Delta_{\mathcal{D}'}^* \qquad \text{whenever} \quad \mathcal{D} \neq \mathcal{D}'.$$

Taking $\mathcal{D}_1$ to be set empty set, so that $B = B_{\mathcal{D}_1}$, we find $\Delta_{\mathcal{D}_1}^* = \mathbf{D}^\sharp$. Taking $\mathcal{D}_0 = \{D \subset C \mid D \neq \emptyset\}$, so that $B_{\mathcal{D}_0}$ is of k-dimension 1 (generated by the class of $e_\emptyset$), we find $\Delta_{\mathcal{D}_0}^* = \mathbf{D}^\natural$, cf. Lemma 2.5.

In the following examples we choose (as we may) the coordinate t such that $\Phi(t) = \pi t + t^q$. We assume $|D| = 1$ and drop subscripts $(.)_d$. We describe various Δ's as quotients $\Delta = (k[[t]][\varphi] \otimes_k M)/\nabla$ with finite dimensional k-vector spaces M, and where always ∇ is generated as a $k[[t]][\varphi]$-submodule by elements in $k[[t]]\varphi \otimes_k M + k[[t]] \otimes_k M$ only (i.e., no higher powers of φ occur in these generators).

In all these examples, φ acts injectively in Δ (hence ψ acts surjectively on Δ^*) and t acts surjectively on Δ (hence Δ^* is t-torsion free).

(b) We describe a Δ defining an extension between two rank one étale (φ, Γ)-modules.

Fix $\alpha \in k$. Let $\langle e_1, e_2, f\rangle_k$ denote the k-vector space with basis $\{e_1, e_2, f\}$. In $k[[t]][\varphi] \otimes_k \langle e_1, e_2, f\rangle_k$ consider the subset[6]

[6] In writing the elements of $\mathcal{R}$ we suppress the symbol $\otimes$.

$$\begin{aligned}\mathcal{R} = \{&te_1 - e_2, \quad te_2, \quad tf, \\ &t^{q-1}\varphi f - f, \quad t^{q-1}\varphi e_1 - e_1 - \alpha t^{q-2}\varphi f\}\end{aligned}$$

and let Δ denote the quotient of $k[[t]][\varphi] \otimes_k \langle e_1, e_2, f\rangle_k$ by the $k[[t]][\varphi]$-submodule generated by the elements in $\mathcal{R}$. Let $\langle f\rangle_k$ denote the k-sub vector space of $\langle e_1, e_2, f\rangle_k$ spanned by f and define the k-vector space $\overline{\langle e_1, e_2\rangle_k}$ by the exact sequence

$$0 \longrightarrow \langle f\rangle_k \longrightarrow \langle e_1, e_2, f\rangle_k \longrightarrow \overline{\langle e_1, e_2\rangle_k} \longrightarrow 0.$$

We identify $\{e_1, e_2\}$ with a k-basis of $\overline{\langle e_1, e_2\rangle_k}$. We obtain an exact sequence

$$0 \longrightarrow \frac{k[[t]][\varphi] \otimes_k \langle f\rangle_k}{\nabla_1} \longrightarrow \Delta \longrightarrow \frac{k[[t]][\varphi] \otimes_k \overline{\langle e_1, e_2\rangle_k}}{\nabla_2} \longrightarrow 0$$

where ∇_1 (resp. ∇_2) is the respective $k[[t]][\varphi]$-submodule generated by tf and $t^{q-1}\varphi f - f$ (resp. by $te_1 - e_2, te_2$ and $t^{q-1}\varphi e_1 - e_1$).

Next, fix $a \in \mathbb{Z}$ and let $\gamma \in \Gamma$ act on $\langle e_1, e_2, f\rangle_k$ by means of

$$\gamma \cdot f = \gamma^{2+a} f \qquad \text{and} \qquad \gamma \cdot e_i = \gamma^{i+a} e_i.$$

(Here, in the expression $\gamma^{2+a} f$ resp. $\gamma^{i+a} e_i$ the γ refers to the scalar in $\mathbb{F}_q^\times$ to which $\gamma \in \Gamma = \mathcal{O}_F^\times$ is projected.) Then

$$k[[t]][\varphi] \otimes_k \langle e_1, e_2, f\rangle_k \cong k[[t]][\varphi, \Gamma] \otimes_{k[\Gamma]} \langle e_1, e_2, f\rangle_k$$

so that $k[[t]][\varphi] \otimes_k \langle e_1, e_2, f\rangle_k$ receives a $k[[t]][\varphi, \Gamma]$-action. One checks that all elements in $\mathcal{R}$ are eigenvectors for the action of Γ. (For the elements $t^{q-1}\varphi f - f$ and $t^{q-1}\varphi e_1 - e_1 - \alpha t^{q-2}\varphi f$ this computation uses that $[\gamma](t) \equiv \gamma t$ modulo $t^q k[[t]]$, as is implied by our assumption $\Phi(t) = \pi t + t^q$, see Lemma 0.1 in [4].) It follows that Δ, as well as the above exact sequence are in fact $k[[t]][\varphi, \Gamma]$-equivariant.

In view of Proposition 1.5 we get an induced exact sequence (3) of étale (φ, Γ)-modules, with $\dim_{k((t))}(\mathbf{D}_1) = \dim_{k((t))}(\mathbf{D}_2) = 1$. If $F = \mathbb{Q}_p$ then the Galois character attached (by Theorem 1.2) to $\mathbf{D}_1$ is obtained from the one attached to $\mathbf{D}_2$ by multiplying with the cyclotomic character. We have $\Delta = \mathbf{D}^\natural$. Neither the sequence (4) nor the sequence (5) is exact.

(c) In contrast to what one might be tempted to think in view of Proposition 2.6, the possible failure of exactness of the sequences (4) or (5) can *not* exclusively be reduced to the non-uniqueness of (ψ, Γ)-stable lattices with surjective ψ-operator inside étale (φ, Γ)-module of rank one.

Let $\langle e_1, e_2, \tilde{e}, f_1, f_2\rangle_k$ denote the k-vector space with basis $\{e_1, e_2, \tilde{e}, f_1, f_2\}$. Let $0 \leq s \leq q - 1$. In $k[[t]][\varphi] \otimes_k \langle e_1, e_2, \tilde{e}, f_1, f_2\rangle_k$ consider the subset

$$\begin{aligned}\mathcal{R} = \{&t^{q-1}\varphi e_1 - e_2 - t^s\varphi f_2, \quad \varphi e_2 - e_1, \quad t^{q-2-s}\varphi f_1 - f_2, \quad t^{1+s}\varphi f_2 - f_1, \\ &te_1 - \tilde{e}, \quad t\tilde{e}, \quad te_2, \quad tf_1, \quad tf_2\}\end{aligned}$$

and let Δ denote the quotient of $k[[t]][\varphi]\otimes_k \langle e_1, e_2, \tilde{e}, f_1, f_2\rangle_k$ by the $k[[t]][\varphi]$-submodule generated by the elements in $\mathcal{R}$. The exact sequence $0\to\langle f_1, f_2\rangle_k\to\langle e_1, e_2, \tilde{e}, f_1, f_2\rangle_k\to\langle e_1, e_2, \tilde{e}\rangle_k\to 0$ gives rise to an exact sequence

$$0\longrightarrow\frac{k[[t]][\varphi]\otimes_k\langle f_1, f_2\rangle_k}{\nabla_1}\longrightarrow\Delta\longrightarrow\frac{k[[t]][\varphi]\otimes_k\langle e_1, e_2, \tilde{e}\rangle_k}{\nabla_2}\longrightarrow 0,$$

where ∇_1 (resp. ∇_2) is the respective $k[[t]][\varphi]$-submodule generated by $t^{q-2-s}\varphi f_1 - f_2$ and $t^{1+s}\varphi f_2 - f_1$ and tf_1, tf_2 (resp. by $t^{q-1}\varphi e_1 - e_2$ and $\varphi e_2 - e_1$ and $te_1 - \tilde{e}$, $t\tilde{e}, te_2$).

Next, fix $a\in\mathbb{Z}$ and let $\gamma\in\Gamma$ act on $\langle e_1, e_2, \tilde{e}, f_1, f_2\rangle_k$ by means of

$$\gamma\cdot e_1=\gamma^a e_1,\quad\gamma\cdot e_2=\gamma^a e_2,\quad\gamma\cdot\tilde{e}=\gamma^{a+1}\tilde{e},\qquad\gamma\cdot f_1=\gamma^{1+a}f_1,\quad\gamma\cdot f_2=\gamma^{a-s}f_2.$$

Then

$$k[[t]][\varphi]\otimes_k\langle e_1, e_2, \tilde{e}, f_1, f_2\rangle_k\cong k[[t]][\varphi,\Gamma]\otimes_{k[\Gamma]}\langle e_1, e_2, \tilde{e}, f_1, f_2\rangle_k$$

so that $k[[t]][\varphi]\otimes_k\langle e_1, e_2, \tilde{e}, f_1, f_2\rangle_k$ receives a $k[[t]][\varphi,\Gamma]$-action. One checks that all elements in $\mathcal{R}$ are eigenvectors for the action of Γ. It follows that Δ, as well as the above exact sequence are in fact $k[[t]][\varphi,\Gamma]$-equivariant. In view of Proposition 1.5 we get an induced exact sequence (3) of étale (φ,Γ)-modules, with $\dim_{k((t))}(\mathbf{D}_1)=\dim_{k((t))}(\mathbf{D}_2)=2$. We have $\mathbf{D}_1^{\sharp}=\mathbf{D}_1^{\natural}$ and $\mathbf{D}_2^{\sharp}=\mathbf{D}_2^{\natural}$, with both $\mathbf{D}_1$ and $\mathbf{D}_2$ being irreducible, but the sequence (5) is not exact.

(d) Let $\langle e_1, e_2, f_1, f_2\rangle_k$ denote the k-vector space with basis $\{e_1, e_2, f_1, f_2\}$. We view it as a $k[[t]]$-module with trivial action by t. Fix $0\le s\le k\le q-1$. In $k[[t]][\varphi]\otimes_{k[[t]]}\langle e_1, e_2, f_1, f_2\rangle_k$ consider the subset

$$\mathcal{R}=\{t^k\varphi e_1-e_2+t^s\varphi f_2,\quad t^{q-1-k}\varphi e_2-e_1,\quad t^{k-s}\varphi f_1-f_2,\quad t^{q-1-k+s}\varphi f_2-f_1\}$$

and let Δ denote the quotient of $k[[t]][\varphi]\otimes_{k[[t]]}\langle e_1, e_2, f_1, f_2\rangle_k$ by the $k[[t]][\varphi]$-submodule generated by the elements in $\mathcal{R}$. One first checks that there is a natural exact sequence

$$0\longrightarrow\frac{k[[t]][\varphi]\otimes_{k[[t]]}\langle f_1, f_2\rangle_k}{\nabla_1}\longrightarrow\Delta\longrightarrow\frac{k[[t]][\varphi]\otimes_{k[[t]]}\langle e_1, e_2\rangle_k}{\nabla_2}\longrightarrow 0,$$

where ∇_1 (resp. ∇_2) is the respective $k[[t]][\varphi]$-submodule generated by $t^{k-s}\varphi f_1 - f_2$ and $t^{q-1-k+s}\varphi f_2 - f_1$ (resp. by $t^k\varphi e_1 - e_2$ and $t^{q-1-k}\varphi e_2 - e_1$).

Next, let Γ act on $\langle e_1, e_2, f_1, f_2\rangle_k$ by means of

$$\gamma\cdot e_2=\gamma^k e_2,\quad\gamma\cdot f_2=\gamma^{k-s}f_2,\quad\gamma\cdot f_1=f_1,\quad\gamma\cdot e_1=e_1$$

(understanding $\gamma^k e_2$ and $\gamma^{k-s}f_2$ similarly as before). Then

$$k[[t]][\varphi] \otimes_{k[[t]]} \langle e_1, e_2, f_1, f_2 \rangle_k \cong k[[t]][\varphi, \Gamma] \otimes_{k[[t]][\Gamma]} \langle e_1, e_2, f_1, f_2 \rangle_k$$

so that $k[[t]][\varphi] \otimes_{k[[t]]} \langle e_1, e_2, f_1, f_2 \rangle_k$ receives a $k[[t]][\varphi, \Gamma]$-action. Now one checks that all elements in $\mathcal{R}$ are eigenvectors for the action of Γ. It follows that Δ, as well as the above exact sequence are in fact $k[[t]][\varphi, \Gamma]$-equivariant. In view of Proposition 1.5 we get an induced exact sequence (3) of étale (φ, Γ)-modules, with $\dim_{k((t))}(\mathbf{D}_1) = \dim_{k((t))}(\mathbf{D}_2) = 2$. It does not split. The étale (φ, Γ)-module $\mathbf{D}$ lies in the essential image of the functor from supersingular modules over the pro-p Iwahori Hecke algebra of $\mathrm{GL}_2(F)$ to étale (φ, Γ)-modules constructed in [4] if and only if $s = 0$.

Acknowledgements I am very grateful for the invitation to the superb event dedicated to perfectoid spaces at the ICTS Bangalore in 2019, and for the opportunity to contribute to its aftermath by means of the present note. I thank the anonymous referee for his helpful remarks on the text.

References

1. J..M.. Fontaine, Repréesentations p-adiques des corps locaux, in *The Grothendieck Festschrift (P. Cartier et al., ed.), vol. II, Progress in Math*, vol. 87, (Birkhäuser, Boston, 1990), pp. 249–309
2. P. Colmez, Représentations de $\mathrm{GL}_2(\mathbb{Q}_p)$ et (φ, Γ)-modules. Astérisque **330**, 281–509 (2010)
3. G. Zábrádi, Multivariable (φ, Γ)-modules and products of Galois groups. Math. Res. Lett. **25**(2), 687–721 (2018)
4. E. Grosse-Klönne, Supersingular Hecke modules as Galois representations, Algebra and Number Theory. Alg. Number Th. **14**, 67–118 (2020)
5. E. Grosse-Klönne, A note on multivariable (φ, Γ)-modules. Res. Number Theory **5**, 6 (2019)
6. M. Kisin, W. Ren, Galois representations and Lubin tate groups. Doc. Math. **14**, 441–461 (2009)
7. P. Schneider, *Galois representations and (φ, Γ) -modules, Cambridge Studies in Advanced Mathematics* (Cambridge University Press, Cambridge, 2017)
8. G. Pupazan, Multivariable (φ, Γ)-modules and representations of products of Galois groups. Ph.D-thesis, Humboldt University, Berlin (2021)
9. P. Colmez, (φ, Γ)-modules et représentations du mirabolique de $\mathrm{GL}_2(\mathbb{Q}_p)$. Astérisque **330**, 61–153 (2010)

The Relative (de-)Perfectoidification Functor and Motivic p-Adic Cohomologies

Alberto Vezzani

1 Introduction

The categories of (derived, abelian) motives arise naturally by imposing homotopy-invariance onto the (infinity) category of sheaves of Λ-vector spaces on the category of smooth spaces over a base S. Depending on the choice of the topology (typically: the Nisnevich topology or the étale topology), the choice of S (a scheme, a rigid analytic variety [1]...) the choice of the interval over which homotopies are defined (typically the affine line, but there are log-variants [2]) and the choice of the coefficient ring Λ (which may even be omitted [3] or replaced with a ring spectrum [4]) such categories may enjoy different properties and may be useful for the inspection of the various invariants and constructions related to Weil cohomology theories such as periods, Chow groups, the six functor formalism, nearby cycles or even automorphic forms, etc.

The aim of this paper is to make a quick survey on some particular applications of the formalism of motives in the realm of p-adic Hodge theory. More specifically, we consider perfectoid $\mathrm{PerfDA}_{\acute{e}t}(S) = \mathrm{PerfDA}_{\acute{e}t}(S, \mathbb{Q})$ and rigid analytic étale motives $\mathrm{RigDA}_{\acute{e}t}(S) = \mathrm{RigDA}_{\acute{e}t}(S, \mathbb{Q})$. That is, we consider the homotopy invariant infinity-étale sheaves of $\mathbb{Q}$-vector spaces on smooth perfectoid resp. rigid analytic varieties over an adic space S, where homotopies are defined over the perfectoid (closed) ball resp. the rigid analytic (closed) ball.

The author is partially supported by the *Agence Nationale de la Recherche* (ANR), projects ANR-14-CE25-0002 and ANR-18-CE40-0017

A. Vezzani (✉)
Dipartimento di Matematica F. Enriques, Università degli Studi di Milano, Via Cesare Saldini 50, 20133 Milan, Italy

LAGA, Université Sorbonne Paris Nord, 99 Avenue Jean-Baptiste Clément, 93430 Villetaneuse, France
e-mail: alberto.vezzani@unimi.it

© The Author(s), under exclusive license to Springer Nature Singapore Pte Ltd. 2022

D. Banerjee et al. (eds.), *Perfectoid Spaces*, Infosys Science Foundation Series in Mathematical Sciences https://doi.org/10.1007/978-981-16-7121-0_2

In particular, we focus on the *equivalence* between the two categories introduced above:

$$\mathrm{RigDA}_{\text{ét}}(S) \cong \mathrm{PerfDA}_{\text{ét}}(S) \qquad (\clubsuit)$$

that is shown in [5]. Such an equivalence can be considered as a method to "(de-)perfectoidify" functorially and canonically an adic space over a base, up to homotopy. We remark that whenever S is perfectoid, there is a canonical equivalence $\mathrm{PerfDA}_{\text{ét}}(S) \cong \mathrm{PerfDA}_{\text{ét}}(S^\flat)$ induced by the classic tilting functor of perfectoid spaces, which preserves homotopies and the étale sites. This leads to an equivalence

$$\mathrm{RigDA}_{\text{ét}}(S) \cong \mathrm{RigDA}_{\text{ét}}(S^\flat) \qquad (\spadesuit)$$

that can be interpreted as a way to "(un-)tilt" canonically and functorially even rigid analytic spaces, up to homotopy. It is expected (see [5]) to give the following generalization of ($\spadesuit$) which should hold for an arbitrary adic space S over $\mathbb{Q}_p$, using the language of diamonds:

$$\mathrm{RigDA}_{\text{ét}}(S) \cong \mathrm{RigDA}_{\text{ét}}(S^\diamond). \qquad (\diamondsuit)$$

In this paper, we give a full proof of ($\clubsuit$) in the case of a perfectoid base S in characteristic p, generalizing the statement of [6] that only deals with the case of a perfectoid field $S = \mathrm{Spa}(K, K^\circ)$.

Moreover, we make a survey on how the language of motives can be used to define and prove some fundamental properties of de Rham-like p-adic cohomologies on adic spaces and algebraic variety in characteristic p (that is, Große-Klönne's overconvergent de Rham cohomology, and Berthelot's rigid cohomology). We then recall how to merge such constructions with the (un-)tilting and (de-)perfection procedures of ($\clubsuit$)-($\spadesuit$)-($\diamondsuit$) and obtain new de Rham like cohomology theories for perfectoid varieties and rigid spaces in positive characteristic. Finally, we cite further cohomology theories that have been introduced using rigid motives by other authors (such as Ayoub and Le Bras) and a Betti-like cohomology in the spirit of Berkovich. We insist on the fact that, in all these procedures, the role of homotopies is crucial, and that consequently, motivic categories provide a natural framework where such definitions and proofs can be made.

2 Definitions and Main Properties of Adic Motives

Once and for all, we fix a cardinal κ and we consider only adic spaces that have a κ-small covering by affinoid subspaces. The categories of motives that we will introduce are easily seen not to depend on κ, but this choice allows one to prove that they are compactly generated, under suitable hypotheses (see [7, Proposition 2.4.20]).

Definition 2.1 Let K be a non-archimedean field, and S be a stably uniform adic space over it.

(1) We let $\mathbb{B}^1_K$ be the rigid analytic variety $\operatorname{Spa}(K\langle T\rangle, \mathcal{O}_K\langle T\rangle)$ and $\mathbb{B}^1_S$ be the fiber product $S \times_K \mathbb{B}^1_K$, for any adic space S over K.
(2) If K is perfectoid, we let $\widehat{\mathbb{B}}^1_K$ be the perfectoid space $\operatorname{Spa}(K\langle T^{1/p^\infty}\rangle, \mathcal{O}_K\langle T^{1/p^\infty}\rangle)$ and $\widehat{\mathbb{B}}^1_S$ be the fiber product $S \times_K \widehat{\mathbb{B}}^1_K$, for any perfectoid space S over K.
(3) We also let $\mathbb{T}^1_S$ [resp. $\widehat{\mathbb{T}}^1_S$] be the rational open $U(1/T)$ of $\mathbb{B}^1_K$ [resp. of $\widehat{\mathbb{B}}^1_K$].
(4) We let RigSm/S be the full subcategory of adic spaces over S whose objects are locally étale over a poly-disc $\mathbb{B}^N_S$ (in case S is a rigid analytic variety, this recovers the usual notion of smooth rigid analytic varieties over S) and equip it with the étale topology.
(5) In case K is a perfectoid field and S is perfectoid, we also consider the full subcategory PerfSm/S of adic spaces over S whose objects are locally étale over the perfectoid poly-disc $\widehat{\mathbb{B}}^N_S$, and equip it with the étale topology.

Definition 2.2 Let K and S be as above, and Λ be a (commutative, unital) $\mathbb{Q}$-algebra.

(1) We let $\operatorname{Sh}_{\text{ét}}(\operatorname{RigSm}/S, \Lambda)$ [resp. $\operatorname{Sh}_{\text{ét}}(\operatorname{PerfSm}/S, \Lambda)$] be the monoidal DG-category of complexes of étale sheaves of Λ-vector spaces on RigSm/S [resp. PerfSm/S].
(2) We let $\operatorname{RigDA}^{\text{eff}}_{\text{ét},\mathbb{B}^1}(S, \Lambda)$ (or $\operatorname{RigDA}^{\text{eff}}_{\text{ét}}(S)$ for short) be the monoidal DG-subcategory of $\operatorname{Sh}_{\text{ét}}(\operatorname{RigSm}/S, \Lambda)$ spanned by those objects $\mathcal{F}$ that are $\mathbb{B}^1$-local, meaning that the natural map $\mathcal{F}(\mathbb{B}^1_X) \to \mathcal{F}(X)$ is an equivalence, for all $X \in \operatorname{RigSm}/S$. We recall that there is a left adjoint $\operatorname{Sh}_{\text{ét}}(\operatorname{RigSm}/S, \Lambda) \to \operatorname{RigDA}^{\text{eff}}_{\text{ét}}(S)$ to the natural inclusion.
(3) Similarly, we let $\operatorname{PerfDA}^{\text{eff}}_{\text{ét},\widehat{\mathbb{B}}^1}(S, \Lambda)$ (or $\operatorname{PerfDA}^{\text{eff}}_{\text{ét}}(S)$ for short) be the DG-subcategory of $\operatorname{Sh}_{\text{ét}}(\operatorname{PerfSm}/S, \Lambda)$ spanned by those objects $\mathcal{F}$ that are $\widehat{\mathbb{B}}^1$-local.
(4) We will use the same notation $\operatorname{RigDA}^{\text{eff}}_{\text{ét}}(S)$, $\operatorname{PerfDA}^{\text{eff}}_{\text{ét}}(S)$ for the associated monoidal stable infinity-categories.

We remark that Yoneda defines a functor

$$h\colon \operatorname{RigSm}/S \to \operatorname{Psh}(\operatorname{RigSm}/S) \to \operatorname{Psh}(\operatorname{RigSm}/S, \Lambda) \to \operatorname{RigDA}^{\text{eff}}_{\text{ét}}(S)$$

and for any X we will let $\Lambda_S(X)$ be the image of X under h. We use the same notation for perfectoid spaces and $\operatorname{PerfDA}^{\text{eff}}_{\text{ét}}(S)$.

Definition 2.3 We let K, S and Λ be as above.

(1) We let T_S be the quotient of the split inclusion $\Lambda_S(S) \to \Lambda_S(\mathbb{T}^1)$ given by the unit.
(2) Similarly, if S is perfectoid, we define an object $\widehat{T}_S$ in $\operatorname{PerfDA}^{\text{eff}}_{\text{ét}}(S)$ as the quotient of the split inclusion $\Lambda_S(S) \to \Lambda_S(\widehat{\mathbb{T}}^1)$ given by the unit.
(3) We introduce $\operatorname{RigDA}_{\text{ét}}(S)$ and $\operatorname{PerfDA}_{\text{ét}}(S)$ as the targets of the universal left adjoint DG-functors

$$\mathrm{RigDA}^{\mathrm{eff}}_{\text{ét}}(S) \to \mathrm{RigDA}_{\text{ét}}(S) \quad \mathrm{PerfDA}^{\mathrm{eff}}_{\text{ét}}(S) \to \mathrm{PerfDA}_{\text{ét}}(S)$$

to DG-categories in which the endofunctor $- \otimes T_S$ [resp. $- \otimes \widehat{T}_S$] becomes invertible. They are endowed with a monoidal structure for which the functors above are monoidal.

(4) We use the notation $\mathrm{RigDA}_{\text{ét}}(S)$ and $\mathrm{PerfDA}_{\text{ét}}(S)$ also for the associated monoidal stable infinity-categories.

(5) When we write $\mathrm{RigDA}^{(\mathrm{eff})}_{\text{ét}}(S)$ we mean that one can consider either the category $\mathrm{RigDA}^{\mathrm{eff}}_{\text{ét}}(S)$ (eff standing for *effective* motives) or the category $\mathrm{RigDA}_{\text{ét}}(S)$, and similarly for $\mathrm{PerfDA}^{(\mathrm{eff})}_{\text{ét}}(S)$.

All in all, in the category $\mathrm{RigDA}^{\mathrm{eff}}_{\text{ét}}(S)$ one can find objects of the form $\Lambda_S(X)$ where X is any smooth rigid analytic variety over S coming from the Yoneda functor, as well as any complex of sheaves Ω that represents a Weil cohomology theory (with Λ-coefficients, in our situation). The homology of the mapping complexes $\mathrm{Map}(\Lambda_S(X), \Omega)$ coincide with the cohomology theory associated to Ω. Almost by construction, we point out that the objects $\Lambda_S(X)$ are isomorphic to $\Lambda_S(\mathbb{B}^1_X)$ and coincide with the homotopy colimit of any diagram of the form $\Lambda_S(U^\bullet)$ with $U^\bullet$ being an étale hypercover of X. This translates in terms of cohomology theories into $\mathbb{B}^1$-invariance, and the existence of some exact sequences à la Mayer-Vietoris.

By means of the six-functor formalism (see [8]) it is possible to define motives $\Lambda_S(X)$ attached to *any* rigid analytic variety X over S (not necessarily smooth). In particular, the definition of a well-behaved cohomology theory on smooth varieties extends automatically to all varieties.

It is also possible to consider (co-)homology theories which are equipped with a richer structure than the one of a mere Λ-module: as soon as one has a functor $H\colon \mathrm{RigSm}/S \to \mathbf{C}$ with $\mathbf{C}$ being a Λ-linear DG-category such that H satisfies étale descent and is homotopy invariant [and for which the Tate twist is invertible] then by construction one can (Kan) extend it to motives

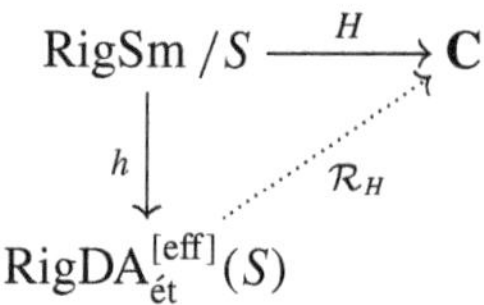

obtaining a so-called *realization functor* $\mathcal{R}_H$. In Sect. 5 we will try to convince the reader that it is sometimes *easier* to define a motivic realization $\mathcal{R}_H$ and hence *deduce* an interesting (co-)homology theory H on RigSm/S.

Remark 2.4 One is free to replace the sites, the interval objects, and the rings of coefficients with any other choice and define corresponding categories of motives. Classically, the categories of étale motives over a scheme S are denoted by $\mathrm{DA}^{(\mathrm{eff})}_{\text{ét}}(S, \Lambda)$ (here, $\mathrm{DA}^{(\mathrm{eff})}_{\text{ét}}(S)$ for short). One may also consider non-commutative variants where the category of Λ-modules is replaced by the infinity-category of spectra, or the category of modules of any commutative ring spectrum.

Remark 2.5 For the categories of algebraic motives $\mathrm{DA}_{\text{ét}}(S)$ the realizations functors induced by Betti, de Rham and ℓ-adic cohomologies have been widely studied in different articles: see [9–11].

Rather than making a full recollection of all the formal properties of motives and their variants, for which there are already staple references such as [9, 10, 12, 13], we focus on two peculiar properties of the categories of rigid motives which are proved in [7].

The first property is the so-called (effective) semi-separatedness.

Theorem 2.6 *([7, Corollary 2.9.10]) Let $S' \to S$ be a universal homeomorphism. The base change functor induces an equivalence of categories*

$$RigDA_{ét}^{(\mathrm{eff})}(S) \cong RigDA_{ét}^{(\mathrm{eff})}(S')$$

Remark 2.7 As noted in [7], the effective part of the statement is not known for the usual algebraic motives $\mathrm{DA}_{\text{ét}}^{\mathrm{eff}}(S)$.

Corollary 2.8 *Let $X' \to X$ be a universal homeomorphism between smooth rigid analytic varieties over a base S. The induced map of motives $\Lambda_S(X') \to \Lambda_S(X)$ is invertible in $RigDA_{ét}^{(\mathrm{eff})}(S)$.*

Proof The motive $\Lambda(X)$ is the image of $\Lambda_X(X) = \Lambda$ under the functor $p_\sharp$ which is the left adjoint to the functor $\mathrm{RigDA}_{\text{ét}}^{\mathrm{eff}}(S) \to \mathrm{RigDA}_{\text{ét}}^{\mathrm{eff}}(X)$ induced by the pullback p^* along the map $p\colon X \to S$. By the previous theorem, we deduce $p_\sharp\Lambda_X(X) \cong p'_\sharp\Lambda_{X'}(X')$ as wanted. □

The second property is referred to as "continuity" in [7].

Theorem 2.9 *([7, Theorem 2.8.14]) Let $\{S_i\}$ be a cofiltered system of stably uniform adic spaces over a non-archimedean field K with qcqs transition maps and let S be a uniform adic space such that $S \sim \varprojlim S_i$ in the sense of Huber [14, (2.4.1)]. Then the base change functors induce an equivalence of categories*

$$RigDA_{ét}^{(\mathrm{eff})}(S) \cong \varinjlim RigDA_{ét}^{(\mathrm{eff})}(S_i)$$

where the homotopy colimit is computed in the category of presentable infinity-categories, and colimit-preserving functors.

Remark 2.10 The analogous statement for algebraic motives also holds: in case S is the limit of a diagram of schemes $\{S_i\}$ with affine transition maps, then $\mathrm{DA}_{\text{ét}}^{(\mathrm{eff})}(S) \cong \varinjlim \mathrm{DA}_{\text{ét}}^{(\mathrm{eff})}(S_i)$. However, the proof of the analytic version is much more involved, and uses homotopies in a crucial way: this is related to the fact that (in case all spaces S, S_i are affinoid) the ring $\varinjlim \mathcal{O}(S_i)$ does not coincide with $\mathcal{O}(S)$, but it is only dense in it. In particular, the "continuity" statement for étale sheaves (on the big sites), before performing the $\mathbb{B}^1$-localization, is false.

A special case of continuity gives the following computation of "stalks" for $\mathrm{RigDA}_{\text{ét}}(-)$.

Corollary 2.11 *Let $s = \mathrm{Spa}(C, C^+) \to S$ be an étale point of a stably uniform adic space S. Let $U \to S$ vary among étale neighborhoods of s in S. The base-change functors induce an equivalence in the category of presentable infinity-categories and colimit-preserving functors:*

$$\mathit{RigDA}^{\mathrm{eff}}_{\acute{e}t}(\mathrm{Spa}(C, C^+)) \cong \varinjlim_{s\in U\to S} \mathit{RigDA}^{(\mathrm{eff})}_{\acute{e}t}(U).$$

3 De-Perfectoidification

The aim of this section is to prove the following.

Theorem 3.1 *Let S be a rigid space over a non-archimedean field K of characteristic p. Then the base change along $S^{\mathrm{Perf}} \to S$ and the relative perfection functor define equivalences:*

$$\mathit{RigDA}^{(\mathrm{eff})}_{\acute{e}t}(\mathrm{S}) \cong \mathit{RigDA}^{(\mathrm{eff})}_{\acute{e}t}(\mathrm{S}^{\mathrm{Perf}}) \cong \mathit{PerfDA}^{(\mathrm{eff})}_{\acute{e}t}(\mathrm{S}^{\mathrm{Perf}}).$$

The first half of the statement follows from the "separatedness" and the "continuity" properties of $\mathrm{RigDA}_{\text{ét}}$.

Proposition 3.2 *Let S be a rigid space over a non-archimedean field K of characteristic p. Then the base change along $S^{\mathrm{Perf}} \to S$ defines an equivalence:*

$$\mathit{RigDA}^{(\mathrm{eff})}_{\acute{e}t}(S) \cong \mathit{RigDA}^{(\mathrm{eff})}_{\acute{e}t}(S^{\mathrm{Perf}}).$$

Proof The space S^{Perf} is a weak projective limit of the diagram $\cdots \to S \overset{\varphi}{\to} S \overset{\varphi}{\to} S$ with φ being the Frobenius. We note that by Theorem 2.6 the motivic functor φ^* is an equivalence, and the claim then follows from Theorem 2.9. □

We now move to the second part of the statement. From now on, we will therefore assume that $S = S^{\mathrm{Perf}}$ is a perfectoid space of characteristic $p > 0$. The second half is a refinement of [6, Theorem 6.9] in two different directions: on the one hand we get rid of the Frobét-localization (or, equivalently, of correspondences see [15]) proving an effective claim that holds for $\mathrm{RigDA}^{\mathrm{eff}}_{\text{ét}}$; on the other hand, we promote the equivalence from the case of a base field of height one $\mathrm{Spa}(K^\flat, K^{\flat\circ})$ to a general (perfectoid) base S.

It is worth noting that, because of the computation of stalks for $\mathrm{RigDA}^{(\mathrm{eff})}_{\text{ét}}(-)$ (which can be generalized easily to $\mathrm{PerfDA}^{(\mathrm{eff})}_{\text{ét}}(-)$) the missing crucial case is the one of a base $S = \mathrm{Spa}(C, C^+)$ which is a complete algebraically closed valued field, with a valuation of height $n \in \mathbb{N}_{\geq 2}$. As this case is not essentially easier than the one of a general base S we do not restrict to this case in what follows.

In order to extend the result of [6], we follow the blueprint given by the proof of loc. cit., and we simplify it at the same time. We try to highlight here the main differences with respect to the original approach. We first introduce some notation.

Definition 3.3 Let $\mathcal{F}$ be in $\mathrm{Psh}(\mathrm{RigSm}/S, \Lambda)$.

(1) We let $\mathbb{L}_{\varphi}\mathcal{F}$ be the presheaf $X \mapsto \varinjlim_n \mathcal{F}(X^{(p^{-n})})$ where we let $X^{(p^{-n})}$ be $X \times_{S,\varphi^{-n}} S$ and $X^{(p^{-n-1})} \to X^{(p^{-n})}$ be the map induced by Frobenius.
(2) We let $\mathbb{L}_{\mathbb{B}^1}\mathcal{F}$ be the normalized complex associated to the cubical presheaf of complexes of abelian groups $\underline{\mathrm{Hom}}(\mathbb{B}^{\bullet}, \mathcal{F})$ where $\underline{\mathrm{Hom}}(\mathbb{B}^r, \mathcal{F})(-) = \mathcal{F}((-)\langle u_1, \dots, u_r\rangle)$.

Proposition 3.4 *Let $\mathcal{F}$ be in* $\mathrm{Psh}(\mathrm{RigSm}/S, \Lambda)$. *The natural map $\mathcal{F} \to \mathbb{L}_{\mathbb{B}^1}\mathbb{L}_{\varphi}\mathcal{F}$ is an equivalence in $\mathrm{RigDA}^{\mathrm{eff}}_{\acute{e}t}(S, \Lambda)$.*

Proof It is well known that the maps $\mathcal{F} \to \mathbb{L}_{\mathbb{B}^1}\mathcal{F}$ are $\mathbb{B}^1$-equivalences, see [16]. By construction, the complex $\mathbb{L}_{\varphi}\mathcal{F}$ is local with respect to the relative Frobenius maps $\Lambda_S(X^{(p^{-1})}) \to \Lambda_S(X)$ (we will refer to this property as "being Frob-local") and doesn't alter those which are already Frob-local. We deduce that the map $\mathcal{F} \to \mathbb{L}_{\varphi}\mathcal{F}$ is a Frob-local equivalence, that is an equivalence with respect to the localization over relative Frobenius maps. In particular, the map of the statement is a ($\mathbb{B}^1$, Frob)-local equivalence hence a ($\mathbb{B}^1$, Frob, ét)-local equivalence, but the latter are simply ($\mathbb{B}^1$, ét)-local equivalences as shown in Corollary 2.8. □

Proposition 3.5 *Locally with respect to the analytic topology, any space $X \in$* RigSm/S *[resp.* PerfSm/S*] is given by* $\mathrm{Spa}(R, R^+)$ *with* (R, R^+) *given by [the completed perfection of] the following adic pair*

$$\left(\mathcal{O}(U)\langle\underline{x}, \underline{y}\rangle/(P_1, \dots, P_m), \mathcal{O}(U)\langle\underline{x}, \underline{y}\rangle/(P_1, \dots, P_m)^+\rangle\right) \qquad (\star)$$

where $U \subset S$ is an affinoid subspace, $\underline{x} := (x_1, \dots, x_n)$ is a n-tuple of variables with $n \in \mathbb{N}$, $\underline{y} := (y_1, \dots, y_m)$ is a m-tuple of variables with $m \in \mathbb{N}$, and $\underline{P} := (P_1, \dots, P_m)$ is a m-tuple of polynomials in $\mathcal{O}(U)[\underline{x}, \underline{y}]$ such that $\det(\frac{\partial P_i}{\partial y_j})$ *is invertible in $\mathcal{O}(U)\langle\underline{x}, \underline{y}\rangle/(P_1, \dots, P_m)$.*

Proof Any étale space over the perfectoid relative poly-disc $\widehat{\mathbb{B}}^n_S \sim \varprojlim_{\varphi} \mathbb{B}^n_S$ is locally defined over $\mathbb{B}^n_S$ so the claim on RigSm/S, which follows from [14, Proposition 1.7.1(iii)] immediately implies the claim on PerfSm/S. □

Proposition 3.6 *Let* $X = \mathrm{Spa}(R, R^+)$ *and* $X' = \mathrm{Spa}(R', R'^+)$ *be spaces in* RigSm/S *of the form* ($\star$) *and let $f : X'^{\mathrm{Perf}} \to X$ be a morphism. There exists a homotopy $H : X'^{\mathrm{Perf}} \times \widehat{\mathbb{B}}^1 \to X$ such that $H_0 = f$ and H_1 has a (unique) model $X'^{(p^{-n})} \to X$ for some $n \gg 0$.*

Proof We let R'^+_n be the image in $\widehat{R'^+} := \mathcal{O}^+(X'^{\mathrm{Perf}})$ of the injective map $\mathcal{O}^+(X'^{(p^{-n})}) \to \mathcal{O}^+(X'^{\mathrm{Perf}})$ (we recall that X' is reduced as it is smooth over the reduced space S which is perfectoid) and we remark that $\widehat{R}'^{+}$ is the π-adic completion of $R'^{+}_{\infty} := \bigcup R'^{+}_{n}$. The morphism f is determined by some mapping $x_i \mapsto s_i \in \widehat{R}'^{+}$ and $y_j \mapsto t_j \in \widehat{R}'^{+}$. By means of [6, Corollary A.2] we may find a unique array of power series $F_1, \dots, F_m \in \widehat{R}'[\![\underline{\sigma} - \underline{s}]\!]$ such that $P(\sigma, F(\sigma)) = 0$, $F(\underline{s}) = \underline{t}$. Moreover, they are in $\widehat{R}'^{+}[\![\pi^{-N}(\underline{\sigma} - \underline{s})]\!]$ for a sufficiently big $N \gg 0$. For any $\underline{\tilde{s}} \in R'^{+}_{\infty}$ which is sufficiently close to $\underline{s}$ we may then define a homotopy H as the map determined by

$$(\underline{x}, \underline{y}) \mapsto (\underline{s} + (\underline{s} - \underline{\tilde{s}}) \cdot \tau, \underline{F}((\underline{s} - \underline{\tilde{s}}) \cdot \tau))$$

and remark that, by definition, we have $H_0 = f$. In order to show that H_1 factors (uniquely, as the maps $R'_n \to \widehat{R}'$ are injective) over some $X'^{(p^{-n})}$, we are left to show that the elements $\underline{\tilde{t}} := \underline{F}(\underline{s} - \underline{\tilde{s}})$ lie in R'_{∞}.

Suppose without loss of generality that $\underline{\tilde{s}}$ lie in $R' = R'_0$. We consider the R'-algebra E defined as $E = R'\langle \underline{y} \rangle / (\underline{P}(\underline{\tilde{s}}, \underline{y}))$ which is étale over R', and over which the map $R' \to \widehat{R}'$ factors. In particular, the étale morphism $\mathrm{Spa}(E, E^{+}) \times_{X'} X'^{\mathrm{Perf}} \to X'^{\mathrm{Perf}}$ splits. In light of the equivalence between the étale topos of X'^{Perf} and X_0 we conclude that $\mathrm{Spa}(E, E^{+}) \to X'$ splits proving that $\tilde{t}$ is a m-tuple in R'_0 as wanted. □

Proposition 3.7 *Let* $X = \mathrm{Spa}(R, R^{+})$ *and* $X' = \mathrm{Spa}(R', R'^{+})$ *be spaces in* RigSm/S *of the form* ($\star$). *The canonical map*

$$(\mathbb{L}_{\mathbb{B}^1}\mathbb{L}_{\varphi}\Lambda(X))(X') \to (\mathbb{L}_{\mathbb{B}^1}\mathbb{L}_{\varphi}\,\mathrm{Perf}_*\,\mathrm{Perf}^*\,\Lambda(X))(X')$$

is a quasi-isomorphism.

Proof By direct inspection, we may rewrite the two complexes above as follows:

$$\varinjlim_{n} N\Lambda((X' \times \mathbb{B}^{\bullet})^{(p^{-n})}, X) \to N\Lambda(X'^{\mathrm{Perf}} \times \widehat{\mathbb{B}}^{\bullet}, X)$$

with N denoting the normalized complex associated to the cubical complex of abelian groups. The claim then follows from (the proof of) Proposition 3.6 by arguing as in [6, Proposition 4.2]. □

Proof of Theorem 3.1 The effective part of the theorem easily implies the stable version, so we stick to it for simplicity. By means of Proposition 3.5 and the equivalence fo the étale site of X^{Perf} and of X we see that Perf^* sends a class of compact generators to a class of compact generators, and that Perf_* commutes with ét-sheafification, preserving then the ét-local equivalences. The multiplication μ on $\widehat{\mathbb{B}}^1$ defines a morphism

$$\mathrm{Perf}^*(\mathrm{Perf}_*\,\Lambda_S(\widehat{\mathbb{B}}^1_X) \otimes \Lambda_S(\mathbb{B}^1_S)) \cong \mathrm{Perf}^*\,\mathrm{Perf}_*\,\Lambda_S(\widehat{\mathbb{B}}^1_X) \otimes \Lambda_S(\widehat{\mathbb{B}}^1_S) \to \Lambda_S(\widehat{\mathbb{B}}^1_X) \otimes \widehat{\mathbb{B}}^1_S \xrightarrow{\mu} \Lambda_S(\widehat{\mathbb{B}}^1_X)$$

which induces a homotopy between the identity and the zero-map on $\mathrm{Perf}_*(\Lambda(\widehat{\mathbb{B}}^1_X))$, showing that Perf_* sends $\widehat{\mathbb{B}}^1$-local equivalences to $\mathbb{B}^1$-local equivalences.

We deduce that in order to prove the claim, it suffices to show that $\mathcal{F} \to \operatorname{Perf}_* \operatorname{Perf}^* \mathcal{F}$ is a $(\mathbb{B}^1, \text{ét})$-local equivalence for any $\mathcal{F}$ and we may actually restrict to the case where $\mathcal{F}$ is $\Lambda_S(X)$ with X as in $(\star)$ as such motives are a class of compact generators (by Proposition 3.5). Using Proposition 3.4 we may alternatively prove that $\mathbb{L}_{\mathbb{B}^1}\mathbb{L}_{\varphi}\Lambda_S(X) \to \mathbb{L}_{\mathbb{B}^1}\mathbb{L}_{\varphi} \operatorname{Perf}_* \operatorname{Perf}^* \Lambda_S(X)$ is a weak-equivalence, and this follows from Proposition 3.7. □

When trying to generalize the second half of Theorem 3.1 to the case of a perfectoid S in characteristic 0, one is immediately stopped by the lack of a canonical map $\operatorname{RigDA}_{\text{ét}}(S) \to \operatorname{PerfDA}_{\text{ét}}(S)$ which is as "geometric" as the one given by the perfection in positive characteristic. As in [6] we now give an alternative route to constructing such a map, in a compatible way with the characteristic p case.

Definition 3.8 We let $s\operatorname{PerfSm}/S$ be the full subcategory of Rig/S whose objects are spaces X that are locally étale over $\mathbb{B}^N_S \times_S \widehat{\mathbb{B}}^M_S$ for some $M, N \in \mathbb{N}$. This category obviously contains Sm/S (by taking $M = 0$) and PerfSm/S (by taking $N = 0$). We let $s\operatorname{PerfDA}^{\text{eff}}_{\text{ét}}(S)$ be the category of $\widehat{\mathbb{B}}^1$-invariant étale (hyper)sheaves on $s\operatorname{PerfSm}/S$ with values in Λ-modules. The continuous inclusions $\alpha\colon \operatorname{Sm}/S \to s\operatorname{PerfSm}/S$ and $\beta\colon \operatorname{PerfSm}/S \to s\operatorname{PerfSm}/S$ induce adjoint pairs

$$\alpha^*\colon \operatorname{RigDA}^{\text{eff}}_{\text{ét}}(S) \rightleftarrows s\operatorname{PerfDA}^{\text{eff}}_{\text{ét}}(S) :\alpha_*$$

and

$$\beta^*\colon \operatorname{PerfDA}^{\text{eff}}_{\text{ét}}(S) \rightleftarrows s\operatorname{PerfDA}^{\text{eff}}_{\text{ét}}(S) :\beta_*.$$

In particular, there is a functor $\beta_*\alpha^*\colon \operatorname{RigDA}^{\text{eff}}_{\text{ét}}(S) \to \operatorname{PerfDA}^{\text{eff}}_{\text{ét}}(S)$.

We remark that the functor above is the same as the one of Theorem 3.1 in case $\operatorname{char} S = p$. Indeed, under this hypothesis, we may consider the relative perfection functor also at the level of semi-perfectoid spaces $s\operatorname{PerfSm}/S \to \operatorname{PerfSm}/S$, $X \mapsto X^{\operatorname{Perf}}$. It induces an adjoint pair

$$\operatorname{Perf}'^*\colon \operatorname{RigDA}^{\text{eff}}_{\text{ét}}(S) \rightleftarrows s\operatorname{PerfDA}^{\text{eff}}_{\text{ét}}(S) :\operatorname{Perf}'_*$$

and we note that the functor Perf^* is nothing more than the composition $\operatorname{Perf}'^*\alpha^*$. Our claim then follows from the following:

Proposition 3.9 *Suppose that S has characteristic p. The functor β^* is a left adjoint to* Perf'^*. *In particular,* $\beta_*\alpha^* \cong \operatorname{Perf}^*$.

Proof We remark that $\beta\colon \operatorname{PerfSm}/S \to s\operatorname{PerfSm}/S$ is a left adjoint to $\operatorname{Perf}'\colon s\operatorname{PerfSm}/S \to \operatorname{PerfSm}/S$. By the Yoneda lemma, we deduce that they extend to an adjoint pair

$$\beta^*, \colon \operatorname{Psh}(\operatorname{PerfSm}/S, \Lambda) \rightleftarrows \operatorname{Psh}(s\operatorname{PerfSm}/S, \Lambda)\colon \operatorname{Perf}'^*$$

between the (infinity) categories of (complexes of) presheaves. Both functors preserve étale (hyper)covers, fiber products and the object $\widehat{\mathbb{B}}^1_S$ so they both preserve (ét, $\widehat{\mathbb{B}}^1_S$)-equivalences. As Perf'^* has a left adjoint that preserves these equivalences, we deduce that it also preserves (ét, $\widehat{\mathbb{B}}^1_S$)-local objects. We then conclude that the adjunction (β^*, Perf'^*) extends to an adjunction on the motivic categories, as wanted. □

It is a non-trivial endeavor to prove the following generalization of Theorem 3.1 whose proof we won't comment here.

Theorem 3.10 *([5]) Let S be a perfectoid space over some field. The functor $\beta_*\alpha^*$ defines an equivalence $RigDA^{(\mathrm{eff})}_{\acute{e}t}(S) \cong PerfDA^{(\mathrm{eff})}_{\acute{e}t}(S)$.*

By putting together Theorem 3.1 and the previous result, we obtain the following:

Corollary 3.11 *Let S be a perfectoid space. There is an equivalence $RigDA^{(\mathrm{eff})}_{\acute{e}t}(S) \cong PerfDA^{(\mathrm{eff})}_{\acute{e}t}(S) \cong RigDA^{(\mathrm{eff})}_{\acute{e}t}(S^\flat)$.*

Proof The tilting equivalence translates motivically into an equivalence $\mathrm{PerfDA}^{(\mathrm{eff})}_{\mathrm{\acute{e}t}}(S) \cong \mathrm{PerfDA}^{(\mathrm{eff})}_{\mathrm{\acute{e}t}}(S^\flat)$. The equivalence of the statement is then obtained by putting together Theorems 3.1 and 3.10. □

4 Classic De Rham-Like Cohomologies via Motives

In this section, we make a survey on the "classic" de Rham-like cohomology theories for rigid analytic varieties and perfectoid spaces, revisited in the language of motives, based on [17] and [18] which is further expanded by [5].

Remark 4.1 Though we won't comment on them in the present article, also ℓ-adic realizations for analytic motives have been defined in [19] and [7, Sect. 2.10].

We start by a recollection of standard facts on the rigid and the overconvergent de Rham cohomologies, that will be necessarily imprecise and incomplete. All the details can be found in [20–23].

Let's fix a field k of characteristic $p > 0$ that we will assume to be perfect (for simplicity). The approach of Berthelot [24] and Monsky-Washnitzer [25] for the definition of a de Rham-like p-adic cohomology for varieties over k can be summarized (somehow a posteriori, following Große-Klönne) as follows: a smooth variety $\bar{X}$ over k can be lifted locally as a smooth variety $\mathcal{X}$ over $W(k)$ (the DVR given by the Witt ring). The choice of such lifts is unique, étale-locally on the special fiber, "up to homotopy", and even canonical "up to automorphisms" if we consider smooth formal lifts $\mathfrak{X}$ over $W(k)$. A precise statement can be found in [26, Théorèmes 2.2.2, 3.3.2]. It is therefore possible, "somehow canonically" to associate locally a smooth rigid analytic variety (the generic fiber X of $\mathfrak{X}$) to the smooth variety $\bar{X}$.

This way, we have changed the base field: from k (of positive characteristic) to $K = \mathrm{Frac}\, W(k)$ which has characteristic 0, but with a major drawback: we now have to consider *rigid analytic* varieties rather than algebraic varieties. Without further structure, a (non-proper) smooth rigid analytic variety doesn't give rise to a well-behaved de Rham-like cohomology theory. One needs to do a choice of a "thickening" (what we will call an *overconvergent* structure following [27]) $X^\dagger := (X \Subset X')$ of X into a strictly larger (i.e., containing the absolute compactification over K) smooth rigid analytic variety X' and consider the subcomplex $\Omega_{X^\dagger/K}$ of $\Omega_{X/K}$ of those differential forms that extend to a strict neighborhood of X inside X'. Once again, such local choices are sufficiently canonical, "up to homotopy" (see [22]). It is therefore possible "somehow canonically" to associate locally to the smooth variety X a smooth overconvergent variety $X^\dagger$ and a de Rham-like complex $\Omega_{X^\dagger/K}$ which is used to define a cohomology theory for X, and a posteriori for $\bar{X}$ by combining the two procedures above. It is a non-trivial task to prove that these cohomology theories are well-defined and functorial, and enjoy the expected properties of a de Rham-like cohomology theory (for example, being finite dimensional on qcqs varieties): see [23, 28–30] etc.

We now give an alternative way to describe the above phenomena. Since the eventual aim is to define a Weil cohomology theory for varieties over k [resp. analytic varieties over K] it is quite natural to consider the motivic categories associated to these objects. As expected, they form a convenient setting where to state and study lifts and thickenings "up to homotopy". We collect the principal motivic facts in the following statement.

Theorem 4.2 *Let K be a complete non-archimedean field of characteristic 0 with valuation ring $\mathcal{O}_K$ and a perfect residue field k, and let Λ be a $\mathbb{Q}$-subalgebra of K. We also let $\mathbb{B}^{1\dagger}_K$ be the overconvergent variety given by the strict embedding $\mathbb{B}^1_K \Subset \mathbb{P}^{1\,\mathrm{an}}_K$ and $RigDA^{\dagger(\mathrm{eff})}_{\acute{e}t}(K)$ be the (effective) DG-category of $\mathbb{B}^{1\dagger}$-invariant étale sheaves of Λ-vector spaces on smooth overconvergent varieties over K.*

(1) The complex of presheaves $\Omega^\dagger\colon X^\dagger \mapsto \Omega_{X^\dagger/K}$ is a $(\mathbb{B}^{1\dagger}, \text{ét})$-local object of $RigDA^{\dagger\,\mathrm{eff}}_{\acute{e}t}(K)$. In particular, for any overconvergent smooth rigid variety $X^\dagger$ one has

$$\mathrm{Map}(\Lambda_K(X^\dagger), \Omega^\dagger) \cong \Omega^\dagger_{X^\dagger/K}.$$

(2) The forgetful functor $l\colon X^\dagger = (X \Subset X') \mapsto X$ induces an equivalence of monoidal compactly generated stable infinity categories $RigDA^{\dagger(\mathrm{eff})}_{\acute{e}t}(K) \cong RigDA^{(\mathrm{eff})}_{\acute{e}t}(K)$.

(3) The analytification functor $\tilde{X} \mapsto \tilde{X}^{\mathrm{an}}$ induces a compact-preserving, colimit-preserving map of monoidal compactly generated stable infinity categories $\mathrm{DA}^{(\mathrm{eff})}_{\acute{e}t}(K) \to RigDA^{(\mathrm{eff})}_{\acute{e}t}(K)$.

(4) The special fiber functor induces an equivalence of monoidal compactly generated stable infinity categories $\mathrm{FDA}^{(\mathrm{eff})}_{\acute{e}t}(\mathcal{O}_K) \cong \mathrm{DA}^{(\mathrm{eff})}_{\acute{e}t}(k)$ where $\mathrm{FDA}^{(\mathrm{eff})}_{\acute{e}t}(\mathcal{O}_K)$ is the category of (effective, étale, with Λ-coefficients) motives of formal schemes over $\mathcal{O}_K$ (see [7, Remark 3.1.5(2)]).

(5) *The generic fiber functor induces a compact-preserving, colimit-preserving map of monoidal compactly generated stable infinity categories* $\mathrm{FDA}^{(\mathrm{eff})}_{\acute{e}t}(\mathcal{O}_K) \to RigDA^{(\mathrm{eff})}_{\acute{e}t}(K)$.

(6) *In particular, we obtain the following compact-preserving, colimit-preserving, monoidal functor:*

$$\mathrm{DA}_{\acute{e}t}(k) \cong \mathrm{FDA}_{\acute{e}t}(\mathcal{O}_K) \to RigDA_{\acute{e}t}(K) \cong RigDA^{\dagger}_{\acute{e}t}(K)$$

and a monoidal contravariant realization functor on the last category, with values in K-modules:

$$\mathcal{R}^{\dagger}_{dR} \colon RigDA^{\dagger}_{\acute{e}t}(\mathrm{K}) \to \mathrm{D}(\mathrm{K})^{\mathrm{op}}$$

induced by $M \mapsto \mathrm{Map}(M, \Omega^{\dagger})$. *The associated cohomology theory on* $\mathrm{DA}_{\acute{e}t}(k)$ *coincides with Berthelot's rigid cohomology* H^*_{rig}, *the one on* $RigDA_{\acute{e}t}(K)$ *coincides with Große-Klönne's overcovergent de Rham cohomology* $H^*_{\mathrm{dR}^{\dagger}}$ *and the one on* $\mathrm{DA}_{\acute{e}t}(K)$ *(via analytification) coincides with the usual algebraic de Rham cohomology* H^*_{dR}.

(7) *In case K is perfectoid, for any fixed embedding* $k \to K^{\flat}$ *we can define, in light of (♠) a functor*

$$\mathrm{DA}_{\acute{e}t}(k) \to \mathrm{DA}_{\acute{e}t}(K^{\flat}) \to RigDA_{\acute{e}t}(K^{\flat}) \cong RigDA_{\acute{e}t}(K) \cong RigDA^{\dagger}_{\acute{e}t}(K).$$

which is equivalent to the one in Point (6).

(8) *Compact motives of* $RigDA_{\acute{e}t}(K)$ *are fully dualizable. In particular, the overconvergent de Rham cohomology is finite dimensional on any compact motive in* $RigDA_{\acute{e}t}(K)$ *such as motives of smooth quasi-compact rigid varieties over* K, *or analytifications of quasi-projective (not necessarily smooth) varieties over* K.

(9) *Compact motives of* $\mathrm{DA}_{\acute{e}t}(k)$ *are fully dualizable. In particular, rigid cohomology is finite dimensional on any compact motive in* $\mathrm{DA}_{\acute{e}t}(k)$ *such as motives of any quasi-projective (not necessarily smooth) variety* $\bar{X}$ *over k.*

Proof Points (1) and (6) are shown in [17, Proposition 5.12], point (2) is [17, Theorem 4.23] and point (4) is [1, Corollaire 1.4.29]. Point (7) is the content of [18]. The functors of points (3) and (5) are left adjoint functors, hence colimit-preserving. As they preserve direct products, they induce monoidal functors on motives. Moreover, they send affine smooth varieties to affinoid smooth varieties. Motives of such spaces are a class of compact generators (by [1, Proposition 1.2.34]) so the functors are also compact-preserving. The fact that compact motives in $\mathrm{DA}_{\acute{e}t}(k)$ and $\mathrm{DA}_{\acute{e}t}(K)$ are dualizable follows from [31] and the same is true for $\mathrm{RigDA}_{\acute{e}t}(K)$ by [1, Théorème 2.5.35]. Points (8) and (9) then follows from the classic description of compact objects (perfect complexes) in $D(K)$. □

Remark 4.3 The content of Theorem 4.2(4) is the most precise way to state the following: it is possible to associate canonically a smooth rigid analytic motive over

K to any variety $\bar{X}$ over k. Similarly, the content of Theorem 4.2(2) can be rephrased by saying that it is possible to associate canonically an overconvergent rigid analytic motive over to any rigid analytic variety X over K.

Remark 4.4 Let $\mathcal{X}$ be an algebraic variety over $\mathcal{O}_K$. The special fiber of its π-adic completion is just the special fiber $\mathcal{X}_k$ of $\mathcal{X}$. In light of Theorem 4.2(4) we conclude that cohomologically speaking, the act of π-adically completing $\mathcal{X}$ gives the same information as the act of taking its special fiber. More precisely: the following triangle commutes.

$$\begin{array}{ccc} & & \mathrm{FDA}_{\text{ét}}(\mathcal{O}_K) \\ & \nearrow^{\widehat{(-)}^*} & \Big\| \wr\; (-)_k^* \\ \mathrm{DA}_{\text{ét}}(\mathcal{O}_K) & \searrow_{\iota^*} & \\ & & \mathrm{DA}_{\text{ét}}(k) \end{array}$$

Remark 4.5 Following [32, Sects. 0.2–0.3] there are two possible ways to "analitify" a smooth algebraic variety $\mathcal{X}$ over $\mathcal{O}_K$: on the one hand one can consider the formal scheme given by its π-adic completion $\mathfrak{X}$ and then the generic fiber $\mathfrak{X}_\eta$ of it; on the other hand one can first take the generic fiber $\mathcal{X}_K$ (an algebraic variety over K) and then its analytification $\mathcal{X}_K^{\mathrm{an}}$. It is well-known (see [32, Proposition 0.3.5]) that the first rigid analytic space is canonically embedded as an open subvariety of the second, and that they coincide whenever $\mathcal{X}$ is smooth and proper (they differ in general: for example whenever $\mathcal{X}$ is lives on the generic fiber Spec K of Spec $\mathcal{O}_K$, the first space is empty). It is easy to see that such functors preserve étale covers and homotopies, therefore defining the following (non-commutative) square of monoidal colimit-preserving maps:

$$\begin{array}{ccccccc} & & \mathrm{DA}_{\text{ét}}(\mathcal{O}_K) & \longrightarrow & \mathrm{DA}_{\text{ét}}(K) & & \\ & & \downarrow & \overset{\alpha}{\nearrow\!\!\!\nearrow} & \downarrow & & \\ \mathrm{DA}_{\text{ét}}(k) & \overset{\sim}{=\!=\!=} & \mathrm{FDA}_{\text{ét}}(\mathcal{O}_K) & \longrightarrow & \mathrm{RigDA}_{\text{ét}}(K) & \xrightarrow{\mathcal{R}_{dR}} & D(K)^{\mathrm{op}} \end{array}$$

where the natural transformation α is induced by the functorial open immersion $\mathfrak{X}_\eta \subset \mathcal{X}_K^{\mathrm{an}}$ and $\mathcal{R}_{dR}$ is the (overconvergent) de Rham realization. We then obtain two monoidal realizations $\mathrm{DA}_{\text{ét}}(\mathcal{O}_K) \rightrightarrows D(K)$. When applied to some motive $\Lambda_{\mathcal{O}_K}(\mathcal{X})$, one gives the rigid cohomology of the special fiber $\mathcal{X}_k$ and the other gives the de Rham cohomology of the generic fiber $\mathcal{X}_K$, respectively. Moreover, α defines a canonical natural transformation between the two which is invertible on the full monoidal subcategory with sums generated by (the motives of) smooth and proper varieties.

Remark 4.6 Even though the motivic categories are defined as sheaves on *smooth* varieties (or *smooth* formal schemes, or *smooth* rigid varieties etc.) it is possible to define motives attached to an arbitrary quasi-projective variety (or arbitrary rigid varieties) using the 6 functor formalism: the (homological) motive $\Lambda(X)$ attached to such a variety X is given by $f_! f^! \Lambda$ with $f\colon X \to \operatorname{Spa} K$ being the structural morphism. This formalism is fully developed in [8, 33] (in the algebraic case) and in [7] (in the analytic case).

5 New De Rham-Like Cohomologies via Motives

In this section, we make a survey on the "new" de Rham-like cohomology theories for rigid analytic varieties and perfectoid spaces whose construction is based on the properties of motivic categories.

We start by a de Rham cohomology for perfectoid spaces introduced in [17, 18, 34], which is further expanded by [5]. Simply by combining Theorems 3.1 and 4.2 we deduce the following.

Theorem 5.1 *Let K be a perfectoid field.*

(1) Suppose that $\operatorname{char} K = 0$. *Let $\widehat{X} \sim \varprojlim X_h$ be a smooth perfectoid space obtained by relative perfection of an étale map $X_0 \to \mathbb{B}^N_K$. For any i, the system $H^i_{dR^\dagger}(X_h)$ is eventually constant, and the association $\widehat{X} \mapsto H^i_{dR}(X_h)$, $h \gg 0$ induces a well-defined functorial cohomology theory $H^*_{\mathrm{dR}}(\widehat{X}, K)$ on smooth perfectoid motives over K. It has étale descent, a Künneth formula, and finite dimension whenever X is quasi-compact.*

(2) Suppose that $\operatorname{char} K = p > 0$. *For any fixed un-tilt $K^\sharp$ of K the association $\widehat{X} \mapsto H^i_{dR}(\widehat{X}^\sharp, K^\sharp)$ is a well-defined functorial cohomology theory $H^i_{dR^\dagger}(\widehat{X}, K^\sharp)$ on smooth perfectoid motives over K. It has étale descent, a Künneth formula, and finite dimension whenever X is quasi-compact.*

(3) Suppose that $\operatorname{char} K = p > 0$. *For any fixed un-tilt $K^\sharp$ of K the association $X \mapsto H^i_{dR}(X^{\mathrm{Perf}}, K^\sharp)$ is a well-defined functorial cohomology theory on smooth rigid analytic varieties over K which extends to arbitrary rigid analytic varieties and is compatible with rigid cohomology with coefficients is $K^\sharp$ whenever X is of good reduction. Moreover, it has étale descent, a Künneth formula, and finite dimension whenever X is smooth and quasi-compact or the analytification of a quasi-projective algebraic variety.*

Remark 5.2 In [5] also a relative version of the (overconvergent) de Rham cohomology for rigid analytic spaces is introduced. It is also shown that it enjoys many properties which are common to the archimedean/algebraic analogue, such as the fact that $H^*_{dR}(X/S)$ is a vector bundle on the base whenever $X \to S$ is smooth and proper.

Suppose that K is a perfectoid field of characteristic $p > 0$. The need of choosing an un-tilt of K in order to define a de Rham-like cohomology theory for (rigid analytic) varieties over K can be considered unnatural and unsatisfactory for some purposes. To remedy this, in [5] the various cohomology theories $X \mapsto H^*_{dR}(X, K^\sharp)$

are "pasted together" into a vector bundle over the Fargues-Fontaine curve of K by means of the following

Theorem 5.3 *([5]) Let K be a perfectoid field of characteristic $p > 0$. We let $\mathcal{X}_K$ be the analytic space given by the adic Fargues-Fontaine curve associated to it. There is a monoidal realization functor*

$$\mathcal{R}_{dRFF}\colon RigDA_{\acute{e}t}(K) \to \mathrm{QCoh}(\mathcal{X}_K)^{\mathrm{op}}$$

giving rise to a cohomology theory $H^*_{dR}(-, \mathcal{X}_K)$ with values in quasi-coherent $\mathcal{X}_K$-modules (defined as in [35]). Moreover, whenever M is compact (eg. M is the motive of a quasi-compact smooth rigid variety, or the analytification of a quasi-projective algebraic variety) then the modules $H^*_{dR}(M, \mathcal{X}_K)$ are vector bundles, and equal to zero if $|i| \gg 0$.

Remark 5.4 The previous result gives a canonical analytic de Rham cohomology in positive characteristic, and answers positively to a conjecture of Fargues [36, Conjecture 1.13] and Scholze [37, Conjecture 6.4].

Remark 5.5 In [5] also a relative version of the previous theorem is shown, building on Remark 5.2.

The cohomology $H^*_{dR}(-, \mathcal{X}_K)$ above is not the only motivic cohomology theory with values on vector bundles on a Fargues-Fontaine curve. Fix an algebraically closed complete valued field C over $\mathbb{Q}_p$. In [34], Le Bras gives a motivic, over-convergent and rational version of the $\mathbb{A}_{\mathrm{inf}}$-cohomology introduced for smooth and proper formal schemes over $\mathcal{O}_C$ defined as follows: consider the pro-étale sheaf $\mathbb{A}_{\mathrm{inf},X}$ defined on affinoid perfectoid spaces over $X := \mathfrak{X}_C$ as $\mathrm{Spa}(P, P^+) \mapsto W(P^{\flat +})$ and its pull-back $R\nu_*\mathbb{A}_{\mathrm{inf}}$ to the Zariski site of $\mathfrak{X}$. Take the complex obtained by décalage $L\eta_\mu R\nu_*\mathbb{A}_{\mathrm{inf}}$ (μ being $[\varepsilon] - 1$) and, finally, the complex $R\Gamma_{\mathrm{Zar}}(\mathfrak{X}, L\eta_\mu R\nu_*\mathbb{A}_{\mathrm{inf}})$. This complex is known to be related to the various p-adic integral cohomologies defined on $\mathcal{X}$ (see [38]).

Theorem 5.6 *([34]) Let C be an algebraically closed complete valued field over $\mathbb{Q}_p$.*

(1) Let X be a smooth rigid analytic variety over C endowed with a dugger structure $X^\dagger = (X \Subset X')$. Consider the association

$$X \mapsto \varinjlim_{X \Subset X_h \subset X'} R\Gamma_{\acute{e}t}(X_h, L\eta_t R\nu_*\mathbb{B})$$

where $R\nu_*\mathbb{B}$ is the pull-back to the étale topos of X_h of the pro-étale sheaf $\mathbb{B}$ defined on affinoid perfectoid spaces over X_h as $\mathrm{Spa}(P, P^+) \mapsto \mathcal{O}(\mathcal{Y}_{(P,P^+)})$ and $L\eta_t$ is the décalage functor with respect to a generator t of $\ker(\theta\colon W(P^{\flat +}) \to P^+)$. It gives rise to a well-defined functor

$$\mathcal{R}_{\mathcal{FF}}\colon RigDA^{(\mathrm{eff})}_{\acute{e}t}(C) \to \mathrm{QCoh}(\mathcal{X}_{C^\flat})^{\mathrm{op}}$$

(2) *If $C = \mathbb{C}_p$ and $M \in RigDA^{(\mathrm{eff})}_{\acute{e}t}(C)$ is compact (for example, it is the motive of a quasi-compact smooth rigid variety, or the analytification of a quasi-projective algebraic variety) then the cohomology groups $H^i(\mathcal{R}_{\mathcal{FF}}M)$ are vector bundles on the curve $\mathcal{X}_{C^\flat}$ and equal to 0 for $|i| \gg 0$.*

Remark 5.7 It is not hard to see that $\mathcal{R}_{\mathcal{FF}}$ is the rational, overconvergent analogue of the $\mathbb{A}_{\mathrm{inf}}$-cohomology, and it is also possible to relate it to the de Rham cohomology, see [34].

Finally, we sketch briefly the construction of Ayoub of a "new motivic Weil cohomology" for varieties over a field k of positive characteristic. The aim of this construction is somehow different from the previous ones: we have mentioned that the $\mathbb{A}_{\mathrm{inf}}$-cohomology specializes to the various p-adic cohomology theories, and is therefore intimately linked to p-adic Hodge theory and p-adic periods. The constructions above are aimed to generalizations and extensions of this idea.

On a different direction, one can try to build a realization which specializes to the various ℓ-adic realizations (including $\ell = p$): such an approach would be interesting, for example, to inspect the independence on ℓ for ℓ-adic cohomologies. Choose a (non necessarily complete!) valued field K of mixed characteristic, with k as residue field and let $\widehat{K}$ be its completion. We already considered the following two adjoint pairs

$$\xi\colon \mathrm{DA}_{\acute{e}t}(k) \rightleftarrows \mathrm{RigDA}_{\acute{e}t}(\widehat{K})\colon \chi$$

$$\mathrm{Rig}^*\colon \mathrm{DA}_{\acute{e}t}(K) \rightleftarrows \mathrm{RigDA}_{\acute{e}t}(\widehat{K})\colon \mathrm{Rig}_*$$

The fact that ξ and Rig^* are monoidal induces formally a decomposition of the functors above:

$$\mathrm{DA}_{\acute{e}t}(k) \rightleftarrows \mathrm{DA}_{\acute{e}t}(k, \chi 1) \underset{\chi'}{\overset{\xi'}{\rightleftarrows}} \mathrm{RigDA}_{\acute{e}t}(\widehat{K}) \qquad (A)$$

$$\mathrm{DA}_{\acute{e}t}(K) \rightleftarrows \mathrm{DA}_{\acute{e}t}(K, \mathrm{Rig}_* 1) \underset{\mathrm{Rig}'_*}{\overset{\mathrm{Rig}'^*}{\rightleftarrows}} \mathrm{RigDA}_{\acute{e}t}(\widehat{K}) \qquad (B)$$

where the category $\mathrm{DA}_{\acute{e}t}(k, \chi)$ [resp. $\mathrm{DA}_{\acute{e}t}(K, \mathrm{Rig}_* 1)$] in the middle denotes the category of modules over the motive $\chi 1$ [resp. $\mathrm{Rig}_* 1$]. This object inherits a natural algebra structure deduced from the monoidality of ξ [resp. Rig^*]. The adjunction on the left is simply given by the free module structure/forgetful pair, while the adjunction on the right is built out of the natural $\chi 1$-module structure [resp. $\mathrm{Rig}_* 1$-module structure] which can be given to the objects of the form χM [resp. $\mathrm{Rig}_* M$].

The main theorem of [39] is then the following.

Theorem 5.8 *([39]) Let K be a subfield of $\mathbb{C}$ equipped with a rank-1 valuation with residue field k of characteristic $p > 0$ and completion $\widehat{K}$.*

(1) *The functor* Rig'^* *of (B) gives an equivalence of monoidal ∞-categories*

$$\mathrm{DA}_{\text{ét}}(K, \mathrm{Rig}_* 1) \cong RigDA_{\acute{e}t}(\widehat{K}).$$

(2) The (homological) algebraic de Rham realization

$$\mathcal{R}_{dR}\colon \mathrm{DA}_{\text{ét}}(K) \to D(K)$$

induces a monoidal functor

$$\mathcal{R}'_{dR}\colon RigDA_{\acute{e}t}(\widehat{K}) \cong \mathrm{DA}_{\text{ét}}(K, \mathrm{Rig}_* 1) \to D(K, \mathcal{A}_K)$$

where the category on the right denotes the category of modules over the object $\mathcal{A}_K := \mathcal{R}_{dR}\,\mathrm{Rig}_* 1$ equipped with its natural DG-algebra structure.

(3) The complex $\mathcal{A}_K$ is in $D_{\geq 0}(K)$. In particular $A_K := H_0(\mathcal{A}_K)$ has a K-algebra structure, there exists a map of DG-algebras $\mathcal{A}_K \to A_K$ and one can define a realization for $\mathrm{DA}_{\text{ét}}(k)$ as follows:

$$\mathcal{R}_{new}\colon \mathrm{DA}_{\text{ét}}(k) \xrightarrow{\xi} RigDA_{\acute{e}t}(\widehat{K}) \cong \mathrm{DA}_{\text{ét}}(K, \mathrm{Rig}_* 1) \xrightarrow{\mathcal{R}'_{dR}} D(K, \mathcal{A}_K) \to D(A_K).$$

(4) The algebra A_K can be explicitly computed in terms of generators and relations.

(5) There are ring maps $A_K \to \widehat{K}$ and $A_K \to \overline{\mathbb{Q}}_\ell$ for any $\ell \neq p$ (depending on a choice of isomorphism $\overline{\mathbb{Q}}_\ell \cong \mathbb{C}$) such that the realizations obtained by base change

$$\mathrm{DA}_{\text{ét}}(k) \xrightarrow{\mathcal{R}_{new}} D(A_K) \to D(\widehat{K}) \qquad \mathrm{DA}_{\text{ét}}(k) \xrightarrow{\mathcal{R}_{new}} D(A_K) \to D(\overline{\mathbb{Q}}_\ell)$$

are equivalent to the rigid realization, and the ℓ-adic realization, respectively.

Remark 5.9 One of the main results of [39] is actually the explicit computation of the algebra A_K that we only vaguely mentioned in the theorem above. It turns out that the description of the ring A_K is the non-archimedean analogue of the construction of the ring of complex periods considered in [40]. In spite of this explicit presentation, a full understanding of the algebraic properties of A_K (eg. being an integral domain) seem to be out of reach, and any progress in this direction would be of much interest as explained in [39].

The construction above is based on the equivalence $\mathrm{RigDA}_{\text{ét}}(\widehat{K}) \cong \mathrm{DA}_{\text{ét}}(K, \mathrm{Rig}_* 1)$ arising from the monoidal left adjoint functor Rig^*. In [7] the other monoidal functor ξ is analyzed, obtaining the following analogue of Theorem 5.8((1)).

Theorem 5.10 ([7, Theorem 3.3.3]) *Suppose that $\widehat{K}$ is algebraically closed. The functor ξ' of (A) defines an equivalence of monoidal ∞-categories*

$$\mathrm{DA}_{\text{ét}}(k, \chi 1) \cong RigDA_{\acute{e}t}(\widehat{K}).$$

Remark 5.11 In [7] a more general statement is shown: one can extend the equivalence above even in the case of a higher rank valuation, defined by some valuation subring $\widehat{K}^+ \subset \widehat{K}$, obtaining an equivalence

$$\mathrm{DA}_{\text{ét}}(\mathrm{Spec}(\widehat{K}^+/\pi), \chi 1) \cong \mathrm{RigDA}_{\text{ét}}(\mathrm{Spa}(\widehat{K}, \widehat{K}^+))$$

where $\pi \in \widehat{K}^+$ is a non-zero topologically nilpotent element.

6 A Betti-Like Cohomology via Motives

The aim of this section is to define another motivic cohomology theory for rigid analytic varieties and perfectoid spaces. Contrarily to the de Rham version considered above, this won't be a Weil cohomology and can't be expected to compare to ℓ-adic cohomologies. Nonetheless, Berkovich showed ([41]) that it contains some interesting information, and the motivic language can be used to extend his results. The main theorems of this section are taken from [42].

We recall that the Berkovich topological space $|X|_{\mathrm{Berk}}$ underlying a rigid analytic variety X is the maximal Hausdorff quotient of the (locally spectral) topological space $|X|$ It coincides with the topological space defined by the partially proper topology on X, or equivalently, to the topological space introduced by Berkovich.

Theorem 6.1 *([42]) Let K be a complete non-archimedean valued field, and let ℓ be a prime which is invertible in the residue field k. Let also C be a fixed complete algebraic closure of K.*

(1) Put $\Lambda = \mathbb{Q}$. There is an ℓ-adic realization functor

$$\mathcal{R}_\ell \colon RigDA_{\acute{e}t}^{(\mathrm{eff})}(K) \to \mathrm{Sh}_{\mathrm{pro\acute{e}t}}(K, \mathbb{Q}_\ell)$$

which is monoidal, and which sends compact objects to constructible complexes. For any smooth variety X over K, the homology groups $H_*(\mathcal{R}_\ell \Lambda(X))$ compute the ℓ-adic homology of X.

(2) The canonical functor

$$\iota^* \colon \mathrm{Sh}_{\text{ét}}(K, \Lambda) \to RigDA_{\acute{e}t}^{\mathrm{eff}}(K)$$

induced by the inclusion of the small étale site into the big one, has a left adjoint

$$\mathcal{R}_B \colon RigDA_{\acute{e}t}^{\mathrm{eff}}(K) \to \mathrm{Sh}_{\text{ét}}(K, \Lambda)$$

that can be described explicitly as the functor induced by mapping a variety X to the singular complex $\Lambda[\mathrm{Sing}(|X_C|_{\mathrm{Berk}})]$ of the topological space $|X_C|_{\mathrm{Berk}}$ with coefficients in Λ.

(3) *Suppose that K is a finite extension of $\mathbb{Q}_p$ and let $F\colon \mathrm{Gal}(k) \to \mathrm{Gal}(K)$ be a fixed lift of Frobenius. The following diagram*

$$\begin{array}{ccc}
\mathrm{Sh}^{\mathrm{ct}}_{\acute{e}t}(K,\mathbb{Q}) & \xrightarrow{\iota^*} & RigDA^{\mathrm{eff,ct}}_{\acute{e}t}(K,\mathbb{Q}) \\
\downarrow{\scriptstyle -\otimes_{\mathbb{Q}}\mathbb{Q}_\ell=\mathcal{R}_\ell} & & \downarrow{\scriptstyle \mathcal{R}_\ell} \\
\mathrm{Sh}^{\mathrm{ct}}_{\acute{e}t}(K,\mathbb{Q}_\ell) & & \mathrm{Sh}^{\mathrm{ct}}_{\mathrm{proet}}(K,\mathbb{Q}_\ell) \\
\downarrow{\scriptstyle F^*} & & \downarrow{\scriptstyle F^*} \\
\mathrm{Sh}^{\mathrm{ct}}_{\acute{e}t}(k,\mathbb{Q}_\ell) & \xrightarrow{\iota^*} & \mathrm{Sh}^{\mathrm{ct}}_{\mathrm{proet}}(k,\mathbb{Q}_\ell)
\end{array}$$

is commutative and left adjointable, in the sense that there are left adjoint functors $\mathcal{R}_B$ to the functors ι^* and the canonical natural transformation $\mathcal{R}_B F^* \mathcal{R}_\ell \Rightarrow F^* \mathcal{R}_\ell \mathcal{R}_B$ is invertible.

Proof Only the last point does not appear as stated in [42], but it is easily seen to be equivalent to [42, Corollary 5.5]. □

Remark 6.2 In [42] we showed that the theorem above can be used to have a concrete generalization of a result of Berkovich [41] for which whenever K is a local field, the Betti cohomology of the underlying Berkovich space with $\mathbb{Q}_\ell$-coefficients of a variety X coincides with the smooth part of the Galois ℓ-adic representation given by the associated étale cohomology.

As we focus in this paper on the role of motivic tilting and de-perfectoidification, we point out that the Berkovich realization given above can be equivalently defined for perfectoid motives, in a compatible way with the equivalences of Theorem 3.1.

Proposition 6.3 *([42]) Let K be a perfectoid field and C be a complete algebraic closure of it.*

(1) *The functor $X \mapsto \Lambda[\mathrm{Sing}(|X_C|_{\mathrm{Berk}})]$, where $\Lambda[\mathrm{Sing}(T)]$ is the singular complex of a topological space T with coefficients in Λ, induces a colimit-preserving monoidal functor*

$$\mathcal{R}_B\colon PerfDA^{\mathrm{eff}}_{\acute{e}t}(K) \to \mathrm{Sh}_{\acute{e}t}(K,\Lambda).$$

(2) *The following diagram is commutative.*

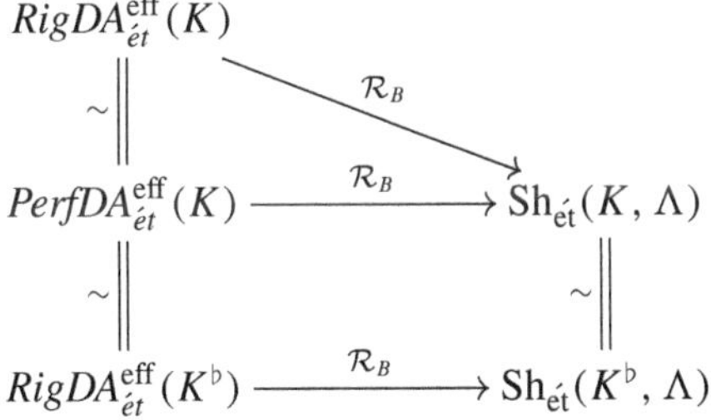

Proof In light of [42, Sect. 4] only the formula for the perfectoid version of $\mathcal{R}_B$ needs to be justified. As the tilting equivalence gives rise to homeomorphisms on the Berkovich spaces attached to perfectoid spaces, we may assume char $C = p$. By the equivalence $\mathrm{Perf}^*\colon \mathrm{RigDA}^{\mathrm{eff}}_{\mathrm{\acute{e}t}}(C) \cong \mathrm{PerfDA}^{\mathrm{eff}}_{\mathrm{\acute{e}t}}(C)$ the formula then follows from the analogous formula for rigid analytic varieties and the homeomorphism $|X^{\mathrm{Perf}}| \cong |X|$. □

As previously anticipated, this cohomology theory is not a Weil cohomology. It does not even extend to the stable categories of motives as indeed it "kills" Tate twists (this is compatible with Remark 6.2).

Remark 6.4 If X is a geometrically connected rigid variety of good reduction over K, then $\mathcal{R}_B\Lambda(X) \cong \Lambda[0]$. In particular, we have $\mathcal{R}_B(T_K) \cong 0$.

As a matter of fact, because of the functor $\xi\colon \mathrm{DA}_{\mathrm{\acute{e}t}}(k) \to \mathrm{RigDA}_{\mathrm{\acute{e}t}}(K)$ that we introduced above, it is impossible to define any Weil realization functor from $\mathrm{RigDA}_{\mathrm{\acute{e}t}}(K)$ with values in $\mathbb{Q}$-vector spaces, whenever K is a local field of mixed characteristic. Any such realization would otherwise violate Serre's counterexample to the existence of a rational Weil cohomology theory for varieties over a finite field ([43, Page 315]).

Acknowledgements We thank the organizers and the scientific committee of the conference "p-adic Automorphic Forms and Perfectoid Spaces" held in ICTS, Bangalore in 2019 for their kind invitation.

References

1. J. Ayoub, Motifs des variétés analytiques rigides. Mém. Soc. Math. Fr. (N. S.) **140–141**, 386 (2015)
2. F. Binda, D. Park, and P. Arne Østvær, Triangulated categories of logarithmic motives over a field. arXiv:2004.12298 [math.AG] (2020)
3. Fabien Morel, Vladimir Voevodsky, A^1-homotopy theory of schemes. Inst. Hautes Études Sci. Publ. Math. **90**, 45–143 (2001)
4. E. Elmanto, H. Kolderup, On modules over motivic ring spectra. Ann. K-Theory **5**(2), 327–355 (2020)
5. A.-C. Le Bras and A. Vezzani, The de Rham-Fargues-Fontaine cohomology. arXiv:2105.13028 [math.AG] (2021)
6. Alberto Vezzani, A motivic version of the theorem of Fontaine and Wintenberger. Compos. Math. **155**(1), 38–88 (2019)
7. J. Ayoub, M. Gallauer, and A. Vezzani, The six-functor formalism for rigid analytic motives. arXiv:2010.15004 [math.AG] (2020)
8. J. Ayoub, Les six opérations de Grothendieck et le formalisme des cycles évanescents dans le monde motivique. I. Astérisque **314**, 466 (2008)
9. J. Ayoub, A guide to (étale) motivic sheaves, in textitProceedings of the ICM 2014 (2014)
10. Denis-Charles. Cisinski, Frédéric. Déglise, étale motives. Compos. Math. **152**(3), 556–666 (2016)
11. Annette Huber, *Mixed motives and their realization in derived categories*, Lecture Notes in Mathematics, vol. 1604. (Springer, Berlin, 1995)

12. Y. André, *Une introduction aux motifs (motifs purs, motifs mixtes, périodes)*, Panoramas et Synthèses [Panoramas and Syntheses], vol. 17. (Société Mathématique de France, Paris, 2004)
13. Carlo Mazza, Vladimir Voevodsky, Charles Weibel, *Lecture notes on motivic cohomology*, Clay Mathematics Monographs, vol. 2. (American Mathematical Society, Providence, RI, 2006)
14. Roland Huber, *Étale cohomology of rigid analytic varieties and adic spaces: E30 (Aspects of Mathematics)* (Friedr. Vieweg & Sohn, Braunschweig, 1996)
15. Alberto Vezzani, Effective motives with and without transfers in characteristic p. Adv. Math. **306**, 852–879 (2017)
16. Joseph Ayoub, L'algèbre de Hopf et le groupe de Galois motiviques d'un corps de caractéristique nulle. I. J. Reine Angew. Math. **693**, 1–149 (2014)
17. Alberto Vezzani, The Monsky-Washnitzer and the overconvergent realizations. Int. Math. Res. Not. IMRN **11**, 3443–3489 (2018)
18. Alberto Vezzani, Rigid cohomology via the tilting equivalence. J. Pure Appl. Algebra **223**(2), 818–843 (2019)
19. Federico Bambozzi, Alberto Vezzani, Rigidity for rigid analytic motives. J. Math. Inst. Jussieu **20**(4), 1341–1369 (2021)
20. P. Berthelot, Géométrie rigide et cohomologie des variétés algébriques de caractéristique p, in *Study group on ultrametric analysis, 9th year: 1981/82, No. 3 (Marseille, 1982)*, pages Exp. No. J2, 18 (Inst. Henri Poincaré, Paris, 1983)
21. Jean Fresnel, Marius van der Put, *Rigid analytic geometry and its applications, Progress in Mathematics*, vol. 218 (Birkhäuser Boston, Boston, MA, 2004)
22. Elmar Große-Klönne, De Rham cohomology of rigid spaces. Math. Z. **247**(2), 223–240 (2004)
23. B. Le Stum, The overconvergent site. Mém. Soc. Math. Fr. (N.S.) **127**, 108 (2012)
24. Pierre Berthelot, *Cohomologie cristalline des schémas de caractéristique* $p > 0$, Lecture Notes in Mathematics, vol. 407. (Springer, New York, 1974)
25. Paul Monsky, Gerard Washnitzer, Formal cohomology. I. Ann. of Math. **2**(88), 181–217 (1968)
26. Alberto Arabia, Relèvements des algèbres lisses et de leurs morphismes. Comment. Math. Helv. **76**(4), 607–639 (2001)
27. Elmar Große-Klönne, Rigid analytic spaces with overconvergent structure sheaf. J. Reine Angew. Math. **519**, 73–95 (2000)
28. Pierre Berthelot, Finitude et pureté cohomologique en cohomologie rigide. Invent. Math. **128**(2), 329–377 (1997). (With an appendix in English by Aise Johan de Jong)
29. Bruno Chiarellotto, Nobuo Tsuzuki, Cohomological descent of rigid cohomology for étale coverings. Rend. Sem. Mat. Univ. Padova **109**, 63–215 (2003)
30. Elmar Große-Klönne, Finiteness of de Rham cohomology in rigid analysis. Duke Math. J. **113**(1), 57–91 (2002)
31. Joël. Riou, Catégorie homotopique stable d'un site suspendu avec intervalle. Bull. Soc. Math. France **135**(4), 495–547 (2007)
32. Pierre Berthelot, *Cohomologie rigide et cohomologie rigide à supports propres* (Première partie, Preprint, 1996)
33. Joseph Ayoub, Les six opérations de Grothendieck et le formalisme des cycles évanescents dans le monde motivique. II. Astérisque **315**, 364 (2007)
34. A.-C. Le Bras, Overconvergent relative de Rham cohomology over the Fargues-Fontaine curve. arXiv:1801.00429 [math.NT] (2018)
35. P. Scholze, Lectures on analytic geometry. Available on the author's webpage (2020)
36. L. Fargues, La courbe, in *Proceedings of the International Congress of Mathematicians. Vol. II. Invited Lectures* (World Scientific Publishing, Hackensack, NJ, 2018), pp. 291–319
37. P. Scholze, p-adic geometry, in *Proceedings of the International Congress of Mathematicians. Vol. I. Plenary Lectures* (2018)
38. Bhargav Bhatt, Matthew Morrow, Peter Scholze, Integral p-adic Hodge theory. Publ. Math. Inst. Hautes Études Sci. **128**, 219–397 (2018)
39. Joseph Ayoub, Nouvelles cohomologies de Weil en caractéristique positive. Algebra Number Theory **14**(7), 1747–1790 (2020)

40. Joseph Ayoub, Une version relative de la conjecture des périodes de Kontsevich-Zagier. Ann. Math. **181**(3), 905–992 (2015)
41. Vladimir G. Berkovich, An analog of Tate's conjecture over local and finitely generated fields. Int. Math. Res. Notices **13**, 665–680 (2000)
42. Alberto Vezzani, The Berkovich realization for rigid analytic motives. J. Algebra **527**, 30–54 (2019)
43. A. Grothendieck, Crystals and the de Rham cohomology of schemes, in Dix exposés sur la cohomologie des schémas, volume 3 of Adv. Stud. Pure Math. (Notes by I. Coates and O. Jussila) (North-Holland, Amsterdam, 1968), pp. 306–358

Diagrams and Mod p Representations of p-adic Groups

Eknath Ghate and Mihir Sheth

2010 Mathematics Subject Classification. 22E50 · 11S37

1 Introduction

Smooth complex representations of reductive p-adic groups play a pivotal role in the global Langlands program as they appear as local factors of automorphic representations. These local representations are admissible. Recall that a representation is smooth if every vector has an open subgroup fixing that vector, and it is admissible if the subspace fixed by any open subgroup is finite-dimensional. The mod p analogue of the local Langlands correspondence makes it necessary to understand smooth mod p representations of reductive p-adic groups. Unlike complex representations, one does not have analytic methods at one's disposal to study smooth mod p representations of p-adic groups because they do not admit a non-zero $\overline{\mathbb{F}}_p$-valued Haar measure. Diagrams give a powerful tool to construct interesting smooth mod p representations of reductive p-adic groups.

Breuil and Paskunas used diagrams attached to certain Galois representations to construct irreducible admissible supercuspidal mod p representations of $\mathrm{GL}_2(\mathbb{Q}_{p^f})$ where $\mathbb{Q}_{p^f}$ is the degree f unramified extension of $\mathbb{Q}_p$ ([1]). The universal supercuspidal representations, i.e., the compact inductions of weights modulo the image of the Hecke operator, classify all irreducible admissible supercuspidal mod p representations of $\mathrm{GL}_2(\mathbb{Q}_p)$, while in general, their irreducible admissible quotients exhaust

E. Ghate (✉) · M. Sheth
School of Mathematics, Tata Institute of Fundamental Research, Homi Bhabha Road, Mumbai 400005, India
e-mail: eghate@math.tifr.res.in

M. Sheth
e-mail: mihir@math.tifr.res.in

© The Author(s), under exclusive license to Springer Nature Singapore Pte Ltd. 2022

D. Banerjee et al. (eds.), *Perfectoid Spaces*, Infosys Science Foundation Series in Mathematical Sciences https://doi.org/10.1007/978-981-16-7121-0_3

all admissible supercuspidal representations of $\mathrm{GL}_2(\mathbb{Q}_{p^f})$ for $f > 1$ ([2], Proposition 4.6). The theory of diagrams can be used to show that, for $f > 1$, the universal supercuspidal representation is not of finite length and is also not admissible ([3], Theorem 3.3). This also follows from [4], Corollary 2.21 and [5], Corollary 4.5. This indicates that the mod p representation theory of $\mathrm{GL}_2(\mathbb{Q}_{p^f})$ is more involved than that of $\mathrm{GL}_2(\mathbb{Q}_p)$ ([6], [1]). For some work on the mod p representation theory of GL_2 over a totally ramified extension of $\mathbb{Q}_p$, see, for example, [7], and for a general finite extension of $\mathbb{Q}_p$, see [8].[1]

By the work of many mathematicians such as Harish-Chandra, Jacquet, and Vignéras, it is known that all smooth irreducible representations of connected reductive[2] p-adic groups over algebraically closed fields of characteristic not equal to p are admissible ([11], II §2.8). The main point is to show that all irreducible supercuspidal representations are admissible, since a general smooth irreducible representation is a subrepresentation of the parabolic induction of an irreducible supercuspidal representation and parabolic induction preserves admissibility. However, it is no longer true that irreducible supercuspidal representations over characteristic p fields are admissible. Recently, Daniel Le constructed non-admissible irreducible (supercuspidal) $\overline{\mathbb{F}}_p$-linear representations of $\mathrm{GL}_2(\mathbb{Q}_{p^f})$ using infinite-dimensional diagrams for all $f > 2$, although only the case $f = 3$ is presented in his paper for simplicity ([12]). Applying Le's method to a diagram attached to a split reducible Galois representation, the authors have constructed non-admissible irreducible representations of $\mathrm{GL}_2(\mathbb{Q}_{p^2})$ ([13]).

This article gives an expository treatment of the theory of diagrams of Breuil and Paskunas, and provides a proof of Le's construction of non-admissible irreducible mod p representations of $\mathrm{GL}_2(\mathbb{Q}_{p^f})$ for all $f > 2$. It is organized as follows. In Sect. 2, we introduce (finite-dimensional) diagrams and describe how they give rise to smooth admissible representations of GL_2 over p-adic fields. Section 3 focuses on diagrams attached to Galois representations and on the irreducible admissible supercuspidal representations of $\mathrm{GL}_2(\mathbb{Q}_{p^f})$ that they give rise to. Finally, we prove Le's theorem for all $f > 2$ in Sect. 4.

1.1 Notation

Let $p > 2$ be a prime number and $\mathbb{Q}_p$ be the field of p-adic numbers. Let $\mathbb{Q}_{p^f}$ denote the unramified extension of $\mathbb{Q}_p$ of degree f with ring of integers $\mathbb{Z}_{p^f}$. The residue field of $\mathbb{Q}_{p^f}$ is the finite field $\mathbb{F}_{p^f}$ with p^f elements. Fix an algebraic closure $\overline{\mathbb{F}}_p$ of $\mathbb{F}_p$ and an embedding $\mathbb{F}_{p^f} \hookrightarrow \overline{\mathbb{F}}_p$.

For an arbitrary but fixed f, let $G = \mathrm{GL}_2(\mathbb{Q}_{p^f})$, $K = \mathrm{GL}_2(\mathbb{Z}_{p^f})$, and $\Gamma = \mathrm{GL}_2(\mathbb{F}_{p^f})$. Let B and U be the subgroups of Γ consisting of the upper triangular

[1] The words *supersingular* and *supercuspidal* are used interchangeably in the literature for mod p representations. These two *a priori* different notions are now known to be equivalent ([9]).

[2] The reductive hypothesis is necessary; see [10].

matrices and the upper triangular unipotent matrices, respectively. Let I and I_1 be the preimages of B and U, respectively, under the natural surjection $K \twoheadrightarrow \Gamma$. The subgroups I and I_1 of K are called the Iwahori and the pro-p Iwahori subgroup of K, respectively. Let K_n denote the nth principal congruence subgroup of K, i.e., the kernel of the reduction map $K \longrightarrow \mathrm{GL}_2(\mathbb{Z}_{p^f}/p^n\mathbb{Z}_{p^f})$ modulo p^n for $n \geq 1$. Write N for the normalizer of I (and of I_1) in G. Then N is generated by I, the center Z of G and by the element $\Pi = \left(\begin{smallmatrix} 0 & 1 \\ p & 0 \end{smallmatrix}\right)$.

Unless stated otherwise, all representations considered in this paper are on $\overline{\mathbb{F}}_p$-vector spaces and are sometimes referred to as mod p representations. A weight is a smooth irreducible representation of K. The K-action on such a representation factors through Γ and thus a weight is an irreducible representation of Γ ([2], Lemma 2.14). For a character χ of I, χ^s denotes its Π-conjugate sending g in I to $\chi(\Pi g \Pi^{-1})$. Given a weight σ, the subspace σ^{I_1} of its I_1-invariants has dimension 1. We denote the corresponding smooth character of I afforded by the space σ^{I_1} by χ_σ. If $\chi_\sigma \neq \chi_\sigma^s$, then there exists a unique weight σ^s such that $\chi_{\sigma^s} = \chi_\sigma^s$ ([14], Theorem 3.1.1). For an I-representation V and an I-character χ, we write V^χ for the χ-isotypic part of V.

2 Diagrams and the Existence Theorem

Diagrams were introduced by Paskunas in [14] to construct smooth admissible representations of G.

Definition 2.1 A diagram is a triple (D_0, D_1, r) where D_0 is a smooth representation of KZ, D_1 is a smooth representation of N, and $r : D_1 \to D_0$ is an IZ-equivariant map. A diagram (D_0, D_1, r) is called a basic diagram if p acts trivially on D_0 and D_1, and r induces an isomorphism $D_1 \xrightarrow{\sim} D_0^{I_1}$ of IZ-representations.

The idea is to use the data of a basic diagram to construct a space Ω admitting actions of both KZ and N which agree on $IZ = KZ \cap N$. Let G^0 be the subgroup of G consisting of matrices whose determinant is a p-adic unit. Since G^0 is an amalgamated product of K and $\Pi K \Pi^{-1}$, and $G = G^0 \rtimes \Pi^{\mathbb{Z}}$, the actions of KZ and N on Ω glue together to give a G-action on Ω. This G-action is unique because KZ and N generate the group G ([2], Theorem 3.3 and Corollary 3.4).

A way to construct Ω is to use injective envelopes of finite-dimensional representations of finite groups. An injective envelope of a representation is the "smallest" injective object containing the representation ([2], Definition 5.12). If the subspace $D_0^{K_1}$ of K_1-invariants of D_0 is finite-dimensional, then the K-socle $\mathrm{soc}_K D_0$ of D_0, i.e., the maximal semi-simple K-subrepresentation of D_0, is finite-dimensional, and therefore the direct limit $\varinjlim_n \mathrm{inj}_{K/K_n}(\mathrm{soc}_K D_0)$ of finite-dimensional injective envelopes exist in the category of smooth K-representations. By [2], Proposition 5.17, this direct limit is the smooth injective envelope $\mathrm{inj}_K(\mathrm{soc}_K D_0)$ of D_0.

Let $\Omega := \mathrm{inj}_K(\mathrm{soc}_K D_0)$ be equipped with the KZ-action such that p acts trivially. The smooth injective I-envelope $\mathrm{inj}_I D_1$ of D_1 appears as an I-direct summand of

Ω via the IZ-equivariant map r. There is a unique N-action on $\text{inj}_I D_1$ compatible with that of I and compatible with the action of N on D_1 ([2], Corollary 6.7). Let e denote the projection of Ω onto $\text{inj}_I D_1$. By [1], Lemma 9.6, there is a non-canonical N-action on $(1-e)(\Omega)$ extending the given I-action. This gives an N-action on Ω whose restriction to IZ is compatible with the action coming from KZ on Ω. Consequently, there is a G-action on Ω as discussed two paragraphs above.

Let π be the G-representation generated by D_0 inside Ω. Then, we see that

$$\text{soc}_K D_0 \subseteq \text{soc}_K \pi \subseteq \text{soc}_K(\text{inj}_K(\text{soc}_K D_0)) = \text{soc}_K D_0$$

so that equality holds throughout.

We summarize the above discussion in the following theorem ([2], Theorem 5.10).

Theorem 2.2 (The existence theorem) *Let (D_0, D_1, r) be a basic diagram such that D_0 is finite-dimensional and K_1 acts trivially on D_0. Then there exists a smooth admissible representation π of G such that*

(1) $(\pi^{K_1}, \pi^{I_1}, \text{can})$ *contains* (D_0, D_1, r)*, where* can *is the canonical inclusion,*
(2) π *is generated by D_0 as a G-representation,*
(3) $\text{soc}_K \pi = \text{soc}_K D_0$.

Note that the representation π in the theorem above is admissible because $\pi^{K_n} \subseteq (\text{inj}_K(\text{soc}_K D_0))^{K_n} = \text{inj}_{K/K_n}(\text{soc}_K D_0)$ which is finite-dimensional (cf. [14], Lemma 6.2.4).

We remark that the discussion in this section, i.e., the notion of a basic diagram and the existence theorem, works for $G = \text{GL}_2(F)$ for any finite extension F of $\mathbb{Q}_p$.

Example 2.3 Let $G = \text{GL}_2(\mathbb{Q}_p)$ and σ be a weight. Take $D_0 = \sigma \oplus \sigma^s$ and $D_1 = D_0^{I_1} = \chi_\sigma \oplus \chi_\sigma^s$. Let Π map a basis vector of the underlying vector space of χ_σ to that of χ_σ^s. By letting p act trivially on D_0 and D_1, we get a basic diagram (D_0, D_1, can) where can is the canonical injection. The existence theorem applied to this diagram gives rise to a G-representation π that is irreducible and supercuspidal, and is *uniquely* determined by the diagram (D_0, D_1, can) ([2], Lemma 5.2). In fact, one obtains all irreducible admissible supercuspidal representations of G up to a smooth twist in this way as σ varies. Under the mod p local Langlands correspondence for GL_2 over $\mathbb{Q}_p$, π is mapped to a continuous 2-dimensional irreducible representation of $\text{Gal}(\overline{\mathbb{Q}_p}/\mathbb{Q}_p)$ whose restriction to the inertia subgroup contains the information of $\text{soc}_K \pi = \text{soc}_K D_0 = \sigma \oplus \sigma^s$.

3 Diagrams Attached to Galois Representations

Let $f > 2$ for the rest of the article.

3.1 Diamond Diagrams

Let $\rho : \mathrm{Gal}(\overline{\mathbb{Q}_p}/\mathbb{Q}_{p^f}) \longrightarrow \mathrm{GL}_2(\overline{\mathbb{F}_p})$ be a continuous irreducible generic Galois representation ([1], Definition 11.7). In [15], Buzzard, Diamond, and Jarvis associate with ρ a finite set $\mathcal{D}(\rho)$ of distinct weights anticipating that it would describe the K-socle of the supercuspidal representation of G corresponding to ρ under the conjectural mod p local Langlands correspondence for GL_2 over $\mathbb{Q}_{p^f}$.[3] As we shall see, the set $\mathcal{D}(\rho)$ can indeed be used to construct irreducible supercuspidal representations with K-socle described by $\mathcal{D}(\rho)$. However, it turns out that there are infinitely many such representations up to isomorphism. The mod p local Langlands correspondence for GL_2 over finite extensions of $\mathbb{Q}_p$ thus still remains puzzling.

The set $\mathcal{D}(\rho)$ has cardinality 2^f. By elementary representation theoretic arguments, there exists a unique finite-dimensional $\overline{\mathbb{F}_p}$-linear representation $D_0(\rho)$ of Γ whose Γ-socle equals $\bigoplus_{\sigma\in\mathcal{D}(\rho)} \sigma$, and is maximal with respect to the property that each $\sigma \in \mathcal{D}(\rho)$ occurs exactly once in $D_0(\rho)$ as a Jordan–Hölder factor. Further, there is an isomorphism of Γ-representations

$$D_0(\rho) \cong \bigoplus_{\sigma\in\mathcal{D}(\rho)} D_{0,\sigma}(\rho)$$

with $\mathrm{soc}_\Gamma D_{0,\sigma}(\rho) = \sigma$ ([1], Proposition 13.1). Viewing $D_0(\rho)$ as a K-representation, let us denote by $D_1(\rho)$ the I-representation $D_0(\rho)^{I_1}$ and by $D_{1,\sigma}(\rho)$ the I-representation $D_{0,\sigma}(\rho)^{I_1}$. If an I-character χ appears in $D_1(\rho)$ then so does χ^s.

While for any finite set of weights, there exists a finite-dimensional Γ-representation D_0 satisfying the same properties listed above, the properties of $D_0(\rho)$ and $D_1(\rho)$ specific to the set of weights $\mathcal{D}(\rho)$ are summarized below.

Proposition 3.1

(1) The Jordan–Hölder factors of $D_0(\rho)$ are multiplicity free.
(2) $D_1(\rho)$ is a multiplicity-free semi-simple I-representation of dimension $3^f - 1$ and thus

$$D_1(\rho) = \bigoplus_{I-\text{character } \chi} \chi \oplus \chi^s.$$

Proof See [1], Corollary 13.5, Corollary 13.6, Lemma 14.1, and Proposition 14.7. □

Proposition 3.1 allows us to define an action of Π on $D_1(\rho)$ by mapping I-characters to their Π-conjugates, thereby giving a family of basic diagrams $D(\rho, r) := (D_0(\rho), D_1(\rho), r)$ parameterized by IZ-equivariant injections $r : D_1(\rho) \hookrightarrow D_0(\rho)$. The diagrams $D(\rho, r)$ attached to Galois representations ρ in this way are called *Diamond diagrams* in [1].

---

[3] They associate a finite set of weights with any continuous semi-simple generic Galois representation ρ. We stick to irreducible ρ in this exposition. However, see Remark 4.5.

3.2 The Map δ

We now introduce the map $\delta : \mathcal{D}(\rho) \longrightarrow \mathcal{D}(\rho)$ which governs the dynamics of the Π-action on $D(\rho, r)$ and plays an important role in proving the irreducibility of representations of G coming from $D(\rho, r)$. There is a natural identification of the set $\mathcal{D}(\rho)$ of weights with the set of subsets of $\mathbb{Z}/f\mathbb{Z} = \{0, 1, \ldots, f-1\}$ ([1], §11). Under this identification, the map δ is defined as follows:

Definition 3.2 For $J \subseteq \mathbb{Z}/f\mathbb{Z}$,

$$\delta(J) := \begin{cases} \{j-1 \mid j \in J\} \cup \{0\} & \text{if } 1 \notin J \\ \{j-1 \mid j \in J\} \setminus \{0\} & \text{if } 1 \in J \end{cases}$$

with the convention $-1 = f - 1$.

Note that δ is a bijection and partitions the set $\mathcal{D}(\rho)$ into δ-orbits.

Example 3.3 We list the four δ-orbits for $f = 5$.

$$\begin{aligned} \Delta_1 &= \big\{\phi \mapsto \{0\} \mapsto \{0,4\} \mapsto \{0,3,4\} \mapsto \{0,2,3,4\} \mapsto \{0,1,2,3,4\} \mapsto \\ &\qquad \{1,2,3,4\} \mapsto \{1,2,3\} \mapsto \{1,2\} \mapsto \{1\}\big\}, \\ \Delta_2 &= \big\{\{2\} \mapsto \{0,1\} \mapsto \{4\} \mapsto \{0,3\} \mapsto \{0,2,4\} \mapsto \{0,1,3,4\} \mapsto \{2,3,4\} \mapsto \\ &\qquad \{0,1,2,3\} \mapsto \{1,2,4\} \mapsto \{1,3\}\big\}, \\ \Delta_3 &= \big\{\{3\} \mapsto \{0,2\} \mapsto \{0,1,4\} \mapsto \{3,4\} \mapsto \{0,2,3\} \mapsto \{0,1,2,4\} \mapsto \{1,3,4\} \mapsto \\ &\qquad \{2,3\} \mapsto \{0,1,2\} \mapsto \{1,4\}\big\}, \\ \Delta_4 &= \big\{\{2,4\} \mapsto \{0,1,3\}\big\}. \end{aligned}$$

The map δ has a nice reinterpretation. Identify the set of subsets of $\mathbb{Z}/f\mathbb{Z}$ as the set of binary numbers (sequences of 0s and 1s) of length f. The subset $J \subseteq \mathbb{Z}/f\mathbb{Z}$ corresponds to the binary number $a_1a_2\ldots a_f$ under the rule $a_j = 1$ if and only if $j \in J$, where we make the identification $f = 0$. Under this identification, δ is the map that moves the first digit of a binary number to the end and changes its parity:

$$\delta(a_1a_2\ldots a_f) = a_2a_3\ldots a_f(a_1+1) \text{ with the convention } 2 = 0.$$

Example 3.4 Let $f = 5$. The subset $\{0, 1, 3\}$ corresponds to the binary number 10101 and $\delta(10101) = 01010$ which corresponds to $\{2, 4\}$.

It follows from the definition that

$$\delta^{2f}(a_1a_2\ldots a_f) = (a_1+2)(a_2+2)\ldots(a_f+2) = a_1a_2\ldots a_f.$$

Hence, the order of δ is at most $2f$. In fact, the order of δ equals $2f$ as one easily sees by considering the δ-orbit of the empty set (= f zeros). It follows that the size of any δ-orbit divides $2f$. Observe that δ changes the size of a subset J by ± 1. So any δ-orbit contains an even number of subsets. Therefore, the size of a δ-orbit is $2f'$ for some f' dividing f. Using the reinterpretation of δ, we can prove the following result which is of independent interest.

Lemma 3.5 *The set $\mathcal{D}(\rho)$ has a δ-orbit of size $2f'$ if and only if $\frac{f}{f'}$ is odd.*

Proof ($\Rightarrow$) Suppose $d := \frac{f}{f'}$ is even. Let $a = a_1 a_2 \dots a_f$ belongs to a δ-orbit of size $2f'$. We write

$$a = a_1 a_2 \dots a_{f'} a_{f'+1} a_{f'+2} \dots a_{2f'} \dots a_{(d-1)f'+1} a_{(d-1)f'+2} \dots a_f.$$

Then

$$\delta^{f'}(a) = a_{f'+1} a_{f'+2} \dots a_{2f'} a_{2f'+1} a_{2f'+2} \dots a_{3f'} \dots (a_1+1)(a_2+1)\dots(a_{f'}+1).$$

Since a_j and $\delta^{f'}(a)_j$ must have opposite parity for all $1 \le j \le f$ and d is even by assumption, we get $a_1 = a_{(d-2)f'+1}$. Comparing the parity of the last block of f' digits in a and $\delta^{f'}(a)$, we also have $a_{(d-1)f'+1} = a_1$. This implies that the first digit $a_{(d-2)f'+1}$ of the second last block of f' digits in a is equal to the first digit $a_{(d-1)f'+1}$ of the second last block of f' digits in $\delta^{f'}(a)$, a contradiction.

($\Leftarrow$) Let a be the f-digit binary number starting with f' 0s, followed by f' 1s, followed by f' 0s, and so on. The number a ends with f' 0s as $\frac{f}{f'}$ is odd. Clearly, $\delta^{f'}(a)$ flips the parity of the digits of a, showing that the δ-orbit of a has size $2f'$.

If $\sigma \in \mathcal{D}(\rho)$ corresponds to a subset J, let $\delta(\sigma)$ denote the weight corresponding to the subset $\delta(J)$. The map δ is characterized by the following property.

Lemma 3.6 *For $\sigma \in \mathcal{D}(\rho)$, $\delta(\sigma) \in \mathcal{D}(\rho)$ is the unique weight such that σ^s is a Jordan–Hölder factor of $D_{0,\delta(\sigma)}(\rho)$.*

Proof See [1], Lemma 15.2.

Using the combinatorics of the Π-action dynamics on $D(\rho, r)$ described by δ, one obtains the following theorem.

Theorem 3.7 *The basic diagram $D(\rho, r)$ is indecomposable, i.e., the KZ-representation $D_0(\rho)$ does not have a proper non-zero KZ-direct summand X such that X^{I_1} is stable under the action of Π.*

Proof See [1], Theorem 15.4.

3.3 Irreducible Admissible Supercuspidal Representations

Let $\tau(\rho, r)$ be a smooth admissible representation of G given by the existence theorem applied to a Diamond diagram $D(\rho, r)$. We briefly sketch the argument of the irreducibility of $\tau(\rho, r)$ using Theorem 3.7. Let $\tau' \subseteq \tau(\rho, r)$ be a non-zero subrepresentation. Since $0 \neq \mathrm{soc}_K \tau' \subseteq \mathrm{soc}_K \tau(\rho, r) = \mathrm{soc}_K D_0(\rho)$, we have $\sigma \in \mathrm{soc}_K \tau'$ for some $\sigma \in \mathcal{D}(\rho)$. Thus, $D_1(\rho)^{\chi_\sigma} \subseteq \tau'$. As τ' is stable under the Π-action, we have $D_1(\rho)^{\chi_\sigma^s} \subseteq \tau'$. By Lemma 3.6, we see that $D_{1,\delta(\sigma)}(\rho)^{\chi_\sigma^s} \subseteq \tau'$. As τ' is clearly a K-representation, it follows that τ' contains the unique K-subrepresentation $I(\delta(\sigma), \sigma^s)$ of $D_{0,\delta(\sigma)}(\rho)$ with quotient σ^s. It is a non-trivial fact that the embedding $I(\delta(\sigma), \sigma^s) \hookrightarrow \tau'$ extends uniquely to an embedding $D_{0,\delta(\sigma)}(\rho) \hookrightarrow \tau'$. This requires delicate analysis of non-split extensions between weights (cf. [1], §17 and 18). Repeating the argument for $\delta(\sigma)$, we get $D_{0,\delta^2(\sigma)}(\rho) \subseteq \tau'$ and so on. Since the map δ has finite order, we get $D_{0,\sigma}(\rho) \subseteq \tau'$. It then follows that

$$\bigoplus_{\sigma \in \mathrm{soc}_K \tau'} D_{0,\sigma}(\rho) = \tau' \cap D_0(\rho).$$

Since the space of I_1-invariants of the right-hand side in the above is stable under the action of Π, the same is true for the left-hand side which is a non-zero direct summand of $D_0(\rho)$. This contradicts Theorem 3.7 unless $\tau' = \tau(\rho, r)$. Hence, $\tau(\rho, r)$ is irreducible.

As $\mathrm{soc}_K \tau(\rho, r) = \mathrm{soc}_K D_0(\rho)$, the number of weights in the K-socle of $\tau(\rho, r)$ is equal to the size of $\mathcal{D}(\rho)$ which is $2^f > 2$. Any subquotient of a principal series representation of G has at most two weights in its K-socle ([2], Remark 4.9). It follows that $\tau(\rho, r)$ is supercuspidal.

Finally, we remark that if $D(\rho, r)$ and $D(\rho, r')$ are two non-isomorphic basic diagrams, then any two smooth admissible G-representations $\tau(\rho, r)$ and $\tau(\rho, r')$ are non-isomorphic ([1], Theorem 19.8 (ii)). In fact, even the representation $\tau(\rho, r)$ is not uniquely determined by $D(\rho, r)$ ([6]).

3.4 Extra Characters

Let us now fix a diagram $D = (D_0, D_1, r)$ in the family $\{D(\rho, r)\}_r$ for the rest of the article. We have $D_0 = \bigoplus_{\sigma \in \mathcal{D}(\rho)} D_{0,\sigma}$. Write $D_{1,\sigma} = (D_{0,\sigma})^{I_1}$. For any δ-orbit Δ, we write

$$D_{0,\Delta} := \bigoplus_{\sigma \in \Delta} D_{0,\sigma} \text{ and } D_{1,\Delta} := (D_{0,\Delta})^{I_1}.$$

We call an I-character $\chi \subset D_1$ *extra* if $\chi \neq \chi_\sigma$ and $\chi \neq \chi_\sigma^s$ for any $\sigma \in \mathcal{D}(\rho)$. There are a total of $3^f - 1$ characters in D_1 (Proposition 3.1). Of these, at most 2^{f+1} characters correspond to the socle weights and their Π-conjugates. Therefore, the set of extra characters is non-empty because $3^f - 1 > 2^{f+1}$ as $f > 2$. We remark that Lemma 3.6 together with Theorem 3.7 implies that for a given δ-orbit Δ, there is an extra character χ such that $(D_{1,\Delta})^\chi \neq 0$.

Let n be the number of δ-orbits of $\mathcal{D}(\rho)$. As the set $\mathcal{D}(\rho)$ has cardinality 2^f and $f > 2$, we have $n > 1$. The existence of the set of extra characters established in the following lemma is used crucially by Le in his construction of non-admissible irreducible G-representations.

Lemma 3.8 *There exists a set S of $2(n-1)$ extra characters closed under Π-conjugation such that given a δ-orbit Δ, there is a $\chi \in S$ satisfying $(D_{1,\Delta})^\chi \neq 0$.*

Proof Choose any δ-orbit, call it Δ_1, and pick an extra character, say χ_1, such that

$$(D_{1,\Delta_1})^{\chi_1} \neq 0 \text{ and } \Big(\bigoplus_{\sigma \in \mathcal{D}(\rho)\setminus \Delta_1} D_{1,\sigma} \Big)^{\chi_1^s} \neq 0.$$

The existence of such a χ_1 is guaranteed by Theorem 3.7. Call the orbit Δ_2 for which $(D_{1,\Delta_2})^{\chi_1^s} \neq 0$. Using Theorem 3.7 again, there is an extra character χ_2 such that

$$\big(D_{1,\Delta_1} \bigoplus D_{1,\Delta_2}\big)^{\chi_2} \neq 0 \text{ and } \Big(\bigoplus_{\sigma \in \mathcal{D}(\rho)\setminus(\Delta_1 \sqcup \Delta_2)} D_{1,\sigma} \Big)^{\chi_2^s} \neq 0.$$

Note that $\chi_2 \notin \{\chi_1, \chi_1^s\}$. Call the orbit Δ_3 for which $(D_{1,\Delta_3})^{\chi_2^s} \neq 0$. Proceeding in this way, we find n δ-orbits $\Delta_1, \Delta_2, \ldots, \Delta_n$ of $\mathcal{D}(\rho)$ and $(n-1)$ extra characters $\chi_1, \chi_2, \ldots, \chi_{n-1}$ such that $(D_{1,\Delta_{j+1}})^{\chi_j^s} \neq 0$ for all $1 \leq j \leq n-1$. Take $S = \{\chi_1, \chi_1^s, \chi_2, \chi_2^s, \ldots, \chi_{n-1}, \chi_{n-1}^s\}$. □

4 Infinite-dimensional Diagrams and Non-admissible Representations

We now explain Le's method of constructing infinite-dimensional diagrams from Diamond diagrams to produce non-admissible irreducible representations. Let $D_0(\infty) := \bigoplus_{i \in \mathbb{Z}} D_0(i)$ be the smooth KZ-representation with componentwise KZ-action, where there is a fixed isomorphism $D_0(i) \cong D_0$ of KZ-representations for every $i \in \mathbb{Z}$. Denote the natural inclusion $D_0 \xrightarrow{\sim} D_0(i) \hookrightarrow D_0(\infty)$ by ι_i, and write $v_i := \iota_i(v)$ for $v \in D_0$ for every $i \in \mathbb{Z}$. Let $D_1(\infty) := D_0(\infty)^{I_1}$.

We make use of the δ-orbits and the set S of extra characters from the proof of Lemma 3.8 to define a Π-action on $D_1(\infty)$ which is different from the componentwise Π-action. Pick a pair of extra characters $\{\psi, \psi^s\}$ not belonging to the set S. To justify the existence of such a pair, note that it is enough to show the inequality $2(n-1) < 3^f - 1 - 2^{f+1}$ for all $f > 2$. Since the size of any δ-orbit is even, we have $n \leq 2^{f-1}$. Thus $2(n-1) \leq 2^f - 2$. It is now easy to check that $2^f - 2 < 3^f - 1 - 2^{f+1}$ for all $f > 2$.

Let us choose a weight $\sigma_k \in \Delta_k$ for all $1 \le k \le n$ and let $\lambda = (\lambda_i) \in \prod_{i\in\mathbb{Z}} \overline{\mathbb{F}_p}^{\times}$. For all integers $i \in \mathbb{Z}$, define

$$\Pi v_i := \begin{cases} (\Pi v)_i & \text{if } v \in D_1^{\chi} \text{ for } \chi \notin \{\chi_{\sigma_1}, \chi_{\sigma_1}^s, \dots, \chi_{\sigma_n}, \chi_{\sigma_n}^s, \psi, \psi^s\}, \\ (\Pi v)_{i+1} & \text{if } v \in D_1^{\chi} \text{ for } \chi \in \{\chi_{\sigma_1}, \dots, \chi_{\sigma_{n-1}}\}, \\ (\Pi v)_{i-1} & \text{if } v \in D_1^{\chi_{\sigma_n}}, \\ \lambda_i (\Pi v)_i & \text{if } v \in D_1^{\psi}. \end{cases}$$

This uniquely determines a smooth N-action on $D_1(\infty)$ such that $p = \Pi^2$ acts trivially on it. Thus, we get a basic diagram $D(\lambda) := (D_0(\infty), D_1(\infty), \text{can})$ with the above actions where can is the canonical inclusion $D_1(\infty) \hookrightarrow D_0(\infty)$.

Theorem 4.1 (Le) *There exists a smooth representation π of G such that*

(1) $(\pi|_{KZ}, \pi|_N, \text{id})$ *contains* $D(\lambda)$,
(2) π *is generated by* $D_0(\infty)$ *as a* G*-representation, and*
(3) $\text{soc}_K \pi = \text{soc}_K D_0(\infty)$.

Proof The idea is to consider the infinite direct sum $\bigoplus_{i\in\mathbb{Z}} \Omega(i)$ where each $\Omega(i)$ is isomorphic to the smooth injective K-envelope Ω of D_0, and equip this direct sum with an N-action extending the N-action on $D_1(\infty)$ defined above. The proof is the same as that of [12], Theorem 3.2, presented for $f = 3$.

Theorem 4.2 (Le) *If $\lambda_i \neq \lambda_0$ for all $i \neq 0$, then any smooth representation π of G satisfying the properties (1), (2), and (3) of Theorem 4.1 is irreducible and non-admissible.*

Proof Let $\pi' \subseteq \pi$ be a non-zero subrepresentation of G. By property (3), we have $\text{Hom}_K(\sigma, \pi') \neq 0$ for some $\sigma \in \text{soc}_K D_0$. Considering that σ could be embedded diagonally in π', there exists a non-zero $(c_i) \in \bigoplus_{i\in\mathbb{Z}} \overline{\mathbb{F}}_p$ such that

$$\Big(\sum_i c_i \iota_i\Big)(\sigma) \subset \pi',$$

or equivalently

$$\Big(\sum_i c_i \iota_i\Big)(D_{0,\sigma}) \cap \pi' \neq 0,$$

because the K-socle of $\big(\sum_i c_i \iota_i\big)(D_{0,\sigma})$ is $\big(\sum_i c_i \iota_i\big)(\sigma)$, which is irreducible.

We claim that

$$\Big(\sum_i c_i \iota_{i+j}\Big)(D_0) \subset \pi' \text{ for all } j \in \mathbb{Z}. \tag{4.3}$$

We prove the claim (4.3) assuming $\sigma \in \Delta_n$. The cases where σ is in an orbit other than Δ_n are proved similarly. If $\sigma \in \Delta_n$, then σ is in the same δ-orbit Δ_n as σ_n is. So it follows from the discussion in Sect. 3.3 that

$$(\sum_i c_i \iota_i)(\sigma) \subset \pi' \Longrightarrow (\sum_i c_i \iota_i)(\sigma_n) \subset \pi'.$$

Note that the indices i are unchanged since the action of Π on $\iota_i(D_1^{\chi_\sigma})$ fixes the index i of the embedding ι_i for all $\sigma \in \Delta_n$ except σ_n. Since the Π-action takes $\iota_i(D_1^{\chi_{\sigma_n}})$ to $\iota_{i-1}(D_1^{\chi^s_{\sigma_n}})$, we have

$$(\sum_i c_i \iota_{i-1})(D_{0,\delta(\sigma_n)}) \subset \pi'.$$

Therefore, again from the discussion in Sect. 3.3, we get that

$$(\sum_i c_i \iota_{i-1})(D_{0,\Delta_n}) \subset \pi'.$$

Continuing in this fashion, we obtain

$$(\sum_i c_i \iota_{i+j})(D_{0,\Delta_n}) \subset \pi' \text{ for all } j < 0.$$

Making use of the extra character χ^s_{n-1} in the proof of Lemma 3.8, we have in particular,

$$(\sum_i c_i \iota_{i+j})(D_1^{\chi^s_{n-1}}) \subset \pi' \text{ for all } j < 0.$$

Therefore,

$$(\sum_i c_i \iota_{i+j})(D_1^{\chi_{n-1}}) \subset \pi' \text{ for all } j < 0.$$

We know from the proof of Lemma 3.8 that $(D_{1,\Delta_k})^{\chi_{n-1}} \neq 0$ for some $1 \leq k < n$. Since the Π-action takes $\iota_i(D_1^{\chi_{\sigma_k}})$ to $\iota_{i+1}(D_1^{\chi^s_{\sigma_k}})$, we obtain

$$(\sum_i c_i \iota_{i+j})(D_{0,\Delta_k}) \subset \pi' \text{ for all } j \in \mathbb{Z} \text{ for some } 1 \leq k < n.$$

Making use of the extra character χ^s_{k-1}, by the same arguments as above, we obtain

$$(\sum_i c_i \iota_{i+j})(D_{0,\Delta_{k'}}) \subset \pi' \text{ for all } j \in \mathbb{Z} \text{ for some } 1 \leq k' < k.$$

Continuing in this fashion, we finally get that

$$(\sum_i c_i \iota_{i+j})(D_{0,\Delta_1}) \subset \pi' \text{ for all } j \in \mathbb{Z}.$$

Recall from the proof of Lemma 3.8 that

$$\Big(\bigoplus_{m=1}^{l} D_{1,\Delta_m}\Big)^{\chi_l} \neq 0 \text{ and } (D_{1,\Delta_{l+1}})^{\chi_l^s} \neq 0 \text{ for all } 1 \leq l \leq n-1. \tag{4.4}$$

Using (4.4) with $l = 1$, we get

$$\Big(\sum_i c_i \iota_{i+j}\Big)((D_{1,\Delta_1})^{\chi_1}) \subset \pi' \text{ and } \Big(\sum_i c_i \iota_{i+j}\Big)((D_{1,\Delta_2})^{\chi_1^s}) \subset \pi' \text{ for all } j \in \mathbb{Z}.$$

This implies

$$\Big(\sum_i c_i \iota_{i+j}\Big)(D_{0,\Delta_2}) \subset \pi \text{ for all } j \in \mathbb{Z}.$$

Similarly, using (4.4) successively for $l = 2, \ldots, n-1$, we obtain

$$\Big(\sum_i c_i \iota_{i+j}\Big)(D_{0,\Delta_r}) \subset \pi' \text{ for all } j \in \mathbb{Z} \text{ and for all } 1 \leq r \leq n.$$

Hence, $\big(\sum_i c_i \iota_{i+j}\big)(D_0) \subset \pi'$ for all $j \in \mathbb{Z}$ as desired.

For $(d_i) \in \bigoplus_{i\in\mathbb{Z}} \overline{\mathbb{F}}_p$, let $\#(d_i)$ denote the number of non-zero d_i's. Among all the non-zero elements (c_i) of $\bigoplus_{i\in\mathbb{Z}} \overline{\mathbb{F}}_p$ for which $\big(\sum_i c_i \iota_i\big)(D_0) \subset \pi'$, we pick one with $\#(c_i)$ minimal. We may also assume that $c_0 \neq 0$ using (4.3). We now show that $\#(c_i) = 1$. Assume to the contrary that $\#(c_i) > 1$. Since $\big(\sum_i c_i \iota_i\big)(D_1^{\psi}) \subset \pi'$ and π' is stable under the Π-action, we have

$$\Big(\sum_i \lambda_i c_i \iota_i\Big)(D_1^{\psi^s}) \subset \pi'.$$

Since $\big(\sum_i \lambda_0 c_i \iota_i\big)(D_1^{\psi^s})$ is also clearly in π', subtracting it from the above, we get

$$\Big(\sum_i (\lambda_i - \lambda_0) c_i \iota_i\Big)(D_1^{\psi^s}) \subset \pi'.$$

Let $\nu \in \mathcal{D}(\rho)$ be the weight for which $D_{1,\nu}^{\psi^s} \neq 0$. Writing $(c_i') := ((\lambda_i - \lambda_0)c_i)$, we see that

$$\Big(\sum_i c_i' \iota_i\Big)(D_{0,\nu}) \cap \pi' \neq 0.$$

Following the same arguments as in the previous paragraph proving the claim (4.3), we get that $\big(\sum_i c_i' \iota_i\big)(D_0) \subset \pi'$. However, the hypothesis $\lambda_i \neq \lambda_0$ for all $i \neq 0$, and the assumption $\#(c_i) > 1$ imply that (c_i') is non-zero and $\#(c_i') = \#(c_i) - 1$ contradicting the minimality of $\#(c_i)$. Therefore, we have $c_0 \iota_0(D_0) \subset \pi'$. So $\iota_0(D_0) \subset \pi'$. Using (4.3) again, we get that $\bigoplus_{j\in\mathbb{Z}} \iota_j(D_0) = D_0(\infty) \subset \pi'$. By property (2) of Theorem 4.1, we have $\pi' = \pi$.

The non-admissibility of π is clear because $\pi^{K_1} \supseteq \mathrm{soc}_K \pi$ and $\mathrm{soc}_K \pi$ is not finite-dimensional by property (3) of Theorem 4.1. □

Remark 4.5 The strategy to construct non-admissible irreducible representations explained above fails for the group $\mathrm{GL}_2(\mathbb{Q}_{p^2})$ because of the absence of extra characters in $D_1(\rho)$ when $f = 2$. However, it turns out that a Diamond diagram attached to a *reducible split* mod p Galois representation of $\mathrm{Gal}(\overline{\mathbb{Q}_p}/\mathbb{Q}_{p^2})$ does have enough extra characters to employ Le's strategy to produce non-admissible irreducible representations of $\mathrm{GL}_2(\mathbb{Q}_{p^2})$ (cf. [13]).

Remark 4.6 Note that the smooth irreducible non-admissible representations π in Theorem 4.2 and in [13], Theorem 3.2, have a central character because the action of p on π is trivial. By [16], Theorem 33 (1), π is a quotient of $\text{c-Ind}_{KZ}^G\sigma/(T - \lambda)(\text{c-Ind}_{KZ}^G\sigma)$ for some $\sigma \in \mathrm{soc}_K\pi$ and $\lambda \in \overline{\mathbb{F}}_p$. If $\lambda \neq 0$, by [16], Corollary 31, π is the unique irreducible quotient and by [16], Lemma 28 (1) and Theorem 33, all such quotients are admissible. It follows that $\lambda = 0$ and π is a quotient of $\text{c-Ind}_{KZ}^G\sigma/T(\text{c-Ind}_{KZ}^G\sigma)$, i.e., π is supercuspidal. Since quotients of admissible representations are admissible, by [17], Theorem 1, we deduce that the universal supercuspidal representation $\text{c-Ind}_{KZ}^G\sigma/T(\text{c-Ind}_{KZ}^G\sigma)$ is not admissible. This was already known, as mentioned in the introduction.

Acknowledgements We acknowledge the support of the Department of Atomic Energy, Government of India, under project number 12-R&D-TFR-5.01-0500. We thank the organizers of the ICTS (International Centre for Theoretical Sciences) program on "Perfectoid spaces" (Code: ICTS/perfectoid2019/09) for the discussion meeting on "p-adic automorphic forms and perfectoid spaces". The research talks on mod p representation theory of p-adic groups during the meeting provided an impetus to our joint work mentioned in this article.

References

1. C. Breuil, V. Paškūnas, *Towards a modulo p Langlands correspondence for* GL_2, Mem. Amer. Math. Soc. **216** (2012)
2. C. Breuil, *Representations of Galois and of* GL_2 *in characteristic* p, Lecture notes of a graduate course at Columbia University (2007)
3. C. Breuil, The emerging p-adic Langlands programme, in *Proceedings of the International Congress of Mathematicians Volume II* (Hindustan Book Agency, New Delhi, 2010), pp. 203–230
4. B. Schraen, Sur la présentation des représentations supersingulières de $\mathrm{GL}_2(F)$. J. Reine Angew. Math. **704**, 187–208 (2015)
5. Z. Wu, *A note on presentations of supersingular representations of* $\mathrm{GL}_2(F)$, Manuscripta Math. 165(3–4), 583–596 (2021)
6. Y. Hu, Sur quelques représentations supersinguliéres de $\mathrm{GL}_2(\mathbb{Q}_{p^f})$. J. Algebra **324**, 1577–1615 (2010)
7. M. Schein, On the universal supersingular mod p representations of $\mathrm{GL}_2(F)$. J. Number Theory **141**, 242–277 (2014)
8. Y. Hendel, On the universal mod p supersingular quotients for $\mathrm{GL}_2(F)$ over $\bar{\mathbb{F}}_p$ for a general $F/\mathbb{Q}_p$. J. Algebra **591**, 1–38 (2019)

9. N. Abe, G. Henniart, F. Herzig, M.-F. Vignéras, A classification of irreducible admissible mod p representations of p-adic reductive groups. J. Amer. Math. Soc. **30**(2), 495–559 (2017)
10. G.F. Helminck, An irreducible smooth non-admissible representation. Indag. Math. (N.S.) **1**(4), 435–438 (1990)
11. M.-F. Vignéras, *Représentations l-modulaires d'un groupe réductif p-adique avec $l \neq p$, Progress in Mathematics 137* (Birkhäuser, Boston, MA, 1996)
12. D. Le, On some non-admissible smooth representations of GL_2, Math. Res. Lett. **26**(6), 1747–1758 (2019)
13. E. Ghate, M. Sheth, On non-admissible irreducible modulo p representations of $\mathrm{GL}_2(\mathbb{Q}_{p^2})$. C. R. Math. Acad. Sci. Paris **358**(5), 627–632 (2020)
14. V. Paškūnas, Coefficient systems and supersingular representations of $\mathrm{GL}_2(F)$, Mém. Soc. Math. Fr. (N.S.) **99** (2004)
15. K. Buzzard, F. Diamond, F. Jarvis, On Serre's conjecture for mod l Galois representations over totally real fields. Duke Math. J. **55**, 105–161 (2010)
16. L. Barthel, R. Livné, Irreducible modular representations of GL_2 of a local field. Duke Math. J. **75**(2), 261–292 (1994)
17. G. Henniart, Sur les représentations modulo p de groupes réductifs p-adiques. Contemp. Math. **489**, 41–55 (2009)

A Short Review on Local Shtukas and Divisible Local Anderson Modules

Urs Hartl and Rajneesh Kumar Singh

Mathematics Subject Classification: 11G09 · 13A35 · 14L05

1 Introduction

The theory of p-adic Galois representations is concerned with the continuous representations

$$\rho\colon \operatorname{Gal}(L^{\mathrm{alg}}/L) \longrightarrow \operatorname{GL}_r(\mathbb{Q}_p) \tag{1.1}$$

of the absolute Galois group $\operatorname{Gal}(L^{\mathrm{alg}}/L)$ of a finite field extension L of $\mathbb{Q}_p$. It started with Tate's introduction of *p-divisible groups* in [33]. These are also called *Barsotti-Tate groups*. The Tate module T_pX of a p-divisible group X of height r over L induces Galois representations $V_pX := T_pX \otimes_{\mathbb{Z}_p} \mathbb{Q}_p$ and $\mathrm{H}^1_{\acute{e}t}(X, \mathbb{Q}_p) := \operatorname{Hom}_{\mathbb{Z}_p}(T_pX, \mathbb{Q}_p)$ as in (1.1). If X extends to a p-divisible group over $\mathcal{O}_L$, one says that X has *good reduction*. In this case, the special fiber $X_0 := X \otimes_{\mathcal{O}_L} \kappa$ of X over the residue field κ of $\mathcal{O}_L$ can be described by its *crystalline cohomology* $H^1_{\mathrm{cris}}\big(X_0/W(\kappa)\big)$, where $W(\kappa)$ is the ring of p-typical Witt vectors with coefficients in κ. The p-divisible group X, which can be viewed as a lift of X_0 to $\mathcal{O}_L$, is described by the F-crystal $H^1_{\mathrm{cris}}\big(X_0/W(\kappa)\big)$ together with its Hodge filtration. All this was proved by Messing [27]. Grothendieck [15] reformulated this as a functor relating the p-adic étale cohomology $\mathrm{H}^1_{\acute{e}t}(X, \mathbb{Q}_p)$ to the crystalline cohomology $\mathrm{H}^1_{\mathrm{cris}}(X_0/L_0)$ with its Hodge filtration, where $L_0 := W(\kappa)[\frac{1}{p}]$ and $\mathrm{H}^1_{\mathrm{cris}}(X_0/L_0)$ is a *filtered*

U. Hartl
Fachbereich Mathematik und Informatik, Westfalische Wilhelms-Universitat Munster,,
Einsteinstrasse 62, 48149, Munster, Germany

R. K. Singh (✉)
Quant Risk Management Associate, CME Group, Bangalore 560093, India
e-mail: rajneeshkumar.s@gmail.com

© The Author(s), under exclusive license to Springer Nature Singapore Pte Ltd. 2022
D. Banerjee et al. (eds.), *Perfectoid Spaces*, Infosys Science Foundation Series
in Mathematical Sciences https://doi.org/10.1007/978-981-16-7121-0_4

isocrystal; see Remark 6.6 below. Grothendieck then posed the problem to extend this functor, which he called the *mysterious functor*, to general proper smooth schemes X over L with good reduction. For those X, the problem was solved by Fontaine [10–13], who defined the notion of *crystalline p-adic Galois representations* and constructed a functor from crystalline p-adic Galois representations to filtered isocrystals. Fontaine conjectured that $\mathrm{H}^i_{\text{ét}}(X \times_L L^{\mathrm{alg}}, \mathbb{Q}_p)$ is crystalline when X is a proper smooth scheme over $\mathcal{O}_L$. After contributions by Grothendieck, Tate, Fontaine, Lafaille, Messing, Hyodo, Kato, and many others, Fontaine's conjecture was proved independently by Faltings [8], Niziol [28], and Tsuji [34].

Our goal in this survey is to describe the function field analog of the above. In this analog, p-divisible groups are replaced by *divisible local Anderson modules* which we discuss in Sect. 4. The analog of Messing's [27] theory of crystalline Dieudonné-modules for p-divisible groups is Theorem 4.2. In it, Messings F-crystals are replaced by *local shtukas*, which we treat first in Sect. 2. The anti-equivalence between divisible local Anderson modules and local shtukas passes through finite flat group schemes and finite shtukas. We review it in Sect. 3. Analogous to the étale and crystalline cohomology we mentioned for p-divisible groups in the previous paragraph, local shtukas possess cohomology realizations as described in Sect. 5. In the final Sect. 6, we explain how the theory of local shtukas provides the function field analog of Fontaine's theory of p-adic Galois representations (1.1).

2 Local Shtukas

The theory of local shtukas is the function field analog of Fontaine's theory of p-adic Galois representations. Let A_ε be a complete discrete valuation ring with finite residue field $\mathbb{F}_\varepsilon$ of characteristic p such that the fraction field Q_ε of A_ε also has characteristic p. The rings A_ε and Q_ε are the function field analogs of $\mathbb{Z}_p$ and $\mathbb{Q}_p$. We choose a uniformizing parameter $z \in A_\varepsilon$. Then A_ε is canonically isomorphic to $\mathbb{F}_\varepsilon[\![z]\!]$. Let $\hat{q} = \#\mathbb{F}_\varepsilon$ be the cardinality of $\mathbb{F}_\varepsilon$. As base rings R over which our objects are defined, we are interested in this article in two kinds of A_ε-algebras:

(a) The first kind are A_ε-algebras in which the image ζ of the uniformizer z of A_ε is nilpotent. We denote the category of these A_ε-algebras by $\mathcal{N}ilp_{A_\varepsilon}$.
(b) Let K be a field which is complete with respect to a non-trivial, non-Archimedean absolute value $|\,.\,|\colon K \to \mathbb{R}_{\geq 0}$ and let $\mathcal{O}_K = \{x \in K : |x| \leq 1\}$ be the valuation ring of K. We make $\mathcal{O}_K$ into an A_ε-algebra via an *injective* ring homomorphism $\gamma\colon A_\varepsilon \hookrightarrow \mathcal{O}_K$ such that $\zeta := \gamma(z) \neq 0$ lies in the maximal ideal $\mathfrak{m}_K \subset \mathcal{O}_K$.

The relation between the two kinds of base rings is that $\mathcal{O}_K/(\zeta^n) \in \mathcal{N}ilp_{A_\varepsilon}$ for all positive integers n.

Let R be a base ring as in (a) or (b). To define local shtukas over R, we consider modules $\hat{M}$ over the power series ring $R[\![z]\!]$, which Zariski locally on $\operatorname{Spec} R$ are free over $R[\![z]\!]$. We call such a module a *locally free* $R[\![z]\!]$*-module of rank* r. We set $\hat{M}[\frac{1}{z-\zeta}] := \hat{M} \otimes_{R[\![z]\!]} R[\![z]\!][\frac{1}{z-\zeta}]$, and $\hat{M}[\frac{1}{z}] := \hat{M} \otimes_{R[\![z]\!]} R(\!(z)\!)$ where

$R((z)) := R[[z]][\frac{1}{z}]$, and $\hat{\sigma}^*\hat{M} := \hat{M} \otimes_{R[[z]],\hat{\sigma}} R[[z]]$ where $\hat{\sigma}$ is the endomorphism of $R[[z]]$ with $\hat{\sigma}(z) = z$ and $\hat{\sigma}(b) = b^{\hat{q}}$ for $b \in R$. Note that $R[[z]][\frac{1}{z-\zeta}] = R((z))$ if $R \in \mathcal{N}ilp_{A_\varepsilon}$ as in (a), but $R[[z]][\frac{1}{z-\zeta}] \neq R((z))$ if R is a valuation ring as in (b). There is a natural $\hat{\sigma}$-semilinear map $\hat{M} \to \hat{\sigma}^*\hat{M}$, $m \mapsto \hat{\sigma}^*_{\hat{M}} m := m \otimes 1$. For a morphism of $R[[z]]$-modules $f\colon \hat{M} \to \hat{M}'$, we set $\hat{\sigma}^* f := f \otimes \mathrm{id}\colon \hat{\sigma}^*\hat{M} \to \hat{\sigma}^*\hat{M}'$.

Definition 2.1 A *local $\hat{\sigma}$-shtuka* (or *local shtuka*) *of rank r* over R is a pair $\underline{\hat{M}} = (\hat{M}, \tau_{\hat{M}})$ consisting of a locally free $R[[z]]$-module $\hat{M}$ of rank r, and an isomorphism $\tau_{\hat{M}}\colon \hat{\sigma}^*\hat{M}[\frac{1}{z-\zeta}] \xrightarrow{\sim} \hat{M}[\frac{1}{z-\zeta}]$. If $\tau_{\hat{M}}(\hat{\sigma}^*\hat{M}) \subset \hat{M}$ then $\underline{\hat{M}}$ is called *effective*, and if $\tau_{\hat{M}}(\hat{\sigma}^*\hat{M}) = \hat{M}$ then $\underline{\hat{M}}$ is called *étale*. We say that $\tau_{\hat{M}}$ is *topologically nilpotent*, if $\underline{\hat{M}}$ is effective and there is an integer n such that $\mathrm{im}(\tau^n_{\hat{M}}) \subset z\hat{M}$, where $\tau^n_{\hat{M}} := \tau_{\hat{M}} \circ \hat{\sigma}^*\tau_{\hat{M}} \circ \ldots \circ \hat{\sigma}^{(n-1)*}\tau_{\hat{M}}\colon \hat{\sigma}^{n*}\hat{M} \to \hat{M}$.

A *morphism* of local shtukas $f\colon (\hat{M}, \tau_{\hat{M}}) \to (\hat{M}', \tau_{\hat{M}'})$ over R is a morphism of $R[[z]]$-modules $f\colon \hat{M} \to \hat{M}'$ which satisfies $\tau_{\hat{M}'} \circ \hat{\sigma}^* f = f \circ \tau_{\hat{M}}$. We denote the set of morphisms from $\underline{\hat{M}}$ to $\underline{\hat{M}}'$ by $\mathrm{Hom}_R(\underline{\hat{M}}, \underline{\hat{M}}')$.

A *quasi-morphism* between local shtukas $f\colon (\hat{M}, \tau_{\hat{M}}) \to (\hat{M}', \tau_{\hat{M}'})$ over R is a morphism of $R((z))$-modules $f\colon \hat{M}[\frac{1}{z}] \to \hat{M}'[\frac{1}{z}]$ with $\tau_{\hat{M}'} \circ \hat{\sigma}^* f = f \circ \tau_{\hat{M}}$. It is called a *quasi-isogeny* if it is an isomorphism of $R((z))$-modules. We denote the set of quasi-morphisms from $\underline{\hat{M}}$ to $\underline{\hat{M}}'$ by $\mathrm{QHom}_R(\underline{\hat{M}}, \underline{\hat{M}}')$.

For any local shtuka $(\hat{M}, \tau_{\hat{M}})$ over $R \in \mathcal{N}ilp_{A_\varepsilon}$, the homomorphism $\hat{M} \to \hat{M}[\frac{1}{z-\zeta}]$ is injective by the flatness of $\hat{M}$ and the following.

Lemma 2.2 *([21, Lemma 2.2]) Let R be an A_ε-algebra as in (a) or (b). Then the sequence of $R[[z]]$-modules*

$$0 \longrightarrow R[[z]] \longrightarrow R[[z]] \longrightarrow R \longrightarrow 0$$

$$1 \longmapsto z - \zeta\,, \quad z \longmapsto \zeta$$

is exact. In particular, $R[[z]] \subset R[[z]][\frac{1}{z-\zeta}]$.

Of fundamental importance is the following.

Example 2.3 Let $\mathbb{F}_q$ be a finite field with q elements, let C be a smooth projective geometrically irreducible curve over $\mathbb{F}_q$, and let $Q := \mathbb{F}_q(C)$ be the function field of C. Fix a closed point ∞ of C, and let $A := \Gamma(C \smallsetminus \{\infty\}, \mathcal{O}_C)$ be the ring of regular functions on C outside ∞. The rings A and Q are the function field analogs of $\mathbb{Z}$ and $\mathbb{Q}$.

Let $\varepsilon \subset A$ be a maximal ideal and let A_ε be the completion of A at ε. Then $\mathbb{F}_\varepsilon$ is a field extension of $\mathbb{F}_q$ with $\hat{q} := \#\mathbb{F}_\varepsilon = q^{[\mathbb{F}_\varepsilon:\mathbb{F}_q]}$. Let R be a base A_ε-algebra as in (a) or (b) and denote its structure morphism by $\gamma\colon A_\varepsilon \to R$. Set $A_R := A \otimes_{\mathbb{F}_q} R$ and let $\sigma := \mathrm{id}_A \otimes \mathrm{Frob}_{q,R}$ be the endomorphism of A_R with $\sigma(a \otimes b) = a \otimes b^q$ for $a \in A$ and $b \in R$. An *effective A-motive of rank r over R* is a pair

$\underline{M} = (M, \tau_M)$ consisting of a locally free A_R-module M of rank r and an injective A_R-homomorphism $\tau_M : \sigma^* M \hookrightarrow M$ whose cokernel is a finite free R-module and is annihilated by a power of the ideal $\mathcal{J} := (a \otimes 1 - 1 \otimes \gamma(a) : a \in A) = \ker(\gamma \otimes \mathrm{id}_R : A_R \twoheadrightarrow R) \subset A_R$.

More generally, an *A-motive of rank r over R* is a pair $\underline{M} = (M, \tau_M)$ consisting of a locally free A_R-module M of rank r and an isomorphism $\tau_M : \sigma^* M|_{\mathrm{Spec}\, A_R \smallsetminus \mathrm{V}(\mathcal{J})} \xrightarrow{\sim} M|_{\mathrm{Spec}\, A_R \smallsetminus \mathrm{V}(\mathcal{J})}$ of the associated sheaves outside $\mathrm{V}(\mathcal{J}) \subset \mathrm{Spec}\, A_R$. Note that if $A = \mathbb{F}_q[t]$, then $\mathcal{J} = (t - \gamma(t))$ and $\mathrm{Spec}\, A_R \smallsetminus \mathrm{V}(\mathcal{J}) = \mathrm{Spec}\, R[t][\frac{1}{t-\gamma(t)}]$.

Let $\underline{M}$ be an (effective) A-motive over R. We consider the ε-adic completions $A_{\varepsilon,R} = \varprojlim A_R/\varepsilon^n A_R$ of A_R and $\underline{M} \otimes_{A_R} A_{\varepsilon,R} := (M \otimes_{A_R} A_{\varepsilon,R}\,,\ \tau_M \otimes \mathrm{id})$ of $\underline{M}$. If $\mathbb{F}_\varepsilon = \mathbb{F}_q$, and hence $\hat{q} = q$ and $\hat{\sigma} = \sigma$, we have $A_{\varepsilon,R} = R[\![z]\!]$ and $\mathcal{J} \cdot A_{\varepsilon,R} = (z - \zeta)$ because $R \otimes_{A_R} A_{\varepsilon,R} = R$. So $\underline{M} \otimes_{A_R} A_{\varepsilon,R}$ is an (effective) local shtuka over R which we denote by $\underline{\hat{M}}_\varepsilon(\underline{M})$ and call the *local $\hat{\sigma}$-shtuka at ε associated with $\underline{M}$*. If $f := [\mathbb{F}_\varepsilon : \mathbb{F}_q] > 1$, the construction is slightly more complicated; compare the discussion in [4, after Proposition 8.4]. Namely, we consider the canonical isomorphism $\mathbb{F}_\varepsilon[\![z]\!] \xrightarrow{\sim} A_\varepsilon$ and the ideals $\mathfrak{a}_i = (a \otimes 1 - 1 \otimes \gamma(a)^{q^i} : a \in \mathbb{F}_\varepsilon) \subset A_{\varepsilon,R}$ for $i \in \mathbb{Z}/f\mathbb{Z}$, which satisfy $\prod_{i \in \mathbb{Z}/f\mathbb{Z}} \mathfrak{a}_i = (0)$, because $\prod_{i \in \mathbb{Z}/f\mathbb{Z}} (X - a^{q^i}) \in \mathbb{F}_q[X]$ is a multiple of the minimal polynomial of a over $\mathbb{F}_q$ and even equal to it when $\mathbb{F}_\varepsilon = \mathbb{F}_q(a)$. By the Chinese remainder theorem, $A_{\varepsilon,R}$ decomposes

$$A_{\varepsilon,R} \;=\; \prod_{i \in \mathbb{Z}/f\mathbb{Z}} A_{\varepsilon,R}/\mathfrak{a}_i\,. \tag{2.1}$$

Each factor is canonically isomorphic to $R[\![z]\!]$. The factors are cyclically permuted by σ because $\sigma(\mathfrak{a}_i) = \mathfrak{a}_{i+1}$. In particular, σ^f stabilizes each factor. The ideal $\mathcal{J}$ decomposes as follows: $\mathcal{J} \cdot A_{\varepsilon,R}/\mathfrak{a}_0 = (z - \zeta)$ and $\mathcal{J} \cdot A_{\varepsilon,R}/\mathfrak{a}_i = (1)$ for $i \neq 0$. We define the *local $\hat{\sigma}$-shtuka at ε associated with $\underline{M}$* as $\underline{\hat{M}}_\varepsilon(\underline{M}) := (\hat{M}, \tau_{\hat{M}}) := \big(M \otimes_{A_R} A_{\varepsilon,R}/\mathfrak{a}_0\,,\ (\tau_M \otimes 1)^f\big)$, where $\tau_M^f := \tau_M \circ \sigma^* \tau_M \circ \ldots \circ \sigma^{(f-1)*} \tau_M$. Of course, if $f = 1$ we get back the definition of $\underline{\hat{M}}_\varepsilon(\underline{M})$ given above. Also note if $\underline{M}$ is effective, then $M/\tau_M(\sigma^* M) = \hat{M}/\tau_{\hat{M}}(\hat{\sigma}^* \hat{M})$.

The local shtuka $\underline{\hat{M}}_\varepsilon(\underline{M})$ allows to recover $\underline{M} \otimes_{A_R} A_{\varepsilon,R}$ via the isomorphism

$$\bigoplus_{i=0}^{f-1} (\tau_M \otimes 1)^i \bmod \mathfrak{a}_i : \Big(\bigoplus_{i=0}^{f-1} \sigma^{i*}(M \otimes_{A_R} A_{\varepsilon,R}/\mathfrak{a}_0),\ (\tau_M \otimes 1)^f \oplus \bigoplus_{i \neq 0} \mathrm{id}\Big) \xrightarrow{\sim} \underline{M} \otimes_{A_R} A_{\varepsilon,R}\,,$$

because for $i \neq 0$ the equality $\mathcal{J} \cdot A_{\varepsilon,R}/\mathfrak{a}_i = (1)$ implies that $\tau_M \otimes 1$ is an isomorphism modulo $\mathfrak{a}_i$; see [4, Propositions 8.8 and 8.5] for more details. Note that $\underline{M} \mapsto \underline{\hat{M}}_\varepsilon(\underline{M})$ is a functor.

We quote the next lemma from [21, Lemma 2.3].

Lemma 2.4 *Let $(\hat{M}, \tau_{\hat{M}})$ be a local shtuka over R. Then there are $e, e' \in \mathbb{Z}$ such that $(z - \zeta)^{e'} \hat{M} \subset \tau_{\hat{M}}(\hat{\sigma}^* \hat{M}) \subset (z - \zeta)^{-e} \hat{M}$. For any such e, the map $\tau_{\hat{M}}$:*

$\hat\sigma^*\hat M \to (z-\zeta)^{-e}\hat M$ is injective, and the quotient $(z-\zeta)^{-e}\hat M/\tau_M(\hat\sigma^*\hat M)$ is a locally free R-module of finite rank.

Example 2.5 We discuss the case of the Carlitz module [5]. We keep the notation from Example 2.3 and set $A=\mathbb{F}_q[t]$. Let $\mathbb{F}_q(\theta)$ be the rational function field in the variable θ and let $\gamma\colon A\to\mathbb{F}_q(\theta)$ be given by $\gamma(t)=\theta$. The *Carlitz motive over* $\mathbb{F}_q(\theta)$ is the A-motive $\underline{M}=\big(\mathbb{F}_q(\theta)[t],\, t-\theta\big)$.

Now let $\varepsilon=(z)\subset A$ be a maximal ideal generated by a monic prime element $z=z(t)\in\mathbb{F}_q[t]$. Then $\mathbb{F}_\varepsilon=A/(z)$ and A_ε is canonically isomorphic to $\mathbb{F}_\varepsilon[[z]]$. Let $\mathcal{O}_K\supset\mathbb{F}_\varepsilon[[\zeta]]$ be a valuation ring as in (b) and let $\theta=\gamma(t)\in\mathcal{O}_K$. The Carlitz motive has *good reduction* in the sense that it has a model over $\mathcal{O}_K$ given by the A-motive $\underline{M}=(\mathcal{O}_K[t],\, t-\theta)$ over $\mathcal{O}_K$.

If $\deg_t z(t)=1$, that is, $z(t)=t-a$ for $a\in\mathbb{F}_q$, then $\mathbb{F}_\varepsilon=\mathbb{F}_q$, $\zeta=\theta-a$, and $z-\zeta=t-\theta$. So $\underline{\hat M}_\varepsilon(\underline{M})=(\mathcal{O}_K[[z]],\, z-\zeta)$.

If $\deg_t z(t)=f>1$, then $\underline{\hat M}_\varepsilon(\underline{M})=\big(\mathcal{O}_K[[z]],\,(t-\theta)(t-\theta^q)\cdots(t-\theta^{q^{f-1}})\big)$. Here, the product $(t-\theta)(t-\theta^q)\cdots(t-\theta^{q^{f-1}})=(z-\zeta)u$ for a unit $u\in\mathbb{F}_\varepsilon[[\zeta]][[z]]^\times$, because $\tau_M(\sigma^*M)=(t-\theta)M$ implies that $\underline{\hat M}_\varepsilon(\underline{M})$ is effective and $\hat M/\tau_{\hat M}(\hat\sigma^*\hat M)=M/\tau_M(\sigma^*M)$ is free over $\mathcal{O}_K$ of rank 1. In order to get rid of u, we denote the image of t in $\mathbb{F}_\varepsilon$ by λ. Then $\mathbb{F}_\varepsilon=\mathbb{F}_q(\lambda)$ and $z(t)$ equals the minimal polynomial $(t-\lambda)\cdots(t-\lambda^{q^{f-1}})$ of λ over $\mathbb{F}_q$. Moreover, $t\equiv\lambda \bmod zA_\varepsilon$ and $\theta\equiv\lambda \bmod \zeta\mathbb{F}_\varepsilon[[\zeta]]$. We compute in $\mathbb{F}_\varepsilon[[\zeta]][[z]]/(\zeta)$

$$z(t) \;=\; (t-\lambda)\cdots(t-\lambda^{q^{f-1}}) \;\equiv\; (t-\theta)\cdots(t-\theta^{q^{f-1}}) \;\equiv\; (z-\zeta)u \;\equiv\; zu \mod \zeta\,.$$

Since z is a non-zero-divisor in $\mathbb{F}_\varepsilon[[\zeta]][[z]]/(\zeta)$, it follows that $u\equiv 1 \bmod \zeta\,\mathbb{F}_\varepsilon[[\zeta]][[z]]$. We write $u=1+\zeta u'$ and observe that the product

$$w \;:=\; \prod_{n=0}^{\infty}\hat\sigma^n(u) \;=\; \prod_{n=0}^{\infty}\hat\sigma^n(1+\zeta u') \;=\; \prod_{n=0}^{\infty}\big(1+\zeta^{\hat q^n}\hat\sigma^n(u')\big)$$

converges in $\mathbb{F}_\varepsilon[[\zeta]][[z]]^\times$ because $\mathbb{F}_\varepsilon[[\zeta]][[z]]$ is ζ-adically complete. It satisfies $w=u\cdot\hat\sigma(w)$ and so multiplication with w defines a canonical isomorphism $(\mathcal{O}_K[[z]],\, z-\zeta)\xrightarrow{\sim}\underline{\hat M}_\varepsilon(\underline{M})$.

We conclude that $\underline{\hat M}_\varepsilon(\underline{M})=(\mathcal{O}_K[[z]],\, z-\zeta)$, regardless of $\deg_t z(t)$.

3 Finite Shtukas

In this section, let R be an arbitrary $\mathbb{F}_\varepsilon$-algebra. For an R-module $\hat M$ we set $\hat\sigma^*\hat M:=\hat M\otimes_{R,\,\mathrm{Frob}_{\hat q}}R$ where $\mathrm{Frob}_{\hat q}$ is the $\hat q$-Frobenius endomorphism of R with $\mathrm{Frob}_{\hat q}(b)=b^{\hat q}$ for $b\in R$.

Definition 3.1 A *finite* $\mathbb{F}_\varepsilon$*-shtuka* over R is a pair $\underline{\hat{M}} = (\hat{M}, \tau_{\hat{M}})$ consisting of a locally free R-module $\hat{M}$ of finite rank denoted by $\operatorname{rk}\underline{\hat{M}}$, and an R-module homomorphism $\tau_{\hat{M}}\colon \hat{\sigma}^*\hat{M} \to \hat{M}$ satisfying $f \circ \tau_{\hat{M}} = \tau_{\hat{M}'} \circ \hat{\sigma}^* f$. That is, the following diagram is commutative

$$\begin{array}{ccc} \hat{\sigma}^*\hat{M} & \xrightarrow{\hat{\sigma}^* f} & \hat{\sigma}^*\hat{M}' \\ \downarrow{\scriptstyle \tau_{\hat{M}}} & & \downarrow{\scriptstyle \tau_{\hat{M}'}} \\ \hat{M} & \xrightarrow{f} & \hat{M}'\,. \end{array}$$

A finite $\mathbb{F}_\varepsilon$-shtuka over R is called *étale* if $\tau_{\hat{M}}$ is an isomorphism. We say that $\tau_{\hat{M}}$ is *nilpotent* if there is an integer n such that $\tau^n_{\hat{M}} := \tau_{\hat{M}} \circ \hat{\sigma}^*\tau_{\hat{M}} \circ \ldots \circ \sigma^*_{q^{n-1}}\tau_{\hat{M}} = 0$.

Finite $\mathbb{F}_\varepsilon$-shtukas were studied at various places in the literature. They were called "(finite) φ-sheaves" by Drinfeld [7, § 2], Taguchi and Wan [31, 32], and "Dieudonné $\mathbb{F}_q$-modules" by Laumon [25]. Finite $\mathbb{F}_\varepsilon$-shtukas over a field admit a canonical decomposition.

Proposition 3.2 *([25, Lemma B.3.10]) If $R = L$ is a field, every finite $\mathbb{F}_\varepsilon$-shtuka $\underline{\hat{M}} = (\hat{M}, \tau_{\hat{M}})$ is canonically an extension of finite $\mathbb{F}_\varepsilon$-shtukas*

$$0 \longrightarrow (\hat{M}_{\text{ét}}, \tau_{\text{ét}}) \longrightarrow (\hat{M}, \tau_{\hat{M}}) \longrightarrow (\hat{M}_{\text{nil}}, \tau_{\text{nil}}) \longrightarrow 0$$

where $\tau_{\text{ét}}$ is an isomorphism and τ_{nil} is nilpotent. $\underline{\hat{M}}_{\text{ét}} = (\hat{M}_{\text{ét}}, \tau_{\text{ét}})$ is the largest étale finite $\mathbb{F}_q$-sub-shtuka of $\underline{\hat{M}}$ and equals $\operatorname{im}(\tau_{\hat{M}}^{\operatorname{rk}\underline{\hat{M}}})$. If L is perfect, this extension splits canonically.

Example 3.3 Every effective local shtuka $(\hat{M}, \tau_{\hat{M}})$ of rank r over R yields for every $n \in \mathbb{N}$ a finite $\mathbb{F}_\varepsilon$-shtuka $\big(\hat{M}/z^n\hat{M}, \tau_{\hat{M}} \bmod z^n\big)$ of rank rn, and $(\hat{M}, \tau_{\hat{M}})$ equals the projective limit of these finite $\mathbb{F}_\varepsilon$-shtukas.

Thus, from Proposition 3.2 we obtain the following.

Proposition 3.4 *If $R = L$ is a field in $\mathcal{N}ilp_{A_\varepsilon}$, that is, $\zeta = 0$ in L, then every effective local shtuka $(\hat{M}, \tau_{\hat{M}})$ is canonically an extension of effective local shtukas*

$$0 \longrightarrow (\hat{M}_{\text{ét}}, \tau_{\text{ét}}) \longrightarrow (\hat{M}, \tau_{\hat{M}}) \longrightarrow (\hat{M}_{\text{nil}}, \tau_{\text{nil}}) \longrightarrow 0$$

where $\tau_{\text{ét}}$ is an isomorphism and τ_{nil} is topologically nilpotent. $(\hat{M}_{\text{ét}}, \tau_{\text{ét}})$ is the largest étale effective local sub-shtuka of $(\hat{M}, \tau_{\hat{M}})$. If L is perfect, this extension splits canonically. □

Finite $\mathbb{F}_\varepsilon$-shtukas and local shtukas are related to group schemes in the following way. Let $\underline{\hat{M}} = (\hat{M}, \tau_{\hat{M}})$ be a finite $\mathbb{F}_\varepsilon$-shtuka over R. Let

$$E = \operatorname{Spec} \bigoplus_{n \geq 0} \operatorname{Sym}^n_R \hat{M}$$

be the geometric vector bundle corresponding to $\hat{M}$ over Spec R, and let $F_{\hat{q},E}\colon E \to \hat{\sigma}^*E$ be its relative $\hat{q}$-Frobenius morphism over R. On the other hand, the map $\tau_{\hat{M}}$ induces another R-morphism $\operatorname{Spec}(\operatorname{Sym}^\bullet \tau_{\hat{M}})\colon E \to \hat{\sigma}^*E$. Drinfeld defines

$$\mathrm{Dr}_{\hat{q}}(\underline{\hat{M}}) := \ker\bigl(\operatorname{Spec}(\operatorname{Sym}^\bullet \tau_{\hat{M}}) - F_{\hat{q},E}\colon E \to \hat{\sigma}^*E\bigr) = \operatorname{Spec}\Bigl(\bigoplus_{n\geq 0} \operatorname{Sym}^n_R \hat{M}\Bigr)/I$$

where the ideal I is generated by the elements $m^{\otimes q} - \tau_{\hat{M}}(\hat{\sigma}^*m)$ for all elements m of $\hat{M}$. (Here, $m^{\otimes q}$ lives in $\operatorname{Sym}^q_R \hat{M}$ and $\tau_{\hat{M}}(\hat{\sigma}^*m)$ in $\operatorname{Sym}^1_R \hat{M}$.) Note that locally on Spec R, we have $\hat{M} = \bigoplus_{i=1}^d R\cdot m_i$ and $E \cong \operatorname{Spec} R[m_1,\ldots,m_d] = \mathbb{G}^d_{a,R}$. The subgroup scheme $\mathrm{Dr}_{\hat{q}}(\underline{\hat{M}})$ is finite locally free over R of order $\hat{q}^{\operatorname{rk}\underline{\hat{M}}}$, that is, the R-algebra $\mathcal{O}_{\mathrm{Dr}_{\hat{q}}(\underline{\hat{M}})}$ is a finite locally free R-module of rank $\hat{q}^{\operatorname{rk}\underline{\hat{M}}}$. It is also an $\mathbb{F}_\varepsilon$-module scheme over R via the comultiplication $\Delta\colon m \mapsto m\otimes 1 + 1\otimes m$ and the $\mathbb{F}_\varepsilon$-action $[a]\colon m \mapsto am$ which it inherits from E. It is even a *strict* $\mathbb{F}_\varepsilon$*-module scheme* in the sense of Faltings [9] and Abrashkin [2]. For a proof, see [2, Theorem 2] or [21, § 5]. This means that $\mathbb{F}_\varepsilon$ acts on the *co-Lie complex of* $\mathrm{Dr}_{\hat{q}}(\underline{\hat{M}})$ *over* R, see Illusie [26, § VII.3.1], via the scalar multiplication through $\mathbb{F}_\varepsilon \subset R$. A detailed explanation of strict $\mathbb{F}_\varepsilon$-module schemes is given in [21, § 4].

Conversely, let $G = \operatorname{Spec} A$ be a finite locally free strict $\mathbb{F}_\varepsilon$-module scheme over R. Note that on the additive group scheme $\mathbb{G}_{a,R} = \operatorname{Spec} R[x]$, the elements $b \in R$ act via endomorphisms $\psi_b\colon \mathbb{G}_{a,R} \to \mathbb{G}_{a,R}$ given by $\psi_b^*\colon R[x] \to R[x]$, $x \mapsto bx$. This makes $\mathbb{G}_{a,R}$ into an R-module scheme, and in particular, into an $\mathbb{F}_\varepsilon$-module scheme via $\mathbb{F}_\varepsilon \subset R$. We associate with G the R-module of $\mathbb{F}_\varepsilon$-equivariant homomorphisms on R

$$\hat{M}_{\hat{q}}(G) := \operatorname{Hom}_{R\text{-groups},\mathbb{F}_\varepsilon\text{-lin}}(G,\mathbb{G}_{a,R}) = \bigl\{x\in A\colon \Delta(x) = x\otimes 1 + 1\otimes x,\ [a](x) = ax,\ \forall a \in \mathbb{F}_\varepsilon\bigr\},$$

with its action of R via $R \to \operatorname{End}_{R\text{-groups},\mathbb{F}_\varepsilon\text{-lin}}(\mathbb{G}_{a,R})$. It is a finite locally free R-module by [30, Proposition 3.6 and Remark 5.5]; see also [1, $\mathrm{VII_A}$, 7.4.3] in the reedited version of SGA 3 by P. Gille and P. Polo. The composition on the left with the relative $\hat{q}$-Frobenius endomorphism $F_{\hat{q},\mathbb{G}_{a,R}}$ of $\mathbb{G}_{a,R} = \operatorname{Spec} R[x]$ given by $x \mapsto x^{\hat{q}}$ defines a map $\hat{M}_{\hat{q}}(G) \to \hat{M}_{\hat{q}}(G)$, $m \mapsto F_{\hat{q},\mathbb{G}_{a,R}} \circ m$ which is not R-linear, but $\hat{\sigma}$-linear, because $F_{\hat{q},\mathbb{G}_{a,R}} \circ \psi_b = \psi_{b^{\hat{q}}} \circ F_{\hat{q},\mathbb{G}_{a,R}}$. Therefore, $F_{\hat{q},\mathbb{G}_{a,R}}$ induces an R-homomorphism $\tau_{\hat{M}_{\hat{q}}(G)}\colon \hat{\sigma}^*\hat{M}_{\hat{q}}(G) \to \hat{M}_{\hat{q}}(G)$. Then $\underline{\hat{M}}_{\hat{q}}(G) := \bigl(\hat{M}_{\hat{q}}(G), \tau_{\hat{M}_{\hat{q}}(G)}\bigr)$ is a finite $\mathbb{F}_\varepsilon$-shtuka over R. If $f\colon G \to H$ is a morphism of finite locally free strict $\mathbb{F}_\varepsilon$-module schemes over R, then $\underline{\hat{M}}_{\hat{q}}(f)\colon \underline{\hat{M}}_{\hat{q}}(H) \to \underline{\hat{M}}_{\hat{q}}(G)$, $m \mapsto m\circ f$. This defines the functor $\underline{\hat{M}}_{\hat{q}}$ from the category of finite locally free strict $\mathbb{F}_\varepsilon$-module schemes over R to finite $\mathbb{F}_\varepsilon$-shtukas over R. It has the following properties.

Theorem 3.5 ([21, Theorem 5.2])

(a) The contravariant functors $\mathrm{Dr}_{\hat{q}}$ *and* $\underline{\hat{M}}_{\hat{q}}$ *are mutually quasi-inverse anti-equivalences between the category of finite* $\mathbb{F}_\varepsilon$*-shtukas over* R *and the category of finite locally free strict* $\mathbb{F}_\varepsilon$*-module schemes over* R.

(b) *Both functors are $\mathbb{F}_q$-linear and map short exact sequences to short exact sequences. They preserve étale objects and map the canonical decompositions from Propositions 3.2 and 3.6 below to each other.*

Let $\underline{\hat{M}} = (\hat{M}, \tau_{\hat{M}})$ *be a finite* $\mathbb{F}_\varepsilon$*-shtuka over* R *and let* $G = \mathrm{Dr}_{\hat{q}}(\underline{\hat{M}})$*. Then*

(c) *The* $\mathbb{F}_\varepsilon$*-module scheme* $\mathrm{Dr}_{\hat{q}}(\underline{\hat{M}})$ *is radical over* R *if and only if* $\tau_{\hat{M}}$ *is nilpotent.*
(d) *The order of the* R*-group scheme* $\mathrm{Dr}_{\hat{q}}(\underline{\hat{M}})$ *is* $\hat{q}^{\operatorname{rk}\underline{\hat{M}}}$.
(e) *There is a canonical isomorphism between* $\operatorname{coker}\tau_{\hat{M}} = \hat{M}/\tau_{\hat{M}}(\hat{\sigma}^*\hat{M})$ *and the co-Lie module* $\omega_{\mathrm{Dr}_{\hat{q}}(\underline{\hat{M}})} := e^*\Omega^1_{\mathrm{Dr}_{\hat{q}}(\underline{\hat{M}})/R}$ *where* $e\colon \operatorname{Spec} R \to \mathrm{Dr}_{\hat{q}}(\underline{\hat{M}})$ *is the zero section.*

Proposition 3.6 *([21, Proposition 4.2]) If* $R = L$ *is a field, every* $\mathbb{F}_\varepsilon$*-module scheme* G *over* L *is canonically an extension* $0 \to G^\circ \to G \to G^{\text{ét}} \to 0$ *of an étale* $\mathbb{F}_\varepsilon$*-module scheme* $G^{\text{ét}}$ *by a connected* $\mathbb{F}_\varepsilon$*-module scheme* G°*. The* $\mathbb{F}_\varepsilon$*-module scheme* $G^{\text{ét}}$ *is the largest étale quotient of* G*. If* L *is perfect,* $G^{\text{ét}}$ *is canonically isomorphic to the reduced closed* $\mathbb{F}_\varepsilon$*-module subscheme* G^{red} *of* G *and the extension splits canonically,* $G = G^\circ \times_L G^{\text{red}}$.

4 Divisible Local Anderson Modules

Let $R \in \mathcal{N}ilp_{A_\varepsilon}$ and let $\underline{\hat{M}} = (\hat{M}, \tau_{\hat{M}})$ be an effective local shtuka over R. Set $\underline{\hat{M}}_n := (\hat{M}_n, \tau_{\hat{M}_n}) := (\hat{M}/z^n\hat{M}, \tau_{\hat{M}} \bmod z^n)$ and consider the finite locally free strict $\mathbb{F}_\varepsilon$-module scheme $\mathrm{Dr}_{\hat{q}}(\underline{\hat{M}}_n)$ over R from the previous section. $\mathrm{Dr}_{\hat{q}}(\underline{\hat{M}}_n)$ inherits from $\underline{\hat{M}}_n$ an action of $A_\varepsilon/(z^n) = \mathbb{F}_\varepsilon[z]/(z^n)$. The canonical epimorphisms $\underline{\hat{M}}_{n+1} \twoheadrightarrow \underline{\hat{M}}_n$ induce closed immersions $i_n\colon \mathrm{Dr}_{\hat{q}}(\underline{\hat{M}}_n) \hookrightarrow \mathrm{Dr}_{\hat{q}}(\underline{\hat{M}}_{n+1})$. The inductive limit $\mathrm{Dr}_{\hat{q}}(\underline{\hat{M}}) := \varinjlim \mathrm{Dr}_{\hat{q}}(\underline{\hat{M}}_n)$ in the category of sheaves on the big *fppf*-site of $\operatorname{Spec} R$ is a sheaf of A_ε-modules that satisfies the following.

Definition 4.1 A *z-divisible local Anderson module over* R is a sheaf of A_ε-modules G on the big *fppf*-site of $\operatorname{Spec} R$ such that

(a) G is *z-torsion*, that is, $G = \varinjlim G[z^n]$, where $G[z^n] := \ker(z^n\colon G \to G)$;
(b) G is *z-divisible*, that is, $z\colon G \to G$ is an epimorphism;
(c) For every n, the $\mathbb{F}_\varepsilon$-module $G[z^n]$ is representable by a finite locally free strict $\mathbb{F}_\varepsilon$-module scheme over R in the sense of Faltings [9] and Abrashkin [2];
(d) Locally on R, there exists an integer $d \in \mathbb{Z}_{\geq 0}$, such that $(z-\zeta)^d = 0$ on ω_G where $\omega_G := \varprojlim \omega_{G[z^n]}$ and $\omega_{G[z^n]} = e^*\Omega^1_{G[z^n]/R}$ is the pullback under the zero section $e\colon \operatorname{Spec} R \to G[z^n]$. Here, the action of z on ω_G comes from the structure of A_ε-module on G, while the action of ζ on ω_G comes from the structure of R-module on ω_G.

A *morphism of z-divisible local Anderson modules over R* is a morphism of *fppf*-sheaves of $\mathbb{F}_\varepsilon[[z]]$-modules. It is shown in [21, Lemma 8.2 and Theorem 10.8] that ω_G is a finite locally free R-module, and we define the *dimension of G* as $\operatorname{rk}\omega_G$. Moreover, it follows from [21, Proposition 7.5] that there is a locally constant function $h\colon \operatorname{Spec} R \to \mathbb{N}_0,\ s \mapsto h(s)$ such that the order of $G[z^n]$ equals $\hat{q}^{nh}$. We call h the *height* of the z-divisible local Anderson module G.

The category of z-divisible local Anderson modules over R and the category of local shtukas over R are both A_ε-linear. The construction and the equivalence from Sect. 3 extend to an equivalence between the category of effective local shtukas over R and the category of z-divisible local Anderson modules over R.

The quasi-inverse functor to $\underline{\hat{M}} \mapsto \mathrm{Dr}_{\hat{q}}(\underline{\hat{M}})$ is given as follows. Let $G = \varinjlim G[z^n]$ be a z-divisible local Anderson module over R. We set

$$\underline{\hat{M}}_{\hat{q}}(G) \;=\; \big(\hat{M}_{\hat{q}}(G), \tau_{\hat{M}_{\hat{q}}(G)}\big) \;:=\; \varprojlim_n \big(\hat{M}_{\hat{q}}(G[z^n]), \tau_{\hat{M}_{\hat{q}}(G[z^n])}\big)\,.$$

Multiplication with z on G gives $\hat{M}_{\hat{q}}(G)$, the structure of an $R[[z]]$-module. The following theorem was proved in [21, Theorem 8.3].

Theorem 4.2 *Let $R \in \mathcal{N}ilp_{A_\varepsilon}$.*

(a) The two contravariant functors $\mathrm{Dr}_{\hat{q}}$ and $\underline{\hat{M}}_{\hat{q}}$ are mutually quasi-inverse anti-equivalences between the category of effective local shtukas over R and the category of z-divisible local Anderson modules over R.

(b) Both functors are A_ε-linear, map short exact sequences to short exact sequences, and preserve (ind-) étale objects.

Let $\underline{\hat{M}} = (\hat{M}, \tau_{\hat{M}})$ be an effective local shtuka over R, and let $G = \mathrm{Dr}_{\hat{q}}(\underline{\hat{M}})$ be its associated z-divisible local Anderson module. Then

(c) G is a formal A_ε-module, i.e. a formal Lie group equipped with an action of A_ε, if and only if $\tau_{\hat{M}}$ is topologically nilpotent.

(d) The height and dimension of G are equal to the rank and dimension of $\underline{\hat{M}}$.

(e) The $R[[z]]$-modules $\omega_{\mathrm{Dr}_{\hat{q}}(\underline{\hat{M}})}$ and $\operatorname{coker}\tau_{\hat{M}}$ are canonically isomorphic.

Example 4.3 In the notation of Example 2.3, let $R \in \mathcal{N}ilp_{A_\varepsilon}$ and let r be a positive integer. A *Drinfeld A-module of rank r over R* is a pair $\underline{E} = (E, \varphi)$ consisting of a smooth affine group scheme E over $\operatorname{Spec} R$ of relative dimension 1 and a ring homomorphism $\varphi\colon A \to \operatorname{End}_{R\text{-groups}}(E),\ a \mapsto \varphi_a$ satisfying the following conditions:

(a) Zariski-locally on $\operatorname{Spec} R$ there is an isomorphism $\alpha\colon E \xrightarrow{\sim} \mathbb{G}_{a,R}$ of $\mathbb{F}_q$-module schemes such that

(b) the coefficients of $\Phi_a := \alpha \circ \varphi_a \circ \alpha^{-1} = \sum_{i\geq 0} b_i(a)\tau^i \in \operatorname{End}_{R\text{-groups},\mathbb{F}_q\text{-lin}}(\mathbb{G}_{a,R}) = R\{\tau\}$ satisfy $b_0(a) = \gamma(a)$, $b_{r(a)}(a) \in R^\times$ and $b_i(a)$ is nilpotent for all $i > r(a) := -r\,[\mathbb{F}_\infty : \mathbb{F}_q]\operatorname{ord}_\infty(a)$.

Here, $R\{\tau\} := \big\{ \sum\limits_{i=0}^{n} b_i\tau^i : n \in \mathbb{N}_0, b_i \in R \big\}$ is the non-commutative polynomial ring with $\tau b = b^q\tau$, and the isomorphism of rings $R\{\tau\} \xrightarrow{\sim} \mathrm{End}_{R\text{-groups},\mathbb{F}_q\text{-lin}}(\mathbb{G}_{a,R})$ is given by sending τ to the relative $\hat{q}$-Frobenius endomorphism $F_{\hat{q},\mathbb{G}_{a,R}}$ of $\mathbb{G}_{a,R} = \operatorname{Spec} R[x]$ given by $x \mapsto x^{\hat{q}}$ and $b \in R$ to the endomorphism ψ_b given by $\psi_b^* \colon x \mapsto bx$.

For a Drinfeld A-module $\underline{E} = (E, \varphi)$, we consider the set $M := M(\underline{E}) := \mathrm{Hom}_{R\text{-groups},\mathbb{F}_q\text{-lin}}(E, \mathbb{G}_{a,R})$ of $\mathbb{F}_q$-equivariant homomorphisms of R-group schemes. It is a locally free module over $A_R := A \otimes_{\mathbb{F}_q} R$ of rank r under the action given on $m \in M$ by

$$\begin{aligned} A \ni a \colon M &\longrightarrow M,\ m \mapsto m \circ \varphi_a =:\ am \\ R \ni b \colon M &\longrightarrow M,\ m \mapsto \psi_b \circ m =:\ bm \end{aligned}$$

In addition, we consider the map $\tau \colon m \mapsto F_{q,\mathbb{G}_{a,R}} \circ m$ on $m \in M$, where $F_{q,\mathbb{G}_{a,R}}$ is the relative q-Frobenius of $\mathbb{G}_{a,R}$ over R. Since $F_{q,\mathbb{G}_{a,R}} \circ \psi_b = \psi_{b^q} \circ F_{q,\mathbb{G}_{a,R}}$, and hence $\tau(bm) = b^q\tau(m)$, the map τ is σ-semilinear and induces an A_R-linear map $\tau_M \colon \sigma^*M \to M$, which makes $\underline{M}(\underline{E}) := \big(M(\underline{E}), \tau_M\big)$ into an effective A-motive over R in the sense of Example 2.3. The functor $\underline{E} \mapsto \underline{M}(\underline{E})$ is fully faithful and its essential image is described in [18, Theorems 3.5 and 3.9] generalizing Anderson's description [3, Theorem 1].

Now let $\hat{\underline{M}} := \hat{\underline{M}}_\varepsilon(\underline{M}(\underline{E}))$ be the effective local $\hat{\sigma}$-shtuka at ε associated with $\underline{M}(\underline{E})$; see Example 2.3. Let $n \in \mathbb{N}$ and let $\varepsilon^n = (a_1, \ldots, a_s) \subset A$. Then

$$\underline{E}[\varepsilon^n] \ :=\ \ker\big(\varphi_{a_1,\ldots,a_s} := (\varphi_{a_1}, \ldots, \varphi_{a_s}) \colon E \longrightarrow E^s\big)$$

is called the *ε^n-torsion submodule of* $\underline{E}$. It is an A/ε^n-module via $A/\varepsilon^n \to \mathrm{End}_R(\underline{E}[\varepsilon^n])$, $\bar{a} \mapsto \varphi_a$ and independent of the set of generators of ε^n; see [18, Lemma 6.2]. Moreover, by [18, Theorem 7.6] it is a finite locally free R-group scheme and a strict $\mathbb{F}_\varepsilon$-module scheme and there are canonical A/ε^n-equivariant isomorphisms of finite locally free R-group schemes

$$\begin{aligned} \mathrm{Dr}_{\hat{q}}(\hat{\underline{M}}/\varepsilon^n\hat{\underline{M}}) &\xrightarrow{\sim} \underline{E}[\varepsilon^n] \qquad \text{and} \\ \hat{\underline{M}}/\varepsilon^n\hat{\underline{M}} &\xrightarrow{\sim} \mathrm{Hom}_{R\text{-groups},\mathbb{F}_\varepsilon\text{-lin}}\big(\underline{E}[\varepsilon^n], \mathbb{G}_{a,R}\big) \end{aligned}$$

of finite $\mathbb{F}_\varepsilon$-shtukas. In particular, $\underline{E}[\varepsilon^\infty] := \varinjlim \underline{E}[\varepsilon^n] = \mathrm{Dr}_{\hat{q}}(\hat{\underline{M}})$ is a z-divisible local Anderson module over R.

5 Cohomology Realizations of Local Shtukas

In this section, we work over a valuation ring $\mathcal{O}_K$ as in (b). With local shtukas over $\mathcal{O}_K$, one can associate various cohomology realizations, which are related to each other under period isomorphisms. We describe the ε-adic, the de Rham, and the crystalline realizations. These period isomorphisms are used in [20, 22] to study the periods of A-motives with complex multiplication.

Definition 5.1 Let $\underline{\hat{M}} = (\hat{M}, \tau_{\hat{M}})$ be a local shtuka over a valuation ring $\mathcal{O}_K$ as in (b). Then $\tau_{\hat{M}}$ induces an isomorphism $\tau_{\hat{M}} : \hat{\sigma}^*\hat{M} \otimes_{\mathcal{O}_K[\![z]\!]} K[\![z]\!] \xrightarrow{\sim} \hat{M} \otimes_{\mathcal{O}_K[\![z]\!]} K[\![z]\!]$, because $z - \zeta \in K[\![z]\!]^\times$. We define the *(dual) Tate module*

$$\mathrm{H}^1_\varepsilon(\underline{\hat{M}}, A_\varepsilon) := \check{T}_\varepsilon\underline{\hat{M}} := (\hat{M} \otimes_{\mathcal{O}_K[\![z]\!]} K^{\mathrm{sep}}[\![z]\!])^{\hat{\tau}} := \{m \in \hat{M} \otimes_{\mathcal{O}_K[\![z]\!]} K^{\mathrm{sep}}[\![z]\!] : \tau_{\hat{M}}(\hat{\sigma}^*_{\hat{M}} m) = m\}$$

and the *rational (dual) Tate module*

$$\mathrm{H}^1_\varepsilon(\underline{\hat{M}}, Q_\varepsilon) := \check{V}_\varepsilon\underline{\hat{M}} := \{m \in \hat{M} \otimes_{\mathcal{O}_K[\![z]\!]} K^{\mathrm{sep}}(\!(z)\!) : \tau_{\hat{M}}(\hat{\sigma}^*_{\hat{M}} m) = m\} = \check{T}_\varepsilon\underline{\hat{M}} \otimes_{A_\varepsilon} Q_\varepsilon .$$

By [19, Proposition 4.2], the Tate modules are free over A_ε, resp. Q_ε of rank equal to $\operatorname{rk} \underline{\hat{M}}$ and carry a continuous action of $\mathrm{Gal}(K^{\mathrm{sep}}/K)$. They are also called the *ε-adic realizations of $\underline{\hat{M}}$.*

Theorem 5.2 *([19, Theorem 4.20]) Assume that $\mathcal{O}_K$ is discretely valued. Then the functor $\check{T}_\varepsilon : \underline{\hat{M}} \mapsto \check{T}_\varepsilon\underline{\hat{M}}$ from the category of local shtukas over $\mathcal{O}_K$ to the category* $\mathrm{Rep}_{A_\varepsilon} \mathrm{Gal}(K^{\mathrm{sep}}/K)$ *of continuous representations of* $\mathrm{Gal}(K^{\mathrm{sep}}/K)$ *on finite free A_ε-modules and the functor $\check{V}_\varepsilon : \underline{\hat{M}} \mapsto \check{V}_\varepsilon\underline{\hat{M}}$ from the category of local shtukas over $\mathcal{O}_K$ with quasi-morphisms to the category* $\mathrm{Rep}_{Q_\varepsilon} \mathrm{Gal}(K^{\mathrm{sep}}/K)$ *of continuous representations of* $\mathrm{Gal}(K^{\mathrm{sep}}/K)$ *on finite-dimensional Q_ε-vector spaces are fully faithful.*

Definition 5.3 Let $\mathcal{O}_K$ be discretely valued. The full subcategory of $\mathrm{Rep}_{Q_\varepsilon} \mathrm{Gal}(K^{\mathrm{sep}}/K)$ which is the essential image of the functor $\check{V}_\varepsilon$ from Theorem 5.2 is called the *category of equal characteristic crystalline representations.*

We will explain the motivation for this definition in Sect. 6.

Example 5.4 We describe the ε-adic (dual) Tate module $\check{T}_\varepsilon\underline{M} = \check{T}_\varepsilon\underline{\hat{M}}_\varepsilon(\underline{M})$ of the Carlitz motive $\underline{M} = (\mathcal{O}_K[t], t - \theta)$ from Example 2.5 by using the local shtuka $\underline{\hat{M}} := \underline{\hat{M}}_\varepsilon(\underline{M}) = (\mathcal{O}_K[\![z]\!], z - \zeta)$ computed there. For all $i \in \mathbb{N}_0$, let $\ell_i \in K^{\mathrm{sep}}$ be solutions of the equations $\ell_0^{\hat{q}-1} = -\zeta$ and $\ell_i^{\hat{q}} + \zeta\ell_i = \ell_{i-1}$. This implies $|\ell_i| = |\zeta|^{\hat{q}^{-i}/(\hat{q}-1)} < 1$. Define the power series $\ell_+ = \sum_{i=0}^{\infty} \ell_i z^i \in \mathcal{O}_{K^{\mathrm{sep}}}[\![z]\!]$. It satisfies $\hat{\sigma}(\ell_+) = (z - \zeta)\cdot\ell_+$, but depends on the choice of the ℓ_i. A different choice yields a different power series $\tilde{\ell}^+$ which satisfies $\tilde{\ell}^+ = u\ell_+$ for a unit $u \in (K^{\mathrm{sep}}[\![z]\!]^\times)^{\hat{\sigma}=\mathrm{id}} = A_\varepsilon^\times$, because $\hat{\sigma}(u) = \frac{\hat{\sigma}(\tilde{\ell}^+)}{\hat{\sigma}(\ell_+)} = \frac{\tilde{\ell}^+}{\ell_+} = u$. The field extension $\mathbb{F}_\varepsilon(\!(\zeta)\!)(\ell_i : i \in \mathbb{N}_0)$ of $\mathbb{F}_\varepsilon(\!(\zeta)\!)$ is the function field analog of the cyclotomic tower $\mathbb{Q}_p(\sqrt[p^i]{1} : i \in \mathbb{N}_0)$; see [16, § 1.3 and § 3.4]. There is an isomorphism of topological groups called the *ε-adic cyclotomic character*

$$\chi_\varepsilon : \mathrm{Gal}\big(\mathbb{F}_\varepsilon(\!(\zeta)\!)(\ell_i : i \in \mathbb{N}_0)/\mathbb{F}_\varepsilon(\!(\zeta)\!)\big) \xrightarrow{\sim} A_\varepsilon^\times ,$$

which satisfies $g(\ell_+) := \sum_{i=0}^{\infty} g(\ell_i)z^i = \chi_\varepsilon(g)\cdot\ell_+$ in $K^{\text{sep}}[[z]]$ for g in the Galois group. It is independent of the choice of the ℓ_i. The ε-adic (dual) Tate module $\check{T}_\varepsilon\underline{\hat{M}}$ of $\underline{\hat{M}}$ and $\underline{M}$ is generated by ℓ_+^{-1} on which the Galois group acts by the inverse of the cyclotomic character.

Definition 5.5 Let $\underline{\hat{M}}$ be a local shtuka over a valuation ring $\mathcal{O}_K$ as in (b). We denote by $K[[z-\zeta]]$ the power series ring over K in the "variable" $z-\zeta$ and by $K((z-\zeta))$ its fraction field. We consider the ring homomorphism $\mathcal{O}_K[[z]] \hookrightarrow K[[z-\zeta]]$, $z \mapsto z = \zeta + (z-\zeta)$ and define the *de Rham realization of* $\underline{\hat{M}}$ as

$$\begin{aligned} \mathrm{H}^1_{\mathrm{dR}}\big(\underline{\hat{M}}, K[[z-\zeta]]\big) &:= \hat{\sigma}^*\hat{M} \otimes_{\mathcal{O}_K[[z]]} K[[z-\zeta]]\,, \\ \mathrm{H}^1_{\mathrm{dR}}\big(\underline{\hat{M}}, K((z-\zeta))\big) &:= \hat{\sigma}^*\hat{M} \otimes_{\mathcal{O}_K[[z]]} K((z-\zeta)) \qquad \text{and} \\ \mathrm{H}^1_{\mathrm{dR}}(\underline{\hat{M}}, K) &:= \hat{\sigma}^*\hat{M} \otimes_{\mathcal{O}_K[[z]],\, z\mapsto\zeta} K \\ &= \mathrm{H}^1_{\mathrm{dR}}\big(\underline{\hat{M}}, K[[z-\zeta]]\big) \otimes_{K[[z-\zeta]]} K[[z-\zeta]]/(z-\zeta)\,. \end{aligned}$$

The de Rham realization $\mathrm{H}^1_{\mathrm{dR}}\big(\underline{\hat{M}}, K((z-\zeta))\big)$ contains a full $K[[z-\zeta]]$-lattice

$$\mathfrak{q}^{\underline{\hat{M}}} := \tau_{\hat{M}}^{-1}(\hat{M} \otimes_{\mathcal{O}_K[[z]]} K[[z-\zeta]]), \tag{5.1}$$

which is called the *Hodge-Pink lattice of* $\underline{\hat{M}}$. The de Rham realization $\mathrm{H}^1_{\mathrm{dR}}(\underline{\hat{M}}, K)$ carries a descending separated and exhausting filtration $F^\bullet$ by K-subspaces called the *Hodge-Pink filtration of* $\underline{\hat{M}}$. It is defined via $\mathfrak{p} := \mathrm{H}^1_{\mathrm{dR}}(\underline{\hat{M}}, K[[z-\zeta]])$ and (for $i \in \mathbb{Z}$)

$$F^i\,\mathrm{H}^1_{\mathrm{dR}}(\underline{\hat{M}}, K) := \big(\mathfrak{p} \cap (z-\zeta)^i\mathfrak{q}^{\underline{\hat{M}}}\big)\big/\big((z-\zeta)\mathfrak{p} \cap (z-\zeta)^i\mathfrak{q}^{\underline{\hat{M}}}\big) \subset \mathrm{H}^1_{\mathrm{dR}}(\underline{\hat{M}}, K). \tag{5.2}$$

If we equip $\mathrm{H}^1_{\mathrm{dR}}\big(\underline{\hat{M}}, K((z-\zeta))\big)$ with the descending filtration $F^i\,\mathrm{H}^1_{\mathrm{dR}}\big(\underline{\hat{M}}, K((z-\zeta))\big) := (z-\zeta)^i\mathfrak{q}^{\underline{\hat{M}}}$ by $K[[z-\zeta]]$-submodules, then $F^i\,\mathrm{H}^1_{\mathrm{dR}}(\underline{\hat{M}}, K)$ is the image of $\mathrm{H}^1_{\mathrm{dR}}\big(\underline{\hat{M}}, K[[z-\zeta]]\big) \cap F^i\,\mathrm{H}^1_{\mathrm{dR}}\big(\underline{\hat{M}}, K((z-\zeta))\big)$ in $\mathrm{H}^1_{\mathrm{dR}}(\underline{\hat{M}}, K)$. Since $z = \zeta + (z-\zeta)$ is invertible in $K[[z-\zeta]]$, the de Rham realization with Hodge-Pink lattice and filtration is a functor on the category of local shtukas over $\mathcal{O}_K$ with quasi-morphisms.

Note, however, that the Hodge-Pink filtration on $\mathrm{H}^1_{\mathrm{dR}}(\underline{\hat{M}}, K)$ does not behave well under tensor products, as opposed to the Hodge-Pink lattice; see Remark 6.3 below. Therefore, the more important concept is the Hodge-Pink lattice $\mathfrak{q}^{\underline{\hat{M}}}$.

Theorem 5.6 *([19, Theorem 4.15]) Let $\overline{K}$ be the completion of an algebraic closure K^{alg} of K. There is a canonical functorial comparison isomorphism*

$$h_{\varepsilon,\mathrm{dR}}\colon\ \mathrm{H}^1_\varepsilon(\underline{\hat{M}}, Q_\varepsilon) \otimes_{Q_\varepsilon} \overline{K}((z-\zeta)) \xrightarrow{\ \sim\ } \mathrm{H}^1_{\mathrm{dR}}\big(\underline{\hat{M}}, K((z-\zeta))\big) \otimes_{K((z-\zeta))} \overline{K}((z-\zeta))\,,$$

which satisfies $h_{\varepsilon,\mathrm{dR}}\big(\mathrm{H}^1_\varepsilon(\underline{\hat{M}}, Q_\varepsilon) \otimes_{Q_\varepsilon} \overline{K}[[z-\zeta]]\big) = \mathfrak{q}^{\underline{\hat{M}}} \otimes_{K[[z-\zeta]]} \overline{K}[[z-\zeta]]$ and which is equivariant for the action of $\mathrm{Gal}(K^{\text{sep}}/K)$*, where on the source of $h_{\varepsilon,\mathrm{dR}}$*

this group acts on both factors of the tensor product and on the target of $h_{\varepsilon,\mathrm{dR}}$ it acts only on $\overline{K}$.

Definition 5.7 Let $k = \mathcal{O}_K/\mathfrak{m}_K$ be the residue field of $\mathcal{O}_K$. A *z-isocrystal* over k is a pair (D, τ_D) consisting of a finite-dimensional $k(\!(z)\!)$-vector space together with a $k(\!(z)\!)$-isomorphism $\tau_D\colon \hat{\sigma}^*D \xrightarrow{\sim} D$. A *morphism* $(D, \tau_D) \to (D', \tau_{D'})$ is a $k(\!(z)\!)$-homomorphism $f\colon D \to D'$ satisfying $\tau_{D'} \circ \hat{\sigma}^* f = f \circ \tau_D$.

Definition 5.8 Let $\underline{\hat{M}} = (\hat{M}, \tau_{\hat{M}})$ be local shtuka over a valuation ring $\mathcal{O}_K$ as in (b). Then the *crystalline realization* of $\underline{\hat{M}}$ is defined as the z-isocrystal over $k = \mathcal{O}_K/\mathfrak{m}_K$

$$\mathrm{H}^1_{\mathrm{cris}}\big(\underline{\hat{M}}, k(\!(z)\!)\big) \;:=\; \hat{\sigma}^*(\hat{M}, \tau_{\hat{M}}) \otimes_{\mathcal{O}_K[\![z]\!]} k(\!(z)\!)\,. \tag{5.3}$$

It only depends on the special fiber $\underline{\hat{M}} \otimes_{\mathcal{O}_K} k$ of $\underline{\hat{M}}$ and defines a functor $\underline{\hat{M}} \mapsto \mathrm{H}^1_{\mathrm{cris}}\big(\underline{\hat{M}}, k(\!(z)\!)\big)$ from the category of local shtukas over $\mathcal{O}_K$ with quasi-morphism to the category of z-isocrystals. This functor is faithful by [19, Lemma 4.24] if $\bigcap_n \hat{\sigma}^n(\mathfrak{m}_K) = (0)$.

To formulate the comparison between the de Rham and the crystalline realization, we assume that there exists a fixed section $k \hookrightarrow \mathcal{O}_K$. Then there is a ring homomorphism

$$k(\!(z)\!) \hookrightarrow K[\![z-\zeta]\!]\,, \quad z \longmapsto \zeta + (z-\zeta)\,, \quad \sum_i b_i z^i \longmapsto \sum_{j=0}^{\infty} (z-\zeta)^j \cdot \sum_i \tbinom{i}{j} b_i \zeta^{i-j}\,. \tag{5.4}$$

We always consider $K[\![z-\zeta]\!]$ and its fraction field $K(\!(z-\zeta)\!)$ as $k(\!(z)\!)$-vector spaces via (5.4).

Theorem 5.9 *([19, Theorem 5.18]) Let $\underline{\hat{M}}$ be a local shtuka over $\mathcal{O}_K$. Assume that $\mathcal{O}_K$ is discretely valued or that $\underline{\hat{M}} = \underline{\hat{M}}_\varepsilon(\underline{M})$ for an A-motive $\underline{M}$ over $\mathcal{O}_K$ as in Example 2.3. Then there are canonical functorial comparison isomorphisms between the de Rham and crystalline realizations*

$$\begin{aligned} h_{\mathrm{dR,cris}}\colon \mathrm{H}^1_{\mathrm{dR}}(\underline{\hat{M}}, K[\![z-\zeta]\!]) &\xrightarrow{\sim} \mathrm{H}^1_{\mathrm{cris}}\big(\underline{\hat{M}}, k(\!(z)\!)\big) \otimes_{k(\!(z)\!)} K[\![z-\zeta]\!] \quad \textit{and} \\ h_{\mathrm{dR,cris}}\colon \quad \mathrm{H}^1_{\mathrm{dR}}(\underline{\hat{M}}, K) &\xrightarrow{\sim} \mathrm{H}^1_{\mathrm{cris}}\big(\underline{\hat{M}}, k(\!(z)\!)\big) \otimes_{k(\!(z)\!),\, z\mapsto\zeta} K\,. \end{aligned}$$

To formulate the comparison between the crystalline and the ε-adic realizations, we introduce the $\mathcal{O}_K$-algebra

$$\mathcal{O}_K[\![z, z^{-1}\} := \Big\{ \sum_{i=-\infty}^{\infty} b_i z^i :\; b_i \in \mathcal{O}_K\,,\; |b_i|\,|\zeta|^{ri} \to 0 \; (i \to -\infty) \text{ for all } r > 0 \Big\}\,. \tag{5.5}$$

It is a subring of $K[[z-\zeta]]$ via the expansion $\sum\limits_{i=-\infty}^{\infty} b_i z^i = \sum\limits_{j=0}^{\infty} \zeta^{-j}\Big(\sum\limits_{i=-\infty}^{\infty} \binom{i}{j} b_i \zeta^i\Big)(z-\zeta)^j$. The homomorphism (5.4) factors through $\mathcal{O}_K[[z,z^{-1}\}$. We view the elements of $\mathcal{O}_K[[z,z^{-1}\}$ as functions that converge on the punctured open unit disk $\{0<|z|<1\}$. An example of such a function is

$$\ell_- := \prod_{i\in\mathbb{N}_0} (1-\tfrac{\zeta^{\hat{q}^i}}{z}) \in \mathbb{F}_\varepsilon[[\zeta]][[z,z^{-1}\} \subset \mathcal{O}_K[[z,z^{-1}\}, \tag{5.6}$$

which satisfies $\ell_- = (1-\frac{\zeta}{z})\cdot\hat{\sigma}(\ell_-)$. In addition, we let $\overline{K}$ be the completion of an algebraic closure K^{alg} of K and recall the element $\ell_+ \in \mathcal{O}_{\overline{K}}[[z]]$ from Example 5.4, which satisfies $\hat{\sigma}(\ell_+) = (z-\zeta)\cdot\ell_+$. We set

$$\ell := \ell_+\ell_- \in \mathcal{O}_{\overline{K}}[[z,z^{-1}\}. \tag{5.7}$$

Then $\hat{\sigma}(\ell) = z\cdot\ell$ and $g(\ell) = \chi_\varepsilon(g)\cdot\ell$ for $g\in \mathrm{Gal}(K^{\mathrm{sep}}/K)$ where χ_ε is the cyclotomic character from Example 5.4.

Theorem 5.10 *([19, Theorem 5.20]) Let $\underline{\hat{M}}$ be a local shtuka over $\mathcal{O}_K$. Assume that $\mathcal{O}_K$ is discretely valued or that $\underline{\hat{M}} = \underline{\hat{M}}_\varepsilon(\underline{M})$ for an A-motive $\underline{M}$ over $\mathcal{O}_K$ as in Example 2.3. Then there is a canonical functorial comparison isomorphism between the ε-adic and crystalline realizations*

$$h_{\varepsilon,\mathrm{cris}}\colon\ \mathrm{H}^1_\varepsilon(\underline{\hat{M}},Q_\varepsilon)\otimes_{Q_\varepsilon}\mathcal{O}_{\overline{K}}[[z,z^{-1}\}[\ell^{-1}] \xrightarrow{\ \sim\ } \mathrm{H}^1_{\mathrm{cris}}(\underline{\hat{M}},k((z)))\otimes_{k((z))}\mathcal{O}_{\overline{K}}[[z,z^{-1}\}[\ell^{-1}].$$

The isomorphism $h_{\varepsilon,\mathrm{cris}}$ is $\mathrm{Gal}(K^{\mathrm{sep}}/K)$- and $\hat{\tau}$-equivariant, where on the left module $\mathrm{Gal}(K^{\mathrm{sep}}/K)$ acts on both factors and $\hat{\tau}$ is $\mathrm{id}\otimes\hat{\sigma}$, and on the right module $\mathrm{Gal}(K^{\mathrm{sep}}/K)$ acts only on $\mathcal{O}_{\overline{K}}[[z,z^{-1}\}[\ell^{-1}]$ and $\hat{\tau}$ is $(\tau_D\circ\hat{\sigma}_D^*)\otimes\hat{\sigma}$. In other words, $h_{\varepsilon,\mathrm{cris}} = \tau_D\circ\hat{\sigma}^*h_{\varepsilon,\mathrm{cris}}$. Moreover, $h_{\varepsilon,\mathrm{cris}}$ satisfies $h_{\varepsilon,\mathrm{dR}} = (h^{-1}_{\mathrm{dR,cris}}\otimes \mathrm{id}_{\overline{K}((z-\zeta))})\circ(h_{\varepsilon,\mathrm{cris}}\otimes\mathrm{id}_{\overline{K}((z-\zeta))})$. It allows to recover $\mathrm{H}^1_\varepsilon(\underline{\hat{M}},Q_\varepsilon)$ from $\mathrm{H}^1_{\mathrm{cris}}(\underline{\hat{M}},k((z)))$ as the intersection inside $\mathrm{H}^1_{\mathrm{cris}}(\underline{\hat{M}},k((z)))\otimes_{k((z))}\overline{K}((z-\zeta))$

$$h_{\varepsilon,\mathrm{cris}}(\mathrm{H}^1_\varepsilon(\underline{\hat{M}},Q_\varepsilon)) = (\mathrm{H}^1_{\mathrm{cris}}(\underline{\hat{M}},k((z)))\otimes_{k((z))}\mathcal{O}_{\overline{K}}[[z,z^{-1}\}[\ell^{-1}])^{\hat{\tau}=\mathrm{id}} \cap \mathfrak{q}_D\otimes_{K[[z-\zeta]]}\overline{K}[[z-\zeta]],$$

where $\mathfrak{q}_D\subset \mathrm{H}^1_{\mathrm{cris}}(\underline{\hat{M}},k((z)))\otimes_{k((z))}K((z-\zeta))$ is the Hodge-Pink lattice of $\underline{\hat{M}}$.

6 Crystalline Representations over Function Fields

We explain the motivation for Definition 5.3; compare [19, Remarks 5.13 and 6.17].

Let $\mathcal{O}_K$ be discretely valued and let $\underline{\hat{M}}$ be a local shtuka over $\mathcal{O}_K$. Theorem 5.9 allows to define a Hodge-Pink lattice and a Hodge-Pink filtration on $\mathrm{H}^1_{\mathrm{cris}}\big(\underline{\hat{M}}, k(\!(z)\!)\big)$. More precisely, we equip the finite-dimensional $k(\!(z)\!)$-vector space $D := \mathrm{H}^1_{\mathrm{cris}}\big(\underline{\hat{M}}, k(\!(z)\!)\big)$ with the Hodge-Pink lattice

$$\mathfrak{q}_D \;:=\; (h_{\mathrm{dR,cris}} \otimes \mathrm{id}_{K(\!(z-\zeta)\!)})(\mathfrak{q}^{\underline{\hat{M}}}) \;\subset\; D \otimes_{k(\!(z)\!)} K(\!(z-\zeta)\!)\,,$$

where $\mathfrak{q}^{\underline{\hat{M}}} \subset \mathrm{H}^1_{\mathrm{dR}}\big(\underline{\hat{M}}, K(\!(z-\zeta)\!)\big)$ is the Hodge-Pink lattice from (5.1). Together with the Frobenius $\tau_D := \hat{\sigma}^*\tau_{\hat{M}} \otimes \mathrm{id}_{k(\!(z)\!)}$ on $D = \mathrm{H}^1_{\mathrm{cris}}\big(\underline{\hat{M}}, k(\!(z)\!)\big)$ from (5.3), the triple $\underline{D}(\underline{\hat{M}}) := \underline{D} = (D, \tau_D, \mathfrak{q}_D)$ forms a z-isocrystal with a Hodge-Pink structure as in the following.

Definition 6.1 A *z-isocrystal with Hodge-Pink structure over $\mathcal{O}_K$* is a triple $\underline{D} = (D, \tau_D, \mathfrak{q}_D)$ consisting of a z-isocrystal (D, τ_D) over k and a $K[\![z-\zeta]\!]$-lattice $\mathfrak{q}_D$ in $D \otimes_{k(\!(z)\!)} K(\!(z-\zeta)\!)$ of full rank, which is called the *Hodge-Pink lattice of* $\underline{D}$. The dimension of D is called the *rank of* $\underline{D}$ and is denoted by $\mathrm{rk}\,\underline{D}$.

A *morphism* $(D, \tau_D, \mathfrak{q}_D) \to (D', \tau_{D'}, \mathfrak{q}_{D'})$ is a $k(\!(z)\!)$-homomorphism $f\colon D \to D'$ satisfying $\tau_{D'} \circ \hat{\sigma}^* f = f \circ \tau_D$ and $(f \otimes \mathrm{id})(\mathfrak{q}_D) \subset \mathfrak{q}_{D'}$.

A *strict subobject* $\underline{D}' \subset \underline{D}$ is a z-isocrystal with Hodge-Pink structure of the form $\underline{D}' = \big(D', \tau_D|_{\hat{\sigma}^*D'}, \mathfrak{q}_D \cap D' \otimes_{k(\!(z)\!)} K(\!(z-\zeta)\!)\big)$ where $D' \subset D$ is a $k(\!(z)\!)$-subspace with $\tau_D(\hat{\sigma}^* D') = D'$.

On a z-isocrystal with Hodge-Pink structure $\underline{D}$, there always is the tautological $K[\![z-\zeta]\!]$-lattice $\mathfrak{p}_D := D \otimes_{k(\!(z)\!)} K[\![z-\zeta]\!]$. Since $K[\![z-\zeta]\!]$ is a principal ideal domain, the elementary divisor theorem provides basis vectors $v_i \in \mathfrak{p}_D$ such that $\mathfrak{p}_D = \bigoplus_{i=1}^r K[\![z-\zeta]\!] \cdot v_i$ and $\mathfrak{q}_D = \bigoplus_{i=1}^r K[\![z-\zeta]\!] \cdot (z-\zeta)^{\mu_i} \cdot v_i$ for integers $\mu_1 \geq \ldots \geq \mu_r$. We call $(\mu_1, \ldots, \mu_r)$ the *Hodge-Pink weights of* $\underline{D}$. Alternatively, if e is large enough such that $\mathfrak{q}_D \subset (z-\zeta)^{-e}\mathfrak{p}_D$ or $(z-\zeta)^e\mathfrak{p}_D \subset \mathfrak{q}_D$, then the Hodge-Pink weights are characterized by

$$(z-\zeta)^{-e}\mathfrak{p}_D/\mathfrak{q}_D \cong \bigoplus_{i=1}^n K[\![z-\zeta]\!]/(z-\zeta)^{e+\mu_i}\,,$$

$$\text{or}\quad \mathfrak{q}_D/(z-\zeta)^e\mathfrak{p}_D \cong \bigoplus_{i=1}^n K[\![z-\zeta]\!]/(z-\zeta)^{e-\mu_i}\,.$$

Like in (5.2), the Hodge-Pink lattice $\mathfrak{q}_D$ induces a descending filtration of $D_K := D \otimes_{k(\!(z)\!),\, z\mapsto\zeta} K$ by K-subspaces as follows. Consider the natural projection

$$\mathfrak{p}_D \twoheadrightarrow \mathfrak{p}_D/(z-\zeta)\mathfrak{p}_D \;=\; D_K\,.$$

The *Hodge-Pink filtration* $F^\bullet D_K = (F^i D_K)_{i\in\mathbb{Z}}$ is defined by letting $F^i D_K$ be the image in D_K of $\mathfrak{p}_D \cap (z-\zeta)^i\mathfrak{q}_D$ for all $i \in \mathbb{Z}$. This means, $F^i D_K = \big(\mathfrak{p}_D \cap (z-\zeta)^i\mathfrak{q}_D\big)\big/\big((z-\zeta)\mathfrak{p}_D \cap (z-\zeta)^i\mathfrak{q}_D\big)$.

Definition 6.2 Let $\underline{D} = (D, \tau_D, \mathfrak{q}_D)$ be a z-isocrystal with Hodge-Pink structure over $\mathcal{O}_K$ and set $r = \dim_{k((z))} D$.

(a) Choose a $k((z))$-basis of D and let $\det \tau_D$ be the determinant of the matrix representing τ_D with respect to this basis. The number $t_N(\underline{D}) := \mathrm{ord}_z(\det \tau_D)$ is independent of this basis and is called the *Newton slope of* $\underline{D}$.
(b) The integer $t_H(\underline{D}) := -\mu_1 - \ldots - \mu_r$, where $\mu_1, \ldots, \mu_r$ are the Hodge-Pink weights of $\underline{D}$ from Definition 6.1, satisfies $\wedge^r \mathfrak{q}_D = (z-\zeta)^{-t_H(\underline{D})} \wedge^r \mathfrak{p}_D$ and is called the *Hodge slope of* $\underline{D}$.
(c) $\underline{D}$ is called *weakly admissible* if

$$t_H(\underline{D}) = t_N(\underline{D}) \quad \text{and} \quad t_H(\underline{D}') \le t_N(\underline{D}') \quad \text{for every strict subobject } \underline{D}' \subset \underline{D}.$$

Remark 6.3 One can show that the tensor product

$$\underline{D} \otimes \underline{D}' = \left(D \otimes_{k((z))} D', \tau_D \otimes \tau_{D'}, \mathfrak{q}_D \otimes_{K[\![z-\zeta]\!]} \mathfrak{q}_{D'}\right)$$

of two weakly admissible z-isocrystals with Hodge-Pink structures $\underline{D}$ and $\underline{D}'$ over $\mathcal{O}_K$ is again weakly admissible. It was Pink's insight that for this result the *Hodge-Pink filtration* does not suffice, but one needs the finer information present in the *Hodge-Pink lattice*. The problem arises if the field extension $K/\mathbb{F}_q((\zeta))$ is inseparable; see [29, Example 5.16]. This is Pink's ingenious discovery.

Proposition 6.4 *([19, Corollary 6.11]) Let $\underline{\hat{M}}$ be a local shtuka over $\mathcal{O}_K$. Assume that $\mathcal{O}_K$ is discretely valued or that $\underline{\hat{M}} = \underline{\hat{M}}_\varepsilon(\underline{M})$ for an A-motive $\underline{M}$ over $\mathcal{O}_K$ as in Example 2.3. Then the z-isocrystal with Hodge-Pink structure $\underline{D}(\underline{\hat{M}})$ constructed at the beginning of this section is weakly admissible. The functor $\underline{\hat{M}} \longmapsto \underline{D}(\underline{\hat{M}})$ from the category of local shtukas over $\mathcal{O}_K$ with quasi-morphisms to the category of weakly admissible z-isocrystals with Hodge-Pink structure is fully faithful.*

There is a converse to this proposition.

Theorem 6.5 *([14, Théorème 7.3], [17, Theorem 2.5.3]) If $\mathcal{O}_K$ is discretely valued, then every weakly admissible z-isocrystal with Hodge-Pink structure $\underline{D}$ over $\mathcal{O}_K$ is of the form $\underline{D}(\underline{\hat{M}})$ for a local shtuka $\underline{\hat{M}}$ over $\mathcal{O}_K$.*

Remark 6.6 The theory presented here has as analog, the theory of p-adic Galois representations. There L is a discretely valued extension of $\mathbb{Q}_p$ with perfect residue field κ and $L_0 := W(\kappa)[\frac{1}{p}]$ is the maximal, absolutely unramified subfield of L. Let $\hat{\sigma} := W(\mathrm{Frob}_p)$ be the lift to L_0 of the p-Frobenius on κ which fixes the uniformizer p of L_0. Crystalline p-adic Galois representations are described by *filtered isocrystals* $\underline{D} = (D, \tau_D, F^\bullet D_L)$ over L, where D is a finite-dimensional L_0-vector space, $\tau_D \colon \hat{\sigma}^* D \xrightarrow{\sim} D$ is an L_0-isomorphism, and $F^\bullet D_L$ is a descending filtration on $D_L := D \otimes_{L_0} L$ by L-subspaces. More precisely, the Theorem of Colmez and Fontaine [6, Théorème A] says that a continuous representation of $\mathrm{Gal}(L^{\mathrm{sep}}/L)$ in a finite-dimensional $\mathbb{Q}_p$-vector space is crystalline if and only if it isomorphic to $F^0(\underline{D} \otimes_{L_0} \widetilde{\mathbf{B}}_{\text{rig}})^{\tau=\text{id}}$ for a *weakly admissible* filtered isocrystal $\underline{D} = (D, \tau_D, F^\bullet D_L)$ over L. Here, $\widetilde{\mathbf{B}}_{\text{rig}}$ is a certain period ring from Fontaine's theory of p-adic Galois representations, which carries a filtration and a Frobenius endomorphism Frob_p. The function field analog of $\widetilde{\mathbf{B}}_{\text{rig}}$ is the Q_ε-algebra $\mathcal{O}_{\overline{K}}[\![z, z^{-1}\}[\ell^{-1}]$; see [16, §§ 2.5 and 2.7]. In the function field case, when K is discretely valued, we could therefore define the *category of equal characteristic crystalline representations of* $\text{Gal}(K^{\text{sep}}/K)$ as the essential image of the functor

$$\underline{D} = (D, \tau_D, \mathfrak{q}_D) \longmapsto \big(\underline{D} \otimes_{k(\!(z)\!)} \mathcal{O}_{\overline{K}}[\![z, z^{-1}\}[\ell^{-1}]\big)^{\tau=\text{id}} \cap \mathfrak{q}_D \otimes_{K[\![z-\zeta]\!]} \overline{K}[\![z-\zeta]\!] \qquad (6.1)$$

from weakly admissible z-isocrystals with Hodge-Pink structure $\underline{D}$ to continuous representations of $\text{Gal}(K^{\text{sep}}/K)$ in finite-dimensional Q_ε-vector spaces. By Theorems 6.5, 5.10, and 5.2 and Proposition 6.4, this functor is fully faithful and this definition coincides with our Definition 5.3 above.

References

1. M. Demazure, A. Grothendieck: *SGA 3: Schémas en Groupes I, II, III*, LNM **151**, **152**, **153** (Springer, Berlin, 1970); also available at http://library.msri.org/books/sga/sga/pdf/
2. V. Abrashkin, Galois modules arising from Faltings's strict modules. Compos. Math. **142**(4), 867–888 (2006)
3. G. Anderson, t Motives. Duke Math. J. **53**, 457–502 (1986)
4. M. Bornhofen, U. Hartl, Pure Anderson motives and abelian τ-sheaves. Math. Z. **268**, 67–100 (2011)
5. L. Carlitz, On certain functions connected with polynomials in a Galois field. Duke Math. J. **1**(2), 137–168 (1935)
6. P. Colmez, J.-M. Fontaine, Construction des représentations p-adiques semi-stables. Invent. Math. **140**(1), 1–43 (2000)
7. V.G. Drinfeld, Moduli variety of F-sheaves. Funct. Anal. Appl. **21**(2), 107–122 (1987)
8. G. Faltings, *Crystalline cohomology and p-adic Galois-representations, algebraic analysis, geometry, and number theory (Baltimore 1988)* (Johns Hopkins University Press, Baltimore, 1989), pp. 25–80
9. G. Faltings, Group schemes with strict $\mathcal{O}$-action. Mosc. Math. J. **2**(2), 249–279 (2002)
10. J.-M. Fontaine, Modules galoisiens, modules filtrés et anneaux de Barsotti-Tate, Journées de Géométrie Algébrique de Rennes (Rennes, 1978), Vol. III, Astérisque **65** (Soc. Math. France, Paris, 1979), pp. 3–80
11. J.-M. Fontaine, Sur certains types de représentations p-adiques du groupe de Galois d'un corps local; construction d'un anneau de Barsotti-Tate. Ann. of Math. **115**(3), 529–577 (1982)
12. J.-M. Fontaine: Représentations p-adiques des corps locaux I, in *The Grothendieck Festschrift*, Vol. II, Progr. Math. 87 (Birkhäuser Boston, Boston, MA, 1990), pp. 249–309
13. J.-M. Fontaine, Le corps des périodes p-adiques, in Périodes p-adiques (Bures-sur-Yvette, 1988). Astérisque **223**, 59–111 (1994)
14. A. Genestier, V. Lafforgue, Théorie de Fontaine en égales charactéristiques. Ann. Sci. École Norm. Supér. **44**(2), 263–360 (2011)
15. A. Grothendieck, *Groupes de Barsotti-Tate et cristaux de Dieudonné, Séminaire de Mathématiques Supérieures 45* (Les Presses de l'Université de Montréal, Montreal, 1974)

16. U. Hartl, A Dictionary between Fontaine-Theory and its Analogue in Equal Characteristic. J. Number Theory **129**(7), 1734–1757 (2009)
17. U. Hartl, Period Spaces for Hodge Structures in Equal Characteristic. Annals of Math. **173**(3), 1241–1358 (2011)
18. U. Hartl, Isogenies of abelian Anderson A-modules and A-motives, to appear in Annali della Scuola Normale Superiore di Pisa, Classe di Scienze. Preprint as arXiv:math/1706.06807
19. U. Hartl, and W. Kim, Local Shtukas, Hodge-Pink Structures and Galois Representations, in *Proceedings of the conference on "t-motives: Hodge structures, transcendence and other motivic aspects*, BIRS, Banff, Canada 2009, eds. G. Böckle, D. Goss, U. Hartl, M. Papanikolas (EMS, 2019)
20. U. Hartl, and R.K. Singh, Product Formulas for Periods of CM Abelian Varieties and the Function Field Analog. Preprint at arXiv:math/1809.02990
21. U. Hartl, R.K. Singh, Local Shtukas and divisible local Anderson modules. Can. J. Math. **71**, 1163–1207 (2019)
22. U. Hartl, R.K. Singh, Periods of Drinfeld modules and local shtukas with complex multiplication. J. Inst. Math. Jussieu **19**(1), 175–208 (2020)
23. R. Hartshorne, *Algebraic Geometry, GTM 52* (Springer, Berlin, 1977)
24. U. Hartl, E. Viehmann, The Newton stratification on deformations of local G-shtukas. J. Reine Angew. Math. (Crelle) **656**, 87–129 (2011)
25. G. Laumon, *Cohomology of Drinfeld Modular Varieties I, Cambridge Studies in Advanced Mathematics 41* (Cambridge University Press, Cambridge, 1996)
26. L. Illusie, *Complexe cotangent et déformations II*, Lecture Notes in Mathematics, vol. 283. (Springer, Berlin, 1972)
27. W. Messing, *The Crystals Associated to Barsotti-Tate Groups, LNM 264* (Springer, Berlin, 1972)
28. W. Niziol, Crystalline conjecture via K-theory. Ann. Sci. École Norm. Sup. (4) **31**(5), 659–681 (1998)
29. R. Pink, Hodge structures over function fields. Preprint 1997 at http://www.math.ethz.ch/~pinkri/ftp/HS.pdf
30. T. Poguntke, Group schemes with $\mathbb{F}_q$-action. Bull. Soc. Math. France **145**(2), 345–380 (2017)
31. Y. Taguchi, A duality for finite t-modules. J. Math. Sci. Univ. Tokyo **2**, 563–588 (1995)
32. Y. Taguchi, D. Wan, L-functions of φ-sheaves and Drinfeld modules. J. Amer. Math. Soc. **9**(3), 755–781 (1996)
33. J. Tate, p-divisible groups, in *Proc. Conf. Local Fields* (Springer, Berlin, 1967), pp. 158–183
34. T. Tsuji, p-adic étale cohomology and crystalline cohomology in the semi-stable reduction case. Invent. Math. **137**(2), 233–411 (1999)

An Introduction to p-Adic Hodge Theory

Denis Benois

Mathematics Subject Classification 11F85 · 11S15 · 11S31 · 11G20

Introduction

0.1 These notes grew out of author's lectures at the International Center of Theoretical Sciences of Tata Institute in Bangalore in September 2019. Their aim is to provide a self-contained introduction to p-adic Hodge theory with minimal prerequisties. The reader should be familiar with valuations, complete fields and basic results in the theory of local fields, including ramification theory as, for example, the first four chapters of Serre's book [142]. In Sects. 3 and 4, we use the language of continuous cohomology. Sects. 15 and 16 require the knowledge of Galois cohomology and local class field theory, as in [142] or [140].

0.2 Section 1 is utilitarian. For the convenience of the reader, it assembles basic definitions and results from the theory of local fields repeatedly used in the text. In Sect. 2, we discuss the structure of the absolute Galois group of a local field. Although only a portion of this material is used in the remainder of the text, we think that it is important in its own right. In Sect. 3, we illustrate the ramification theory by the example of $\mathbf{Z}_p$-extensions. Following Tate, we define the normalized trace map and compute continuous cohomology of Galois groups of such extensions.

D. Benois (✉)
Institut de Mathématiques, Université de Bordeaux, 351, cours de la Libération,
33405 Talence, France
e-mail: denis.benois@math.u-bordeaux.fr

© The Author(s), under exclusive license to Springer Nature Singapore Pte Ltd. 2022
D. Banerjee et al. (eds.), *Perfectoid Spaces*, Infosys Science Foundation Series
in Mathematical Sciences https://doi.org/10.1007/978-981-16-7121-0_5

Krasner [100] was probably the first to remark that local fields of characteristic p appear as "limits" of totally ramified local fields of characteristic 0.[1] In Sects. 4–6, we study three important manifestations of this phenomenon. In Sect. 4, we introduce Tate's method of almost étale extensions. We consider deeply ramified extensions of local fields and prove that finite extensions of a deeply ramified field are almost étale. The main reference here is the paper of Coates and Greenberg [37]. The book of Gabber and Ramero [78] provides a new conceptual approach to this theory in a very general setting, but uses the tools which are beyond the scope of these notes. As an application, we compute continuous Galois cohomology of the absolute Galois group of a local field.

In Sect. 5, we study perfectoid fields following Scholze [130] and Fargues–Fontaine [60]. The connection of this notion with the theory of deeply ramified extensions is given by a theorem of Gabber–Ramero. Again, we limit our study to the arithmetic case and refer the interested reader to [130] for the general treatment. In Sect. 6, we review the theory of field of norms of Fontaine–Wintenberger and discuss its relation with perfectoid fields.

Sections 7–13 are devoted to the general theory of p-adic representations. In Sect. 7, we introduce basic notions and examples and discuss Grothendieck's ℓ-adic monodromy theorem. Next, we turn to the case $\ell = p$. Section 8 gives an introduction to Fontaine's theory of (φ, Γ)-modules [69]. Here, we classify p-adic representations of local fields using the link between the fields of characteristic 0 and p studied in Sects. 5–6. In Sects. 9–13, we introduce and study special classes of p-adic representations. The general formalism of admissible representations is reviewed in Sect. 9. In Sect. 10, we discuss the notion of a Hodge–Tate representation and put it in the frame of Sen's theory of $\mathbf{C}_K$-representations. Here, the computation of the continuous Galois cohomology from Sect. 4 plays a fundamental role. In Sects. 11–13, we define the rings of p-adic periods $\mathbf{B}_{\mathrm{dR}}$, $\mathbf{B}_{\mathrm{cris}}$, and $\mathbf{B}_{\mathrm{st}}$ and introduce Fontaine's hierarchy of p-adic representations. Its relation with p-adic comparison isomorphisms is quickly discussed at the end of Sect. 13.

In the remainder of the text, we study p-adic representations arising from formal groups. In this case, the main constructions of the theory have an explicit description, and p-adic representations can be studied without an extensive use of algebraic geometry. In Sect. 14, we review the p-adic integration on formal groups following Colmez [38]. A completely satisfactory exposition of this material should cover the general case of p-divisible groups, which we decided not to include in these notes. For this material, we refer the reader to [30, 39, 64, 66]. In Sects. 15 and 16, we illustrate the p-adic Hodge theory of formal groups by two applications: complex multiplication of abelian varieties and Hilbert pairings on formal groups. In Sect. 17, we prove the theorem "weakly admissible $\Rightarrow$ admissible" in the case of dimension one by the method of Laffaille [102]. This implies the surjectivity of the Gross–

[1] See [52] for a modern exposition of Krasner's results.

Hopkins period map. Finally, we apply the theory of formal groups to the study of the spaces $(\mathbf{B}_{\mathrm{cris}}^{+})^{\varphi^h=p}$, which play an important role in the theory of Fargues–Fontaine. For further detail and applications of these results, we refer the reader to [60].

0.3 These notes should not be viewed as a survey paper. Several important aspects of p-adic Hodge theory are not even mentioned. As a partial substitute, we propose some references for further reading in the body of the text.

1 Local Fields: Preliminaries

1.1 Non-Archimedean Fields

1.1.1 We recall basic definitions and facts about non-Archimedean fields.

Definition A non-Archimedean field is a field K equipped with a non-Archimedean absolute value that is, an absolute value $|\cdot|_K$ satisfying the ultrametric triangle inequality:

$$|x+y|_K \leqslant \max\{|x|_K, |y|_K\}, \qquad \forall x, y \in K.$$

We will say that K is complete if it is complete for the topology induced by $|\cdot|_K$.

To any non-Archimedean field K, one associates its ring of integers

$$O_K = \{x \in K \mid |x|_K \leqslant 1\}.$$

The ring O_K is local, with the maximal ideal

$$\mathfrak{m}_K = \{x \in K \mid |x|_K < 1\}.$$

The group of units of O_K is

$$U_K = \{x \in K \mid |x|_K = 1\}.$$

The residue field of K is defined as

$$k_K = O_K/\mathfrak{m}_K.$$

Theorem 1.1.2 *Let K be a complete non-Archimedean field and let L/K be a finite extension of degree $n = [L : K]$. Then the absolute value $|\cdot|_K$ has a unique continuation $|\cdot|_L$ to L, which is given by*

$$|x|_L = \left|N_{L/K}(x)\right|_K^{1/n},$$

where $N_{L/K}$ is the norm map.

Proof See, for example, [10, Chap. 2, Theorem 7]. □

1.1.3 We fix an algebraic closure $\overline{K}$ of K and denote by K^{sep} the separable closure of K in $\overline{K}$. If $\mathrm{char}(K) = p > 0$, we denote by $K^{\mathrm{rad}} := K^{1/p^{\infty}}$ the purely inseparable closure of K. Thus $\overline{K} = K^{\mathrm{sep}}$ if $\mathrm{char}(K) = 0$, and $\overline{K} = (K^{\mathrm{rad}})^{\mathrm{sep}}$ if $\mathrm{char}(K) = p > 0$. Theorem 1.1.2 allows to extend $|\cdot|_K$ to $\overline{K}$. To simplify notation, we denote again by $|\cdot|_K$ the extension of $|\cdot|_K$ to $\overline{K}$.

Proposition 1.1.4 (Krasner's Lemma) *Let K be a complete non-Archimedean field. Let $\alpha \in K^{\mathrm{sep}}$ and let $\alpha_1 = \alpha, \alpha_2, \dots, \alpha_n$ denote the conjugates of α over K. Set*

$$d_\alpha = \min\left\{|\alpha - \alpha_i|_K \mid 2 \leqslant i \leqslant n\right\}.$$

If $\beta \in K^{\mathrm{sep}}$ is such that $|\alpha - \beta| < d_\alpha$, then $K(\alpha) \subset K(\beta)$.

Proof We recall the proof (see, for example, [119, Proposition 8.1.6]). Assume that $\alpha \notin K(\beta)$. Then $K(\alpha, \beta)/K(\beta)$ is a non-trivial extension, and there exists an embedding $\sigma : K(\alpha, \beta)/K(\beta) \to \overline{K}/K(\beta)$ such that $\alpha_i := \sigma(\alpha) \neq \alpha$. Hence,

$$|\beta - \alpha_i|_K = |\sigma(\beta - \alpha)|_K = |\beta - \alpha|_K < d_\alpha,$$

and

$$|\alpha - \alpha_i|_K = |(\alpha - \beta) + (\beta - \alpha_i)|_K \leqslant \max\left\{|\alpha - \beta|_K, |\beta - \alpha_i|_K\right\} < d_\alpha.$$

This gives a contradiction. □

Proposition 1.1.5 (Hensel's Lemma) *Let K be a complete non-Archimedean field. Let $f(X) \in O_K[X]$ be a monic polynomial such that*

(a) the reduction $\bar{f}(X) \in k_K[X]$ of $f(X)$ modulo $\mathfrak{m}_K$ has a root $\bar{\alpha} \in k_K$;
(b) $\bar{f}'(\bar{\alpha}) \neq 0$.

Then there exists a unique $\alpha \in O_K$ such that $f(\alpha) = 0$ and $\bar{\alpha} = \alpha \pmod{\mathfrak{m}_K}$.

Proof See, for example, [106, Chap. 2, §2]. □

1.1.6 Recall that a valuation on K is a function $v_K : K \to \mathbf{R} \cup \{+\infty\}$ satisfying the following properties:

(1) $v_K(xy) = v_K(x) + v_K(y), \quad \forall x, y \in K^*$;
(2) $v_K(x+y) \geqslant \min\{v_K(x), v_K(y)\}, \quad \forall x, y \in K^*$;
(3) $v_K(x) = \infty \Leftrightarrow x = 0$.

For any $\rho \in]0,1[$, the function $|x|_\rho = \rho^{v_K(x)}$ defines an ultrametric absolute value on K. Conversely, if $|\cdot|_K$ is an ultrametric absolute value, then for any $\rho \in]0,1[$ the function $v_\rho(x) = \log_\rho |x|_K$ is a valuation on K. This establishes a one to one correspondence between equivalence classes of non-Archimedean absolute values and equivalence classes of valuations on K.

Definition A discrete valuation field is a field K equipped with a valuation v_K such that $v_K(K^*)$ is a discrete subgroup of $\mathbf{R}$. Equivalently, K is a discrete valuation field if it is equipped with an absolute value $|\cdot|_K$ such that $|K^*|_K \subset \mathbf{R}_+$ is discrete.

Let K be a discrete valuation field. In the equivalence class of discrete valuations on K, we can choose the unique valuation v_K such that $v_K(K^*) = \mathbf{Z}$. An element $\pi_K \in K$ such that $v_K(\pi_K) = 1$ is called a uniformizer of K. Every $x \in K^*$ can be written in the form $x = \pi_K^{v_K(x)} u$ with $u \in U_K$, and one has

$$K^* \simeq \langle \pi_K \rangle \times U_K, \qquad \mathfrak{m}_K = (\pi_K).$$

1.1.7 Let K be a complete non-Archimedean field. We finish this section by discussing the Galois action on the completion $\mathbf{C}_K$ of $\overline{K}$.

Theorem 1.1.8 (Ax–Sen–Tate) *Let K be a complete non-Archimedean field. The following statements hold true:*

(i) The completion $\mathbf{C}_K$ of $\overline{K}$ is an algebraically closed field, and K^{sep} is dense in $\mathbf{C}_K$.
(ii) The absolute Galois group $G_K = \mathrm{Gal}(K^{\mathrm{sep}}/K)$ acts continuously on $\mathbf{C}_K$, and this action identifies G_K with the group of all continuous automorphisms of $\mathbf{C}_K$ that act trivially on K.
(iii) For any closed subgroup $H \subset G_K$, the field $\mathbf{C}_K^H$ coincides with the completion of the purely inseparable closure of $(K^{\mathrm{sep}})^H$ in $\overline{K}$.

Proof The statement (i) follows easily from Krasner's Lemma, and (ii) is an immediate consequence of continuity of the Galois action. The last statement was first proved by Tate [151] for local fields of characteristic 0. In full generality, the theorem was proved by Ax [11]. Tate's proof is based on the ramification theory and leads to the notion of an almost étale extension, which is fundamental for p-adic Hodge theory. We review it in Sect. 4. □

1.2 Local Fields

1.2.1 In these notes, we adopt the following convention.

Definition 1.2.2 A local field is a complete discrete valuation field K whose residue field k_K is finite.

Note that many (but not all) results and constructions of the theory are valid under the weaker assumption that the residue field k_K is perfect.

We will always assume that the discrete valuation

$$v_K : K \to \mathbf{Z} \cup \{+\infty\}$$

is surjective. Let $p = \mathrm{char}(k_K)$. The following well-known classification of local fields can be easily proved using Ostrowski's theorem:

- If $\mathrm{char}(K) = p$, then K is isomorphic to the field $k_K((x))$ of Laurent power series, where k_K is the residue field of K and x is transcendental over k. The discrete valuation on K is given by

$$v_K(f(x)) = \mathrm{ord}_x f(x).$$

  Note that x is a uniformizer of K and $O_K \simeq k_K[[x]]$.
- If $\mathrm{char}(K) = 0$, then K is isomorphic to a finite extension of the field of p-adic numbers $\mathbf{Q}_p$. The absolute value on K is the extension of the p-adic absolute value

$$\left| \frac{a}{b} p^k \right|_p = p^{-k}, \qquad p \nmid a, b.$$

In all cases, set $f_K = [k_K : \mathbf{F}_p]$ and denote by $q_K = p^{f_K}$ the cardinality of k_K. The group of units U_K is equipped with the exhaustive descending filtration:

$$U_K^{(n)} = 1 + \pi_K^n O_K, \qquad n \geqslant 0.$$

For the factors of this filtration, one has

$$U_K / U_K^{(1)} \simeq k_K^*, \qquad U_K^{(n)} / U_K^{(n+1)} \simeq \mathfrak{m}_K^n / \mathfrak{m}_K^{n+1}. \quad \text{if } n \geqslant 1. \tag{1}$$

1.2.3 If L/K is a finite extension of local fields, the ramification index $e(L/K)$ and the inertia degree $f(L/K)$ of L/K are defined as follows:

$$e(L/K) = v_L(\pi_K), \qquad f(L/K) = [k_L : k_K].$$

Recall the fundamental formula:

$$f(L/K)\, e(L/K) = [L : K]$$

(see, for example, [10, Chap. 3, Theorem 6]).

Definition 1.2.4 One says that L/K is
(i) unramified if $e(L/K) = 1$ (and therefore $f(L/K) = [L : K]$);
(ii) totally ramified if $e(L/K) = [L : K]$ (and therefore $f(L/K) = 1$).

The following useful proposition follows easily from Krasner's lemma.

Proposition 1.2.5 *Let K be a local field of characteristic* 0*. For any $n \geqslant 1$, there exists only a finite number of extensions of K of degree $\leqslant n$.*

Proof See [106, Chap. 2, Proposition 14]. □

We remark that, looking at Artin–Schreier extensions, it's easy to see that a local field of characteristic p has infinitely many separable extensions of degree p.

1.2.6 The unramified extensions can be described entirely in terms of the residue field k_K. Namely, there exists a one-to-one correspondence

$$\{\text{finite extensions of } k_K\} \longleftrightarrow \{\text{finite unramified extensions of } K\},$$

which can be explicitly described as follows. Let k/k_K be a finite extension of k_K. Write $k = k_K(\alpha)$ and denote by $f(X) \in k_K[X]$ the minimal polynomial of α. Let $\widehat{f}(X) \in O_K[X]$ denote any lift of $f(X)$. Then we associate to k the extension $L = K(\widehat{\alpha})$, where $\widehat{\alpha}$ is the unique root of $\widehat{f}(X)$ whose reduction modulo $\mathfrak{m}_L$ is α. An easy argument using Hensel's lemma shows that L doesn't depend on the choice of the lift $\widehat{f}(X)$.

Unramified extensions form a distinguished class of extensions in the sense of [104]. In particular, for any finite extension L/K, one can define its maximal unramified subextension L_{ur} as the compositum of all its unramified subextensions. Then

$$f(L/K) = [L_{\mathrm{ur}} : K], \qquad e(L/K) = [L : L_{\mathrm{ur}}].$$

The extension L/L_{ur} is totally ramified.

1.2.7 Assume that L/K is totally ramified of degree n. Let π_L be any uniformizer of L, and let

$$f(X) = X^n + a_{n-1}X^{n-1} + \cdots + a_1X + a_0 \in O_K[X]$$

be the minimal polynomial of π_L. Then $f(X)$ is an Eisenstein polynomial, namely

$$v_K(a_i) \geqslant 1 \qquad \text{for } 0 \leqslant i \leqslant n-1, \text{ and } v_K(a_0) = 1.$$

Conversely, if α is a root of an Eisenstein polynomial of degree n over K, then $K(\alpha)/K$ is totally ramified of degree n, and α is an uniformizer of $K(\alpha)$.

Definition 1.2.8 One says that an extension L/K is
(i) tamely ramified if $e(L/K)$ is coprime to p.
(ii) totally tamely ramified if it is totally ramified and $e(L/K)$ is coprime to p.

Using Krasner's lemma, it is easy to give an explicit description of totally tamely ramified extensions.

Proposition 1.2.9 *If L/K is totally tamely ramified of degree n, then there exists a uniformizer $\pi_K \in K$ such that*

$$L = K(\pi_L), \qquad \pi_L^n = \pi_K.$$

Proof Assume that L/K is totally tamely ramified of degree n. Let Π be a uniformizer of L and $f(X) = X^n + \cdots + a_1X + a_0$ its minimal polynomial. Then $f(X)$ is Eisenstein, and $\pi_K := -a_0$ is a uniformizer of K. Let $\alpha_i \in \overline{K}$ $(1 \leqslant i \leqslant n)$ denote the roots of $g(X) := X^n + a_0$. Then

$$|g(\Pi)|_K = |g(\Pi) - f(\Pi)|_K \leqslant \max_{1\leqslant i\leqslant n-1} |a_i\Pi^i|_K < |\pi_K|_K$$

Since $|g(\Pi)|_K = \prod_{i=1}^{n}(\Pi - \alpha_i)$, and $\Pi = (-1)^n \prod_{i=1}^{n}\alpha_i$, we have

$$\prod_{i=1}^{n} |\Pi - \alpha_i|_K < \prod_{i=1}^{n} |\alpha_i|_K.$$

Therefore, there exists i_0 such that

$$|\Pi - \alpha_{i_0}|_K < |\alpha_{i_0}|_K. \tag{2}$$

Set $\pi_L = \alpha_{i_0}$. Then

$$\prod_{i\neq i_0}(\pi_L - \alpha_i) = g'(\pi_L) = n\pi_L^{n-1}.$$

Since $(n, p) = 1$ and $|\pi_L - \alpha_i|_K \leqslant |\pi_L|_K$, the previous equality implies that

$$d := \min_{i\neq i_0}|\pi_L - \alpha_i|_K = |\pi_L|_K.$$

Together with (2), this gives

$$|\Pi - \alpha_{i_0}|_K < d.$$

Applying Krasner's lemma, we find that $K(\pi_L) \subset L$. Since $[L : K] = [K(\pi_L) : K] = n$, we obtain that $L = K(\pi_L)$, and the proposition is proved. □

1.2.10 Let L/K be a finite separable extension of local fields. Consider the bilinear non-degenerate form

$$t_{L/K} : L \times L \to K, \qquad t_{L/K}(x, y) = \mathrm{Tr}_{L/K}(xy), \tag{3}$$

where $\mathrm{Tr}_{L/K}$ is the trace map. The set

$$O_L' := \left\{ x \in L \mid t_{L/K}(x, y) \in O_K, \quad \forall y \in O_L \right\}$$

is a fractional ideal, and

$$\mathfrak{D}_{L/K} := O_L'^{-1} := \left\{ x \in L \mid x O_L' \subset O_L \right\}$$

is an ideal of O_L.

Definition The ideal $\mathfrak{D}_{L/K}$ is called the different of L/K.

If $K \subset L \subset M$ is a tower of separable extensions, then

$$\mathfrak{D}_{M/K} = \mathfrak{D}_{M/L}\mathfrak{D}_{L/K}. \tag{4}$$

(see, for example, [106, Chap. 3, Proposition 5]).

Set

$$v_L(\mathfrak{D}_{L/K}) = \inf\{v_L(x) \mid x \in \mathfrak{D}_{L/K}\}.$$

Proposition 1.2.11 *Let L/K be a finite separable extension of local fields and $e = e(L/K)$ the ramification index. The following assertions hold true:*

(i) If $O_L = O_K[\alpha]$, and $f(X) \in O_K[X]$ is the minimal polynomial of α, then $\mathfrak{D}_{L/K} = (f'(\alpha))$.

(ii) $\mathfrak{D}_{L/K} = O_L$ if and only if L/K is unramified.

(iii) $v_L(\mathfrak{D}_{L/K}) \geqslant e - 1$.

(iv) $v_L(\mathfrak{D}_{L/K}) = e - 1$ if and only if L/K is tamely ramified.

Proof The first statement holds in the more general setting of Dedekind rings (see, for example, [106, Chap. 3, Proposition 2]). We prove ii-iv) for reader's convenience (see also [106, Chap. 3, Proposition 8]).

(a) Let L/K be an unramified extension of degree n. Write $k_L = k_K(\bar{\alpha})$ for some $\bar{\alpha} \in k_L$. Let $\bar{f}(X) \in k_K[X]$ denote the minimal polynomial of $\bar{\alpha}$. Then $\deg(\bar{f}) = n$. Take any lift $f(X) \in O_K[X]$ of $\bar{f}(X)$ of degree n. By Proposition 1.1.5 (Hensel's lemma), there exists a unique root $\alpha \in O_L$ of $f(X)$ such that $\bar{\alpha} = \alpha \pmod{\mathfrak{m}_K}$. It's easy to see that $O_L = O_K[\alpha]$. Since $\bar{f}(X)$ is separable, $\bar{f}'(\bar{\alpha}) \neq 0$, and therefore $f'(\alpha) \in U_L$. Applying (i), we obtain

$$\mathfrak{D}_{L/K} = (f'(\alpha)) = O_L.$$

Therefore, $\mathfrak{D}_{L/K} = O_L$ if L/K is unramified.

(b) Assume that L/K is totally ramified. Then $O_L = O_K[\pi_L]$, where π_L is any uniformizer of O_L. Let $f(X) = X^e + a_{e-1}X^{e-1} + \cdots + a_1X + a_0$ be the minimal polynomial of π_L. Then

$$f'(\pi_L) = e\pi_L^{e-1} + (e-1)a_{e-1}\pi_L^{e-2} + \cdots + a_1.$$

Since $f(X)$ is Eisenstein, $v_L(a_i) \geqslant e$, and an easy estimation shows that $v_L(f'(\pi_L)) \geqslant e-1$. Thus,

$$v_L(\mathfrak{D}_{L/K}) = v_L(f'(\alpha)) \geqslant e-1.$$

This proves (iii). Moreover, $v_L(f'(\alpha)) = e-1$ if and only if $(e,p)=1$, i.e. if and only if L/K is tamely ramified. This proves iv).

(c) Assume that $\mathfrak{D}_{L/K} = O_L$. Then $v_L(\mathfrak{D}_{L/K}) = 0$. Let $L_{\rm ur}$ denote the maximal unramified subextension of L/K. By (4), a) and b) we have

$$v_L(\mathfrak{D}_{L/K}) = v_L(\mathfrak{D}_{L/L_{\rm ur}}) \geqslant e-1.$$

Thus, $e=1$, and we showed that each extension L/K such that $\mathfrak{D}_{L/K} = O_L$ is unramified. Together with a), this proves (i). □

1.3 Ramification Filtration

1.3.1 Let L/K be a finite Galois extension of local fields. Set $G = \mathrm{Gal}(L/K)$. For any integer $i \geqslant -1$, define

$$G_i = \{g \in G \mid v_L(g(x)-x) \geqslant i+1, \quad \forall x \in O_L\}.$$

Then G_i are normal subgroups of G, called ramification subgroups. We have a descending chain

$$G = G_{-1} \supset G_0 \supset G_1 \supset \cdots \supset G_m = \{1\}$$

called the ramification filtration on G (in low numbering). From definition, it easily follows that

$$G_0 = \mathrm{Gal}(L/L_{\rm ur}), \qquad G/G_0 \simeq \mathrm{Gal}(k_L/k_K).$$

Below, we summarize some basic results about the factors of the ramification filtration. First remark that for each $i \geqslant 0$, one has

$$G_i = \left\{ g \in G_0 \mid v_L\left(1 - \frac{g(\pi_L)}{\pi_L}\right) \geqslant i \right\}.$$

Proposition 1.3.2 *(i) For all* $i \geqslant 0$, *the map*

$$s_i \,:\, G_i/G_{i+1} \to U_L^{(i)}/U_L^{(i+1)}, \tag{5}$$

which sends $\bar{g} = g \mod G_{i+1}$ *to* $s_i(\bar{g}) = \dfrac{g(\pi_L)}{\pi_L} \pmod{U_L^{(i+1)}}$, *is a well-defined monomorphism which doesn't depend on the choice of the uniformizer* π_L *of* L.

(ii) The composition of s_i *with the maps (1) gives monomorphisms:*

$$\delta_0 : G_0/G_1 \to k^*, \qquad \delta_i : G_i/G_{i+1} \to \mathfrak{m}_K^i/\mathfrak{m}_K^{i+1}, \quad \textit{for all } i \geqslant 1. \tag{6}$$

Proof The proof is straightforward. See [142, Chapitre IV, Propositions 5-7]. □

An important corollary of this proposition is that the Galois group G is solvable for any Galois extension. Also, since $\mathrm{char}(k_K) = p$, the order of G_0/G_1 is coprime to p, and the order of G_1 is a power of p. Therefore, $L_{\mathrm{tr}} = L^{G_1}$ is the maximal tamely ramified subextension of L. From this, one can easily deduce that the class of tamely ramified extensions is distinguished. To sup up, we have the tower of extensions:

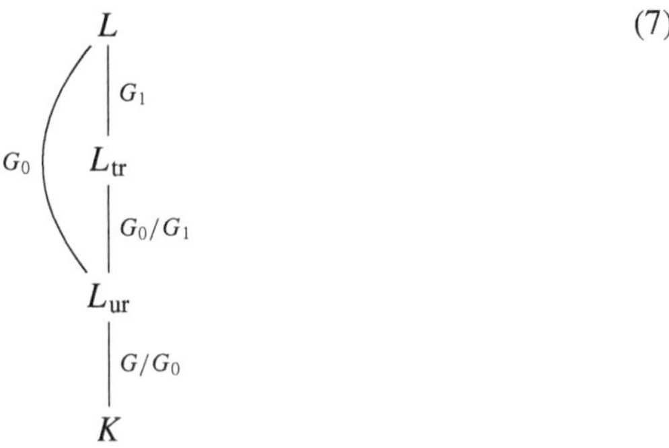

(7)

Definition 1.3.3 The groups $I_{L/K} := G_0$ and $P_{L/K} := G_1$ are called the inertia subgroup and the wild inertia subgroup, respectively.

1.3.4 The different $\mathfrak{D}_{L/K}$ of a finite Galois extension can be computed in terms of the ramification subgroups.

Proposition 1.3.5 *Let L/K be a finite Galois extension of local fields. Then*

$$v_L(\mathfrak{D}_{L/K}) = \sum_{i=0}^{\infty}(|G_i| - 1). \tag{8}$$

Proof Let $O_L = O_K[\alpha]$, and let $f(X)$ be the minimal polynomial of α. For any $g \in G$, set $i_{L/K}(g) = v_L(g(\alpha) - \alpha)$. From the definition of ramification subgroups, it follows that $g \in G_i$ if and only if $i_{L/K}(g) \geqslant i + 1$. Since

$$f'(\alpha) = \prod_{g \neq 1}(\alpha - g(\alpha)),$$

we have

$$v_L(\mathfrak{D}_{L/K}) = v_L(f'(\alpha)) = \sum_{g\neq 1} v_L(\alpha - g(\alpha)) = \sum_{g\neq 1} i_{L/K}(g) = \sum_{i=0}^{\infty}(i+1)(|G_i| - |G_{i+1}|)$$
$$= \sum_{i=0}^{\infty}(i+1)\big((|G_i|-1) - (|G_{i+1}|-1)\big) = \sum_{i=0}^{\infty}(|G_i|-1).$$

□

1.3.6 We review Hasse–Herbrand's theory of upper ramification. It is convenient to define G_u for all *real* $u \geqslant -1$ setting

$$G_t = G_i, \quad \text{where } i \text{ is the smallest integer } \geqslant u.$$

For any finite Galois extension, the Hasse–Herbrand functions are defined as follows:

$$\begin{aligned} \varphi_{L/K}(u) &= \int_0^u \frac{dt}{(G_0 : G_t)}, \\ \psi_{L/K}(v) &= \varphi_{L/K}^{-1}(v) \qquad \text{(the inverse of } \varphi_{L/K}). \end{aligned} \tag{9}$$

Proposition 1.3.7 *Let $K \subset F \subset L$ be a tower of finite Galois extensions. Set $G = \mathrm{Gal}(L/K)$ and $H = \mathrm{Gal}(L/F)$. Then the following holds true:*
(i) $\varphi_{L/K} = \varphi_{F/K} \circ \varphi_{L/F}$ and $\psi_{L/K} = \psi_{L/F} \circ \psi_{F/K}$.
ii) (Herbrand's theorem) For any $u \geqslant 0$,

$$G_u H/H \simeq (G/H)_{\varphi_{M/L}(u)}.$$

Proof See [142, Chap. IV, §3]. □

Definition The ramification subgroups in upper numbering $G^{(v)}$ are defined by

$$G^{(v)} = G_{\psi_{L/K}(v)},$$

or, equivalently, by $G^{(\varphi_{L/K}(u))} = G_u$.

Therefore, Herbrand's theorem can be stated as follows:

$$(G/H)^{(v)} = G^{(v)}/G^{(v)} \cap H, \qquad \forall v \geqslant 0. \tag{10}$$

The Hasse–Herbrand function $\psi_{L/K}$ can be written as

$$\psi_{L/K}(v) = \int_0^v (G^{(0)} : G^{(t)})dt.$$

1.3.8 Hebrand's theorem allows to define the ramification filtration for infinite Galois extensions. Namely, for any (finite or infinite) Galois extension of local fields L/K define

$$\mathrm{Gal}(L/K)^{(v)} = \varprojlim_{F} \mathrm{Gal}(F/K)^{(v)},$$

where F runs through finite Galois subextensions of L/K. In particular, we can consider the ramification filtration on the absolute Galois group G_K of K. This filtration contains fundamental information about the field K. We discuss it in more detail in Sect. 2.3.

Definition A real number $v \geqslant 0$ is a ramification jump of a Galois extension L/K if

$$\mathrm{Gal}(L/K)^{(v+\varepsilon)} \neq \mathrm{Gal}(L/K)^{(v)} \qquad \text{for any } \varepsilon > 0.$$

1.3.9 Formula (8) can be written in terms of upper ramification subgroups:

$$v_K(\mathfrak{D}_{L/K}) = \int_{-1}^{\infty} \left(1 - \frac{1}{|G^{(v)}|}\right) dv.$$

In this form, it can be generalized to arbitrary finite extensions as follows. For any $v \geqslant 0$, define

$$\overline{K}^{(v)} = \overline{K}^{G_K^{(v)}}.$$

Then for any finite extension L/K, one has

$$v_K(\mathfrak{D}_{L/K}) = \int_{-1}^{\infty} \left(1 - \frac{1}{[L : L \cap \overline{K}^{(v)}]}\right) dv \qquad (11)$$

(see [37, Lemma 2.1]).

1.3.10 The description of the ramification filtration for general Galois extensions is a difficult problem (see Sect. 2.3 below). It is completely solved for abelian extensions (see Sect. 2.2). In particular, the ramification jumps of an abelian extension are rational integers (theorem of Hasse–Arf). For non-abelian extensions, we have the following result.

Theorem 1.3.11 (Sen) *Let K_∞/K be an infinite totally ramified Galois extension whose Galois group $G = \mathrm{Gal}(K_\infty/K)$ is a p-adic Lie group. Fix a Lie filtration $(G(n))_{n \geqslant 0}$ on G. Then there exists a constant $c \geqslant 0$ such that*

$$G^{(ne_K+c)} \subset G(n) \subset G^{(ne_K-c)}, \qquad \forall n \geqslant 0.$$

In particular, $(G : G^{(v)}) < +\infty$ for all $v \geqslant 0$.

Proof This is the main result of [134]. □

1.4 Norms and Traces

1.4.1 The results proved in this section are technical by the nature, but they play a crucial role in our discussion of deeply ramified extensions and the field of norms functor. Assume that L/K is a finite extension of local fields of characteristic 0.

Lemma 1.4.2 *One has*

$$\mathrm{Tr}_{L/K}(\mathfrak{m}_L^n) = \mathfrak{m}_K^r,$$

where $r = \left[\frac{v_L(\mathfrak{D}_{L/K})+n}{e(L/K)}\right]$.

Proof From the definition of the different, it follows immediately that

$$\mathrm{Tr}_{L/K}(\mathfrak{D}_{L/K}^{-1}) = O_K.$$

Set $\delta = v_L(\mathfrak{D}_{L/K})$ and $e = e(L/K)$. Then:

$$\mathfrak{m}_K^r = \mathrm{Tr}_{L/K}\left(\mathfrak{m}_K^r \mathfrak{D}_{L/K}^{-1}\right) = \mathrm{Tr}_{L/K}\left(\mathfrak{m}_L^{re-\delta}\right) \subset \mathrm{Tr}_{L/K}\left(\mathfrak{m}_L^{(\delta+n)-\delta}\right) = \mathrm{Tr}_{L/K}\left(\mathfrak{m}_L^n\right).$$

Conversely, one has

$$\mathrm{Tr}_{L/K}(\mathfrak{m}_L^n \mathfrak{m}_K^{-r}) = \mathrm{Tr}_{L/K}(\mathfrak{m}_L^n \mathfrak{m}_L^{-er}) \subset \mathrm{Tr}_{L/K}(\mathfrak{m}_L^{n-(\delta+n)}) = \mathrm{Tr}_{L/K}(\mathfrak{D}_{L/K}^{-1}) = O_K,$$

Therefore, $\mathrm{Tr}_{L/K}(\mathfrak{m}_L^n) \subset \mathfrak{m}_K^r$, and the lemma is proved. □

1.4.3 Assume that L/K is a totally ramified Galois extension of degree p. Set $G = \mathrm{Gal}(L/K)$ and denote by t the maximal natural number such that $G_t = G$ (and therefore $G_{t+1} = \{1\}$). Formula (8) reads:

$$v_L(\mathfrak{D}_{L/K}) = (p-1)(t+1). \qquad (12)$$

Lemma 1.4.4 *For any* $x \in \mathfrak{m}_L^n$,

$$N_{L/K}(1+x) \equiv 1 + N_{L/K}(x) + \mathrm{Tr}_{L/K}(x) \pmod{\mathfrak{m}_K^s},$$

where $s = \left[\frac{(p-1)(t+1)+2n}{p}\right]$.

Proof Set $G = \mathrm{Gal}(L/K)$, and for each $1 \leqslant n \leqslant p$ denote by C_n the set of all n-subsets $\{g_1, \dots, g_n\}$ of G (note that $g_i \neq g_j$ if $i \neq j$). Then:

$$N_{L/K}(1+x) = \prod_{g \in G}(1+g(x)) = 1 + N_{L/K}(x) + \mathrm{Tr}_{L/K}(x)$$
$$+ \sum_{\{g_1,g_2\} \in C_2} g_1(x)g_2(x) + \cdots + \sum_{\{g_1,\dots g_{p-1}\} \in C_{p-1}} g_1(x)\cdots g_{p-1}(x).$$

It's clear that the rule

$$g \star \{g_1, \ldots, g_n\} = \{gg_1, \ldots, gg_n\}$$

defines an action of G on C_n. Moreover, from the fact that $|G| = p$ is a prime number, it follows that all stabilizers are trivial, and therefore each orbit has p elements. This implies that each sum

$$\sum_{\{g_1,\ldots g_n\}\in C_n} g_1(x)\cdots g_n(x), \qquad 2 \leqslant n \leqslant p-1$$

can be written as the trace $\mathrm{Tr}_{L/K}(x_n)$ of some $x_n \in \mathfrak{m}_L^{2n}$. From (12) and Lemma 1.4.2, it follows that $\mathrm{Tr}_{L/K}(x_n) \in \mathfrak{m}_K^s$. The lemma is proved. □

Corollary 1.4.5 *Let L/K is a totally ramified Galois extension of degree p. Then*

$$v_K(N_{L/K}(1+x) - 1 - N_{L/K}(x)) \geqslant \frac{t(p-1)}{p}.$$

Proof From Lemmas 1.4.2 and 1.4.4, it follows that

$$v_K(N_{L/K}(1+x) - 1 - N_{L/K}(x)) \geqslant \left\lceil \frac{(p-1)(t+1)}{p} \right\rceil,$$

Since
$$\left\lceil \frac{(p-1)(t+1)}{p} \right\rceil = \left\lceil \frac{(p-1)t}{p} + 1 - \frac{1}{p} \right\rceil \geqslant \frac{t(p-1)}{p},$$

the corollary is proved. □

1.5 Witt Vectors

1.5.1 In this subsection, we review the theory of Witt vectors. Consider the sequence of polynomials $w_0(x_0), w_1(x_0, x_1), \ldots$ defined by

$$\begin{aligned}
&w_0(x_0) = x_0,\\
&w_1(x_0, x_1) = x_0^p + px_1,\\
&w_2(x_0, x_1, x_2) = x_0^{p^2} + px_1^p + p^2x_2,\\
&\ldots\ldots\ldots\ldots\ldots\ldots\ldots\ldots\ldots\ldots\\
&w_n(x_0, x_1, \ldots x_n) = x_0^{p^n} + px_1^{p^{n-1}} + p^2x_2^{p^{n-2}} + \cdots + p^nx_n,\\
&\ldots\ldots\ldots\ldots\ldots\ldots\ldots\ldots\ldots\ldots\ldots\ldots\ldots\ldots
\end{aligned}$$

Proposition 1.5.2 *Let $F(x,y)\in \mathbf{Z}[x,y]$ be a polynomial with coefficients in $\mathbf{Z}$ such that $F(0,0)=0$. Then there exists a unique sequence of polynomials*

$$\Phi_0(x_0,y_0)\in \mathbf{Z}[x_0,y_0],$$
$$\Phi_1(x_0,y_0,x_1,y_1)\in \mathbf{Z}[x_0,y_0,x_1,y_1],$$
$$\cdots\cdots\cdots\cdots\cdots\cdots\cdots\cdots\cdots\cdots$$
$$\Phi_n(x_0,y_0,x_1,y_1,\dots,x_n,y_n)\in \mathbf{Z}[x_0,y_0,x_1,y_1,\dots,x_n,y_n],$$
$$\cdots\cdots\cdots\cdots\cdots\cdots\cdots\cdots\cdots\cdots\cdots\cdots\cdots\cdots$$

such that

$$w_n(\Phi_0,\Phi_1,\dots,\Phi_n)=F(w_n(x_0,x_1,\dots,x_n),w_n(y_0,y_1,\dots,y_n)),\qquad \textit{for all } n\geqslant 0. \tag{13}$$

To prove this proposition, we need the following elementary lemma.

Lemma 1.5.3 *Let $f\in \mathbf{Z}[x_0,\dots,x_n]$. Then*

$$f^{p^m}(x_0,\dots,x_n)\equiv f^{p^{m-1}}(x_0^p,\dots,x_n^p)\pmod{p^m},\qquad \textit{for all } m\geqslant 1.$$

Proof The proof is left to the reader. □

1.5.4 *Proof of Proposition* 1.5.2 We prove the proposition by induction on n. For $n=0$, we have $\Phi_0(x_0,y_0)=F(x_0,y_0)$. Assume that $\Phi_0,\Phi_1,\dots,\Phi_{n-1}$ are constructed. From (13), it follows that

$$\Phi_n=\frac{1}{p^n}\left(F(w_n(x_0,x_1,\dots,x_n),w_n(y_0,y_1,\dots,y_n))-(\Phi_0^{p^n}+\cdots+p^{n-1}\Phi_{n-1}^p)\right). \tag{14}$$

This proves the uniqueness. It remains to prove that Φ_n has coefficients in $\mathbf{Z}$. Since

$$w_n(x_0,\dots,x_{n-1},x_n)\equiv w_{n-1}(x_0^p,\dots,x_{n-1}^p)\pmod{p^n},$$

we have

$$\begin{aligned}F(w_n(x_0,\dots,x_{n-1},x_n),\,&w_n(y_0,\dots,y_{n-1},y_n))\\ &\equiv F(w_{n-1}(x_0^p,\dots,x_{n-1}^p),w_{n-1}(y_0^p,\dots,y_{n-1}^p))\pmod{p^n}.\end{aligned} \tag{15}$$

On the other hand, applying Lemma 1.5.3 and the induction hypothesis, we have

$$\begin{aligned}\Phi_0^{p^n}+\cdots+p^{n-1}\Phi_{n-1}^p&\equiv w_{n-1}\left(\Phi_0(x_0^p,y_0^p),\dots,\Phi_{n-1}(x_0^p,y_0^p,\dots,x_{n-1}^p,y_{n-1}^p)\right)\\ &\equiv F(w_{n-1}(x_0^p,\dots,x_{n-1}^p),w_{n-1}(y_0^p,\dots,y_{n-1}^p))\pmod{p^n}.\end{aligned} \tag{16}$$

From (15) and (16), we obtain that

$$F(w_n(x_0, \ldots, x_{n-1}, x_n), w_n(y_0, \ldots, y_{n-1}, y_n)) \equiv \Phi_0^{p^n} + \cdots + p^{n-1}\Phi_{n-1}^p \pmod{p^n}.$$

Together with (14), this whows that Φ_n has coeffiients in $\mathbf{Z}$. The proposition is proved.

1.5.5 Let $(S_n)_{n\geqslant 0}$ denote the polynomials $(\Phi_n)_{n\geqslant 0}$ for $F(x, y) = x + y$ and $(P_n)_{n\geqslant 0}$ denote the polynomials $(\Phi_n)_{n\geqslant 0}$ for $F(x, y) = xy$. In particular,

$$S_0(x_0, y_0) = x_0 + y_0, \quad S_1(x_0, y_0, x_1, y_1) = x_1 + y_1 + \frac{x_0^p + y_0^p - (x_0 + y_0)^p}{p},$$
$$P_0(x_0, y_0) = x_0 y_0, \qquad P_1(x_0, y_0, x_1, y_1) = x_0^p y_1 + x_1 y_0^p + p x_1 y_1.$$

1.5.6 For any commutative ring A, we denote by $W(A)$ the set of infinite vectors $a = (a_0, a_1, \ldots) \in A^{\mathbf{N}}$ equipped with the addition and multiplication defined by the formulas:

$$a + b = (S_0(a_0, b_0), S_1(a_0, b_0, a_1, b_1), \ldots),$$
$$a \cdot b = (P_0(a_0, b_0), P_1(a_0, b_0, a_1, b_1), \ldots).$$

Theorem 1.5.7 (Witt) *With addition and multiplication defined as above, $W(A)$ is a commutative unitary ring with the identity element*

$$1 = (1, 0, 0, \ldots).$$

Proof (a) We show the associativity of addition. From construction, it is clear that there exist polynomials $(u_n)_{n\geqslant 0}$, and $(v_n)_{n\geqslant 0}$ with integer coefficients such that $u_n, v_n \in \mathbf{Z}[x_0, y_0, z_0, \ldots, x_n, y_n, z_n]$ and for any $a, b, c \in W(A)$

$$(a + b) + c = (u_0(a_0, b_0, c_0), \ldots, u_n(a_0, b_0, c_0, \ldots, a_n, b_n, c_n), \ldots),$$
$$a + (b + c) = (v_0(a_0, b_0, c_0), \ldots, v_n(a_0, b_0, c_0, \ldots, a_n, b_n, c_n), \ldots).$$

Moreover,

$$\begin{aligned} w_n(u_0, \ldots, u_n) &= w_n(f_0(x_0, y_0), f_1(x_0, y_0, x_1, y_1), \ldots) + w_n(z_0, \ldots, z_n) \\ &= w_n(x_0, \ldots, x_n) + w_n(y_0, \ldots, y_n) + w_n(z_0, \ldots, z_n), \end{aligned}$$

and

$$\begin{aligned} w_n(v_0, \ldots, v_n) &= w_n(x_0, \ldots, x_n) + w_n(f_0(y_0, z_0), f_1(y_0, z_0, y_1, z_1), \ldots) \\ &= w_n(x_0, \ldots, x_n) + w_n(y_0, \ldots, y_n) + w_n(z_0, \ldots, z_n). \end{aligned}$$

Therefore,

$$w_n(u_0, \ldots, u_n) = w_n(v_0, \ldots, v_n), \qquad \forall n \geqslant 0,$$

and an easy induction shows that $u_n = v_n$ for all n. This proves the associativity of addition.

(b) We will show the formula:

$$(x_0, x_1, x_2, \ldots) \cdot (y_0, 0, 0, \ldots) = (x_0 y_0, x_1 y_0^p, x_1 y_0^{p^2}, \ldots). \tag{17}$$

In particular, it implies that $1 = (1, 0, 0, \ldots)$ is the identity element of $W(A)$. We have

$$(x_0, x_1, x_2, \ldots) \cdot (y_0, 0, 0, \ldots) = (h_0, h_1, \ldots),$$

where $h_0, h_1, \ldots$ are some polynomials in $y_0, x_0, x_1 \ldots$. We prove by induction that $h_n = x_n y_0^n$. For $n = 0$, we have $h_0 = g_0(x_0, y_0) = x_0 y_0$. Assume that the formula is proved for all $i \leqslant n-1$. We have

$$w_n(h_0, h_1, \ldots, h_n) = w_n(x_0, x_1, \ldots, x_n) w_n(y_0, 0, \ldots, 0).$$

Hence:

$$h_0^{p^n} + p h_1^{p^{n-1}} + \cdots + p^{n-1} h_1 + p^n h_n = (x_0^{p^n} + p x_1^{p^{n-1}} + \cdots + p^{n-1} x_1 + p^n x_n) y_0^{p^n}.$$

By induction hypothesis, $h_i = x_i y_0^{p^i}$ for $0 \leqslant i \leqslant n-1$. Then $h_n = x_n y_0^{p^n}$, and the statement is proved.

Other properties can be proved by the same method. □

1.5.8 Below, we assemble some properties of the ring $W(A)$:

(1) For any homomorphism $\psi : A \to B$, the map

$$W(A) \to W(B), \qquad \psi(a_0, a_1, \ldots) = (\psi(a_0), \psi(a_1), \ldots)$$

is an homomorphism.

(2) If p is invertible in A, then there exists an isomorphism of rings $W(A) \simeq A^{\mathbf{N}}$.

Proof The map

$$w : W(A) \to A^{\mathbf{N}}, \qquad w(a_0, a_1, \ldots) = (w_0(a_0), w_1(a_0, a_1), w_2(a_0, a_1, a_2), \ldots)$$

is an homomorphism by the definition of the addition and multiplication in $W(A)$. If p is invertible, then for any $(b_0, b_1, b_2, \ldots)$, the system of equations

$$w_0(x_0) = b_0, \quad w_1(x_0, x_1) = b_1, \quad w_2(x_0, x_1, x_2) = b_2, \ldots$$

has a unique solution in A. Therefore, w is an isomorphism. □

(3) For any $a \in A$, define its Teichmüller lift $[a] \in W(A)$ by

$$[a] = (a, 0, 0, \ldots).$$

Then $[ab] = [a][b]$ for all $a, b \in A$. This follows from (17).

(4) The shift map (Verschiebung)

$$V : W(A) \to W(A), \qquad (a_0, a_1, 0, \ldots) \mapsto (0, a_0, a_1, \ldots)$$

is additive, i.e. $V(a + b) = V(a) + V(b)$. This can be proved by the same method as for Theorem 1.5.7.

(5) For any $n \geqslant 0$, define:

$$I_n(A) = \{(a_0, a_1, \ldots) \in W(A) \mid a_i = 0 \text{ for all } 0 \leqslant i \leqslant n\}.$$

Then $(I_n(A))_{n \geqslant 0}$ is a descending chain of ideals, which defines a separable filtration on $W(A)$. Set:

$$W_n(A) := W(A)/I_n(A).$$

Then

$$W(A) = \varprojlim W(A)/I_n(A).$$

We equip $W(A)/I_n(A)$ with the discrete topology and define the *standard topology* on $W(A)$ as the topology of the projective limit. It is clearly Hausdorff. This topology coincides with the topology of the direct product on $W(A)$:

$$W(A) = A \times A \times A \times \cdots,$$

where each copy of A is equipped with the discrete topology. The ideals $I_n(A)$ form a base of neighborhoods of 0 (each open neighborhood of 0 contains $I_n(A)$ for some n).

(6) For any $a = (a_0, a_1, \ldots) \in W(A)$, one has

$$(a_0, a_1, a_2, \ldots) = \sum_{n=0}^{\infty} V^n[a_n].$$

This can be proved by the same method as for Theorem 1.5.7.

(7) If A is a ring of characteristic p, then the map

$$\varphi : W(A) \to W(A), \qquad (a_0, a_1, \ldots) \mapsto (a_0^p, a_1^p, \ldots),$$

is a ring endomomorphism. In addition,

$$\varphi V = V\varphi = p.$$

Proof The map φ is induced by the absolute Frobenius

$$\varphi : A \to A, \qquad \varphi(x) = x^p.$$

We should show that

$$p(a_0, a_1, \ldots) = (0, a_0^p, a_1^p, \ldots).$$

By definition of Witt vectors, the multiplication by p is given by

$$p(a_0, a_1, \ldots) = (\bar{h}_0(a_0), \bar{h}_1(a_0, a_1), \ldots),$$

where $\bar{h}_n(x_0, x_1, \ldots, x_n)$ is the reduction $\mod p$ of the polynomials defined by the relations:

$$w_n(h_0, h_1, \ldots, h_n) = p w_n(x_0, x_1, \ldots, x_n), \qquad n \geqslant 0.$$

An easy induction shows that $h_n \equiv x_{n-1}^p \pmod{p}$ and the formula is proved. □

Definition Let A be a ring of charactersitic p. We say that A is perfect if φ is an isomorphism. We will say that A is semiperfect if φ is surjective.

Proposition 1.5.9 *Assume that A is an integral perfect ring of characteristic p. The following holds true:*

(i) $p^{n+1}W(A) = I_n(A)$.

(ii) The standard topology on $W(A)$ *coincides with the* p*-adic topology.*

(iii) Each $a = (a_0, a_1, \ldots) \in W(A)$ *can be written as*

$$(a_0, a_1, a_2, \ldots) = \sum_{n=0}^{\infty} [a_n^{p^{-n}}] p^n.$$

Proof (i) Since φ is bijective on A (and therefore on $W(A)$), we can write:

$$p^{n+1} W(A) = V^{n+1} \varphi^{-(n+1)} W(A) = V^{n+1} W(A) = I_n(A).$$

(ii) This follows directly from (i).

(iii) One has

$$(a_0, a_1, a_2, \ldots) = \sum_{n=0}^{\infty} V^n([a_n]) = \sum_{n=0}^{\infty} p^n \varphi^{-n}([a_n]) = \sum_{n=0}^{\infty} [a_n^{p^{-n}}] p^n.$$

□

Theorem 1.5.10 *(i) Let A be an integral perfect ring of characteristic p. Then there exists a unique, up to an isomorphism, ring R such that:*

(a) R is integral of characteristic 0*;*

(b) $R/pR \simeq A$*;*

(c) R is complete for the p-adic topology, namely

$$R \simeq \varprojlim_n R/p^n R;$$

(ii) The ring $W(A)$ satisfies properties a–c).

Proof (i) See [142, Chapitre II, Théorème 3].

(ii) This follows from Proposition 1.5.9. □

Example 1.5.11 (1) $W(\mathbf{F}_p) \simeq \mathbf{Z}_p$.

(2) Let $\overline{\mathbf{F}}_p$ be the algebraic closure of $\mathbf{F}_p$. Then $W(\overline{\mathbf{F}}_p)$ is isomorphic to the ring of integers of the p-adic completion $\widehat{\mathbf{Q}}_p^{\mathrm{ur}}$ of $\mathbf{Q}_p^{\mathrm{ur}}$.

1.6 Non-abelian Cohomology

1.6.1 In this section, we review basic results about non-abelian cohomology. We refer the reader to [119, Chap. 2, §2 and Theorem 6.2.1] for further detail.

Let G be a topological group. One says that a (not necessarily abelian) topological group M is a G-group if it is equipped with a continuous action of G, i.e. a continuous map

$$G \times M \to M, \qquad (g, m) \mapsto gm$$

such that

$$\begin{aligned} g(m_1m_2) &= g(m_1)\, g(m_2), \qquad &&\text{if} \quad g \in G,\ m_1, m_2 \in M, \\ (g_1g_2)(m) &= g_1(g_2(m)), &&\text{if} \quad g_1, g_2 \in G,\ m \in M. \end{aligned}$$

Let M be a G-group. A 1-cocycle with values in M is a continuous map $f : G \to M$ which satisfies the cocycle condition

$$f(g_1g_2) = f(g_1)\,(g_1 f(g_2)), \qquad g_1, g_2 \in G.$$

Two cocycles f_1 and f_2 are said to be homologous if there exists $m \in M$ such that

$$f_2(g) = m\, f_1(g)\, g(m)^{-1}, \qquad g \in G.$$

This defines an equivalence relation $\sim$ on the set $Z^1(G, M)$ of 1-cocycles. The first cohomology $H^1(G, M)$ of G with coefficients in M is defined to be the quotient set $Z^1(G, M)/\sim$. It is easy to see that if M is abelian, this construction coincides with the usual definition of the first continuous cohomology. In general, $H^1(G, M)$ is not a group but it has a distinguished element which is the class of the trivial cocycle. This allows to consider $H^1(G, M)$ as a pointed set. The following properties of the non-abelian H^1 are sufficient for our purposes:

(1) *Inflation–restriction exact sequence.* Let H be a closed normal subgroup of G. Then there exists an exact sequence of pointed sets:

$$0 \to H^1(G/H, M^H) \xrightarrow{\text{inf}} H^1(G, M) \xrightarrow{\text{res}} H^1(H, M)^{G/H}.$$

(2) *Hilbert's Theorem* 90. Let E be a field, and F/E be a finite Galois extension. Then $\mathrm{GL}_n(F)$ is a discrete $\mathrm{Gal}(F/E)$-group, and

$$H^1(\mathrm{Gal}(F/E), \mathrm{GL}_n(F)) = 0, \qquad n \geqslant 1.$$

1.6.2 A direct consequence of the non-abelian Hilbert's Theorem 90 is the following fact. Let V be a finite-dimensional F-vector space equipped with a *semi-linear* action of $\mathrm{Gal}(F/E)$:

$$\begin{aligned} g(x+y) &= g(x) + g(y), \qquad \forall x, y \in V, \\ g(ax) &= g(a)g(x), \qquad \forall a \in F, \forall x \in V. \end{aligned}$$

Let $\{e_1, \dots, e_n\}$ be a basis of V. For any $g \in \mathrm{Gal}(F/E)$, let $A_g \in \mathrm{GL}_n(F)$ denote the unique matrix such that

$$g(e_1, \dots, e_n) = (e_1, \dots, e_n)A_g.$$

Then the map

$$f : \mathrm{Gal}(F/E) \to \mathrm{GL}_n(F), \qquad f(g) = A_g$$

is a 1-cocyle. Hilbert's Theorem 90 shows that there exists a matrix B such that the $(e_1, \dots, e_n)B$ is $\mathrm{Gal}(F/E)$-invariant. To sum up, V always has a $\mathrm{Gal}(F/E)$-invariant basis.

Passing to the direct limit, we obtain the following result.

Proposition 1.6.3 *(i)* $H^1(G_E, GL_n(E^{\mathrm{sep}})) = 0$ *for all* $n \geqslant 1$.

(ii) Each finite-dimensional E^{sep}*-vector space* V *equipped with a semi-linear discrete action of* G_E *has a* G_E*-invariant basis.*

1.6.4 Let E be a field of characteristic p, and let $\mathscr{E}$ be a complete unramified field with residue field E. Let $\mathscr{E}^{\mathrm{ur}}$ denote the maximal unramified extension of $\mathscr{E}$. The residue field of $\mathscr{E}^{\mathrm{ur}}$ is isomorphic to E^{sep}, and we have an isomorphism of Galois groups:

$$\mathrm{Gal}(\mathscr{E}^{\mathrm{ur}}/\mathscr{E}) \simeq G_E.$$

Let $\widehat{\mathscr{E}}^{\mathrm{ur}}$ denote the p-adic completion of $\mathscr{E}^{\mathrm{ur}}$ and $\widehat{O}_{\mathscr{E}}^{\mathrm{ur}}$ its ring of integers. The following version of Hilbert's Theorem 90 can be deduced from Proposition 1.6.3 by devissage.

Proposition 1.6.5 *(i)* $H^1(\mathrm{Gal}(\mathscr{E}^{\mathrm{ur}}/\mathscr{E}), GL_n(\widehat{O}_{\mathscr{E}}^{\mathrm{ur}})) = 0$ *for all* $n \geqslant 1$.

(ii) Each free $\widehat{O}_{\mathscr{E}}^{\mathrm{ur}}$*-module equipped with a semi-linear continuous action of* G_E *has a* G_E*-invariant basis.*

2 Galois Groups of Local Fields

2.1 Unramified and Tamely Ramified Extensions

2.1.1 In this section, we review the structure of Galois groups of local fields. Let K be a local field. Fix a separable closure K^{sep} of K, and set $G_K = \mathrm{Gal}(K^{\mathrm{sep}}/K)$. Set $q = |k_K|$. Since the compositum of two unramified (respectively, tamely ramified) extensions of K is unramified (respectively, tamely ramified) we have the well defined notions of the maximal unramified (respectively, maximal tamely ramified) extension of K. We denote these extensions by K^{ur} and K^{tr} respectively.

2.1.2 The maximal unramified extension K^{ur} of K is procyclic and its Galois group is generated by the Frobenius automorphism Fr_K:

$$\begin{aligned} \mathrm{Gal}(K^{\mathrm{ur}}/K) &\xrightarrow{\sim} \widehat{\mathbf{Z}}, \\ \mathrm{Fr}_K &\longleftrightarrow 1, \\ \mathrm{Fr}_K(x) &\equiv x^q \pmod{\pi_K}, \qquad \forall x \in O_{K^{\mathrm{ur}}}. \end{aligned}$$

2.1.3 Passing to the direct limit in the diagram (7), we have:

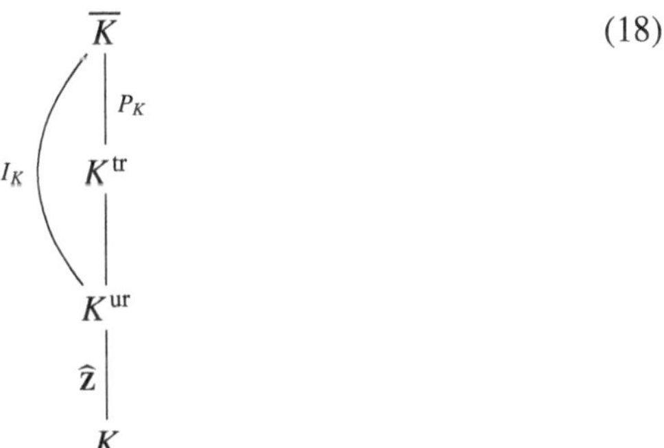

(18)

Consider the exact sequence:

$$1 \to \mathrm{Gal}(K^{\mathrm{tr}}/K^{\mathrm{ur}}) \to \mathrm{Gal}(K^{\mathrm{tr}}/K) \to \mathrm{Gal}(K^{\mathrm{ur}}/K) \to 1. \tag{19}$$

Here $\mathrm{Gal}(K^{\mathrm{ur}}/K) \simeq \widehat{\mathbf{Z}}$. From the explicit description of tamely ramified extensions, it follows that K^{tr} is generated over K^{ur} by the roots $\pi_K^{1/n}$, $(n, p) = 1$ of any uniformizer π_K of K. This immediately implies that

$$\mathrm{Gal}(K^{\mathrm{tr}}/K^{\mathrm{ur}}) \simeq \prod_{\ell \neq p} \mathbf{Z}_\ell. \tag{20}$$

Let τ_K be a topological generator of $\mathrm{Gal}(K^{\mathrm{tr}}/K^{\mathrm{ur}})$. Fix a lift of the Frobenius automorphism Fr_K to an element $\widehat{\mathrm{Fr}}_K \in \mathrm{Gal}(K^{\mathrm{tr}}/K)$. Analyzing the action of these elements on the elements $\pi_K^{1/n}$, one can easily determine the structure of $\mathrm{Gal}(K^{\mathrm{tr}}/K)$.

Proposition 2.1.4 (Iwasawa) *The group* $\mathrm{Gal}(K^{\mathrm{tr}}/K)$ *is topologically generated by the automorphisms* $\widehat{\mathrm{Fr}}_K$ *and* τ_K *with the only relation:*

$$\widehat{\mathrm{Fr}}_K\,\tau_K\,\widehat{\mathrm{Fr}}_K^{-1}=\tau_K^q. \tag{21}$$

Proof See [89] or [119, Theorem 7.5.3]. From (19), it follows that $\mathrm{Gal}(K^{\mathrm{tr}}/K)$ is topologically generated by $\widehat{\mathrm{Fr}}_K$ and τ_K. The relation (21) follows from the explicit action of τ_K and $\widehat{\mathrm{Fr}}_K$ on $\pi_K^{1/n}$ for $(n,p)=1$. □

2.2 Local Class Field Theory

2.2.1 Let K^{ab} denote the maximal abelian extension of K. Then $\mathrm{Gal}(K^{\mathrm{ab}}/K)$ is canonically isomorphic to the abelianization $G_K^{\mathrm{ab}}=G_K/[G_K,G_K]$ of G_K. Local class field theory gives an explicit description of $\mathrm{Gal}(K^{\mathrm{ab}}/K)$ in terms of K. Namely, there exists a canonical injective homomorphism (called the reciprocity map) with dense image

$$\theta_K\,:\,K^*\to \mathrm{Gal}(K^{\mathrm{ab}}/K)$$

such that:

(i) For any finite abelian extension L/K, the homomorphism θ_K induces an isomorphism

$$\theta_{L/K}\,:\,K^*/N_{L/K}(L^*)\xrightarrow{\sim}\mathrm{Gal}(L/K),$$

where $N_{L/K}$ is the norm map;

(ii) If L/K is unramified, then for any uniformizer $\pi\in K^*$ the automorphism $\theta_{L/K}(\pi)$ coincides with the arithmetic Frobenius $\mathrm{Fr}_{L/K}$;

(iii) For any $x\in K^*$, the automorphism $\theta_K(x)$ acts on K^{ur} as

$$\theta_K(x)|_{K^{\mathrm{ur}}}=\mathrm{Fr}_K^{v_K(x)}.$$

The reciprocity map is compatible with the canonical filtrations of K^* and $\mathrm{Gal}(K^{\mathrm{ab}}/K)^{(v)}$. Namely, for any real $v\geqslant 0$ set $U_K^{(v)}=U_K^{(n)}$, where n is the smallest integer $\geqslant v$. Then

$$\theta_K\left(U_K^{(v)}\right)=\mathrm{Gal}(K^{\mathrm{ab}}/K)^{(v)},\qquad \forall v\geqslant 0. \tag{22}$$

For the classical proof of this result, see [142, Chap. XV].

2.2.2 The theory of Lubin–Tate [111] (see also [140]) gives an explicit construction of K^{ab} in terms of torsion points of formal groups with a "big" endomorphism ring, and describes the action of the Galois group $\mathrm{Gal}(K^{\mathrm{ab}}/K)$ on these points. In particular, it gives a simple and natural proof of (22). This theory can be seen as a local analog of the theory of complex multiplication, providing the solution of Hilbert's twelfth problem for local fields. We review it in Sect. 15 below.

2.2.3 Local class field theory was generalized to the infinite residue field case by Serre, Hazewinkel and Suzuki–Yoshida [53, 138, 149]. In another direction, Parshin and Kato developed the class field theory of higher-dimensional local fields [91, 122, 123]. We refer the reader to [63] for survey articles and further references.

2.3 *The Absolute Galois Group of a Local Field*

2.3.1 First, we review the structure of the Galois group of the maximal p-extension of a local field. A finite Galois extension of K is a p-extension if its degree is a power of $p = \mathrm{char}(k_K)$. It is easy to see that p-extensions form a distinguished class, and we can define the maximal pro-p-extension $K(p)$ of K as the compositum of all finite p-extensions. Set $G_K(p) = \mathrm{Gal}(K(p)/K)$.

First assume that $\mathrm{char}(K) = p$. We have the Artin–Schreier exact sequence

$$0 \to \mathbf{F}_p \to K(p) \xrightarrow{\wp} K(p) \to 0,$$

where $\wp(x) = x^p - x$. Taking the associated long exact cohomology sequence and using the fact that $H^i(G_K(p), K(p)) = 0$ for $i \geqslant 1$, we obtain:

$$H^1(G_K(p), \mathbf{F}_p) = K(p)/\wp(K(p)), \qquad H^2(G_K(p), \mathbf{F}_p) = 0.$$

General results about pro-p-groups (see, for example, [99, Chap. 6]) say that

$$\begin{aligned} \dim_{\mathbf{F}_p} H^1(G_K(p), \mathbf{F}_p) &= \text{cardinality of a minimal system of generators of } G_K(p); \\ \dim_{\mathbf{F}_p} H^2(G_K(p), \mathbf{F}_p) &= \text{cardinality of a minimal relation system of } G_K(p). \end{aligned} \tag{23}$$

This leads to the following theorem:

Theorem 2.3.2 *If* $\mathrm{char}(K) = p$, *then* $G_K(p)$ *is a free pro-p-group of countable infinite rank.*

The situation is more complicated in the inequal characteristic case. Let K be a finite extension of $\mathbf{Q}_p$ of degree N. For any n, let μ_n denote the group of nth roots of unity.

Theorem 2.3.3 (Shafarevich, Demushkin) *Assume that* $\mathrm{char}(K) = 0$.

(i) If K doesn't contain the group μ_p, then $G_K(p)$ is a free pro-p-group of rank $N+1$.

(ii) If K contains μ_p, then $G_K(p)$ is a pro-p-group of rank $N+2$, and there exists a system of generators $g_1, g_2, \dots, g_{N+2}$ of $G_K(p)$ with the only relation:

$$g_1^{p^s}[g_1, g_2][g_3, g_4]\cdots[g_{N+1}, g_{N+2}] = 1, \tag{24}$$

where p^s denotes the highest p-power such that K contains μ_{p^s}

Comments on the proof. The Poincaré duality in local class field theory gives perfect pairings:

$$H^i(G_K(p), \mathbf{F}_p) \times H^{2-i}(G_K(p), \mu_p) \to H^2(G_K(p), \mu_p) \simeq \mathbf{F}_p, \qquad 0 \leqslant i \leqslant 2.$$

Therefore, we have:

$$H^1(G_K(p), \mathbf{F}_p) \simeq (K^*/K^{*p})^\vee, \qquad H^2(G_K(p), \mathbf{F}_p) \simeq \mu_p(K)^\vee,$$

where $^\vee$ denotes the duality of $\mathbf{F}_p$-vector spaces. Assume that K doesn't contain the group μ_p. Then these isomorphisms give:

$$\begin{aligned} &\dim_{\mathbf{F}_p} H^1(G_K(p), \mathbf{F}_p) = N+1, \\ &H^2(G_K(p), \mathbf{F}_p) = 0. \end{aligned}$$

Now from (23) we obtain that $G_K(p)$ is free of rank $N+1$. Note that this was first proved by Shafarevich [145] by another method.

Assume now that K contains μ_p. In this case, we have:

$$\begin{aligned} &\dim_{\mathbf{F}_p} H^1(G_K(p), \mathbf{F}_p) = N+2, \\ &H^2(G_K(p), \mathbf{F}_p) = 1. \end{aligned}$$

Therefore, $G_K(p)$ can be generated by $N+2$ elements $g_1, \dots, g_{N+2}$ with only one relation. In [54], Demushkin proved that $g_1, \dots, g_{N+2}$ can be chosed in such a way that (24) holds. See also [101] and [139].

2.3.4 The structure of the absolute Galois group in the characteristic p case can be determined using the above arguments. One easily sees that the wild inertia subgroup P_K is pro-p-free with a countable number of generators. This allows to describe G_K as an explicit semi-direct product of the tame Galois group $\mathrm{Gal}(K^{\mathrm{tr}}/K)$ and P_K (see [98] or [119, Theorem 7.5.13]). The characteristic 0 case is much more difficult. If K is a finite extension of $\mathbf{Q}_p$, the structure of the G_K in terms of generators and relations was first described by Yakovlev [163] under additional assumption $p \neq 2$. A simpler description was found by Jannsen and Wingberg in [90]. For the case $p = 2$, see [164, 165].

2.3.5. The ramification filtration $(G_K^{(v)})$ on G_K has a highly non-trivial structure. We refer the reader to [1, 2, 4, 7, 79] for known results in this direction. Abrashkin [5] and Mochizuki [113] proved that a local field can be completely determined by its absolute Galois group together with the ramification filtration. In another direction, Weinstein [157] interpreted $G_{\mathbf{Q}_p}$ as the fundamental group of some "perfectoid" object.

3 $\mathbf{Z}_p$-Extensions

3.1 *The Different in* $\mathbf{Z}_p$*-Extensions*

3.1.1 The results of this section were proved by Tate [151]. We start with illustrating the ramification theory with the example of $\mathbf{Z}_p$-extensions. Let K be a local field of characteristic 0. Set $e = e(K/\mathbf{Q}_p)$. Let $v_K : \overline{K} \to \mathbf{Q} \cup \{+\infty\}$ denote the extension of the discrete valuation on K to $\overline{K}$.

Definition A $\mathbf{Z}_p$-extension is a Galois extension whose Galois group is topologically isomorphic to $\mathbf{Z}_p$.

Let K_∞/K be a $\mathbf{Z}_p$-extension. Set $\Gamma = \mathrm{Gal}(K_\infty/K)$. For any n, $p^n\mathbf{Z}_p$ is the unique open subgroup of $\mathbf{Z}_p$ of index p^n, and we denote by $\Gamma(n)$ the corresponding subgroup of Γ. Set $K_n = K_\infty^{\Gamma(n)}$. Then K_n is the unique subextension of K_∞/K of degree p^n over K, and

$$K_\infty := \bigcup_{n\geqslant 1} K_n, \qquad \mathrm{Gal}(K_n/K) \simeq \mathbf{Z}/p^n\mathbf{Z}.$$

Assume that K_∞/K is totally ramified. Let $(v_n)_{n\geqslant 0}$ denote the increasing sequence of ramification jumps of K_∞/K. Since $\Gamma \simeq \mathbf{Z}_p$, and all quotients $\Gamma^{(v_n)}/\Gamma^{(v_{n+1})}$ are p-elementary, we obtain that

$$\Gamma^{(v_n)} = p^n\mathbf{Z}_p, \qquad \forall n \geqslant 0.$$

Proposition 3.1.2 *Let K_∞/K be a totally ramified $\mathbf{Z}_p$-extension.*
(i) There exists n_0 such that

$$v_{n+1} = v_n + e, \qquad \forall n \geqslant n_0.$$

(ii) There exists a constant c such that

$$v_K(\mathfrak{D}_{K_n/K}) = en + c + p^{-n}a_n,$$

where the sequence $(a_n)_{n\geqslant 0}$ is bounded.

This is [151, Proposition 5]. Below, we reproduce Tate's proof, which uses local class field theory. See also [73, Proposition 1.11].

The following lemma is a classical and well known statement.

Lemma 3.1.3 *(i) The series*

$$\log(1+x)=\sum_{m=1}^{\infty}(-1)^{m+1}\frac{x^m}{m}$$

converges for all $x\in\mathfrak{m}_K$.

(ii) The series

$$\exp(x)=\sum_{m=0}^{\infty}\frac{x^m}{m!}$$

converges for all x *such that* $v_K(x)>\frac{e}{p-1}$.

(iii) For any integer $n>\frac{e}{p-1}$ *we have isomorphisms:*

$$\log\,:\,U_K^{(n)}\to\mathfrak{m}_K^n,\qquad\exp\,:\,\mathfrak{m}_K^n\to U_K^{(n)},$$

which are inverse to each other.

Corollary 3.1.4 *For any integer* $n>\frac{e}{p-1}$, *one has*

$$\left(U_K^{(n)}\right)^p=U_K^{(n+e)}.$$

Proof $\left(U_K^{(n)}\right)^p$ and $U_K^{(n+e)}$ have the same image under $\log$. □

3.1.5 *Proof of Proposition* 3.1.2

(a) Let $\Gamma=\mathrm{Gal}(K_\infty/K)$. By Galois theory, $\Gamma=G_K^{\mathrm{ab}}/H$, where $H\subset G_K^{\mathrm{ab}}$ is a closed subgroup. Consider the exact sequence

$$\{1\}\to\mathrm{Gal}(K^{\mathrm{ab}}/K^{\mathrm{ur}})\to G_K^{\mathrm{ab}}\xrightarrow{s}\mathrm{Gal}(K^{\mathrm{ur}}/K)\to\{1\}.$$

Since K_∞/K is totally ramified, $(K^{\mathrm{ab}})^H\cap K^{\mathrm{ur}}=K$, and $s(H)=\mathrm{Gal}(K^{\mathrm{ur}}/K)$. Therefore

$$\Gamma\simeq\mathrm{Gal}(K^{\mathrm{ab}}/K^{\mathrm{ur}})/(H\cap\mathrm{Gal}(K^{\mathrm{ab}}/K^{\mathrm{ur}})).$$

By local class field theory, $\mathrm{Gal}(K^{\mathrm{ab}}/K^{\mathrm{ur}})\simeq U_K$, and there exists a closed subgroup $N\subset U_K$ such that

$$\Gamma\simeq U_K/N.$$

The order of $U_K/U_K^{(1)}\simeq k_K^*$ is coprime with p. Hence, the index of $U_K^{(1)}/(N\cap U_K^{(1)})$ in U_K/N is coprime with p. On the other hand, $U_K/N\simeq\Gamma$ is a pro-p-group. Therefore,

$$U_K^{(1)}/(N \cap U_K^{(1)}) = U_K/N,$$

and we have an isomorphism:

$$\rho : \Gamma \simeq U_K^{(1)}/(N \cap U_K^{(1)}).$$

(b) To simplify notation, set:

$$\mathscr{U}^{(v)} = U_K^{(v)}/(N \cap U_K^{(v)}), \qquad \forall v \geqslant 1.$$

By (22) and (10), we have:

$$\rho(\Gamma^{(v)}) \simeq \mathscr{U}^{(v)}, \qquad v \geqslant 1.$$

Let γ be a topological generator of Γ. Then $\gamma_n = \gamma^{p^n}$ is a topological generator of $\Gamma(n)$. Let n_0 be an integer such that

$$\rho(\gamma_{n_0}) \in \mathscr{U}^{(m_0)},$$

with some integer $m_0 > \frac{e}{p-1}$. Fix such n_0 and assume that, for this fixed n_0, m_0 is the biggest integer satisfying this condition. Since γ_{n_0} is a generator of $\Gamma(n_0)$, this means that

$$\rho(\Gamma(n_0)) = \mathscr{U}^{(m_0)}, \qquad \text{but} \qquad \rho(\Gamma(n_0)) \neq \mathscr{U}^{(m_0+1)}.$$

Hence, m_0 is the n_0-th ramification jump for K_∞/K, i.e.

$$m_0 = v_{n_0}.$$

We can write $\rho(\gamma_{n_0}) = \overline{x}$, where $\overline{x} = x \pmod{(N \cap U_K^{(m_0)})}$ and $x \in U_K^{(m_0)} \setminus U_K^{(m_0+1)}$. By Corollary 3.1.4,

$$x^{p^n} \in U_K^{(m_0+en)} \setminus U_K^{(m_0+en+1)}, \qquad \forall n \geqslant 0.$$

Since $\rho(\gamma_{n_0+n}) = \overline{x}^{p^n}$, and γ_{n_0+n} is a generator of $\Gamma(m_0+n)$, this implies that

$$\rho(\Gamma(n_0+n)) = \mathscr{U}^{(m_0+ne)}, \quad \text{and} \quad \rho(\Gamma(n_0+n)) \neq \mathscr{U}^{(m_0+ne+1)}.$$

This shows that for each integer $n \geqslant 0$, the ramification filtration has a jump at $m_0 + ne$, and

$$\Gamma^{(m_0+ne)} = \Gamma(n_0+n).$$

In other terms, for any *real* $v \geqslant v_{n_0} = m_0$, we have:

$$\Gamma^{(v)} = \Gamma(n_0+n+1) \qquad \text{if} \qquad v_{n_0} + ne < v \leqslant v_{n_0} + (n+1)e.$$

This shows that $v_{n_0+n} = v_{n_0} + en$ for all $n \geqslant 0$, and assertion i) is proved.

(c) We prove (ii) applying formula (11). For any $n > 0$, set $G(n) = \Gamma/\Gamma(n)$. We have

$$v_K(\mathfrak{D}_{K_n/K}) = \int_{-1}^{\infty} \left(1 - \frac{1}{|G(n)^{(v)}|}\right) dv.$$

By Herbrand's theorem, $G(n)^{(v)} = \Gamma^{(v)}/(\Gamma(n) \cap \Gamma^{(v)})$. Since $\Gamma^{(v_n)} = \Gamma(n)$, the ramification jumps of $G(n)$ are $v_0, v_1, \dots, v_{n-1}$, and we have:

$$|G(n)^{(v)}| = \begin{cases} p^{n-i}, & \text{if } v_{i-1} < v \leqslant v_i, \\ 1, & \text{if } v > v_{n-1} \end{cases} \qquad (25)$$

(for $i = 0$, we set $v_{i-1} := 0$ to uniformize notation). Assume that $n > n_0$. Then

$$v_K(\mathfrak{D}_{K_n/K}) = A + \int_{v_{i_0}}^{v_{n-1}} \left(1 - \frac{1}{|G(n)^{(v)}|}\right) dv,$$

where $A = \displaystyle\int_{-1}^{v_{n_0}} \left(1 - \frac{1}{|G(n)^{(v)}|}\right) dv$. We evaluate the second integral using i) and (25):

$$\int_{v_{n_0}}^{v_{n-1}} \left(1 - \frac{1}{|G(n)^{(v)}|}\right) dv = \sum_{i=n_0+1}^{n-1} (v_i - v_{i-1}) \left(1 - \frac{1}{|G(n)^{(v)}|}\right) = \sum_{i=n_0+1}^{n-1} e \left(1 - \frac{1}{p^{n-i}}\right).$$

Now an easy computation gives:

$$\sum_{i=n_0+1}^{n-1} e \left(1 - \frac{1}{p^{n-i}}\right) = e(n - n_0 - 1) + \frac{e}{p-1} \left(1 - \frac{1}{p^{n-n_0-1}}\right).$$

Setting $c = A - e(n_0 + 1) + \frac{e}{p-1}$, we see that for $n > n_0$,

$$v_K(\mathfrak{D}_{K_n/K}) = c + en - \frac{1}{(p-1)p^{n-n_0-1}}.$$

This implies the proposition. □

Remark 3.1.6 Proposition 3.1.2 shows that the ramified $\mathbf{Z}_p$-extensions are arithmetically profinite in the sense of Sect. 6.1.

3.2 The Normalized Trace

3.2.1 In this section, K_∞/K is a totally ramified $\mathbf{Z}_p$-extension. Fix a topological generator γ of Γ. For any $x \in K_n$, set:

$$\mathrm{T}_{K_\infty/K}(x) = \frac{1}{p^n}\mathrm{Tr}_{K_n/K}(x).$$

It is clear that this definition does not depend on the choice of n. Therefore, we have a well defined homomorphism

$$\mathrm{T}_{K_\infty/K} : K_\infty \to K.$$

Note that $\mathrm{T}_{K_\infty/K}(x) = x$ for $x \in K$. Our first goal is to prove that $\mathrm{T}_{K_\infty/K}$ is continuous. It is probably more natural to state the results of this section in terms of absolute values rather that in terms of valuations. Let $|\cdot|_K$ denote the absolute value on $\overline{K}$ associated to v_K.

Proposition 3.2.2 *(i) There exists a constant $c > 0$ such that*

$$|\mathrm{T}_{K_\infty/K}(x) - x|_K \leqslant c|\gamma(x) - x|_K, \qquad \forall x \in K_\infty.$$

(ii) The map $\mathrm{T}_{K_\infty/K}$ is continuous and extends by continuity to $\widehat{K}_\infty$.

Proof (a) By Proposition 3.1.2, $v_K(\mathfrak{D}_{K_n/K_{n-1}}) = e_K + \alpha_n p^{-n}$, where α_n is bounded. Applying Lemma 1.4.2 to the extension K_n/K_{n-1}, we obtain that

$$|\mathrm{Tr}_{K_n/K_{n-1}}(x)|_K \leqslant |p|_K^{1-b/p^n}|x|_K, \qquad \forall x \in K_n, \tag{26}$$

with some constant $b > 0$ which does not depend on n.

(b) Set $\gamma_n = \gamma^{p^n}$. For any $x \in K_n$ we have:

$$\mathrm{Tr}_{K_n/K_{n-1}}(x) = \sum_{i=0}^{p-1}\gamma_{n-1}^i(x),$$

Therefore

$$\mathrm{Tr}_{K_n/K_{n-1}}(x) - px = \sum_{i=0}^{p-1}(\gamma_{n-1}^i(x) - x) = \sum_{i=1}^{p-1}(1 + \gamma_{n-1} + \cdots + \gamma_{n-1}^{i-1})(\gamma_{n-1}(x) - x),$$

and we obtain that

$$\left|\frac{1}{p}\mathrm{Tr}_{K_n/K_{n-1}}(x) - x\right|_K \leqslant |p|_K^{-1} \cdot |\gamma_{n-1}(x) - x|_K, \qquad \forall x \in K_n.$$

Since $\gamma_{n-1}(x)-x=(1+\gamma+\cdots+\gamma^{p^{n-1}-1})(\gamma(x)-x)$, we also have:

$$\left|\frac{1}{p}\mathrm{Tr}_{K_n/K_{n-1}}(x)-x\right|_K \leqslant |p|^{-1}\cdot|\gamma(x)-x|_K,\qquad \forall x\in K_n. \tag{27}$$

(c) By induction on n, we prove that

$$\left|\mathrm{T}_{K_\infty/K}(x)-x\right|_K \leqslant c_n\cdot|\gamma(x)-x|_K,\qquad \forall x\in K_n, \tag{28}$$

where $c_1=|p|_K$ and $c_n=c_{n-1}\cdot|p|_K^{-b/p^n}$. For $n=1$, this follows from formula (27). For $n\geqslant 2$ and $x\in K_n$, we write:

$$\mathrm{T}_{K_\infty/K}(x)-x=\left(\frac{1}{p}\mathrm{Tr}_{K_n/K_{n-1}}(x)-x\right)+(\mathrm{T}_{K_\infty/K}(y)-y),\qquad y=\frac{1}{p}\mathrm{Tr}_{K_n/K_{n-1}}(x).$$

The first term can be bounded using formula (27). For the second term, we have:

$$\begin{aligned}|\mathrm{T}_{K_\infty/K}(y)-y|_K \leqslant c_{n-1}|\gamma(y)-y|_K &= c_{n-1}|p|_K^{-1}\cdot|\mathrm{Tr}_{K_n/K_{n-1}}(\gamma(x)-x)|_K\\ &\leqslant c_{n-1}|p|_K^{-b/p^n}|\gamma(x)-x|_K.\end{aligned}$$

(Here the last inequality follows from (26)). This proves (28).

(d) Set $c=c_1\prod\limits_{n=1}^{\infty}|p|_K^{-b/p^n}=c_1|p|_K^{-b/(p-1)}$. Then $c_n<c$ for all $n\geqslant 1$. From formula (28), we obtain:

$$\left|\mathrm{T}_{K_\infty/K}(x)-x\right|_K\leqslant c\cdot|\gamma(x)-x|_K,\qquad \forall x\in K_\infty.$$

This proves the first assertion of the proposition. The second assertion is its immediate consequence. □

Definition The map $\mathrm{T}_{K_\infty/K}\,:\,\widehat{K}_\infty\to K$ is called the normalized trace.

3.2.3 Since $\mathrm{T}_{K_\infty/K}$ is an idempotent map, we have:

$$\widehat{K}_\infty=K\oplus\widehat{K}_\infty^\circ,$$

where $K_\infty^\circ=\ker(\mathrm{T}_{K_\infty/K})$.

Theorem 3.2.4 (Tate) *(i) The operator $\gamma-1$ is bijective, with a continuous inverse, on $\widehat{K}_\infty^\circ$.*

(ii) For any $\lambda\in U_K^{(1)}$ which is not a root of unity, the map $\gamma-\lambda$ is bijective, with a continuous inverse, on $\widehat{K}_\infty$.

Proof (a) Write $K_n=K\oplus K_n^\circ$, where $K_n^\circ=\ker(\mathrm{T}_{K_\infty/K})\cap K_n$. Since $\gamma-1$ is injective on K_n°, and K_n° has finite dimension over K, $\gamma-1$ is bijective on K_n° and on $K_\infty^\circ = \bigcup\limits_{n\geqslant 0} K_n^\circ$. Let $\rho\,:\,K_\infty^\circ \to K_\infty^\circ$ denote its inverse. From Proposition 3.2.2, it follows that

$$|x|_K \leqslant c|(\gamma-1)(x)|_K, \qquad \forall x \in K_\infty^\circ,$$

and therefore

$$|\rho(x)|_K \leqslant c|x|_K, \qquad \forall x \in K_\infty^\circ.$$

Thus ρ is continuous and extends to $\widehat{K}_\infty^\circ$. This proves the theorem for $\lambda = 1$.

(b) Assume that $\lambda \in U_K^{(1)}$ is such that

$$|\lambda - 1|_K < c^{-1}.$$

Then $\rho(\gamma-\lambda) = 1 + (1-\lambda)\rho$, and the series

$$\theta = \sum_{i=0}^{\infty} (\lambda-1)^i \rho^i$$

converges to an operator θ such that $\rho\theta(\gamma-\lambda) = 1$. Thus $\gamma - \lambda$ is invertible on $\widehat{K}_\infty^\circ$. Since $\lambda \neq 1$, it is also invertible on K.

(c) In the general case, we choose n such that $|\lambda^{p^n} - 1|_K < c^{-1}$. By assumptions, $\lambda^{p^n} \neq 1$. Applying part b) to the operator $\gamma^{p^n} - \lambda^{p^n}$, we see that it is invertible on $\widehat{K}_\infty^\circ$. Since

$$\gamma^{p^n} - \lambda^{p^n} = (\gamma - \lambda)\sum_{i=0}^{p^n-1} \gamma^{p^n-i-1}\lambda^i,$$

the operator $\gamma - \lambda$ is also invertible, and the theorem is proved. □

3.3 Application to Continuous Cohomology

3.3.1. We apply the results of the previous section to the computation of some continuous cohomology of Γ. For any continuous character $\eta\,:\,\Gamma \to U_K$, we denote by $\widehat{K}_\infty(\eta)$ the group $\widehat{K}_\infty$ equipped with the natural action of Γ twisted by η:

$$(g, x) \mapsto \eta(g)\cdot g(x), \qquad g \in \Gamma, \quad x \in \widehat{K}_\infty.$$

Let $H^n(\Gamma, -)$ denote the *continuous* cohomology of Γ (see, for example, [119, Chap. II, §7] for definition).

Theorem 3.3.2 (Tate) *(i) $H^0(\Gamma, \widehat{K}_\infty) = K$ and $H^0(\Gamma, \widehat{K}_\infty(\eta)) = 0$ for any continuous character $\eta\,:\,\Gamma \to U_K$ with infinite image.*

(ii) $H^1(\Gamma, \widehat{K}_\infty)$ is a one-dimensional vector space over K, and $H^1(\Gamma, \widehat{K}_\infty(\eta)) = 0$ for any character $\eta\,:\,\Gamma \to U_K$ with infinite image.

Proof (i) The first statement follows directly from Theorem 3.2.4.

(ii) Since Γ is procyclic, any cocycle $f : \Gamma \to \widehat{K}_\infty(\eta)$ is completely determined by $f(\gamma)$. This gives an isomorphism between $H^1(\Gamma, \widehat{K}_\infty(\eta))$ and the cokernel of $\gamma - \eta(\gamma)$. Applying again Theorem 3.2.4, we obtain ii). □

4 Deeply Ramified Extensions

4.1 Deeply Ramified Extensions

4.1.1 In this section, we review the theory of deeply ramified extensions of Coates–Greenberg [37]. This theory goes back to Tate's paper [151], where the case of $\mathbf{Z}_p$-extensions was studied and applied to the proof of the Hodge–Tate decomposition for p-divisible groups.

Let K_∞/K be an infinite algebraic extension of a local field K of characteristic 0. Recall that for each m, the number of algebraic extensions of K of degree m is finite. Hence, we can always write K_∞ in the form

$$K_\infty = \bigcup_{n=0}^{\infty} K_n, \qquad K_0 = K, \qquad K_n \subset K_{n+1}, \qquad [K_n : K] < \infty.$$

Following [75], we define the different of K_∞/K as the intersection of the differents of its finite subextensions:

Definition The different of K_∞/K is defined as

$$\mathfrak{D}_{K_\infty/K} = \bigcap_{n=0}^{\infty} (\mathfrak{D}_{K_n/K} O_{K_\infty}).$$

4.1.2 Let L_∞ be a finite extension of K_∞. Then $L_\infty = K_\infty(\alpha)$, where α is a root of an irreducible polynomial $f(X) \in K_\infty[X]$. The coefficients of $f(X)$ belong to a finite extension K_f of K. Set:

$$n_0 = \min\{n \in \mathbf{N} \mid f(X) \in K_n[X]\}.$$

Let $L_n = K_n(\alpha)$ for all $n \geqslant n_0$. Then

$$L_\infty = \bigcup_{n=n_0}^{\infty} L_n.$$

In what follows, we will assume that $n_0 = 0$ without loss of generality. Note that the degree $[L_n : K_n] = \deg(f)$ does not depend on $n \geqslant 0$.

Proposition 4.1.3 *i) If $m \geqslant n$, then*

$$\mathfrak{D}_{L_n/K_n} O_{L_m} \subset \mathfrak{D}_{L_m/K_m}.$$

ii) One has

$$\mathfrak{D}_{L_\infty/K_\infty} = \bigcup_{n=0}^{\infty} (\mathfrak{D}_{L_n/K_n} O_{L_\infty}).$$

Proof (i) We consider the trace pairing (3):

$$t_{L_n/K_n} \,:\, L_n \times L_n \to K_n.$$

Let $\{e_k\}_{k=1}^s$ be a basis of O_{L_n} over O_{K_n}, and let $\{e_k^*\}_{k=1}^s$ denote the dual basis. Then

$$\mathfrak{D}_{L_n/K_n} = O_{L_n} e_1^* + \cdots + O_{L_n} e_s^*.$$

Since $\{e_k\}_{k=1}^s$ is also a basis of L_m over K_m, any $x \in \mathfrak{D}_{L_m/K_m}^{-1}$ can be written as

$$x = \sum_{k=1}^{s} a_k e_k^*.$$

Then

$$a_k = t_{L_m/K_m}(x, e_k) \in O_{K_m}, \qquad \forall 1 \leqslant k \leqslant s,$$

and we have:

$$x \in O_{K_m} e_1^* + \cdots + O_{K_m} e_s^* \subset \mathfrak{D}_{L_n/K_n}^{-1} O_{L_m}.$$

Hence, $\mathfrak{D}_{L_m/K_m}^{-1} \subset \mathfrak{D}_{L_n/K_n}^{-1} O_{L_m}$, and therefore $\mathfrak{D}_{L_n/K_n} O_{L_m} \subset \mathfrak{D}_{L_m/K_m}$.

(ii) By the same argument as in the proof of (i), the following holds:

$$\bigcup_{n=0}^{\infty} (\mathfrak{D}_{L_n/K_n} O_{L_\infty}) \subset \mathfrak{D}_{L_\infty/K_\infty}.$$

We need to prove that $\mathfrak{D}_{L_\infty/K_\infty} \subset \bigcup_{n=0}^{\infty} (\mathfrak{D}_{L_n/K_n} O_{L_\infty})$ or, equivalently, that

$$\bigcap_{n=0}^{\infty} (\mathfrak{D}_{L_n/K_n}^{-1} O_{L_\infty}) \subset \mathfrak{D}_{L_\infty/K_\infty}^{-1}.$$

Let $x \in \bigcap_{n=0}^{\infty} (\mathfrak{D}_{L_n/K_n}^{-1} O_{L_\infty})$ and $y \in O_{L_\infty}$. Choosing n such that $x \in \mathfrak{D}_{L_n/K_n}^{-1}$ and $y \in O_{L_n}$, we have:

$$t_{L_\infty/K_\infty}(x, y) = t_{L_n/K_n}(x, y) \in O_{K_n} \subset O_{K_\infty}.$$

The proposition is proved. □

4.1.4 For any algebraic extension M/K of local fields (finite or infinite) we set:

$$v_K(\mathfrak{D}_{M/K}) = \inf\{v_K(x) \mid x \in \mathfrak{D}_{M/K}\}.$$

Definition (i) We say that K_∞/K has finite conductor if there exists $v \geqslant 0$ such that $K_\infty \subset \overline{K}^{(v)}$. If that is the case, we call the conductor of K_∞/K the number

$$c(K_\infty) = \inf\{v \mid K_\infty \subset \overline{K}^{(v-1)}\}.$$

(ii) We say that K_∞/K is deeply ramified if it does not have finite conductor.

Below, we give some examples of deeply ramified extensions.

Example 4.1.5 (1) The cyclotomic extension $K(\zeta_{p^\infty})/K$ is deeply ramified. This follows from Proposition 3.1.2.

(2) Fix a uniformizer π of K and set $\pi_n = \pi^{1/p^n}$. Then the infinite Kummer extension $K(\pi^{1/p^\infty}) = \overset{\infty}{\underset{n=1}{\cup}} K(\pi_n)$ is deeply ramified. This can be proved by a direct computation or alternatively computing the different of this extension and using Theorem 4.1.7 below.

(3) Let K_∞/K be a totally ramified infinite Galois extension such that its Galois group $G = \mathrm{Gal}(K_\infty/K)$ is a Lie group. From Theorem 1.3.11, it follows that K_∞/K is deeply ramified. We will come back to this example in Sect. 6.

4.1.6 Now we state our main theorem about deeply ramified extensions.

Theorem 4.1.7 (Coates–Greenberg) *Let K_∞/K be an algebraic extension of local fields. Then the following assertions are equivalent:*

(i) $v_K(\mathfrak{D}_{K_\infty/K}) = +\infty$;

(ii) K_∞/K is deeply ramified;

(iii) For any finite extension L_∞/K_∞, one has

$$v_K(\mathfrak{D}_{L_\infty/K_\infty}) = 0;$$

(iv) For any finite extension L_∞/K_∞, one has

$$\mathrm{Tr}_{L_\infty/K_\infty}(\mathfrak{m}_{L_\infty}) = \mathfrak{m}_{K_\infty}.$$

In sections 4.1.8–4.1.12 below, we prove the implications

$$(i) \Leftrightarrow (ii) \Rightarrow (iii) \Rightarrow (iv).$$

Lemma 4.1.8 *For any finite extension M/K, one has*

$$\frac{c(M)}{2} \leqslant v_K(\mathfrak{D}_{M/K}) \leqslant c(M).$$

Proof We have:

$$[M : M \cap \overline{K}^{(v)}] = 1, \quad \text{for any } v > c(M) - 1;$$
$$[M : M \cap \overline{K}^{(v)}] \geqslant 2, \quad \text{if } -1 \leqslant v < c(M) - 1.$$

Therefore

$$v_K(\mathfrak{D}_{M/K}) = \int_{-1}^{\infty} \left(1 - \frac{1}{[M : M \cap \overline{K}^{(v)}]}\right) dv \leqslant \int_{-1}^{c(M)-1} dv = c(M),$$

and

$$v_K(\mathfrak{D}_{M/K}) = \int_{-1}^{\infty} \left(1 - \frac{1}{[M : M \cap \overline{K}^{(v)}]}\right) dv \geqslant \frac{1}{2} \int_{-1}^{c(M)-1} dv = \frac{c(M)}{2}.$$

The lemma is proved. □

4.1.9 We prove that $(i) \Leftrightarrow (ii)$. First assume that $v_K(\mathfrak{D}_{K_\infty/K}) = +\infty$. For any $c > 0$, there exists $K \subset M \subset K_\infty$ such that $v_K(\mathfrak{D}_{M/K}) \geqslant c$. By Lemma 4.1.8, $c(M) \geqslant c$. This shows that K_∞/K doesn't have finite conductor.

Conversely, assume that K_∞/K doesn't have finite conductor. Then for each $c > 0$, there exists a non-zero element $\beta \in K_\infty \cap \overline{K}^{(c)}$. Let $M = K(\beta)$. Then $v_K(\mathfrak{D}_{M/K}) \geqslant \frac{c}{2}$ by Lemma 4.1.8. Therefore, $v_K(\mathfrak{D}_{K_\infty/K}) = +\infty$.

Lemma 4.1.10 *Assume that w is such that $L \subset \overline{K}^{(w)}$. Then for any $n \geqslant 0$,*

$$[L_n : L_n \cap \overline{K}^{(w)}] = [K_n : K_n \cap \overline{K}^{(w)}].$$

Proof Since $\overline{K}^{(w)}/K$ is a Galois extension, K_n and $\overline{K}^{(w)}$ are linearly disjoint over $K_n \cap \overline{K}^{(w)}$. Therefore K_n and $\overline{K}^{(w)} \cap L_n$ are linearly disjoint over $K_n \cap \overline{K}^{(w)}$. We have:

$$[K_n : K_n \cap \overline{K}^{(w)}] = [K_n \cdot (\overline{K}^{(w)} \cap L_n) : (\overline{K}^{(w)} \cap L_n)]. \tag{29}$$

Clearly $K_n \cdot (\overline{K}^{(w)} \cap L_n) \subset L_n$. Conversely, from $L_n = K_n \cdot L$ and $L \subset \overline{K}^{(w)}$, it follows that $L_n \subset K_n \cdot (\overline{K}^{(w)} \cap L_n)$. Thus

$$L_n = K_n \cdot (\overline{K}^{(w)} \cap L_n).$$

Together with (29), this proves the lemma. □

4.1.11 We prove that $(ii) \Rightarrow (iii)$. By the multiplicativity of the different, for any $n \geqslant 0$, we have:

$$v_K(\mathfrak{D}_{L_n/K_n}) = v_K(\mathfrak{D}_{L_n/K}) - v_K(\mathfrak{D}_{K_n/K}).$$

Let w be such that $L \subset \overline{K}^{(w)}$. Using formula (11) and Lemma 4.1.10, we obtain:

$$v_K(\mathfrak{D}_{L_n/K_n}) = \int_{-1}^{\infty} \left(\frac{1}{[K_n : (K_n \cap \overline{K}^{(v)})]} - \frac{1}{[L_n : (L_n \cap \overline{K}^{(v)})]} \right) dv =$$
$$\int_{-1}^{w} \left(\frac{1}{[K_n : (K_n \cap \overline{K}^{(v)})]} - \frac{1}{[L_n : (L_n \cap \overline{K}^{(v)})]} \right) dv \leqslant \int_{-1}^{w} \frac{dv}{[K_n : (K_n \cap \overline{K}^{(v)})]}.$$

Since $[K_n : (K_n \cap \overline{K}^{(v)})] \geqslant [K_n : (K_n \cap \overline{K}^{(w)})]$ if $v \leqslant w$, this gives the following estimate for the different:

$$v_K(\mathfrak{D}_{L_n/K_n}) \leqslant \frac{w+1}{[K_n : (K_n \cap \overline{K}^{(w)})]}.$$

Since K_∞/K doesn't have finite conductor, for any $c > 0$ there exists $n \geqslant 0$ such that $[K_n : (K_n \cap \overline{K}^{(w)})] > c$, and therefore $v_K(\mathfrak{D}_{L_n/K_n}) \leqslant (w+1)/c$. This proves that $v_K(\mathfrak{D}_{L_\infty/K_\infty}) = 0$.

4.1.12 We prove that $(iii) \Rightarrow (iv)$. We consider two cases.

(a) First assume that the set $\{e(K_n/K) \mid n \geqslant 0\}$ is bounded. Then there exists $n_0 \in I$ such that $e(K_n/K_{n_0}) = 1$ for any $n \geqslant n_0$. Therefore, $e(L_n/L_{n_0}) = 1$ for any $n \geqslant n_0$, and by the mutiplicativity of the different

$$\mathfrak{D}_{L_n/K_n} = \mathfrak{D}_{L_{n_0}/K_{n_0}} O_{L_n}, \qquad \forall n \geqslant n_0.$$

From Proposition 4.1.3 and assumption iii), it follows that $\mathfrak{D}_{L_n/K_n} = O_{L_n}$ for all $n \geqslant n_0$. Therefore, the extensions L_n/K_n are unramified, and Lemma 1.4.2 (or just the well known surjectivity of the trace map for unramified extensions) gives:

$$\mathrm{Tr}_{L_n/K_n}(\mathfrak{m}_{L_n}) = \mathfrak{m}_{K_n}, \qquad \text{for all } n \geqslant n_0.$$

Thus $\mathrm{Tr}_{L_\infty/K_\infty}(\mathfrak{m}_{L_\infty}) = \mathfrak{m}_{K_\infty}$.

(b) Now assume that the set $\{e(K_n/K) \mid n \geqslant 0\}$ is unbounded. Let $x \in \mathfrak{m}_{K_\infty}$. Then there exists n such that $x \in \mathfrak{m}_{K_n}$. By Lemma 1.4.2,

$$\mathrm{Tr}_{L_n/K_n}(\mathfrak{m}_{L_n}) = \mathfrak{m}_{K_n}^{r_n}, \qquad r_n = \left[\frac{v_{L_n}(\mathfrak{D}_{L_n/K_n}) + 1}{e(L_n/K_n)} \right].$$

From our assumptions and Proposition 4.1.3, it follows that we can choose n such that in addition

$$v_K(\mathfrak{D}_{L_n/K_n}) + \frac{1}{e(L_n/K)} \leqslant v_K(x).$$

Then

$$r_n \leqslant \frac{v_{L_n}(\mathfrak{D}_{L_n/K_n}) + 1}{e(L_n/K_n)} = \left(v_K(\mathfrak{D}_{L_n/K_n}) + \frac{1}{e(L_n/K)}\right) e(K_n/K) \leqslant v_{K_n}(x).$$

Since $\mathrm{Tr}_{L_n/K_n}(\mathfrak{m}_{L_n})$ is an ideal in O_{K_n}, this implies that $x \in \mathrm{Tr}_{L_n/K_n}(\mathfrak{m}_{L_n})$, and the inclusion $\mathfrak{m}_{K_\infty} \subset \mathrm{Tr}_{L_\infty/K_\infty}(\mathfrak{m}_{L_\infty})$ is proved. Since the converse inclusion is trivial, we have $\mathfrak{m}_{K_\infty} = \mathrm{Tr}_{L_\infty/K_\infty}(\mathfrak{m}_{L_\infty})$.

4.2 *Almost étale Extensions*

4.2.1 In this section, we introduce, in our very particular setting, the notion of an almost étale extension.

Definition A finite extension E/F of non-Archimedean fields is almost étale if and only if

$$\mathrm{Tr}_{E/F}(\mathfrak{m}_E) = \mathfrak{m}_F.$$

It is clear that an unramified extension of local fields is almost étale. Below, we give two other archetypical examples of almost étale extensions.

Example 4.2.2 (1) Assume that F is a perfect non-Archimedean field of characteristic p. Then any finite extension of F is almost étale.

Proof Let E/F be a finite extension. It is clear that $\mathrm{Tr}_{E/F}(\mathfrak{m}_E) \subset \mathfrak{m}_F$. Moreover, $\mathrm{Tr}_{E/F}(\mathfrak{m}_E)$ is an ideal of O_F, and for any $\alpha \in \mathfrak{m}_E$, one has

$$\lim_{n\to+\infty} |\mathrm{Tr}_{E/F}\varphi^{-n}(\alpha)|_F = 0.$$

This implies that $\mathfrak{m}_F \subset \mathrm{Tr}_{E/F}(\mathfrak{m}_E)$, and the proposition is proved. □

(2) Assume that K_∞ is a deeply ramified extension of a local field K of characteristic 0. By Theorem 4.1.7, any finite extension of K_∞ is almost étale.

4.2.3. Following Tate [151], we apply the theory of almost étale extensions to the proof of the theorem of Ax–Sen–Tate. Let K be a perfect complete non-Archimedean field, and let $\mathbf{C}_K$ denote the completion of $\overline{K}$. For any topological group G, we denote by $H^n(G, -)$ the continuous cohomology of G.

Theorem 4.2.4 *Assume that F is an algebraic extension of K such that any finite extension of F is almost étale. Then*

$$H^0(G_F, \mathbf{C}_K) = \widehat{F}.$$

We first prove the following lemma. Fix an absolute value $|\cdot|_K$ on $\overline{K}$.

Lemma 4.2.5 *Let E/F be an almost étale Galois extension with Galois group G. Then for any $\alpha \in E$ and any $c > 1$, there exists $a \in F$ such that*

$$\left|\alpha - a\right|_K < c \cdot \max_{g \in G} \left|g(\alpha) - \alpha\right|_K.$$

Proof Let $c > 1$. By Theorem 4.1.7 iv), there exists $x \in O_E$ such that $y = \mathrm{Tr}_{E/F}(x)$ satisfies

$$1/c < |y|_K \leqslant 1.$$

Set: $a = \dfrac{1}{y}\sum_{g \in G} g(\alpha x)$. Then

$$|\alpha - a|_K = \left|\frac{\alpha}{y}\sum_{g\in G} g(x) - \frac{1}{y}\sum_{g\in G} g(\alpha x)\right|_K = \left|\frac{1}{y}\sum_{g\in G} g(x)(\alpha - g(\alpha))\right|_K$$
$$\leqslant \frac{1}{|y|_F} \cdot \max_{g\in G}\left|g(\alpha) - \alpha\right|_K.$$

The lemma is proved. □

4.2.6 *Proof of Theorem* 4.2.4 Let $\alpha \in \mathbf{C}_K^{G_F}$. Choose a sequence $(\alpha_n)_{n\in\mathbf{N}}$ of elements $\alpha_n \in \overline{K}$ such that $|\alpha_n - \alpha|_K < p^{-n}$. Then

$$|g(\alpha_n) - \alpha_n|_K = |g(\alpha_n - \alpha) - (\alpha_n - \alpha)|_K < p^{-n}, \qquad \forall g \in G_F.$$

By Lemma 4.2.5, for each n, there exists $\beta_n \in F$ such that $|\beta_n - \alpha_n|_K < p^{-n}$. Then

$$\alpha = \lim_{n\to+\infty} \beta_n \in \widehat{F}.$$

The theorem is proved. □

4.2.7 Now we compute the first cohomology group $H^1(G_F, \mathbf{C}_K)$.

Theorem 4.2.8 *Under the assumptions and notation of Theorem 4.2.4, $\mathfrak{m}_F H^1(G_F, O_{\mathbf{C}_K}) = \{0\}$ and $H^1(G_F, \mathbf{C}_K) = \{0\}$.*

The proof will be given in Sects. 4.2.9–4.2.10 below. For any map $f : X \to O_{\mathbf{C}_K}$, where X is an arbitrary set, we define $|f| := \sup_{x\in X} |f(x)|_K$.

Lemma 4.2.9 *Let E/F be a finite Galois extension with Galois group G. Then for any map $f : G \to O_{\overline{K}}$ and any $y \in \mathfrak{m}_F$, there exists $\alpha \in O_E$ such that*

$$|yf - h_\alpha| < |\partial(f)|_K,$$

where $h_\alpha : G \to O_{\overline{K}}$ is the 1-coboundary $h_\alpha(g) = g(\alpha) - \alpha$ and $\partial(f) : G \times G \to O_{\overline{K}}$ is the 2-coboundary $\partial(f)(g_1, g_2) = g_1 f(g_2) - f(g_1 g_2) + f(g_1)$.

Proof Since E/F is almost étale, there exists $x \in O_E$ such that $y = \mathrm{Tr}_{E/F}(x)$. Set:

$$\alpha := -\sum_{g \in G} g(x) f(g).$$

An easy computation shows that for any $\tau \in G$, one has

$$\tau(\alpha) - \alpha = yf(\tau) - \sum_{g \in G} \tau g(x) \cdot \partial(f)(\tau, g).$$

This proves the lemma. □

4.2.10 *Proof of Theorem* 4.2.8 Let $f : G_F \to O_{\mathbf{C}_K}$ be a 1-cocycle. Fix $y \in \mathfrak{m}_F$. By continuity of f, for any $n \geqslant 0$ there exists a map $\tilde{f} : G_F \to O_{\overline{K}}$ such that $|\tilde{f} - f| < p^{-n}$, and $\tilde{f}$ factors through a finite quotient of G_F. Note that $|\partial(\tilde{f})| < p^{-n}$ because $\partial(f) = 0$. By Lemma 4.2.9, there exists $\alpha \in \mathfrak{m}_{\overline{K}}$ such that

$$|yf - h_\alpha| < |\partial(\tilde{f})| < p^{-n}.$$

Using this argument together with successive approximation, it is easy to see that $y \cdot \mathrm{cl}(f) = 0$. This proves that $\mathfrak{m}_F H^1(G_F, O_{\mathbf{C}_K}) = \{0\}$. Now the vanishing of $H^1(G_F, \mathbf{C}_K)$ is obvious. □

The following corollary should be compared with Theorem 1.1.8.

Corollary 4.2.11 *Let F be a complete perfect non-Archimedean field of characteristic p. Then the following holds true:*

(i) $H^0(G_F, \mathbf{C}_F) = F$;
(ii) $\mathfrak{m}_F \cdot H^1(G_F, O_{\mathbf{C}_F}) = 0$;
(iii) $H^1(G_F, \mathbf{C}_F) = 0$.

4.3 Continuous Cohomology of G_K

4.3.1 Assume that K is a local field of characteristic 0.

Theorem 4.3.2 (Tate) *(i) Let K_∞/K be a deeply ramified extension. Then $H^0(G_{K_\infty}, \mathbf{C}_K) = \widehat{K}_\infty$ and $H^1(G_{K_\infty}, \mathbf{C}_K) = 0$.*

(ii) $H^0(G_K, \mathbf{C}_K) = K$, *and* $H^1(G_K, \mathbf{C}_K)$ *is the one dimensional* K*-vector space generated by any totally ramified additive character* $\eta : G_K \to \mathbf{Z}_p$.

(iii) Let $\eta : G_K \to \mathbf{Z}_p^*$ *be a totally ramified character with infinite image. Then* $H^0(G_K, \mathbf{C}_K(\eta)) = 0$, *and* $H^1(G_K, \mathbf{C}_K(\eta)) = 0$.

Proof (i) The first assertion follows from Theorems 4.1.7 and 4.2.8.

(ii) Let $K_\infty = \overline{K}^{\ker(\eta)}$. Then K_∞/K is a $\mathbf{Z}_p$-extension, and we set $\Gamma = \mathrm{Gal}(K_\infty/K)$. By Proposition 3.1.2, K_∞/K is deeply ramified. Hence, $H^0(G_{K_\infty}, \mathbf{C}_K) = \widehat{K}_\infty$ by Theorem 4.2.4. Applying Theorem 3.3.2, we obtain that $H^0(G_K, \mathbf{C}_K) = H^0(\Gamma, \widehat{K}_\infty) = K$. To compute the first cohomology, consider the inflation-restriction exact sequence:

$$0 \to H^1(\Gamma, \mathbf{C}_K^{G_{K_\infty}}) \to H^1(G_K, \mathbf{C}_K) \to H^1(G_{K_\infty}, \mathbf{C}_K).$$

By assertion i), $\mathbf{C}_K^{G_{K_\infty}} = \widehat{K}_\infty$, and $H^1(G_{K_\infty}, \mathbf{C}_K) = 0$. Hence,

$$H^1(G_K, \mathbf{C}_K) \simeq H^1(\Gamma, \widehat{K}_\infty).$$

Applying Theorem 3.3.2, we see that $H^1(G_K, \mathbf{C}_K)$ is the one-dimensional K-vector space generated by $\eta : G_K \to \mathbf{Z}_p$.

(iii) The last assertion can be proved by the same arguments. □

4.3.3 The group G_K acts on the groups μ_{p^n} of p^n-th roots of unity via the character $\chi_K : G_K \to \mathbf{Z}_p^*$ defined as

$$g(\zeta) = \zeta^{\chi_K(g)}, \qquad \forall g \in G_K,\ \zeta \in \mu_{p^n},\ n \geqslant 1.$$

Definition The character $\chi_K : G_K \to \mathbf{Z}_p^*$ is called the cyclotomic character.

It is clear that $\log \chi_K$ is an additive character of G_K with values in $\mathbf{Z}_p$.

Corollary 4.3.4 $H^1(G_K, \mathbf{C}_K)$ *is the one-dimensional* K*-vector space generated by* $\log \chi_K$.

4.3.5 Let E/K be a finite extension which contains all conjugates τK of K over $\mathbf{Q}_p$. We say that two multiplicative characters $\psi_1, \psi_2 : G_E \to U_K$ are equivalent and write $\psi_1 \sim \psi_2$ if $\mathbf{C}_K(\psi_1) \simeq \mathbf{C}_K(\psi_2)$ as G_E-modules. Theorem 4.3.2 implies the following proposition, which will be used in Sect. 15.

Proposition 4.3.6 *The conditions (a) and b) below are equivalent:*

(a) $\tau \circ \psi_1 \sim \tau \circ \psi_2$ *for all* $\tau \in \mathrm{Hom}(K, E)$.

(b) The characters ψ_1 *and* ψ_2 *coincide on an open subgroup of* I_E.

Proof See [143, Section A2]. □

4.3.7 Using Tate's method, Sen proved the following important result.

Theorem 4.3.8 (Sen) *Assume that K_∞/K is deeply ramified. Then*

$$H^1(G_{K_\infty}, \mathrm{GL}_n(\mathbf{C}_K)) = \{1\}.$$

Proof For deeply ramified $\mathbf{Z}_p$-extensions, it was proved in [136], and the proof is similar in the general case. □

5 From Characteristic 0 to Characteristic p and Vice Versa I: Perfectoid Fields

5.1 *Perfectoid Fields*

5.1.1 The notion of perfectoid field was introduced in Scholze's fundamental paper [130] as a far-reaching generalization of Fontaine's constructions [66, 70]. Fix a prime number p. Let E be a field equipped with a non-Archimedean absolute value $|\cdot|_E : E \to \mathbf{R}_+$ such that $|p|_E < 1$. Note that we don't exclude the case of characteristic p, where the last condition holds automatically. We denote by O_E the ring of integers of E and by $\mathfrak{m}_E$ the maximal ideal of O_E.

Definition Let E be a field equipped with an absolute value $|\cdot|_E : E \to \mathbf{R}_+$ such that $|p|_E < 1$. One says that E is perfectoid if the following holds true:

(i) $|\cdot|_E$ is non-discrete;

(ii) E is complete for $|\cdot|_E$;

(iii) The Frobenius map

$$\varphi : O_E/pO_E \to O_E/pO_E, \qquad \varphi(x) = x^p$$

is surjective.

We give first examples of perfectoid fields, which can be treated directly.

Example 5.1.2 (1) A perfect field of characteristic p, complete for a non-Archimedean valuation, is a perfectoid field.

(2) Let K be a non-Archimedean field. The completion $\mathbf{C}_K$ of its algebraic closure is a perfectoid field.

(3) Let K be a local field. Fix a uniformizer π of K and set $\pi_n = \pi^{1/p^n}$. Then the completion of the Kummer extension $K(\pi^{1/p^\infty}) = \bigcup_{n=1}^{\infty} K(\pi_n)$ is a perfectoid field. This follows from the congruence

$$\left(\sum_{i=0}^{m} [a_i]\pi_n^i\right)^p \equiv \sum_{i=0}^{m} [a_i]^p \pi_{n-1}^i \pmod{p}.$$

5.1.3 The following important result is a particular case of [78, Proposition 6.6.6].

Theorem 5.1.4 (Gabber–Ramero) *Let K be a local field of characteristic* 0. *A complete subfield $K \subset E \subset \mathbf{C}_K$ is a perfectoid field if and only if it is the completion of a deeply ramified extension of K.*

5.2 Tilting

5.2.1 In this section, we describe the tilting construction, which functorially associates to any perfectoid field of characteristic 0 a perfect field of characteristic p. This construction first appeared in the pionnering papers of Fontaine [64, 66]. The tilting of arithmetically profinite extensions is closely related to the field of norms functor of Fontaine–Wintenberger [161]. We will come back to this question in Sect. 6. In the full generality, the tilting was defined in the famous paper of Scholze [130] for perfectoid algebras. This generalization is crucial for geometric application. However, in this introductory paper, we will consider only the arithmetic case.

5.2.2 Let E be a perfectoid field of characteristic 0. Consider the projective limit

$$O_E^\flat := \varprojlim_{\varphi} O_E/pO_E = \varprojlim (O_E/pO_E \overset{\varphi}{\leftarrow} O_E/pO_E \overset{\varphi}{\leftarrow} \cdots), \tag{30}$$

where $\varphi(x) = x^p$. It is clear that $O_E^\flat$ is equipped with a natural ring structure. An element x of $O_E^\flat$ is an infinite sequence $x = (x_n)_{n \geqslant 0}$ of elements $x_n \in O_E/pO_E$ such that $x_{n+1}^p = x_n$ for all n. Below, we summarize first properties of the ring $O_E^\flat$:

(1) For all $m \in \mathbf{N}$, choose a lift $\widehat{x}_m \in O_E$ of x_m. Then for any fixed n, the sequence $(\widehat{x}_{n+m}^{p^m})_{m \geqslant 0}$ converges to an element

$$x^{(n)} = \lim_{m \to \infty} \widehat{x}_{m+n}^{p^m} \in O_E,$$

which does not depend on the choice of the lifts $\widehat{x}_m$. In addition, $\left(x^{(n)}\right)^p = x^{(n-1)}$ fol all $n \geqslant 1$.

Proof Since $x_{m+n}^p = x_{m+n-1}$, we have $\widehat{x}_{m+n}^p \equiv \widehat{x}_{m+n-1} \pmod{p}$, and an easy induction shows that $\widehat{x}_{m+n}^{p^m} \equiv \widehat{x}_{m+n-1}^{p^{m-1}} \pmod{p^m}$. Therefore, the sequence $(\widehat{x}_{n+m}^{p^m})_{m \geqslant 0}$ converges. Assume that $\widetilde{x}_m \in O_E$ are another lifts of $x_m, m \in \mathbf{N}$. Then $\widetilde{x}_m \equiv \widehat{x}_m \pmod{p}$ and therefore $\widetilde{x}_{n+m}^{p^m} \equiv \widehat{x}_{n+m}^{p^m} \pmod{p^{m+1}}$. This implies that the limit doesn't depend on the choice of the lifts. □

(2) For all $x, y \in O_{E^\flat}$ one has

$$(x+y)^{(n)} = \lim_{m \to +\infty} \left(x^{(n+m)} + y^{(n+m)}\right)^{p^m}, \qquad (xy)^{(n)} = x^{(n)}y^{(n)}. \tag{31}$$

Proof It is easy to see that $x^{(n)} \in O_E$ is a lift of x_n. Therefore, $x^{(n+m)} + y^{(n+m)}$ is a lift of $x_{n+m} + y_{n+m}$, and the first formula follows from the definition of $(x+y)^{(n)}$. The same argument proves the second formula. □

(3) The map $x \mapsto (x^{(n)})_{n \geqslant 0}$ defines an isomorphism

$$O_E^\flat \simeq \varprojlim_{x^p \leftarrow x} O_E, \tag{32}$$

where the right hand side is equipped with the addition and the multiplication defined by formula (31).

Proof This follows from from (2). □

Set:

$$| \cdot |_{E^\flat} : O_E^\flat \to \mathbf{R} \cup \{+\infty\},$$
$$|x|_{E^\flat} = |x^{(0)}|_E.$$

Proposition 5.2.3 *(i)* $| \cdot |_{E^\flat}$ *is a non-Archimedean absolute value on* $O_{E^\flat}$.

(ii) $O_E^\flat$ *is a perfect complete valuation ring of characteristic* p, *with maximal ideal* $\mathfrak{m}_E^\flat = \{x \in O_E^\flat \mid |x|_{E^\flat} < 1\}$ *and residue field* k_E.

(iii) Let $E^\flat$ *denote the field of fractions of* $O_E^\flat$. *Then* $|E^\flat|_{E^\flat} = |E|_E$.

Proof (i) Let $x, y \in O_E^\flat$. It is clear that

$$|xy|_{E^\flat} = |(xy)^{(0)}|_E = |x^{(0)}y^{(0)}|_E = |x^{(0)}| \cdot |y^{(0)}|_E = |x|_{E^\flat}|y|_{E^\flat}.$$

One has

$$\begin{aligned}
|x+y|_{E^\flat} = |(x+y)^{(0)}|_E = |\lim_{m \to +\infty} (x^{(m)} + y^{(m)})^{p^m}|_E = \lim_{m \to +\infty} |x^{(m)} + y^{(m)}|_E^{p^m} \\
\leqslant \lim_{m \to +\infty} \max\{|x^{(m)}|_E, |x^{(m)}|_E\}^{p^m} = \lim_{m \to +\infty} \max\left\{\left|(x^{(m)})^{p^m}\right|_E, \left|(x^{(m)})^{p^m}\right|_E\right\} \\
= \max\left\{\left|(x^{(0)})\right|_E, \left|(x^{(0)})\right|_E\right\} = \max\left\{|x|_{E^\flat}, |y|_{E^\flat}\right\}.
\end{aligned}$$

This proves that $| \cdot |_{E^\flat}$ is an non-Archimedean absolute value.

(ii) We prove the completeness of $O_E^\flat$ (other properties follow easily from i) and properties 1-3) above. First remark that if $y = (y_0, y_1, \ldots) \in O_E^\flat$, then

$$y_n = 0 \quad \Leftrightarrow \quad |y|_{E^\flat} \leqslant |p|_E^{p^n}. \tag{33}$$

Let $(x_n)_{n \in \mathbf{N}}$ be a Cauchy sequence in $O_E^\flat$. Then for any $M > 0$, there exist N such that for all $n, m \geqslant N$

$$|x_n - x_m|_{E^\flat} \leqslant |p|_E^{p^M}.$$

Write $x_n = (x_{n,0}, x_{n,1}, \ldots)$ and $x_m = (x_{m,0}, x_{m,1}, \ldots)$. Using formula (33), we obtain that for all $n, m \geqslant N$

$$x_{n,i} = x_{m,i} \quad \text{for all} \quad 0 \leqslant i \leqslant M.$$

Hence, for each $i \geqslant 0$ the sequence $(x_{n,i})_{n \in \mathbf{N}}$ is stationary. Set $a_i = \lim_{n \to +\infty} x_{n,i}$. Then $a = (a_0, a_1, \dots) \in O_E^\flat$, and it is easy to check that $\lim_{n \to +\infty} x_n = a$. □

Definition The field $E^\flat$ will be called the tilt of E.

Proposition 5.2.4 *A perfectoid field E is algebraically closed if and only if $E^\flat$ is.*

Proof The proposition can be proved by successive approximation. We refer the reader to [60, Proposition 2.1.11] for the proof that $E^\flat$ is algebraically closed if E is and to [130, Proposition 3.8], and [60, Proposition 2.2.19, Corollary 3.1.10] for two different proofs of the converse statement. See also [23]. □

5.3 *The Ring* $\mathbf{A}_{\mathrm{inf}}(E)$

5.3.1 Let F be a perfect field, complete for a non-Archimedean absolute value $|\cdot|_F$. The ring of Witt vectors $W(F)$ is equipped with the p-adic (standard) topology defined in Sect. 1.5. Now we equip it with a coarser topology, which will be called the *canonical topology*. It is defined as the topology of the infinite direct product

$$W(F) = F^{\mathbf{N}},$$

where each F is equipped with the topology induced by the absolute value $|\cdot|_F$. For any ideal $\mathfrak{a} \subset O_F$ and integer $n \geqslant 0$, the set

$$U_{\mathfrak{a},n} = \{x = (x_0, x_1, \dots) \in W(F) \mid x_i \in \mathfrak{a} \quad \text{for all } 0 \leqslant i \leqslant n\}$$

is an ideal in $W(F)$. In the canonical topology, the family $(U_{\mathfrak{a},n})$ of these ideals form a base of the fundamental system of neighborhoods of 0.

5.3.2 Let E be a perfectoid field of characteristic 0. Set:

$$\mathbf{A}_{\mathrm{inf}}(E) := W(O_E^\flat).$$

Each element of $\mathbf{A}_{\mathrm{inf}}(E)$ is an infinite vector

$$a = (a_0, a_1, a_2, \dots), \qquad a_n \in O_E^\flat,$$

which also can be written in the form

$$a = \sum_{n=0}^{\infty} [a_n^{p^{-n}}] p^n.$$

Proposition 5.3.3 (Fontaine, Fargues–Fontaine) *(i) The map*

$$\theta_E : \mathbf{A}_{\mathrm{inf}}(E) \to O_E$$

given by

$$\theta_E \left(\sum_{n=0}^{\infty} [a_n] p^n \right) = \sum_{n=0}^{\infty} a_n^{(0)} p^n$$

is a surjective ring homomorphism.

(ii) $\ker(\theta_E)$ *is a principal ideal. An element* $\sum_{n=0}^{\infty}[a_n]p^n \in \ker(\theta_E)$ *is a generator of* $\ker(\theta_E)$ *if and only if* $|a_0|_{E^\flat} = |p|_E$.

Proof (i) For any ring A, set $W_n(A) = W(A)/I_n(A)$. From the definition of Witt vectors, it follows that for any $n \geqslant 0$, the map

$$\begin{aligned} &w_n : W_{n+1}(O_E) \to O_E, \\ &w_n(a_0, a_1, \ldots, a_n) = a_0^{p^n} + pa_1^{p^{n-1}} + \cdots + p^n a_n \end{aligned}$$

is a ring homomorphism. Consider the map:

$$\begin{aligned} &\eta_n : W_{n+1}(O_E/pO_E) \to O_E/p^{n+1}O_E, \\ &\eta_n(a_0, a_1, \ldots, a_n) = \widehat{a}_0^{p^n} + p\widehat{a}_1^{p^{n-1}} + \cdots + p^n\widehat{a}_n, \end{aligned}$$

where $\widehat{a}_i$ denotes any lift of a_i in O_E. It's easy to see that the definition of η_n does not depend on the choice of these lifts. Moreover, the diagram

$$\begin{array}{ccc} W_{n+1}(O_E) & \xrightarrow{w_n} & O_E \\ \downarrow & & \downarrow \\ W_{n+1}(O_E/pO_E) & \xrightarrow{\eta_n} & O_E/p^{n+1}O_E \end{array}$$

commutes by the functoriality of Witt vectors. This shows that η_n is a ring homomorphism. Let $\theta_{E,n} : W_{n+1}(O_E^\flat) \to O_E/p^{n+1}O_E$ denote the reduction of θ_E modulo p^{n+1}. From the definitions of our maps, it follows that $\theta_{E,n}$ coincides with the composition

$$W_{n+1}(O_E^\flat) \xrightarrow{\varphi^{-n}} W_{n+1}(O_E^\flat) \to W_{n+1}(O_E/pO_E) \xrightarrow{\eta_n} O_E/p^{n+1}O_E.$$

This proves that $\theta_{E,n}$ is a ring homomorphism for all $n \geqslant 0$. Therefore, θ_E is a ring homomorphism.

The surjectivity of θ_E follows from the surjectivity of the map

$$\theta_{E,0} \,:\, O_E^{\flat} \to O_E/pO_E.$$

(ii) We refer the reader to [66, Proposition 2.4] for the proof of the following statement: an element $\sum_{n=0}^{\infty}[a_n]p^n \in \ker(\theta_E)$ generates $\ker(\theta_E)$ if and only if $|a_0|_{E^\flat} = |p|_E$.

Since $|E^\flat| = |E|$, there exists $a_0 \in O_{E^\flat}$ such that $|a_0|_{E^\flat} = |p|_E$. Then $\theta_E([a_0])/p \in U_E$, and by the surjectivity of θ_E, there exists $b \in \mathbf{A}_{\mathrm{inf}}(E)$ such that $\theta_E(b) = \theta_E([a_0])/p$. Thus $x = [a_0] - pb \in \ker(\theta_E)$. Since $|a_0|_{E^\flat} = |p|_E$, the above criterion shows that x generates $\ker(\theta_E)$. See [60, Proposition 3.1.9] for further detail. □

5.4 The Tilting Equivalence

5.4.1 We continue to assume that E is a perfectoid field of characteristic 0. Fix an algebraic closure $\overline{E}$ of E and denote by $\mathbf{C}_E$ its completion. By Proposition 5.2.4, $\mathbf{C}_E^{\flat}$ is algebraically closed and we denote by $\overline{E^\flat}$ the algebraic closure of $E^\flat$ in $\mathbf{C}_E^{\flat}$. Let $\mathbf{C}_{E^\flat} := \widehat{\overline{E^\flat}}$ denote the p-adic completion of $\overline{E^\flat}$. We have the following picture, where the horizontal arrows denote the tilting:

$$\begin{array}{ccc} \mathbf{C}_E & \overset{\flat}{\rightsquigarrow} & \mathbf{C}_E^{\flat} \\ | & & | \\ E & \overset{\flat}{\rightsquigarrow} & E^{\flat}. \end{array}$$

Let $\mathfrak{F}$ be a complete intermediate field $E^\flat \subset \mathfrak{F} \subset \mathbf{C}_E^{\flat}$. Fix a generator ξ of $\ker(\theta_E)$. Set:

$$O_{\mathfrak{F}}^{\sharp} := \theta_{\mathbf{C}}(W(O_{\mathfrak{F}})),$$

where we write $\theta_{\mathbf{C}}$ instead $\theta_{\mathbf{C}_E}$ to simplify notation. Consider the diagram:

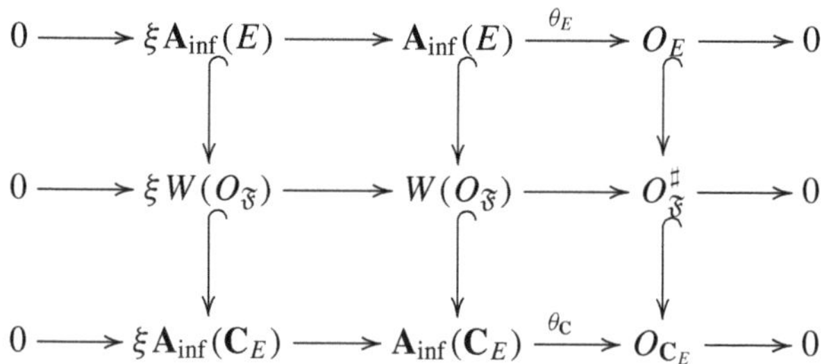

$$\begin{array}{ccccccccc} 0 & \longrightarrow & \xi\mathbf{A}_{\mathrm{inf}}(E) & \longrightarrow & \mathbf{A}_{\mathrm{inf}}(E) & \xrightarrow{\theta_E} & O_E & \longrightarrow & 0 \\ & & \downarrow & & \downarrow & & \downarrow & & \\ 0 & \longrightarrow & \xi W(O_{\mathfrak{F}}) & \longrightarrow & W(O_{\mathfrak{F}}) & \longrightarrow & O_{\mathfrak{F}}^{\sharp} & \longrightarrow & 0 \\ & & \downarrow & & \downarrow & & \downarrow & & \\ 0 & \longrightarrow & \xi\mathbf{A}_{\mathrm{inf}}(\mathbf{C}_E) & \longrightarrow & \mathbf{A}_{\mathrm{inf}}(\mathbf{C}_E) & \xrightarrow{\theta_{\mathbf{C}}} & O_{\mathbf{C}_E} & \longrightarrow & 0 \end{array}$$

Note that $O_{\mathfrak{F}}^{\sharp} = W(O_{\mathfrak{F}})/\xi W(O_{\mathfrak{F}})$. Set:

$$\mathfrak{F}^\sharp = O_{\mathfrak{F}}^\sharp[1/p].$$

Proposition 5.4.2 *$\mathfrak{F}^\sharp$ is a perfectoid field, and $(\mathfrak{F}^\sharp)^\flat = \mathfrak{F}$.*

Proof (a) First prove that $O_{\mathfrak{F}}^\sharp$ is complete. For each $n \geqslant 1$, we have an exact sequence

$$0 \to \xi W_n(O_{\mathfrak{F}}) \to W_n(O_{\mathfrak{F}}) \to O_{\mathfrak{F}}^\sharp/p^n O_{\mathfrak{F}}^\sharp \to 0,$$

where $W_n = W/p^n W$. Since the projection maps $W_{n+1}(O_{\mathfrak{F}}) \to W_n(O_{\mathfrak{F}})$ are surjective, the passage to inverse limits gives an exact sequence

$$0 \to \xi W(O_{\mathfrak{F}}) \to W(O_{\mathfrak{F}}) \to \varprojlim_n O_{\mathfrak{F}}^\sharp/p^n O_{\mathfrak{F}}^\sharp \to 0.$$

Hence, $O_{\mathfrak{F}}^\sharp = \varprojlim_n O_{\mathfrak{F}}^\sharp/p^n O_{\mathfrak{F}}^\sharp$, and $O_{\mathfrak{F}}^\sharp$ is complete.

(b) Fix a valuation v_E on E. We prove that for any $x \in W(O_{\mathfrak{F}})$,

$$v_E(\theta_{\mathrm{C}}(x)) \geqslant n \cdot v_E(p) \Rightarrow x \in p^n W(O_{\mathfrak{F}}) + \xi W(O_{\mathfrak{F}}).$$

It's sufficient to prove this assertion for $n = 1$. Let $x = \sum_{k=0}^{\infty}[x_k]p^k$ be such that $v_E(\theta_{\mathrm{C}}(x)) \geqslant v_E(p)$. If $x_0 = 0$, the assertion is clearly true. Assume that $x_0 \neq 0$. Then $v_E(x_0^{(0)}) \geqslant v_E(p)$. By Proposition 5.3.3, $\xi = \sum_{k=0}^{\infty}[a_k]p^k$ with $v_E(a_0^{(0)}) = v_E(p)$. Hence,

$$x_0 = a_0 y, \qquad \text{for some } y \in O_{\mathfrak{F}},$$

and

$$x = \xi[y] + pz, \qquad \text{for some } z \in W(O_{\mathfrak{F}}).$$

This shows that $x \in pW(O_{\mathfrak{F}}) + \xi W(O_{\mathfrak{F}})$.

(c) Assume that $\alpha \in \mathfrak{F}^\sharp$ belongs to the valuation ring of $\mathfrak{F}^\sharp$. Write $\alpha = \beta/p^n$ with $\beta = \theta_{\mathrm{C}}(x)$, $x \in W(O_{\mathfrak{F}})$. Then $v_E(\theta_{\mathrm{C}}(x)) \geqslant n \cdot v_E(p)$. By part b), there exists $y \in W(O_{\mathfrak{F}})$ such that $\theta_{\mathrm{C}}(x) = p^n\theta_{\mathrm{C}}(y)$. Therefore, $\alpha = \theta_{\mathrm{C}}(y) \in O_{\mathfrak{F}}^\sharp$. This proves that $O_{\mathfrak{F}}^\sharp$ is the valuation ring of $\mathfrak{F}^\sharp$.

(d) From (a) and (c), it follows that $\mathfrak{F}^\sharp$ is a complete field with the valuation ring $O_{\mathfrak{F}}^\sharp$. In addition, the induced valuation on $\mathfrak{F}^\sharp$ is clearly non-discrete. Writing ξ in the form $\xi = \sum_{k=0}^{\infty}[a_k]p^k$, we see that

$$O_{\mathfrak{F}}^\sharp/pO_{\mathfrak{F}}^\sharp \simeq O_{\mathfrak{F}}/a_0 O_{\mathfrak{F}}.$$

This implies that $\mathfrak{F}^\sharp$ is a perfectoid field. Moreover, it is easy to see that the map

$$O_{\mathfrak{F}} \to \varprojlim_{\varphi} O_{\mathfrak{F}}/a_0 O_{\mathfrak{F}}, \qquad z \mapsto \left(\varphi^{-n}(z) \mod (a_0 O_{\mathfrak{F}})\right)_{n \geqslant 0}$$

is an isomorphism. Therefore, $(\mathfrak{F}^\sharp)^\flat = \mathfrak{F}$, and the proposition is proved. □

Proposition 5.4.3 *One has* $\mathbf{C}_E^\flat = \mathbf{C}_{E^\flat}$.

Proof Since $E^\flat \subset \mathbf{C}_E^\flat$, and $\mathbf{C}_E^\flat$ is complete and algebraically closed, we have $\mathbf{C}_{E^\flat} \subset \mathbf{C}_E^\flat$. Set $\mathfrak{F} := \mathbf{C}_{E^\flat}$. By Proposition 5.4.2, $(\mathfrak{F}^\sharp)^\flat = \mathfrak{F}$. Since $\mathfrak{F}$ is complete and algebraically closed, $\mathfrak{F}^\sharp$ is complete and algebraically closed by Proposition 5.2.4. Now from $\mathfrak{F}^\sharp \subset \mathbf{C}_E$, we deduce that $\mathfrak{F}^\sharp = \mathbf{C}_E$. Therefore

$$\mathfrak{F} = (\mathfrak{F}^\sharp)^\flat = \mathbf{C}_E^\flat.$$

The proposition is proved. □

Now we can prove the main result of this section.

Theorem 5.4.4 (Scholze, Fargues–Fontaine) *Let E be a perfectoid field of characteristic* 0. *Then the following holds true:*

(i) One has $G_E \simeq G_{E^\flat}$.

(ii) Each finite extension of E is a perfectoid field.

(iii) The tilt functor $F \mapsto F^\flat$ realizes the Galois correspondence between the categories of finite extensions of E and $E^\flat$ respectively.

(iv) The functor

$$\mathfrak{F} \mapsto \mathfrak{F}^\sharp, \qquad \mathfrak{F}^\sharp := \left(W(O_{\mathfrak{F}})/\xi W(O_{\mathfrak{F}})\right)[1/p]$$

is a quasi-inverse to the tilt functor.

Proof The proof below is due to Fargues and Fontaine [60, Theorem 3.2.1].

(a) We prove assertion (i). The Galois group $G_E = \mathrm{Gal}(\overline{E}/E)$ acts on $\mathbf{C}_E$ and $\mathbf{C}_E^\flat$. To simplify notation, set $\mathbf{F} = \mathbf{C}_{E^\flat}$. By Proposition 5.4.3, $\mathbf{C}_E^\flat = \mathbf{F}$, and we have a map

$$G_E \to \mathrm{Aut}(\mathbf{C}_E^\flat/E^\flat) \xrightarrow{\sim} \mathrm{Aut}(\mathbf{F}/E^\flat) \xrightarrow{\sim} \mathrm{Aut}(\overline{E^\flat}/E^\flat) = G_{E^\flat}. \tag{34}$$

Conversely, again by Proposition 5.4.3, we have an isomorphism

$$W(O_{\mathbf{F}})/\xi W(O_{\mathbf{F}}) \simeq O_{\mathbf{C}_E}, \tag{35}$$

which induces a map

$$G_{E^\flat} \xrightarrow{\sim} \mathrm{Aut}(\mathbf{F}/E^\flat) \to \mathrm{Aut}(\mathbf{C}_E/E) \xrightarrow{\sim} G_E.$$

It is easy to see that the maps (34) and (35) are inverse to each other. Therefore

$$G_E \simeq G_{E^\flat},$$

and by Galois theory we have a one-to-one correspondence

$$\{\text{finite extensions of } E\} \leftrightarrow \{\text{finite extensions of } E^\flat\}. \tag{36}$$

(b) Let $\mathfrak{F}/E^\flat$ be a finite extension. By Proposition 5.4.2, $\mathfrak{F}^\sharp$ is a perfectoid field, and

$$(\mathfrak{F}^\sharp)^\flat = \mathfrak{F}.$$

Consider the exact sequence:

$$0 \to W(O_{\mathbf{F}})[1/p] \xrightarrow{\xi} W(O_{\mathbf{F}})[1/p] \to W(O_{\mathbf{F}})/\xi W(O_{\mathbf{F}})[1/p] \to 0.$$

By Corollary 4.2.11 (Ax–Sen–Tate in characteristic p), one has

$$H^0(G_{\mathfrak{F}}, W(O_{\mathbf{F}})) = W(O_{\mathfrak{F}}).$$

By the same corollary, $\mathfrak{m}_{\mathfrak{F}} \cdot H^1(G_{\mathfrak{F}}, O_{\mathbf{F}}) = 0$. Using successive approximation, one verifies that $[a] \cdot H^1(G_{\mathfrak{F}}, W(O_{\mathbf{F}})) = 0$ for any $a \in \mathfrak{m}_{\mathfrak{F}}$. The generator $\xi \in \ker(\theta_E)$ can be written in the form $\xi = [a] + pu$, where $a \in \mathfrak{m}_{E^\flat}$ and u is invertible in $\mathbf{A}_{\mathrm{inf}}(E)$. If

$$f \in \ker\left(H^1(G_{\mathfrak{F}}, W(O_{\mathbf{F}}))[1/p] \xrightarrow{\xi} H^1(G_{\mathfrak{F}}, W(O_{\mathbf{F}}))[1/p]\right),$$

then $[a]f = 0$, $\xi f = 0$, and therefore $f = 0$. Hence,

$$\ker\big(H^1(G_{\mathfrak{F}}, W(O_{\mathbf{F}}))[1/p] \xrightarrow{\xi} H^1(G_{\mathfrak{F}}, W(O_{\mathbf{F}}))[1/p]\big) = 0.$$

Therefore, the long exact sequence of cohomology associated to the above short exact sequence gives an isomorphism:

$$(W(O_{\mathbf{F}})/\xi W(O_{\mathbf{F}})[1/p])^{G_{\mathfrak{F}}} \simeq W(O_{\mathfrak{F}})/\xi W(O_{\mathfrak{F}})[1/p].$$

The isomorphism $G_E \simeq G_{E^\flat}$ identifies $G_{\mathfrak{F}}$ with an open subgroup of G_E. By Theorem 4.3.2 (Ax–Sen–Tate in characteristic 0), $\mathbf{C}_E^{G_{\mathfrak{F}}} \simeq (\overline{E})^{G_{\mathfrak{F}}}$. Since

$$\mathbf{C}_E \simeq (W(O_{\mathbf{F}})/\xi W(O_{\mathbf{F}}))\,[1/p],$$

one has

$$\overline{E}^{G_{\mathfrak{F}}} \simeq W(O_{\mathfrak{F}})/\xi W(O_{\mathfrak{F}})[1/p] =: \mathfrak{F}^\sharp.$$

We have proved that the Galois correspondence (36) associates to $\mathfrak{F}/E^\flat$ the extension $\mathfrak{F}^\sharp/E$.

(c) Conversely, let F be a finite extension of E. Set $\mathfrak{F} = \left(\overline{E^\flat}\right)^{G_F}$. From part b), it follows that $F = \mathfrak{F}^\sharp$. Applying Proposition 5.4.2, we obtain that F is a perfectoid field and that $F^\flat = \left(\mathfrak{F}^\sharp\right)^\flat = \mathfrak{F}$. This concludes the proof of the theorem. □

Remark 5.4.5 For the theory of almost étale extensions in the geometric setting and Scholze's theory of perfectoid algebras, we refer the reader to [59, 78] and [130]. See also [95]. In another direction, further development of these ideas led to the theory of diamonds [132], closely related to the theory of Fargues–Fontaine [60].

6 From Characteristic 0 to Characteristic p and Vice Versa II: The Field of Norms

6.1 Arithmetically Profinite Extensions

6.1.1 In this section, we review the theory of the arithmetically profinite extensions and the field of norms construction of Fontaine–Wintenberger [161]. Let K be a local field of characteristic 0 with residue field of characteristic p.

Definition An algebraic extension L/K is called arithmetically profinite (APF) if and only if

$$(G_K : G_K^{(v)} G_L) < +\infty, \qquad \forall v \geqslant -1.$$

If L/K is a Galois extension with $G = \mathrm{Gal}(L/K)$, then it is APF if and only if

$$(G : G^{(v)}) < +\infty, \qquad \forall v \geqslant -1.$$

It is clear that any finite extension is APF. Below, we give some archetypical examples of APF extensions.

Example 6.1.2 (1) Any totally ramified $\mathbf{Z}_p$-extension is APF (see Sect. 3.1).

(2) The p-cyclotomic extension $K(\zeta_{p^\infty})/K$ is APF. This easily follows from the fact that $K(\zeta_{p^\infty})/K(\zeta_p)$ is a totally ramified $\mathbf{Z}_p$-extension. See also Proposition 6.1.10 below.

(3) Let π be a fixed uniformizer of K, and let K_π be the maximal abelian extension of K such that π is a universal norm in K_π, namely that

$$\pi \in N_{F/K}(F^*), \qquad \text{for all } K \subset F \subset K_\pi.$$

By local class field theory, K_π/K is totally ramified and one has

$$\mathrm{Gal}(K_\pi/K)^{(v)} \simeq U_K^{(v)}, \qquad \forall v \geqslant 0.$$

Therefore, K_π/K is APF.

(4) More generally, from Sen's Theorem 1.3.11 it follows that any totally ramified p-adic Lie extension is APF. The converse is false in general (see [61] for examples).

(5) Let π be a fixed uniformizer of K. The associated Kummer extension $K(\sqrt[p^\infty]{\pi})$ is an APF extension, which is not Galois. This can be proved by showing first that the Galois extension $K(\zeta_{p^\infty}, \sqrt[p^\infty]{\pi})$ is APF. The last assertion can be either proved by a direct computation or deduced from Sen's theorem. The extension $K(\sqrt[p^\infty]{\pi})$ plays a key role in Abrashkin's approach to the ramification filtration [4, 5, 7] and in integral p-adic Hodge theory [29, 33, 97].

6.1.3 We analyze the ramification jumps of APF extensions. First we extend the definition of a ramification jump to general (not necessarily Galois) extensions.

Definition Let L/K be an algebraic extension. A real number $v \geqslant -1$ is a ramification jump of L/K if and only if

$$G_K^{(v+\varepsilon)}G_L \neq G_K^{(v)}G_L \qquad \forall \varepsilon > 0.$$

If L/K is a Galois extension, this definition coincides with Definition in Sect. 1.3.

Proposition 6.1.4 *Let L/K be an infinite APF extension, and let B denote the set of ramification jumps of K. Then B is a countably infinite unbounded set.*

Proof (a) Let L/K be an APF extension. First we prove that B is discrete. Let $v_2 > v_1 \geqslant -1$ be two ramification jumps. Then

$$(G_K : G_K^{(v_1)}G_L) \leqslant (G_K : G_K^{(v_2)}G_L) < +\infty,$$

and

$$(G_K^{(v_1)}G_L : G_K^{(v_2)}G_L) < +\infty.$$

Therefore, there exists only finitely many subgroups H such that

$$G_K^{(v_2)}G_L \subset H \subset G_K^{(v_1)}G_L.$$

This implies that there are only finitely many ramification jumps in the interval (v_1, v_2).

(b) Assume that B is bounded above by a. Then $G_L G_K^{(a)} = \bigcap\limits_{t\geqslant 0} G_L G_K^{(a+t)}$. Let $g \in G_L G_K^{(a)}$. Then for any $n \geqslant 0$, we can write $g = x_n y_n$ with $x_n \in G_L$ and $y_n \in G_K^{(a+n)}$. Since G_L is compact, we can assume that $(x_n)_{n\geqslant 0}$ converges. Hence, $(y_n)_{n\geqslant 0}$ converges to some $y \in \bigcap\limits_{n\geqslant 0} G_K^{(a+n)}$. From $\bigcap\limits_{n\geqslant 0} G_K^{(a+n)} = \{1\}$, we obtain that $g \in G_L$. This shows that $G_L G_K^{(a)} = G_L$. Therefore

$$(G_K : G_L G_K^{(a)}) = (G_K : G_L) = +\infty,$$

which is in contradiction with the definition of APF extensions. $\square$

6.1.5 Let L/K be an infinite APF extension. We denote by $B^+ = (b_n)_{n\geqslant 1}$ the set of its *strictly positive* ramification jumps. For all $n \geqslant 1$, set:

$$K_n = \overline{K}^{G_L G_K^{(b_n)}}.$$

Proposition 6.1.6 *The following statements hold true:*

(i) $L = \bigcup\limits_{n=1}^{\infty} K_n$.

(ii) K_1 *is the maximal tamely ramified subextension of* L/K.

(iii) For all $n \geqslant 1$, K_{n+1}/K_n *is a non-trivial finite* p*-extension.*

(iv) Assume that L/K *is a Galois extension. Then for all* $n \geqslant 1$, *the group* $\mathrm{Gal}(K_{n+1}/K_n)$ *has a unique ramification jump. In particular,* $\mathrm{Gal}(K_{n+1}/K_n)$ *is a* p*-elementary abelian group.*

Proof We prove assertion (ii). The maximal tamely ramified subextension of L/K is

$$L_{\mathrm{tr}} = \overline{K}^{G_L P_K},$$

where P_K is the wild ramification subgroup. From the definition of the ramification filtration, it follows that P_K is the topological closure of $\bigcup\limits_{v>0} G_K^{(v)}$ in G_K. This implies that $G_L P_K = G_L G_K^{(b_1)}$, and ii) is proved.

The assertions (i), (iii) and (iv) are clear. □

Corollary 6.1.7 *An infinite APF extension is deeply ramified.*

Proof Proposition 6.1.6 shows that such extension does not have finite conductor. □

Remark 6.1.8 The converse of this corollary is clearly wrong. However Fesenko [61] proved that every deeply ramified extension L/K of finite residue degree and with discrete set of ramification jumps is APF.

6.1.9 We record some general properties of APF extensions.

Proposition 6.1.10 *Let* $K \subset F \subset L$ *be a tower of extensions.*

(i) If F/K *is APF and* L/F *is finite, then* L/K *is APF.*

(ii) If F/K *is finite and* L/F *is APF, then* L/K *is APF.*

(iii) If L/K *is APF, then* F/K *is APF.*

Proof See [161, Proposition 1.2.3]. □

6.1.11 The definition of Hasse–Herbrand functions can be extended to APF extensions. Namely, for an APF extension L/K, set:

$$\psi_{L/K}(v) = \begin{cases} v, & \text{if } v \in [-1, 0], \\ \displaystyle\int_0^v (G_K^{(0)} : G_L^{(0)} G_K^{(t)})dt, & \text{if } v \geqslant 0, \end{cases}$$

$$\varphi_{L/K}(u) = \psi_{L/K}^{-1}(u).$$

It is not difficult to check that if $K \subset F \subset L$ with $[F : K] < +\infty$, then one has

$$\psi_{L/K} = \psi_{L/F} \circ \psi_{F/K}, \qquad \varphi_{L/K} = \varphi_{F/K} \circ \varphi_{L/F}.$$

6.2 The Field of Norms

6.2.1 In this section, we review the construction of the field of norms of an APF extension. Let $\mathcal{E}(L/K)$ denote the directed set of finite subextensions of L/K.

Theorem 6.2.2 (Fontaine–Wintenberger) *Let L/F be an infinite APF extension. Set:*

$$\mathscr{X}(L/K) = \varprojlim_{E \in \mathcal{E}(L/K)} E^* \cup \{0\}.$$

Then the following assertions hold true:

(i) Let $\alpha = (\alpha_E)_{E \in \mathcal{E}(L/K)}$ and $\beta = (\beta_E)_{E \in \mathcal{E}(L/K)}$. Set:

$$(\alpha\beta)_E := \alpha_E \beta_E,$$
$$(\alpha + \beta)_E := \lim_{E' \in \mathcal{E}(L/E)} N_{E'/E}(\alpha_{E'} + \beta_{E'}).$$

Then $\alpha\beta := ((\alpha\beta)_E)_{E \in \mathcal{E}(L/K)}$ and $\alpha + \beta := ((\alpha + \beta)_E)_{E \in \mathcal{E}(L/K)}$ are well-defined elements of $\mathscr{X}(L/K)$.

(ii) The above defined addition and multiplication equip $\mathscr{X}(L/K)$ with a structure of a local field of characteristic p with residue field k_L.

(iii) The valuation on $\mathscr{X}(L/K)$ is given by

$$v(\alpha) = v_E(\alpha_E),$$

for any $K_1 \subset E \subset L$. Here K_1 denotes the maximal unramified subextension of L/K.

(iv) For any $\xi \in k_L$, let $[\xi]$ denote its Teichmüller lift. For each $K_1 \subset E \subset L$ set:

$$\xi_E := [\xi]^{1/[E:K_1]}.$$

Then the map

$$k_L \to \mathscr{X}(L/K), \qquad \xi \mapsto (\xi_E)_{E \in \mathcal{E}(L/K_1)}$$

is a canonical embedding.

The proof occupies the remainder of this section. See [161, Sect. 2] for detail.

Definition The field $\mathscr{X}(L/K)$ is called the field of norms of the APF extension L/K.

6.2.3 We start by writing Theorem 6.2.2 in a slightly different form, which also makes more clear its relation to the theory of perfectoid fields.

For any APF extension E/F (finite or infinite), set:

$$i(E/F) = \sup\{v \mid G_E G_F^{(v)} = G_F\}$$

If $E \subset E' \subset E''$ is a tower of finite extensions, then the relation $\psi_{E''/E} = \psi_{E''/E'} \circ \psi_{E'/E}$ implies that

$$i(E''/E) \leqslant \min\{i(E'/E), i(E''/E')\}. \tag{37}$$

Let $B = (b_n)_{n\geqslant 0}$ denote the set of ramification jumps of L/K and let $K_n = \overline{K}^{G_L G_K^{(b_n)}}$. Since

$$\psi_{L/K}(v) = \psi_{K_n/K}(v), \qquad \forall v \in [-1, b_n],$$

from $\psi_{L/K} = \psi_{L/K_n} \circ \psi_{K_n/K}$, it follows that $\psi_{L/K_n}(v) = v$ for $v \in [-1, \psi_{L/K}(b_n)]$, and $\psi_{L/K_n}(v) \neq v$ for $v > \psi_{L/K}(b_n)$. Therefore

$$i(L/K_n) = \psi_{L/K}(b_n), \qquad n \geqslant 1. \tag{38}$$

In particular, $i(L/K_n) \to +\infty$ when $n \to +\infty$.

6.2.4 For any $E \in \mathcal{E}(L/K_1)$, set:

$$r(E) := \text{smallest integer } \geqslant \frac{(p-1)i(L/E)}{p},$$

and

$$\overline{O}_E := O_E/\mathfrak{m}_E^{r(E)}.$$

Theorem 6.2.5 *Let L/K be an infinite APF extension. Then:*

(i) For all finite subextensions $E \subset E'$ of L/K, the norm map induces a ring homomorphism

$$N_{E'/E} : \overline{O}_{E'} \to \overline{O}_E.$$

(ii) The projective limit

$$A(L/K) := \varprojlim_{E \in \mathcal{E}(L/K_1)} \overline{O}_E$$

is a discrete valuation ring of characteristic p with residue field k_L.

(iii) The map

$$k_L \to A(L/K), \qquad \xi \mapsto \left(\xi_E \mod \mathfrak{m}_E^{r(E)}\right)_{E \in \mathcal{E}(L/K_1)}, \qquad \xi_E = [\xi]^{1/[E:K_1]}$$

is a canonical embedding.

6.2.6 The proof of Theorem 6.2.5 relies on the following proposition:

Proposition 6.2.7 *Let E'/E be a finite totally ramified p-extension. Then*
(i) For all $\alpha, \beta \in O_{E'}$,

$$v_E(N_{E'/E}(\alpha+\beta) - N_{E'/E}(\alpha) - N_{E'/E}(\beta)) \geqslant \frac{(p-1)i(E'/E)}{p}.$$

(ii) For any $a \in O_E$, there exists $\alpha \in O_{E'}$ such that

$$v_E(N_{E'/E}(\alpha) - a) \geqslant \frac{(p-1)i(E'/E)}{p}.$$

Proof (a) Assume first that E'/E is a Galois extension of degree p. From Corollary 1.4.5 it follows that for any $x \in O_{E'}$, one has

$$v_E(N_{E'/E}(1+x) - 1 - N_{E'/E}(x)) \geqslant \frac{(p-1)i(E'/E)}{p}.$$

Assume that $v_{E'}(\alpha) \geqslant v_{E'}(\beta)$. Setting $x = \alpha/\beta$, we obtain i).

Let $\pi_{E'}$ be any uniformizer of E'. Set $\pi_E = N_{E'/E}(\pi_{E'})$. Write $a \in O_E$ in the form:

$$a = \sum_{k=0}^{p-1} [\xi_k]\pi_E^k, \qquad \xi_k \in k_E.$$

Then again by Lemma 1.4.5, we have:

$$v_E(N_{E'/E}(\alpha) - a) \geqslant \frac{(p-1)i(E'/E)}{p} \qquad \text{for } \alpha = \sum_{k=0}^{p-1} [\xi_k]^{1/p}\pi_{E'}^k.$$

Therefore, the proposition is proved for Galois extensions of degree p.

(b) Assume that the proposition holds for finite extensions E''/E' and E'/E. Then for $\alpha, \beta \in O_{E''}$ we have:

$$N_{E''/E'}(\alpha+\beta) = N_{E''/E'}(\alpha) + N_{E''/E'}(\beta) + \gamma,$$

and

$$N_{E''/E}(\alpha+\beta) = N_{E''/E}(\alpha) + N_{E''/E}(\beta) + N_{E'/E}(\gamma) + \delta,$$

where $v_{E'}(\gamma) \geqslant \frac{(p-1)i(E''/E')}{p}$ and $v_E(\delta) \geqslant \frac{(p-1)i(E'/E)}{p}$. Since E'/E is totally ramified, one has $v_E(N_{E'/E}(\gamma)) \geqslant \frac{(p-1)i(E''/E')}{p}$, and from (37) it follows that

$$v_E(N_{E''/E}(\alpha+\beta) - N_{E''/E}(\alpha) - N_{E''/E}(\beta)) \geqslant \frac{(p-1)i(E''/E)}{p}.$$

Therefore, the proposition holds for all finite p-extensions.

(c) The general case can be reduced to the case b) by passing to the Galois closure of E'. See [161, Sect. 2.2.2.5] for detail. □

6.2.8 *Sketch of Proof of Theorem* 6.2.5. From Proposition 6.2.7, if follows that $A(L/K)$ is a commutative ring. Let $x = (x_E)_E \in A(L/K)$. If $x \neq 0$, the there exists $E \in \mathcal{E}(L/K_1)$ such that $x_E \neq 0$. For any $E' \in \mathcal{E}(L/E)$, let $\widehat{x}_{E'} \in O_{E'}$ be a lift of $x_{E'}$. Then $v(x) := v_{E'}(\widehat{x}_{E'})$ does not depend on the choice of E' and defines a discrete valuation of $A(L/K)$. It is easy to see that the topology defined by this valuation coincides with the topology of the projective limit of discrete sets on $A(L/K)$. Hence, $A(L/K)$ is complete. Lemma 6.2.9 below shows that the element $x = (x_E)_{E\in\mathcal{E}(L/K_1)}$, with $x_E = p \mod \mathfrak{m}_E^{r(E)}$ for all E, is zero in $A(L/E)$. Therefore, $A(L/E)$ is a ring of characteristic p. For all $\xi_1, \xi_2 \in k_L$, the congruence $[\xi_1 + \xi_2] \equiv [\xi_1] + [\xi_2] \pmod{p}$ together with Lemma 6.2.9 imply that the map

$$k_L \to A(L/K), \qquad \xi \mapsto (\xi_E \mod \pi_E^{r(E)})_{E\in\mathcal{E}(L/K_1)}, \qquad \xi_E = [\xi]^{1/[E:K_1]}$$

is an embedding of fields. Finally, from the definition of the valuation on $A(L/K)$, we see that its residue field is isomorphic to k_L. Theorem 6.2.5 is proved. □

Lemma 6.2.9 *Let L/E be a totally ramified APF pro-p-extension. Then*

$$v_E(p) \geqslant \frac{(p-1)i(L/E)}{p}.$$

Proof First assume that F/E is a Galois extension of degree p. From elementary properties of the ramification filtration, it follows that $G_i = \{1\}$ for all $i > \frac{e_F}{p-1}$, where e_F is the absolute ramification index of F (see [142, Exercise 3, p. 79]). This implies that $v_E(p) \geqslant \frac{(p-1)i(F/E)}{p}$ for such extensions.

Now we consider the general case. Take the Galois closure M of L over E and denote by M_1/E its maximal tamely ramified subextension. It is clear that M_1/E is linearly disjoint with L/E. From Galois theory, it follows that LM_1/M_1 has a Galois subextension F of degree p over M_1. Then the inequality (37) implies that

$$v_E(p) \geqslant \frac{(p-1)i(F/M_1)}{p} \geqslant \frac{(p-1)i(LM_1/M_1)}{p}.$$

Since the extensions M_1/E and LM_1/L are tamely ramified, from $\psi_{LM_1/M_1} \circ \psi_{M_1/E} = \psi_{LM_1/L} \circ \psi_{L/E}$ it follows that $i(LM_1/M_1) = i(F/E)$. The lemma is proved. □

6.2.10 *Sketch of Proof of Theorem* 6.2.2. We will use repeatedly the following inequality: if F/E is a totally ramified p-extension, then for all $x, y \in O_F$, one has

$$v_E(N_{F/E}(x) - N_{F/E}(y)) \geqslant \varphi_{F/E}(t), \qquad \text{if } v_F(x-y) \geqslant t. \tag{39}$$

This estimation can be proved by induction using Corollary 1.4.5. See [142, Chap. V, §6] for the Galois case. The general case can be treated by passing to the Galois closure.

Let $\alpha = (\alpha_E)_{E\in\mathcal{E}(L/K)}$ and $\beta = (\beta_E)_{E\in\mathcal{E}(L/K)} \in \varprojlim_{E\in\mathcal{E}(L/K)} O_E$. From Proposition 6.2.7 and formula (39), it follows that for all intermediate finite subextensions $K \subset E \subset E' \subset E'' \subset L$ one has

$$v_E\left(N_{E''/E}(\alpha_{E''}+\beta_{E''}) - N_{E'/E}(\alpha_{E'}+\beta_{E'})\right) \geqslant \varphi_{E'/E}(r(E')) \geqslant \varphi_{L/K}(r(E')).$$

Since $r(E') \to +\infty$ when E' runs over $\mathcal{E}(L/E)$, this proves the existence of the limit

$$(\alpha+\beta)_E := \lim_{E'\in\mathcal{E}(L/E)} N_{E'/E}(\alpha_{E'}+\beta_{E'}).$$

Therefore, the addition and the multiplication on $\mathscr{X}(L/E)$ are well defined.

Consider the map

$$\varprojlim_{E\in\mathcal{E}(L/K)} O_E \to A(L/K), \qquad (\alpha_E)_{E\in\mathcal{E}(L/K)} \mapsto (\overline{\alpha}_E)_{E\in\mathcal{E}(L/K_1)}, \tag{40}$$

where $\overline{\alpha}_E = \alpha_E \mod \mathfrak{m}_E^{r(E)}$. Proposition 6.2.7 shows that this map is compatible with the addition and the multiplication on the both sets.

Now let $x = (x_E)_E \subset A(L/K)$. For all E, choose a lift $\widehat{x}_E \in O_E$. Applying again the inequality (39), we see that for all E, the sequence $N_{E'/E}(\widehat{x}_{E'})$ converges to some $\alpha_E \in O_E$. From our constructions, it follows that the map

$$A(L/K) \to \varprojlim_{E\in\mathcal{E}(L/K_1)} O_E, \qquad x \mapsto (\alpha_E)_{E\in\mathcal{E}(L/K_1)}$$

is the inverse of the map (40). Now the theorem follows from Theorem 6.2.5. □

6.3 Functorial Properties

6.3.1 In this section, L/K denotes an infinite APF extension. Any finite extension M of L can be written as $M = L(\alpha)$, where α is a root of an irreducible polynomial $f(X) \in L[X]$. The coefficients of $f(X)$ belong to some finite subextension $F \in \mathcal{E}(L/K)$. For any $E \in \mathcal{E}(L/F)$, one has

$$F(\alpha) \cap E = F,$$

and we set:

$$E' = E(\alpha).$$

The system $(E')_{E\in\mathcal{E}(L/K)}$ is cofinal in $\mathcal{E}(M/K)$. Consider the map

$$j_{M/L} : \mathscr{X}(L/K) \to \mathscr{X}(M/K)$$

which sends any $\alpha = (\alpha_E)_{E\in\mathcal{E}(L/K)} \in \mathscr{X}(L/K)$ to the element $\beta = (\beta_{E'})_{E'\in\mathcal{E}(M/K)} \in \mathscr{X}(M/K)$ defined by

$$\beta_{E'} = \alpha_E \qquad \text{if } E' = E(\alpha) \text{ with } E \in \mathcal{E}(L/F).$$

The previous remarks show that $j_{M/L}$ is a well-defined embedding.

The following theorem should be compared with Theorem 5.4.4.

Theorem 6.3.2 (Fontaine–Wintenberger) *(i) Let M/L be a finite extension. Then $\mathscr{X}(M/K)/\mathscr{X}(L/K)$ is a separable extension of degree $[M : L]$. If M/L is a Galois extension, then the natural action of* Gal(M/L) *on $\mathscr{X}(M/L)$ induces an isomorphism*

$$\mathrm{Gal}(M/L) \simeq \mathrm{Gal}\left(\mathscr{X}(M/K)/\mathscr{X}(L/K)\right).$$

(ii) The above construction establishes a one-to-one correspondence

$$\{\textit{finite extensions of } L\} \leftrightarrow \{\textit{finite separable extensions of } \mathscr{X}(L/K)\},$$

which is compatible with the Galois correspondence.

Proof We only explain how to associate to any finite separable extension $\mathscr{M}$ of $\mathscr{X}(L/K)$ a canonical finite extension M of L of the same degree. Let $\mathscr{M} = \mathscr{X}(L/K)(\alpha)$, where α is a root of an irreducible polynomial $f(X)$ with coefficients in the ring of integers of $\mathscr{X}(L/K)$. We can write $f(X)$ as a sequence $f(X) = (f_E(X))_{E\in\mathcal{E}(L/K)}$, where $f_E(X) \in E[X]$. Then $M = L(\widehat{\alpha})$, where $\widehat{\alpha}$ is a root of $f_E(X)$, and E is of "sufficiently big" degree over K. See [161, Sect. 3] for a detailed proof. □

6.3.3 From this theorem, it follows that the separable closure $\overline{\mathscr{X}(L/K)}$ of $\mathscr{X}(L/K)$ can de written as

$$\overline{\mathscr{X}(L/K)} = \bigcup_{[M:L]<\infty} \mathscr{X}(M/K).$$

Corollary 6.3.4 *The field of norms functor induces a canonical isomorphism of absolute Galois groups:*

$$G_{\mathscr{X}(L/K)} \simeq G_L.$$

6.3.5 Let L/K be an infinite totally ramified Galois APF extension. The Galois group Gal(L/K) acts naturally on $\mathscr{X}(L/K)$. Fixing an uniformizer of $\mathscr{X}(L/K)$, we idenfify $\mathscr{X}(L/K)$ with the local field $k_K((x))$ of Laurent power series. Let τ be an automorphism of $k_K((x))$. If τ acts trivially on k_K, then it is completely determined by the power series $\tau(x) = a_1x + a_2x^2 + \cdots \in k_K[[x]]$ with $a_1 \neq 0$. Consider the group of formal power series

$$\mathrm{Aut}\left(k_K((x))\right) = \left\{f(x) = \sum_{i=1}^{\infty} a_i x^i \mid a_1 \neq 0\right\}$$

with respect to the substitution group law $f \circ g(x) = f(g(x))$. We have an injective map

$$\mathrm{Gal}(L/K) \hookrightarrow \mathrm{Aut}\left(k_K((x))\right). \tag{41}$$

This map encodes important information about the ramification filtration on $\mathrm{Gal}(L/K)$. Recall that for any automorphism g of a local field E we defined:

$$i_E(g) = v_E(g(\pi_E) - \pi_E).$$

Now we define this function on the infinite level, setting:

$$i_x(g) = \mathrm{ord}_x(g(x) - x), \qquad g \in \mathrm{Gal}(L/K).$$

Then there exists $F \in \mathcal{E}(L/K)$ such that for any Galois extension $E \in \mathcal{E}(L/F)$, one has

$$i_E(g) = i_x(g)$$

(see [161, Proposition 3.3.2]).

6.3.6 The map (41) can be described explicitly for cyclotomic extensions of unramified local fields. Assume that K is unramified, and set $K_\infty = K(\zeta_{p^\infty})$. Let $\Gamma_K = \mathrm{Gal}(K_\infty/K)$. The action of Γ_K on K_∞ is given by the cyclotomic character:

$$\chi_K : \Gamma_K \to \mathbf{Z}_p^*, \qquad \tau(\zeta_{p^n}) = \zeta_{p^n}^{\chi_K(\tau)}, \qquad \tau \in \Gamma_K.$$

Set:

$$\varepsilon = (\zeta_{p^n})_{n \geqslant 0} \in \mathscr{X}(K_\infty/K). \tag{42}$$

Then $x = \varepsilon - 1$ is a uniformizer of $\mathscr{X}(K_\infty/K)$, and $\mathscr{X}(K_\infty/K) = k_K((x))$. The action of Γ_K on $\mathscr{X}(K_\infty/K)$ is given by

$$\tau(x) = (1+x)^{\chi_K(\tau)} - 1 \pmod{p}, \qquad \tau \in \Gamma_K. \tag{43}$$

This explicit formula can be generalized to the case of maximal abelian totally ramified extensions using the Lubin–Tate theory.

We refer the reader to [62, 107, 108, 133, 159, 160] for further results about the connection between Galois groups and automorphisms of local fields of positive characteristic.

6.3.7 We discuss the compatibility of reciprocity maps in characteristics 0 and p with the field of norms functor. Let L/K be an APF extension. For any $E \in \mathcal{E}(L/K)$

we have the reciprocity map

$$\theta_E : E^* \to G_E^{\mathrm{ab}}.$$

Passing to projective limit, and identifying $\varprojlim_{E\in\mathcal{E}(L/K)} E^*$ with $\mathscr{X}(L/K)^*$, we obtain an injective homomorphism:

$$\theta_\infty : \mathscr{X}(L/K)^* \to G_L^{\mathrm{ab}}.$$

By Corollary 6.3.4, the Galois group G_L^{ab} is canonically isomorphic to $G_{\mathscr{X}(L/K)}^{\mathrm{ab}}$. Let

$$\theta_{\mathscr{X}(L/K)} : \mathscr{X}(L/K)^* \to G_{\mathscr{X}(L/K)}^{\mathrm{ab}}$$

denote the reciprocity map for the field of norms $\mathscr{X}(L/K)$.

Theorem 6.3.8 (Laubie) *The diagram*

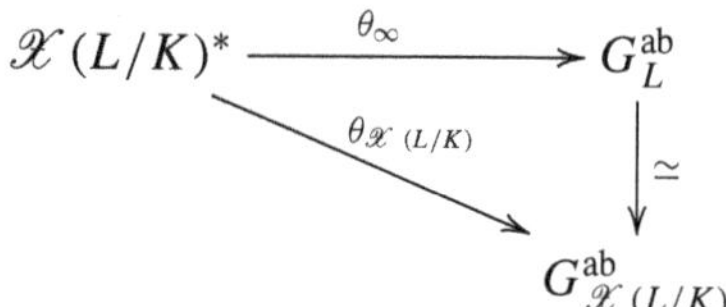

commutes.

Proof See [107, Théorème 3.2.2]. □

6.4 Comparison with the Tilting Equivalence

6.4.1 Recall that an infinite APF extension if deeply ramified, and therefore its completion $\widehat{L}$ is a perfectoid field. We finish this section with comparing the field of norms with the tilting construction. A general result was proved by Fontaine and Wintenberger for APF extensions satisfying some additional condition.

Definition A strictly APF extension is an APF extension satisfying the following property:

$$\liminf_{v\to+\infty} \frac{\psi_{L/K}(v)}{(G_K^{(0)} : G_L^{(0)}G_K^{(v)})} > 0.$$

From Sen's Theorem 1.3.11, it follows that if $\mathrm{Gal}(L/K)$ is a p-adic Lie group, then L/K is strictly APF.

6.4.2 Let L/K be an infinite strict APF extension. Recall that we denote by K_1 the maximal tamely ramified subextension of L/K. For $E \in \mathcal{E}(L/K_1)$, set $d(E) = [E : K_1]$. For each $n \geqslant 1$, set:

$$\mathcal{E}_n = \{E \in \mathcal{E}(L/K_1) \mid p^n \text{ divides } d(E)\}.$$

Let $\alpha = (\alpha_E)_{E \in \mathcal{E}(L/K)} \in \mathscr{X}(L/K)$. It can be proved (see [161, Proposition 4.2.1]) that for any $n \geqslant 1$, the family

$$\alpha_E^{d(E)p^{-n}}, \qquad E \in \mathcal{E}_n$$

converges to some $x_n \in \widehat{L}$. Once the convergence is proved, it's clear that $x_n^p = x_{n-1}^p$ for all n, and therefore $x = (x_n)_{n \geqslant 1} \in \widehat{L}^\flat$. This defines an embedding

$$\mathscr{X}(L/K) \hookrightarrow \widehat{L}^\flat.$$

Theorem 6.4.3 (Fontaine–Wintenberger) *Let L/K be an infinite strict APF extension. Then*

$$\widehat{\mathscr{X}(L/K)}^{\mathrm{rad}} = \widehat{L}^\flat.$$

Proof See [161, Théorème 4.3.2 & Corollaire 4.3.4]. □

Remark 6.4.4 In [61], Fesenko gave examples of deeply ramified extensions which do not contain infinite APF extensions.

7 ℓ-Adic Representations

7.1 Preliminaries

7.1.1 Let E be a complete normed field, and let V be a finite-dimensional E-vector space. Each choice of a basis of V fixes a topological isomorphism $V \simeq E^n$ and equips V with a product topology. Note that this topology does not depend on the choice of the basis.

Definition A representation of a topological group G on V is a continuous homomorphism

$$\rho : G \to \mathrm{Aut}_E V.$$

Fixing a basis of V, one can view a representation of G as a continuous homomorphism $G \to \mathrm{GL}_n(E)$.

Let K be a field and let $\overline{K}$ be a separable closure of K. We denote by G_K the absolute Galois group $\mathrm{Gal}(\overline{K}/K)$ of K. Recall that G_K is equipped with the inverse limit topology and therefore is a compact and totally disconnected topological group.

Definition Let ℓ be a prime number. An ℓ-adic Galois representation is a representation of G_K on a finite dimensional $\mathbf{Q}_\ell$-vector space equipped with the ℓ-adic topology.

Sometimes it is convenient to consider representations with coefficients with a finite extension E of $\mathbf{Q}_\ell$. Below, we give some archetypical examples of ℓ-adic representations.

7.1.2 *One-dimensional representations.* Let V be a one-dimensional Galois representation. Then the action of G_K on V is given by a continuous character $\eta : G_K \to \mathbf{Z}_p^*$, and we will write $\mathbf{Q}_p(\eta)$ instead V.

7.1.3 *Roots of unity.* The following one-dimensional representations are of particular importance for us. Let $\ell \neq \mathrm{char}(K)$. The group G_K acts on the groups μ_{ℓ^n} of ℓ^n-th roots of unity via the ℓ-adic cyclotomic character $\chi_{K,\ell} : G_K \to \mathbf{Z}_\ell^*$:

$$g(\zeta) = \zeta^{\chi_{K,\ell}(g)}, \qquad \forall g \in G_K, \ \zeta \in \mu_{\ell^n}.$$

Set $\mathbf{Z}_\ell(1) = \varprojlim_n \mu_{\ell^n}$ and $\mathbf{Q}_\ell(1) = \mathbf{Z}_\ell(1) \otimes_{\mathbf{Z}_\ell} \mathbf{Q}_\ell$. Then $\mathbf{Q}_\ell(1)$ is a one dimensional $\mathbf{Q}_\ell$-vector space equipped with a continuous action of G_K. The homomorphism $G_K \to \mathrm{Aut}_{\mathbf{Q}_\ell} \mathbf{Q}_\ell(1) \simeq \mathbf{Q}_\ell^*$ concides with $\chi_{K,\ell}$.

7.1.4 *Abelian varieties.* Let A be an abelian variety over K, and let $\ell \neq \mathrm{char}(K)$. The group $A[\ell^n]$ of ℓ^n-torsion points of $A(\overline{K})$ is a Galois module, which is isomorphic (not canonically) to $(\mathbf{Z}/\ell^n\mathbf{Z})^{2d}$ as an abstract group. The ℓ-adic Tate module of A is defined as the projective limit

$$T_\ell(A) = \varprojlim_n A[\ell^n].$$

$T_\ell(A)$ is a free $\mathbf{Z}_\ell$-module of rank $2d$ equipped with a continuous action of G_K. The associated vector space $V_\ell(A) = T_\ell(A) \otimes_{\mathbf{Z}_\ell} \mathbf{Q}_\ell$ gives rise to an ℓ-adic representation

$$\rho_{A,\ell} : G_K \to \mathrm{Aut}_{\mathbf{Q}_\ell} V_\ell(A).$$

Note that $T_\ell(A)$ is a canonical G_K-lattice of $V_\ell(A)$. The reduction of $T_\ell(A)$ modulo ℓ is isomorphic to $A[\ell]$.

7.1.5 *ℓ-Adic Cohomology.* Let X be a smooth projective variety over K. Fix $\ell \neq \mathrm{char}(K)$. The Galois group G_K acts on the étale cohomology $H^n_{\text{ét}}(X \times_K \overline{K}, \mathbf{Z}/\ell^n\mathbf{Z})$. Set:

$$H^n_\ell(X) = \varprojlim_n H^n_{\text{ét}}(X \times_K \overline{K}, \mathbf{Z}/\ell^n\mathbf{Z}) \otimes_{\mathbf{Z}_\ell} \mathbf{Q}_\ell.$$

It is known that the $\mathbf{Q}_\ell$-vector spaces $H^n_\ell(X)$ are finite dimensional and therefore can be viewed as ℓ-adic representations of G_K:

$$G_K \to \mathrm{Aut}_{\mathbf{Q}_\ell} H^n_\ell(X). \tag{44}$$

These representations contain fundamental informations about the arithmetic of algebraic varieties. If X is a smooth proper scheme over a finite field $\mathbf{F}_q$ of characteristic p, then the geometric Frobenius Fr_q acts on $H^n_\ell(X)$, and the zeta-function $Z(X/\mathbf{F}_q, t)$

has the following cohomological interpretation envisioned by Weil and proved by Grothendieck:

$$Z(X/\mathbf{F}_q,t)=\prod_{i=0}^{2d}\left(1-\mathrm{Fr}_q t\mid H^n_\ell(X)\right)^{(-1)^{n+1}}.$$

Katz's survey [93] contains an interesting discussion of what is known and not known about ℓ-adic cohomology over finite fields.

Let now X be a smooth projective variety over a number field K. For any finite place $\mathfrak{p}$ of K, we can consider the restriction of the representation (44) on the decomposition group at $\mathfrak{p}$. This gives a representation of the local Galois group $G_{K_\mathfrak{p}}=\mathrm{Gal}(\overline{K}_\mathfrak{p}/K_\mathfrak{p})$:

$$G_{K_\mathfrak{p}}\to \mathrm{Aut}_{\mathbf{Q}_\ell}H^n_\ell(X).$$

If $\mathfrak{p}\nmid \ell$ and X has a good reduction $X_\mathfrak{p}$ at $\mathfrak{p}$, the base change theorem says that $H^n_\ell(X)$ is isomorphic to $H^n_\ell(X_\mathfrak{p})$. In particular, $H^n_\ell(X)$ is unramified at $\mathfrak{p}$, i.e. $G_{K_\mathfrak{p}}$ acts on $H^n_\ell(X)$ through its maximal unramified quotient $\mathrm{Gal}(K_\mathfrak{p}^{\mathrm{ur}}/K_\mathfrak{p})$. The converse holds for abelian varieties: if $V_\ell(A)$ is unramified, then A has good reduction at $\mathfrak{p}\nmid\ell$ (criterion of Néron–Ogg–Shafarevich [144]).

If $\mathfrak{p}\nmid\ell$, and X has bad reduction at $\mathfrak{p}$, an important information about the action of $G_{K_\mathfrak{p}}$ is provided by Grothendieck's ℓ-adic monodromy theorem (Theorem 7.2.3 below). The case $\mathfrak{p}\mid\ell$ can be studied by the tools of p-adic Hodge theory. This is the main subject of the remainder of these notes.

7.1.6 We denote by $\mathbf{Rep}_{\mathbf{Q}_\ell}(G_K)$ the category of ℓ-adic representations of the absolute Galois group of a field K. Some of its first properties can be summarized in the following proposition:

Proposition 7.1.7 *$\mathbf{Rep}_{\mathbf{Q}_\ell}(G_K)$ is a neutral Tannakian category.*

We refer the reader to [51] for the tannakian formalism. In particular, $\mathbf{Rep}_{\mathbf{Q}_\ell}(G_K)$ is an abelian tensor category. If V_1 and V_2 are ℓ-adic representations, the Galois group G_K acts on $V_1\otimes_{\mathbf{Q}_\ell}V_2$ by

$$g(v_1\otimes v_2)=gv_1\otimes gv_2,\qquad \forall g\in G_K,\quad v_1\in V_1,\quad v_2\in V_2.$$

$\mathbf{Rep}_{\mathbf{Q}_\ell}(G_K)$ is equipped with the internal Hom:

$$\underline{\mathrm{Hom}}(V_1,V_2):=\mathrm{Hom}_{\mathbf{Q}_\ell}(V_1,V_2).$$

The Galois group acts on $\underline{\mathrm{Hom}}(V_1,V_2)$ by

$$g(f)(v_1)=gf(g^{-1}v_1),\qquad \forall g\in G_K,\quad f\in\underline{\mathrm{Hom}}(V_1,V_2),\quad v_1\in V_1.$$

For any ℓ-adic representation V, we denote by V^* its dual representation

$$V^*:=\underline{\mathrm{Hom}}(V,\mathbf{Q}_\ell):=\mathrm{Hom}_{\mathbf{Q}_\ell}(V,\mathbf{Q}_\ell),$$

where $\mathbf{Q}_\ell$ denotes the trivial representation of dimension one.

For any positive n, we set $\mathbf{Q}_\ell(n) = \mathbf{Q}_\ell(1)^{\otimes n}$ and $\mathbf{Q}_\ell(-n) = \mathbf{Q}_\ell(n)^*$.

7.1.8 We will also consider $\mathbf{Z}_\ell$-representations. Namely, a $\mathbf{Z}_\ell$-representation of G_K is a finitely generated free $\mathbf{Z}_\ell$-module equipped with a continuous linear action of G_K. The category $\mathbf{Rep}_{\mathbf{Z}_\ell}(G_K)$ of $\mathbf{Z}_\ell$-representations is abelian. It contains the tannakian subcategory $\mathbf{Rep}_{\mathbf{F}_\ell}(G_K)$ of representations of G_K over the finite field $\mathbf{F}_\ell = \mathbf{Z}/\ell\mathbf{Z}$. We have the reduction-modulo-ℓ functor

$$\begin{aligned} &\mathbf{Rep}_{\mathbf{Z}_\ell}(G_K) \to \mathbf{Rep}_{\mathbf{F}_\ell}(G_K), \\ &T \mapsto T \otimes_{\mathbf{Z}_\ell} \mathbf{F}_\ell. \end{aligned}$$

The following proposition can be easily deduced from the compactness of G_K:

Proposition 7.1.9 *For any ℓ-adic representation V, there exists a $\mathbf{Z}_\ell$-lattice stable under the action of G_K. In particular, the functor*

$$\begin{aligned} &\mathbf{Rep}_{\mathbf{Z}_\ell}(G_K) \to \mathbf{Rep}_{\mathbf{Q}_\ell}(G_K), \\ &T \mapsto T \otimes_{\mathbf{Z}_\ell} \mathbf{Q}_\ell \end{aligned}$$

is essentially surjective.

7.2 ℓ-Adic Representations of Local Fields ($\ell \neq p$)

7.2.1 From now on, we consider ℓ-adic representations of local fields. Let K be a local field with residue field k_K of characteristic p. To distinguish between the cases $\ell \neq p$ and $\ell = p$, we will use in the second case the term p-adic keeping ℓ-adic exclusively for the inequal characteristic case.

7.2.2 We consider the ℓ-adic case. Recall that for the tame quotient of the inertia subgroup we have the isomorphism (20):

$$\mathrm{Gal}(K^{\mathrm{tr}}/K^{\mathrm{ur}}) \simeq \prod_{q \neq p} \mathbf{Z}_q.$$

Let ψ_ℓ denote the projection

$$\psi_\ell : I_K \to \mathrm{Gal}(K^{\mathrm{tr}}/K^{\mathrm{ur}}) \to \mathbf{Z}_\ell.$$

The following general result reflects the Frobenius structure on the tame Galois group.

Theorem 7.2.3 (Grothendieck's ℓ-adic monodromy theorem) *Let*

$$\rho : G_K \to GL(V)$$

be an ℓ-adic representation. Then the following holds true:

(i) There exists an open subgroup H of the inertia group I_K such that the automorphism $\rho(g)$ is unipotent for all $g \in H$.

(ii) More precisely, there exists a nilpotent operator $N : V \to V$ such that

$$\rho(g) = \exp(N\psi_\ell(g)), \qquad \forall g \in H.$$

(iii) Let $\widehat{\mathrm{Fr}}_K \in G_K$ be any lift of the arithmetic Frobenius Fr_K. Set $F = \rho(\widehat{\mathrm{Fr}}_K)$. Then

$$FN = qNF,$$

where $q = |k_K|$.

Proof See [144] for details.

(a) By Proposition 7.1.9, ρ can be viewed as an homomorphism

$$\rho : G_K \to \mathrm{GL}_d(\mathbf{Z}_\ell).$$

Let $U = 1 + \ell^2\mathrm{M}_d(\mathbf{Z}_\ell)$. Then U has finite index in $\mathrm{GL}_d(\mathbf{Z}_\ell)$, and there exists a finite extension K'/K such that $\rho(G_{K'}) \subset U$. Without loss of generality, we may (and will) assume that $K' = K$.

(b) The wild ramification subgroup P_K is a pro-p-group. Since U is a pro-ℓ-group with $\ell \neq p$, we have $\rho(P_K) = \{1\}$, and ρ factors through the tame ramification group $\mathrm{Gal}(K^{\mathrm{tr}}/K)$. Since $\mathrm{Gal}(K^{\mathrm{tr}}/K^{\mathrm{ur}}) \simeq \prod_q \mathbf{Z}_q$, the same argument shows that ρ factors through the Galois group of the extension K_ℓ^{tr}/K, where

$$K_\ell^{\mathrm{tr}} = K^{\mathrm{ur}}(\pi^{1/\ell^\infty}), \qquad \pi \text{ is a uniformizer of } K.$$

Let τ_ℓ be the automorphism that maps to 1 under the isomorphism $\mathrm{Gal}(K_\ell^{\mathrm{tr}}/K^{\mathrm{ur}}) \simeq \mathbf{Z}_\ell$. By Proposition 2.1.4, $\mathrm{Gal}(K_\ell^{\mathrm{tr}}/K)$ is the pro-ℓ-group topologically generated by τ_ℓ and by any lift f_ℓ of the Frobenius automorphism, with the single relation:

$$f_\ell \tau_\ell f_\ell^{-1} = \tau_\ell^q. \tag{45}$$

(c) Set $X = \rho(\tau_\ell) \in U$. The ℓ-adic logarithm map converges on U, and we set:

$$N := \log(X) = \sum_{n=1}^{\infty} (-1)^{n+1} \frac{(X-1)^n}{n}.$$

Then for any $g \in I_K$, we have:

$$\rho(g) = \rho(\tau_\ell^{\psi_\ell(g)}) = \exp(N\psi_\ell(g)).$$

Moreover, applying the identity $\log(BAB^{-1}) = B\log(A)B^{-1}$ to (45) and setting $F = \rho(f_\ell)$, we obtain:

$$FNF^{-1} = qN.$$

(d) From the last formula, it follows that N and qN have the same eigenvalues. Therefore, they are all zero, and N is nilpotent. The theorem is proved. □

8 Classification of p-Adic Representations

8.1 *The Case of Characteristic p*

8.1.1 In this section, we turn to p-adic representations. It turns out, that it is possible to give a full classification of p-adic representations of the Galois group of *any* field K of characteristic p in terms of modules equipped with a semi-linear operator. This can be explained by the existence of the Frobenius structure on K. To simplify the exposition, we will work with the purely inseparable closure $F := K^{\mathrm{rad}}$ of K. However, it is not absolutely necessary (see [69]). On the contrary, it is often preferable to work with non-perfect fields. We will come back to this question in Sect. 8.2.

8.1.2 Consider the ring of Witt vectors

$$O_{\mathscr{F}} = W(F).$$

Recall that $O_{\mathscr{F}}$ is a complete discrete valuation ring of characteristic 0 with maximal ideal $(p) = pO_{\mathscr{F}}$ and residue field F. Its field of fractions $\mathscr{F} = O_{\mathscr{F}}[1/p]$ is an unramified discrete valuation field. The field $\overline{F} = \overline{K}^{\mathrm{rad}}$ is an algebraic closure of F, and the Galois groups of $\overline{K}/K$ and $\overline{F}/F$ are canonically isomorphic. Set:

$$\widehat{O}_{\mathscr{F}}^{\mathrm{ur}} = W(\overline{F}), \qquad \widehat{\mathscr{F}}^{\mathrm{ur}} = \widehat{O}_{\mathscr{F}}^{\mathrm{ur}}[1/p].$$

Then $\widehat{\mathscr{F}}^{\mathrm{ur}}$ is a complete unramified discrete valuation field with residue field $\overline{F}$ and therefore can be identified with the completion of the maximal unramified extension of $\mathscr{F}$. The field $\overline{F}$ is equipped with the following structures:

- The action of the absolute Galois group G_K;
- The absolute Frobenius automorphism $\varphi\,:\,\overline{F} \to \overline{F}$, $\varphi(x) = x^p$.

The actions of G_K and φ commute to each other. One has

$$\overline{F}^{G_K} = F, \qquad \overline{F}^{\varphi=1} = \mathbf{F}_p.$$

These actions extend naturally from $\overline{F}$ to $\widehat{O}_{\mathscr{F}}^{\mathrm{ur}}$ and $\widehat{\mathscr{F}}^{\mathrm{ur}}$, and one has

$$(\widehat{O}_{\mathscr{F}}^{\mathrm{ur}})^{G_K} = O_{\mathscr{F}}, \qquad (\widehat{O}_{\mathscr{F}}^{\mathrm{ur}})^{\varphi=1} = \mathbf{Z}_p.$$

Definition Let $A = F$, $O_{\mathscr{F}}$ or $\mathscr{F}$. A φ-module over A is a finitely generated A-module D equipped with a semi-linear injective operator $\varphi : D \to D$. Namely, φ satisifies the following properties:

$$\begin{aligned} \varphi(x+y) &= \varphi(x) + \varphi(y), && \forall x, y \in D, \\ \varphi(ax) &= \varphi(a)\varphi(x), && \forall a \in A, x \in D. \end{aligned}$$

A morphism of φ-modules is an A-linear map $f : D_1 \to D_2$ which commutes with φ:

$$f(\varphi(d)) = \varphi(f(d)), \qquad \forall d \in D_1.$$

8.1.3 Consider A as an A-module via the Frobenius map $\varphi : A \to A$. For a φ-module D, let $D \otimes_{A,\varphi} A$ denote the tensor product of A-modules D and A. We consider $D \otimes_{A,\varphi} A$ as an A-module:

$$\lambda(d \otimes a) = d \otimes \lambda a, \qquad \lambda \in A, \quad d \otimes a \in D \otimes_{A,\varphi} A.$$

Then the semi-linear map $\varphi : D \to D$ induces an A-linear map

$$\Phi : D \otimes_{A,\varphi} A \to D, \qquad d \otimes a \mapsto a\varphi(d).$$

Definition (i) Let $A = F$ or $O_{\mathscr{F}}$. A φ-module D over A is étale if the map $\Phi : D \otimes_{A,\varphi} A \to D$ is an isomorphism.

(ii) A φ-module over $\mathscr{F}$ is étale if it has an étale $O_{\mathscr{F}}$-lattice.

Let $A = F$ or $O_{\mathscr{F}}$, and assume that D if free over A. Then D is étale if and only if the matrix of $\varphi : D \to D$ is invertible over A. Note that this property does not depend on the choice of the A-base of D.

8.1.4 We denote by $\mathbf{M}_A^{\varphi,\text{ét}}$ the category of étale φ-modules over $A = F, O_{\mathscr{F}}, \mathscr{F}$. We refer the reader to [69] for a detailed study of these categories. All these categories are abelian. They are equipped with the tensor product:

$$D_1 \otimes_A D_2, \qquad \varphi(d_1 \otimes d_2) = \varphi(d_1) \otimes \varphi(d_2)$$

and the internal Hom :

$$\underline{\mathrm{Hom}}(D_1, D_2) := \mathrm{Hom}_A(D_1, D_2).$$

The action of φ on $\underline{\mathrm{Hom}}(D_1, D_2)$ is defined as follows. Let $f : D_1 \to D_2$. Then $\varphi(f)$ is the composition of maps:

$$\varphi(f) : D_1 \xrightarrow{\Phi^{-1}} D_1 \otimes_{A,\varphi} A \xrightarrow{f \otimes \mathrm{id}} D_2 \otimes_{A,\varphi} A \xrightarrow{\Phi} D_2.$$

The categories $\mathbf{M}_F^{\varphi,\text{ét}}$ and $\mathbf{M}_{\mathscr{F}}^{\varphi,\text{ét}}$ are neutral tannakian. If $A = F$ or $\mathscr{F}$, then for any $D \in \mathbf{M}_A^{\varphi,\text{ét}}$, we denote by D^* the dual module:

$$D^* = \mathrm{Hom}_A(D, A).$$

8.1.5 The term *étale* can be explained as follows. Let D be a φ-module over F. Fix a basis $\{e_1, \dots, e_n\}$ of D. Write:

$$\varphi(e_i) = \sum_{i=1}^{n} a_{ij} e_j, \qquad a_{ij} \in F, \qquad 1 \leqslant i \leqslant n.$$

Let $I \subset F[X_1, \dots, X_n]$ denote the ideal generated by

$$X_i^p - \sum_{i=1}^{n} a_{ij} X_j, \qquad 1 \leqslant i \leqslant n.$$

Then the algebra $A := F[X_1, \dots, X_n]/I$ is étale over F if and only if D is an étale φ-module. Consider the $\mathbf{F}_p$-vector space $\mathrm{Hom}_F(D, \overline{F})^{\varphi=1}$. Let $f \in \mathrm{Hom}_F(D, \overline{F})$. Then $\varphi(f) = f$ if and only if the vector $(f(e_1), \dots, f(e_n)) \in \overline{F}^n$ is a solution of the system

$$X_i^p - \sum_{i=1}^{n} a_{ij} X_j = 0, \qquad 1 \leqslant i \leqslant n.$$

Therefore, we have isomorphisms:

$$\mathrm{Hom}_F(D, \overline{F})^{\varphi=1} = \mathrm{Hom}_{F\text{-alg}}(A, \overline{F}) = \mathrm{Spec}(A)(\overline{F}).$$

Note that if D is étale, then the cardinality of $\mathrm{Spec}(A)(\overline{F})$ is p^n, and $\mathrm{Hom}_F(D, \overline{F})^{\varphi=1}$ is a $\mathbf{F}_p$-vector space of dimension n.

8.1.6 For the dual module D^*, we have a canonical isomorphisms:

$$D \otimes_F \overline{F} \simeq \mathrm{Hom}_F(D^*, F) \otimes_F \overline{F} \simeq \mathrm{Hom}_F(D^*, \overline{F}).$$

Then

$$(D \otimes_F \overline{F})^{\varphi=1} \simeq \mathrm{Hom}_F(D^*, \overline{F})^{\varphi=1},$$

and applying the previous remark to D^*, we see that $(D \otimes_F \overline{F})^{\varphi=1}$ is a $\mathbf{F}_p$-vector space of dimension n.

8.1.7 Following Fontaine [69], we construct a canonical equivalence between the category $\mathbf{Rep}_{\mathbf{F}_p}(G_K)$ of modular Galois representations of G_K and $\mathbf{M}_F^{\varphi,\text{ét}}$. For any $V \in \mathbf{Rep}_{\mathbf{F}_p}(G_K)$, set:

$$\mathbf{D}_F(V) = (V \otimes_{\mathbf{F}_p} \overline{F})^{G_K}.$$

Since G_K acts trivially on F, it is clear that $\mathbf{D}_F(V)$ is an F-module equipped with the diagonal action of φ (here φ acts trivially on V). For any $D \in \mathbf{M}_F^{\varphi,\text{ét}}$, set:

$$\mathbf{V}_F(D) = (D \otimes_F \overline{F})^{\varphi=1}.$$

Then $\mathbf{V}_F(D)$ is an $\mathbf{F}_p$-vector space equipped with the diagonal action of G_K (here G_K acts trivially on D).

Theorem 8.1.8 *(i) Let $V \in \mathbf{Rep}_{\mathbf{F}_p}(G_K)$ be a modular Galois representation of dimension n. Then $\mathbf{D}_F(V)$ is an étale φ-module of rank n over F.*

(ii) Let $D \in \mathbf{M}_F^{\varphi,\text{ét}}$ be an étale φ-module of rank n over F. Then $\mathbf{V}_F(D)$ is a modular Galois representation of G_K of dimension n over $\mathbf{F}_p$.

(iii) The functors $\mathbf{D}_F$ and $\mathbf{V}_F$ establish equivalences of tannakian categories

$$\mathbf{D}_F : \mathbf{Rep}_{\mathbf{F}_p}(G_K) \to \mathbf{M}_F^{\varphi,\text{ét}}, \qquad \mathbf{V}_F : \mathbf{M}_F^{\varphi,\text{ét}} \to \mathbf{Rep}_{\mathbf{F}_p}(G_K),$$

which are quasi-inverse to each other. Moreover, for all $T \in \mathbf{Rep}_{\mathbf{F}_p}(G_K)$ and $D \in \mathbf{M}_F^{\varphi,\text{ét}}$, we have canonical and functorial isomorphisms compatible with the actions of G_K and φ on the both sides:

$$\mathbf{D}_F(T) \otimes_F \overline{F} \simeq T \otimes_{\mathbf{F}_p} \overline{F},$$
$$\mathbf{V}_F(D) \otimes_{\mathbf{F}_p} \overline{F} \simeq D \otimes_F \overline{F}.$$

Proof (a) Let $V \in \mathbf{Rep}_{\mathbf{F}_p}(G_K)$ be a modular representation of dimension n. The Galois group G_F acts semi-linearly on $V \otimes_{\mathbf{F}_p} \overline{F}$. From Hilbert's Theorem 90 (Theorem 1.6.3), it follows that $\mathbf{D}_F(V) = (V \otimes_{\mathbf{F}_p} \overline{F})^{G_F}$ has dimension n over F, and that the multiplication in $\overline{F}$ induces an isomorphism

$$(V \otimes_{\mathbf{F}_p} \overline{F})^{G_F} \otimes_F \overline{F} \xrightarrow{\sim} V \otimes_{\mathbf{F}_p} \overline{F}.$$

Hence:

$$\mathbf{D}_F(V) \otimes_F \overline{F} \xrightarrow{\sim} V \otimes_{\mathbf{F}_p} \overline{F}.$$

This isomorphism shows that the matrix of φ is invertible in $\mathrm{GL}_n(\overline{F})$ and therefore in $\mathrm{GL}_n(F)$. This proves that $\mathbf{D}_F(V)$ is étale.

Taking the φ-invariants on the both sides, one has

$$\mathbf{V}_F(\mathbf{D}_F(V)) = (\mathbf{D}_F(V) \otimes_F \overline{F})^{\varphi=1} \xrightarrow{\sim} (V \otimes_{\mathbf{F}_p} \overline{F})^{\varphi=1} = V. \tag{46}$$

(b) Conversely, let $D \in \mathbf{M}_F^{\varphi,\text{ét}}$. We already know (see Sect. 8.1) that $\mathbf{V}_F(D)$ is a $\mathbf{F}_p$-vector space of dimension n. Consider the map

$$\alpha : (D \otimes_F \overline{F})^{\varphi=1} \otimes_{\mathbf{F}_p} \overline{F} \to D \otimes_F \overline{F}, \tag{47}$$

induced by the multiplication in $\overline{F}$. We claim that this map is an isomorpism. Since the both sides have the same dimension over $\overline{F}$, it is sufficient to prove the injectivity. To do that, we use the following argument, known as Artin's trick. Assume that f is not surjective, and take a non-zero element $x \in \ker(\alpha)$ which has a shortest presentation in the form

$$x = \sum_{i=1}^{m} d_i \otimes a_i, \qquad d_i \in \mathbf{V}_F(D), \quad a_i \in \overline{F}.$$

Without loss of generality, we can assume that $a_m = 1$ (dividing by a_m). Note that $\varphi(x) - x \in \ker(\alpha)$. On the other hand, it can be written as

$$\varphi(x) - x = \sum_{i=1}^{m} d_i \otimes (\varphi(a_i) - a_i) = \sum_{i=1}^{m-1} d_i \otimes (\varphi(a_i) - a_i).$$

By our choice of x, this implies that $\varphi(a_i) = a_i$, and therefore $a_i \in \mathbf{F}_p$ for all i. But in this case $x \in \mathbf{V}_F(D)$, and $x = \alpha(x) = 0$. This proves the injectivity of (47).

(c) By part (b), we have an isomorphism:

$$\mathbf{V}_F(D) \otimes_{\mathbf{F}_p} \overline{F} \to D \otimes_F \overline{F}.$$

Taking the Galois invariants on the both sides, we obtain:

$$\mathbf{D}_F(\mathbf{V}_F(D)) = (\mathbf{V}_F(D) \otimes_{\mathbf{F}_p} \overline{F})^{G_F} \xrightarrow{\sim} (D \otimes_F \overline{F})^{G_F} = D. \tag{48}$$

From (46) and (48), it follows that the functors $\mathbf{D}_F$ and $\mathbf{V}_E$ are quasi-inverse to each other. In particular, they are equivalences of categories. Other assertions can be checked easily. □

8.1.9 Now we turn to $\mathbf{Z}_p$-representations. For all $T \in \mathbf{Rep}_{\mathbf{Z}_p}(G_K)$ and $D \in \mathbf{M}_{O_{\mathscr{F}}}^{\varphi,\text{ét}}$, set:

$$\mathbf{D}_{O_{\mathscr{F}}}(T) = (T \otimes_{\mathbf{Z}_p} \widehat{O}_{\mathscr{F}}^{\text{ur}})^{G_K},$$
$$\mathbf{V}_{O_{\mathscr{F}}}(D) = (D \otimes_{O_{\mathscr{F}}} \widehat{O}_{\mathscr{F}}^{\text{ur}})^{\varphi=1}.$$

The following theorem can be deduced from Theorem 8.1.8 by devissage.

Theorem 8.1.10 (Fontaine) *(i) Let $T \in \mathbf{Rep}_{\mathbf{Z}_p}(G_K)$ be a $\mathbf{Z}_p$-representation. Then $\mathbf{D}_{O_{\mathscr{F}}}(T)$ is an étale φ-module over $O_{\mathscr{F}}$.*

(ii) Let $D \in \mathbf{M}_{O_{\mathscr{F}}}^{\varphi,\text{ét}}$ be an étale φ-module over $O_{\mathscr{F}}$. Then $\mathbf{V}_{O_{\mathscr{F}}}(D)$ is a $\mathbf{Z}_p$-representation of G_K.

(iii) The functors $\mathbf{D}_{O_{\mathscr{F}}}$ and $\mathbf{V}_{O_{\mathscr{F}}}$ establish equivalences of categories

$$\mathbf{D}_{O_{\mathscr{F}}} : \mathbf{Rep}_{\mathbf{Z}_p}(G_K) \to \mathbf{M}_{O_{\mathscr{F}}}^{\varphi,\text{ét}}, \qquad \mathbf{V}_{O_{\mathscr{F}}} : \mathbf{M}_{O_{\mathscr{F}}}^{\varphi,\text{ét}} \to \mathbf{Rep}_{\mathbf{Z}_p}(G_K),$$

which are quasi-inverse to each other. Moreover, for all $T \in \mathbf{Rep}_{\mathbf{Z}_p}(G_K)$ and $D \in \mathbf{M}^{\varphi,\text{ét}}_{O_{\mathscr{F}}}$, we have canonical and functorial isomorphisms compatible with the actions of G_K and φ on the both sides:

$$\mathbf{D}_{O_{\mathscr{F}}}(T) \otimes_{O_{\mathscr{F}}} \widehat{O}^{\mathrm{ur}}_{\mathscr{F}} \simeq T \otimes_{\mathbf{Z}_p} \widehat{O}^{\mathrm{ur}}_{\mathscr{F}},$$
$$\mathbf{V}_{O_{\mathscr{F}}}(D) \otimes_{\mathbf{Z}_p} \widehat{O}^{\mathrm{ur}}_{\mathscr{F}} \simeq D \otimes_{O_{\mathscr{F}}} \widehat{O}^{\mathrm{ur}}_{\mathscr{F}}.$$

For p-adic representations, we have the following:

Theorem 8.1.11 *(i) Let V be a p-adic representation of G_K of dimension n. Then $\mathbf{D}_{\mathscr{F}}(V) = (V \otimes_{\mathbf{Q}_p} \widehat{\mathscr{F}}^{\mathrm{ur}})^{G_K}$ is an étale φ-module of dimension n over $\mathscr{F}$.*

(ii) Let $D \in \mathbf{M}^{\varphi,\text{ét}}_{\mathscr{F}}$ be an étale φ-module of dimension n over $\mathscr{F}$. Then $\mathbf{V}_{\mathscr{F}}(D) = (D \otimes_{\mathbf{Q}_p} \widehat{\mathscr{F}}^{\mathrm{ur}})^{\varphi=1}$ is a p-adic Galois representation of G_K of dimension n over $\mathbf{Q}_p$.

(iii) The functors

$$\mathbf{D}_{\mathscr{F}} : \mathbf{Rep}_{\mathbf{Q}_p}(G_K) \to \mathbf{M}^{\varphi,\text{ét}}_{\mathscr{F}},$$
$$\mathbf{V}_{\mathscr{F}} : \mathbf{M}^{\varphi,\text{ét}}_{\mathscr{F}} \to \mathbf{Rep}_{\mathbf{Q}_p}(G_K),$$

are equivalences of tannakian categories, which are quasi-inverse to each other. Moreover, for all $V \in \mathbf{Rep}_{\mathbf{Q}_p}(G_K)$ and $D \in \mathbf{M}^{\varphi,\text{ét}}_{\mathscr{F}}$, we have canonical and functorial isomorphisms compatible with the actions of G_K and φ on the both sides:

$$\mathbf{D}_{\mathscr{F}}(V) \otimes_{\mathscr{F}} \widehat{\mathscr{F}}^{\mathrm{ur}} \simeq V \otimes_{\mathbf{Q}_p} \widehat{\mathscr{F}}^{\mathrm{ur}},$$
$$\mathbf{V}_{\mathscr{F}}(D) \otimes_{\mathbf{Q}_p} \widehat{\mathscr{F}}^{\mathrm{ur}} \simeq D \otimes_{\mathscr{F}} \widehat{\mathscr{F}}^{\mathrm{ur}}.$$

8.2 The Case of Characteristic 0

8.2.1 In this section, K is a local field of characteristic 0 with residual characteristic p. Let $K_\infty = K(\zeta_{p^\infty})$ denote the p-cyclotomic extension of K. Set $H_K = \mathrm{Gal}(\overline{K}/K_\infty)$ and $\Gamma_K = \mathrm{Gal}(K_\infty/K)$. Then K_∞/K is a deeply ramified (even an APF) extension, and we can consider the tilt of its completion:

$$F := \widehat{K}^{\flat}_{\infty}.$$

The field F is perfect, of characteristic p, and we apply to F the contructions of Sect. 8.1. Namely, set $O_{\mathscr{F}} = W(F)$ and $\mathscr{F} = O_{\mathscr{F}}[1/p]$. These rings are equipped with the weak topology, defined in Sect. 5.3. By Proposition 5.4.3, the separable closure $\overline{F}$ of F is dense in $\mathbf{C}^{\flat}_K$ and we have a natural inclusion $\widehat{O}^{\mathrm{ur}}_{\mathscr{F}} \subset W(\mathbf{C}^{\flat}_K)$. The Galois group G_K acts naturally on the maximal unramified extension $\mathscr{F}^{\mathrm{ur}}$ of $\mathscr{F}$ in $W(\mathbf{C}^{\flat}_K)[1/p]$ and on its p-adic completion $\widehat{\mathscr{F}}^{\mathrm{ur}}$. By Theorem 5.4.4, this action induces a canonical isomorphism:

$$H_K \simeq \mathrm{Gal}(\mathscr{F}^{\mathrm{ur}}/\mathscr{F}). \tag{49}$$

In particular, $(\widehat{\mathscr{F}^{\mathrm{ur}}})^{H_K} = \mathscr{F}$. The cyclotomic Galois group Γ_K acts on F and therefore on $O_{\mathscr{F}}$ and $\mathscr{F}$.

Definition Let $A = F, O_{\mathscr{F}}$, or $\mathscr{F}$. A (φ, Γ_K)-module over A is a φ-module over A equipped with a continuous semi-linear action of Γ_K commuting with φ. A (φ, Γ_K)-module is étale if it is étale as a φ-module.

We denote by $\mathbf{M}_A^{\varphi,\Gamma,\text{ét}}$ the category of (φ, Γ_K)-modules over A. It can be easily seen that $\mathbf{M}_A^{\varphi,\Gamma,\text{ét}}$ is an abelian tensor category. Moreover, if $A = F$ or $\mathscr{F}$, it is neutral tannakian.

8.2.2 Now we are in position to introduce the main constructions of Fontaine's theory of (φ, Γ_K)-modules. Let T be a $\mathbf{Z}_p$-representation of G_K. Set:

$$\mathbf{D}_{O_{\mathscr{F}}}(T) = (T \otimes_{\mathbf{Z}_p} \widehat{O}_{\mathscr{F}}^{\mathrm{ur}})^{H_K}.$$

Thanks to the isomorphism (49) and the results of Sect. 8.1, $\mathbf{D}_{O_{\mathscr{F}}}(T)$ is an étale φ-module. In addition, it is equipped with a natural action of Γ_K, and therefore we have a functor

$$\mathbf{D}_{O_{\mathscr{F}}} : \mathbf{Rep}_{\mathbf{Z}_p}(G_K) \to \mathbf{M}_{O_{\mathscr{F}}}^{\varphi,\Gamma,\text{ét}}.$$

Conversely, let D be an étale (φ, Γ_K)-module over $O_{\mathscr{F}}$. Set:

$$\mathbf{V}_{O_{\mathscr{F}}}(D) = (D \otimes_{\mathbf{Z}_p} \widehat{O}_{\mathscr{F}}^{\mathrm{ur}})^{\varphi=1}.$$

By the results of Sect. 8.1, $\mathbf{V}_{O_{\mathscr{F}}}(D)$, is a free $\mathbf{Z}_p$-module of the same rank. Moreover, it is equipped with a natural action of G_K, and we have a functor

$$\mathbf{V}_{O_{\mathscr{F}}} : \mathbf{M}_{O_{\mathscr{F}}}^{\varphi,\Gamma,\text{ét}} \to \mathbf{Rep}_{\mathbf{Z}_p}(G_K).$$

Theorem 8.2.3 (Fontaine) *(i) The functors $\mathbf{D}_{O_{\mathscr{F}}}$ and $\mathbf{V}_{O_{\mathscr{F}}}$ are equivalences of categories, which are quasi-inverse to each other.*

(ii) For all $T \in \mathbf{Rep}_{\mathbf{Z}_p}(G_K)$ and $D \in \mathbf{M}_{O_{\mathscr{F}}}^{\varphi,\text{ét}}$, we have canonical and functorial isomorphisms compatible with the actions of G_K and φ on the both sides:

$$\begin{aligned} \mathbf{D}_{O_{\mathscr{F}}}(T) \otimes_{O_{\mathscr{F}}} \widehat{O}_{\mathscr{F}}^{\mathrm{ur}} &\simeq T \otimes_{\mathbf{Z}_p} \widehat{O}_{\mathscr{F}}^{\mathrm{ur}}, \\ \mathbf{V}_{O_{\mathscr{F}}}(D) \otimes_{\mathbf{Z}_p} \widehat{O}_{\mathscr{F}}^{\mathrm{ur}} &\simeq D \otimes_{O_{\mathscr{F}}} \widehat{O}_{\mathscr{F}}^{\mathrm{ur}}. \end{aligned} \tag{50}$$

Here G_K acts on (φ, Γ_K)-modules through Γ_K.

Proof Theorem 8.1.10 provide us with the canonial and functorial isomorphisms (50), which are compatible with the action of φ and H_K. From construction, it follows that they are compatible with the action of the whole Galois group G_K on the both sides. This implies that the functors $\mathbf{D}_{O_{\mathscr{F}}}$ and $\mathbf{V}_{O_{\mathscr{F}}}$ are quasi-inverse to each other, and the theorem is proved. □

Remark 8.2.4 We invite the reader to formulate and prove the analogous statements for the categories $\mathbf{Rep}_{\mathbf{F}_p}(G_K)$ and $\mathbf{Rep}_{\mathbf{Q}_p}(G_K)$.

8.2.5 One can refine this theory working with the field of norms rather that with the perfectoid field $\widehat{K}_\infty^\flat$. To simplify notation, let $\mathbf{E}_K$ denote the field of norms of K_∞/K. We recall that by Theorem 6.4.3, $\mathbf{E}_K^{\mathrm{rad}}$ is dense in $\widehat{K}_\infty^\flat$. We want to lift $\mathbf{E}_K$ to characteristic 0. First, we consider the maximal unramified subextension K_0 of K. Let $K_{0,\infty}/K_0$ denote its p-cyclotomic extension. Set $\Gamma_{K_0}=\mathrm{Gal}(K_{0,\infty}/K_0)$ and $H_{K_0}=\mathrm{Gal}(\overline{K}/K_{0,\infty})$. Let $\mathbf{E}_{K_0}$ denote the field of norms of $K_{0,\infty}/K_0$. Then $\mathbf{E}_{K_0}=k_K((x))$, where $x=\varepsilon-1$ and $\varepsilon=(\zeta_{p^n})_{n\geqslant 0}$ (see (43)). Take the Teichmüller lift $[\varepsilon]\in O_{\mathscr{F}}$ of ε and set $X=[\varepsilon]-1$. The Galois group and the Frobenius automorphism act on $[\varepsilon]$ and X through Γ_{K_0} as follows:

$$\begin{aligned} &g([\varepsilon])=[\varepsilon]^{\chi_0(g)}, && g\in G_{K_0}, && \varphi([\varepsilon])=[\varepsilon]^p,\\ &g(X)=(1+X)^{\chi_0(g)}-1, && g\in G_{K_0}, && \varphi(X)=(1+X)^p-1, \end{aligned}$$

where $\chi_0 : G_{K_0}\to \mathbf{Z}_p^*$ denotes the p-adic cyclotomic character for K_0. The ring of integers $O_{K_0}=W(k_K)$ is a subring of $O_{\mathscr{F}}$. We define the following subrings of $O_{\mathscr{F}}$:

$$\begin{aligned} &\mathbf{A}_{K_0}^+=O_{K_0}[[X]],\\ &\mathbf{A}_{K_0}=\widehat{\mathbf{A}_{K_0}^+[1/X]}=p\text{-adic completion of}\,\mathbf{A}_{K_0}^+[1/X]. \end{aligned}$$

Note that $\mathbf{A}_{K_0}$ is an unramified discrete valuation ring with residue field $\mathbf{E}_{K_0}$. It can be described explicitely as the ring of power series of the form

$$\sum_{n\in\mathbf{Z}} a_nX^n,\qquad a_n\in O_{K_0}\text{ and }\lim_{n\to-\infty}a_n=0.$$

It is crucial that $\mathbf{A}_{K_0}$ is stable under the actions of Γ_{K_0} and φ. Set $\mathbf{B}_{K_0}=\mathbf{A}_{K_0}[1/p]$. Then $\mathbf{B}_{K_0}$ is an unramified discrete valuation field with the ring of integers $\mathbf{A}_{K_0}$.

8.2.6 By Hensel's lemma, for each finite separable extension $E/\mathbf{E}_0$, there exists a unique complete subring $A\subset \widehat{O}_{\mathscr{F}}^{\mathrm{ur}}$ containing $\mathbf{A}_{K_0}$ and such that its residue field A/pA is isomorphic to E. We denote by $\mathbf{A}_{K_0}^{\mathrm{ur}}$ the compositum of all such extensions in $\widehat{O}_{\mathscr{F}}^{\mathrm{ur}}$ and set $\mathbf{B}_{K_0}^{\mathrm{ur}}=\mathbf{A}_{K_0}^{\mathrm{ur}}[1/p]$. Then $\mathbf{B}_{K_0}^{\mathrm{ur}}$ is the maximal unramified extension of $\mathbf{B}_{K_0}$ and $\mathbf{A}_{K_0}^{\mathrm{ur}}$ is its ring of integers. Let $\mathbf{B}$ and $\mathbf{A}$ denote the p-adic completions of $\mathbf{B}_{K_0}^{\mathrm{ur}}$ and $\mathbf{A}_{K_0}^{\mathrm{ur}}$ respectively. All these rings are stable under the natural action of G_{K_0}. By the theory of fields of norms, this action induces canonical isomorphisms:

$$H_{K_0}\simeq \mathrm{Gal}(\overline{\mathbf{E}}_{K_0}/\mathbf{E}_{K_0})\simeq \mathrm{Gal}(\mathbf{B}_{K_0}^{\mathrm{ur}}/\mathbf{B}_{K_0}).$$

8.2.7 Recall that K is a totally ramified extension of K_0. Set:

$$\mathbf{A}_K=\mathbf{A}^{H_K},\qquad \mathbf{B}_K=\mathbf{A}_K[1/p].$$

Then $\mathbf{B}_K$ is an unramified extension of $\mathbf{B}_{K_0}$ with residue field $\mathbf{E}_K$. One has

$$[\mathbf{B}_K : \mathbf{B}_{K_0}] = [\mathbf{E}_K : \mathbf{E}_{K_0}] = [K_\infty : K_{0,\infty}].$$

These constructions can be summarized in the following diagram, where the horizontal maps are reductions modulo p:

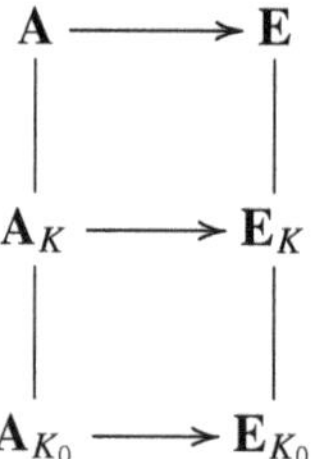

8.2.8 The notion of an (étale) (φ, Γ_K)-module extends verbatim to the case of modules over $\mathbf{A}_K$ (respectively, $\mathbf{B}_K$). We denote by $\mathbf{M}_{\mathbf{A}_K}^{\varphi,\text{ét}}$ and $\mathbf{M}_{\mathbf{B}_K}^{\varphi,\text{ét}}$ the resulting categories. For any $\mathbf{Z}_p$-representation T of G_K, set:

$$\mathbf{D}(T) = (T \otimes_{\mathbf{Z}_p} \mathbf{A})^{H_K}.$$

Conversely, for any étale (φ, Γ_K)-module D over $\mathbf{A}_K$, set:

$$\mathbf{V}(D) = (D \otimes_{\mathbf{Z}_p} \mathbf{A})^{\varphi=1}.$$

Theorem 8.2.9 (Fontaine) *The functors* $\mathbf{D}$ *and* $\mathbf{V}$ *define equivalences of categories*

$$\mathbf{D} : \mathbf{Rep}_{\mathbf{Z}_p}(G_K) \to \mathbf{M}_{\mathbf{A}_K}^{\varphi,\text{ét}}, \qquad \mathbf{V} : \mathbf{M}_{\mathbf{A}_K}^{\varphi,\text{ét}} \to \mathbf{Rep}_{\mathbf{Z}_p}(G_K),$$

which are quasi-inverse to each other.

(ii) For all $T \in \mathbf{Rep}_{\mathbf{Z}_p}(G_K)$ *and* $D \in \mathbf{M}_{\mathbf{A}_K}^{\varphi,\text{ét}}$, *we have canonical and functorial isomorphisms compatible with the actions of* G_K *and* φ *on the both sides:*

$$\mathbf{D}(T) \otimes_{\mathbf{A}_K} \mathbf{A} \simeq T \otimes_{\mathbf{Z}_p} \mathbf{A},$$
$$\mathbf{V}(D) \otimes_{\mathbf{Z}_p} \mathbf{A} \simeq D \otimes_{\mathbf{A}_K} \mathbf{A}.$$

Proof The theorem can be proved by the same arguments as used in the proofs of Theorems 8.1.8 and 8.2.3 above. For details, see [69, Théorème 3.4.3]. □

Remark 8.2.10 We invite the reader to formulate and prove the analogous statements for the categories $\mathbf{Rep}_{\mathbf{F}_p}(G_K)$ and $\mathbf{Rep}_{\mathbf{Q}_p}(G_K)$.

8.2.11 We remark that for all $T \in \mathbf{Rep}_{\mathbf{Z}_p}(G_K)$, one has

$$\mathbf{D}_{O_{\mathscr{F}}}(T) \simeq \mathbf{D}(T) \otimes_{\mathbf{A}_K} O_{\mathscr{F}}.$$

Analogously, for all $D \in \mathbf{M}_{\mathbf{A}_K}^{\varphi,\text{ét}}$, one has

$$\mathbf{V}(D) \simeq \mathbf{V}_{O_{\mathscr{F}}}(D \otimes_{\mathbf{A}_K} O_{\mathscr{F}}).$$

Contrary to $\mathbf{D}_{O_{\mathscr{F}}}(T)$, the module $\mathbf{D}(T)$ is defined over a ring of formal power series. This allows to use the tools of p-adic analysis and relate (φ, Γ_K)-modules to the theory p-adic differential equations (Fontaine's program). See also Sect. 13 for further comments.

9 B-Admissible Representations

9.1 General Approach

9.1.1 The classification of all p-adic representations of local fields of characteristic 0 in terms of (φ, Γ_K)-modules is a powerful result. However, the representations arising in algebraic geometry have very special properties and form some natural subcategories of $\mathbf{Rep}_{\mathbf{Q}_p}(G_K)$. Moreover, as was first observed by Grothendieck in the good reduction case, it should be possible to classify them in terms of some objects of semi-linear algebra, such as filtered Dieudonné modules (Grothendieck's mysterious functor). In this section, we consider Fontaine's general approach to this problem. See [71] for a detailed exposition.

9.1.2 In this section, K is a local field. As usual, we denote by $\overline{K}$ its separable closure and set $G_K = \mathrm{Gal}(\overline{K}/K)$. To simplify notation, in the remainder of this paper we will write $\mathbf{C}$ instead of $\mathbf{C}_K$ for the p-adic completion of $\overline{K}$. Since the field of complex numbers will appear only occasionally, this convention should not lead to confusion.

Let B be a commutative $\mathbf{Q}_p$-algebra without zero divisors, equipped with a $\mathbf{Q}_p$-linear action of G_K. Let C denote the field of fractions of B. Set $E = B^{G_K}$. We adopt the definition of a regular algebra provided by Brinon and Conrad in [32], which differs from the original definition in [71].

Definition The algebra B is G_K-regular if it satisfies the following conditions:

(i) $B^{G_K} = C^{G_K}$;

(ii) Each non-zero $b \in B$ such that the line $\mathbf{Q}_p b$, is stable under the action of G_K, is invertible in B.

If B is a field, these conditions are satisfied automatically.

9.1.3 In the remainder of this section, we assume that B is G_K-regular. From the condition (ii), it follows that E is a field. For any p-adic representation V of G_K we consider the E-module

$$\mathbf{D}_B(V) = (V \otimes_{\mathbf{Q}_p} B)^{G_K}.$$

The multiplication in B induces a natural map

$$\alpha_B : \mathbf{D}_B(V) \otimes_E B \to V \otimes_{\mathbf{Q}_p} B.$$

Proposition 9.1.4 *(i) The map α_B is injective for all $V \in \mathbf{Rep}_{\mathbf{Q}_p}(G_K)$.*
(ii) $\dim_E \mathbf{D}(V) \leqslant \dim_{\mathbf{Q}_p} V$.

Proof See [32, Theorem 5.2.1]. Set $\mathbf{D}_C(V) = (V \otimes_{\mathbf{Q}_p} C)^{G_K}$. Since $B^{G_K} = C^{G_K}$, $\mathbf{D}_C(V)$ is an E-vector space, and we have the following diagram with injective vertical maps:

$$\begin{array}{ccc} \mathbf{D}_B(V) & \xrightarrow{\alpha_B} & V \otimes_{\mathbf{Q}_p} B \\ \downarrow & & \downarrow \\ \mathbf{D}_C(V) & \xrightarrow{\alpha_C} & V \otimes_{\mathbf{Q}_p} B. \end{array}$$

Therefore, it is sufficient to prove that α_C is injective. We prove it applying Artin's trick. Assume that $\ker(\alpha_C) \neq 0$ and choose a non-zero element

$$x = \sum_{i=1}^{m} d_i \otimes c_i \in \ker(\alpha_C)$$

of the shortest length m. It is clear that in this formula, $d_i \in \mathbf{D}_C(V)$ are linearly independent. Moreover, since C is a field, one can assume that $c_m = 1$. Then for all $g \in G_K$

$$g(x) - x = \sum_{i=1}^{m-1} d_i \otimes (g(c_i) - c_i) \in \ker(\alpha_C).$$

This shows that $g(x) = x$ for all $g \in G_K$, and therefore that $c_i \in C^{G_K} = E$ for all $1 \leqslant i \leqslant m$. Thus $x \in \mathbf{D}_C(V)$. From the definition of α_C, it follows that $\alpha_C(x) = x$, hence $x = 0$. □

Definition A p-adic representation V is called B-admissible if

$$\dim_E \mathbf{D}_B(V) = \dim_{\mathbf{Q}_p} V.$$

Proposition 9.1.5 *If V is admissible, then the map α_B is an isomorphism.*

Proof See [71, Proposition 1.4.2]. Let $v = \{v_i\}_{i=1}^n$ and $d = \{d_i\}_{i=1}^n$ be arbitrary bases of V and $\mathbf{D}_B(V)$ respectively. Then $v = Ad$ for some matrix A with coefficients in B. The bases $x = \bigwedge_{i=1}^n d_i \in \bigwedge^n \mathbf{D}_B(V)$ and $y = \bigwedge_{i=1}^n v_i \in \bigwedge^n V$ are related by $x = \det(A)y$. Since G_K acts on $y \in \bigwedge^n V$ as multiplication by a character, the $\mathbf{Q}_p$-vector space generated by $\det(A)$ is stable under the action of G_K. This shows that A is invertible, and α_B is an isomorphism. □

9.1.6 We denote by $\mathbf{Rep}_B(G_K)$ the category of B-admissible representations. The following proposition summarizes some properties of this category.

Proposition 9.1.7 *The category* $\mathbf{Rep}_B(G_K)$ *is a tannakian subcategory of all p-adic representations* $\mathbf{Rep}_{\mathbf{Q}_p}(G_K)$. *In particular, the following holds true:*

(i) If in an exact sequence

$$0 \to V' \to V \to V'' \to 0$$

V *is B-admissible, then* V' *and* V'' *are B-admissible.*

(ii) If V' *and* V'' *are B admissible, then* $V' \otimes_{\mathbf{Q}_p} V''$ *is B admissible.*

(iii) V *is B-admissible if and only if the dual representation* V^* *is B-admissible, and in that case* $\mathbf{D}_B(V^*) = \mathbf{D}_B(V)^*$.

(iv) The functor

$$\mathbf{D}_B : \mathbf{Rep}_B(G_K) \to \mathbf{Vect}_E$$

to the category of finite dimensional E-vector spaces, is exact and faithful.

Proof The proof is formal. See [71, Proposition 1.5.2]. □

9.1.8 We can also work with the contravariant version of the functor $\mathbf{D}_B$:

$$\mathbf{D}_B^*(V) = \mathrm{Hom}_{G_K}(V, B).$$

From definitions, it is clear that

$$\mathbf{D}_B^*(V) = \mathbf{D}_B(V^*).$$

In particular, if V (and therefore V^*) is admissible, then

$$\mathbf{D}_B^*(V) = \mathbf{D}_B(V)^* := \mathrm{Hom}_E(\mathbf{D}_B(V), E).$$

The last isomorphism shows that the covariant and contravariant theories are equivalent. For an admissible V, we have the canonical non-degenerate pairing

$$\langle\, ,\, \rangle : V \times \mathbf{D}^*(V) \to B, \qquad \langle v, f \rangle = f(v),$$

which can be seen as an abstract p-adic version of the canonical duality between singular homology and de Rham cohomology of a complex variety.

9.2 *First Examples*

9.2.1 $B = \overline{K}$, where K is of characteristic 0. The $\overline{K}$-admissible representations are p-adic representations having finite image. Indeed, since the action of G_K is discrete, each $\overline{K}$-admissible representation has finite image. Conversely, if V has finite image, it is $\overline{K}$-admissible by Hilbert's theorem 90.

9.2.2 $B = W(\bar{k}_K)[1/p]$. The B-admissible representations are unramified p-adic representations. This follows from Proposition 1.6.5.

9.2.3 $B = \widehat{\mathscr{F}}^{\mathrm{ur}}$. Let K be a local field of characteristic p, and let $\widehat{\mathscr{F}}^{\mathrm{ur}} = W(K^{\mathrm{rad}})[1/p]$. By Theorem 8.1.11, each p-adic representation of G_K is $\widehat{\mathscr{F}}^{\mathrm{ur}}$-admissible.

9.2.4 $B = \mathbf{C}$, *where* K *is of characteristic* 0. Sen proved (see Corollary 10.2.12 below) that V is $\mathbf{C}$-admissible if and only of I_K acts on V through a finite quotient. The sufficiency of this condition can be proved as follows. Set $n = \dim_{\mathbf{Q}_p} V$. Assume that $\rho(I_K)$ is finite. Let $U \subset I_K$ be a subgroup of finite index such that $\rho(U) = \{1\}$. By the theorem of Ax–Sen–Tate, $(V \otimes_{\mathbf{Q}_p} \mathbf{C})^U = V \otimes_{\mathbf{Q}_p} \widehat{L}$, where $L = \overline{K}^U$. Applying Hilbert's Theorem 90 to the extension $\widehat{L}/\widehat{K}^{\mathrm{ur}}$, we obtain that $(V \otimes_{\mathbf{Q}_p} \mathbf{C})^{I_K}$ is a n-dimensional vector space over $\widehat{K}^{\mathrm{ur}}$ equipped with a semi-linear action of $\mathrm{Gal}(K^{\mathrm{ur}}/K)$. Now from Proposition 1.6.5 it follows that $(V \otimes_{\mathbf{Q}_p} \mathbf{C})^{I_K}$ has a $\mathrm{Gal}(K^{\mathrm{ur}}/K)$-invariant basis, and therefore $\dim_K \mathbf{D}_{\mathbf{C}}(V) = n$.

The necessity is the difficult part of Sen's theorem, and we prove it only for one-dimensional representations.

Proposition 9.2.5 *If the one-dimensional representation* $\mathbf{Q}_p(\eta)$ *is* $\mathbf{C}$*-admissible, then* $\eta(I_K)$ *is finite.*

Proof a) If $\eta(I_K)$ is infinite, then from Theorem 4.3.2, it follows that $\mathbf{C}(\eta)^{G_K} = 0$. Hence, $\mathbf{Q}_p(\eta)$ is not $\mathbf{C}$-admissible. □

9.2.6 Consider the multiplicative group $\mathbb{G}_m$ over the field of complex numbers $\mathbb{C}$. Then $\mathbb{G}_m(\mathbb{C}) = \mathbb{C}^*$ is the punctured complex plane, and the Betti homology $H_1(\mathbb{G}_m)$ is the one-dimensional $\mathbf{Q}$-vector space generated by the counter-clockwise circle centered at 0. The de Rham cohomology $H^1_{\mathrm{dR}}(\mathbb{G}_m)$ is generated over K by the class of the differential form $\frac{dX}{X}$. The integration yields a non-degenerate bilinear map:

$$\begin{aligned} \langle\, , \rangle_{\mathbf{C}} : H_1(\mathbb{G}_m) \times H^1_{\mathrm{dR}}(\mathbb{G}_m) &\to \mathbb{C}, \\ \langle \gamma, \omega \rangle_{\mathbf{C}} &= \int_\gamma \omega. \end{aligned} \tag{51}$$

The p-adic realization of $\mathbb{G}_m$ is its Tate module:

$$T_p(\mathbb{G}_m) := \varprojlim_n \mu_{p^n} \simeq \mathbf{Z}_p(1).$$

The p-adic analog of the pairing (51) should be a non-degenerate bilinear map

$$\langle\, , \rangle : T_p(\mathbb{G}_m) \times H^1_{\mathrm{dR}}(\mathbb{G}_m) \to B,$$

with values in some ring B of "p-adic periods", compatible with the Galois action on $T_p(\mathbb{G}_m)$ and B. Proposition 9.2.5 shows that in the field $\mathbf{C}$, there doesn't exist a non-zero element t such that

$$g(t) = \chi_K(g)t, \qquad g \in G_K.$$

Therefore, the ring of p-adic periods should be in some sense "bigger" that $\mathbf{C}$.

10 Tate–Sen Theory

10.1 Hodge–Tate Representations

10.1.1 We maintain notation and conventions of Sect. 9.1. The notion of a Hodge–Tate representation was introduced in Tate's paper [151]. We use the formalism of admissible representations. Let K be a local field of characteristic 0. Let

$$\mathbf{B}_{\mathrm{HT}} = \mathbf{C}[t, t^{-1}]$$

denote the ring of polynomials in the variable t with integer exponents and coefficients in $\mathbf{C}$. We equip $\mathbf{B}_{\mathrm{HT}}$ with the action of G_K given by

$$g\left(\sum a_i t^i\right) = \sum g(a_i)\, \chi_K^i(g)\, t^i, \qquad g \in G_K,$$

where χ_K denotes the cyclotomic character. Therefore, G_K acts naturally on $\mathbf{C}$, and t can be viewed as the "p-adic $2\pi i$"—the p-adic period of the multiplicative group $\mathbb{G}_m$. For any p-adic representation V of G_K, we set:

$$\mathbf{D}_{\mathrm{HT}}(V) = (V \otimes_{\mathbf{Q}_p} \mathbf{B}_{\mathrm{HT}})^{G_K}.$$

Proposition 10.1.2 *The ring* $\mathbf{B}_{\mathrm{HT}}$ *is* G_K*-regular and* $\mathbf{B}_{\mathrm{HT}}^{G_K} = K$.

Proof (a) The field of fractions $\mathrm{Fr}(\mathbf{B}_{\mathrm{HT}})$ of $\mathbf{B}_{\mathrm{HT}}$ is isomorphic to the field of rational functions $\mathbf{C}(t)$. Embedding it in $\mathbf{C}((t))$, we have:

$$\mathbf{B}_{\mathrm{HT}}^{G_K} \subset \mathrm{Fr}(\mathbf{B}_{\mathrm{HT}})^{G_K} \subset \mathbf{C}((t))^{G_K}.$$

From Theorem 4.3.2, it follows that $(\mathbf{C}t^i)^{G_K} = K$ if $i = 0$, and $(\mathbf{C}t^i)^{G_K} = 0$ otherwise. Hence, $\mathbf{B}_{\mathrm{HT}}^{G_K} = \mathbf{C}((t))^{G_K} = K$. Therefore

$$\mathrm{Fr}(\mathbf{B}_{\mathrm{HT}})^{G_K} = \mathbf{B}_{\mathrm{HT}}^{G_K} = K.$$

(b) Let $b \in \mathbf{B}_{\mathrm{HT}} \setminus \{0\}$. Assume that $\mathbf{Q}_p b$ is stable under the action of G_K. This means that

$$g(b) = \eta(g)b, \qquad \forall g \in G_K \tag{52}$$

for some character $\eta : G_K \to \mathbf{Z}_p^*$. Write b in the form

$$b = \sum_i a_i t^i.$$

We will prove by contradiction that all, except one monomials in this sum are zero. From formula (52), if follows that for all i, one has

$$g(a_i)\chi_K^i(g) = a_i \eta(g), \qquad g \in G_K.$$

Assume that a_i and a_j are both non-zero for some $i \neq j$. Then

$$\frac{g(a_i)\chi_K^i(g)}{a_i} = \frac{g(a_j)\chi_K^j(g)}{a_j}, \qquad \forall g \in G_K.$$

Set $c = a_i/a_j$ and $m = i - j \neq 0$. Then c is a non-zero element of $\mathbf{C}$ such that

$$g(c)\chi_K^m(g) = c, \qquad \forall g \in G_K.$$

This is in contradiction with the fact that $\mathbf{C}(m)^{G_K} = 0$ if $m \neq 0$.

Therefore, $b = a_i t^i$ for some $i \in \mathbf{Z}$ and $a_i \neq 0$. This implies that b is invertible in $\mathbf{B}_{\mathrm{HT}}$. The proposition is proved. □

10.1.3 Let $\mathbf{Grad}_K$ denote the category of finite-dimensional graded K-vector spaces. The morphisms in this category are linear maps preserving the grading. We remark that $\mathbf{D}_{\mathrm{HT}}(V)$ has a natural structure of a graded K-vector space:

$$\mathbf{D}_{\mathrm{HT}}(V) = \underset{i \in \mathbf{Z}}{\oplus} \mathrm{gr}^i \, \mathbf{D}_{\mathrm{HT}}(V), \qquad \mathrm{gr}^i \, \mathbf{D}_{\mathrm{HT}}(V) = \left(V \otimes_{\mathbf{Q}_p} \mathbf{C}t^i\right)^{G_K}.$$

Therefore, we have a functor

$$\mathbf{D}_{\mathrm{HT}} \, : \, \mathbf{Rep}_{\mathbf{Q}_p}(G_K) \to \mathbf{Grad}_K.$$

Note that this functor is clearly left exact but not right exact (see Example 10.2.13 below).

Definition A p-adic representation V is a Hodge–Tate representation if it is $\mathbf{B}_{\mathrm{HT}}$-admissible.

We denote by $\mathbf{Rep}_{\mathrm{HT}}(G_K)$ the category of Hodge–Tate representations. From the general formalism of B-admissible representations, it follows that the restriction of $\mathbf{D}_{\mathrm{HT}}$ on $\mathbf{Rep}_{\mathrm{HT}}(G_K)$ is exact and faithful.

10.1.4 Set:

$$V^{(i)} = \{x \in V \otimes_{\mathbf{Q}_p} \mathbf{C} \mid g(x) = \chi_K(g)^i x, \quad \forall g \in G_K\}, \qquad i \in \mathbf{Z},$$
$$V\{i\} = V^{(i)} \otimes_K \mathbf{C}.$$

It is clear that $V^{(i)} \simeq \mathrm{gr}^{-i}\mathbf{D}_{\mathrm{HT}}(V)$. Moreover, the multiplication in $\mathbf{C}$ induces linear maps of $\mathbf{C}$-vector spaces $V\{i\} \to V \otimes_{\mathbf{Q}_p} \mathbf{C}$. Therefore, one has a $\mathbf{C}$-linear map:

$$\underset{i \in \mathbf{Z}}{\oplus} V\{i\} \to V \otimes_{\mathbf{Q}_p} \mathbf{C}. \tag{53}$$

The following proposition shows that our definition of a Hodge–Tate representation coincides with Tate's original definition:

Proposition 10.1.5 *(i) For any representation V, the map (53) is injective.*
(ii) V is a Hodge–Tate if and only if (53) is an isomorphism.

Proof (i) By Proposition 9.1.4, for *any* p-adic representation V, the map

$$\alpha_{\mathrm{HT}} : \mathbf{D}_{\mathrm{HT}}(V) \otimes_K \mathbf{B}_{\mathrm{HT}} \to V \otimes_{\mathbf{Q}_p} \mathbf{B}_{\mathrm{HT}}$$

is injective. The restriction of α_{HT} on the homogeneous subspaces of degree 0 coincides with the map (53). Therefore (53) is injective.

(ii) By Proposition 9.1.5, V is a Hodge–Tate if and only if α_{HT} is an isomorphism. We remark that α_{HT} is an isomorphism if and only if the map (53) is. This proves the proposition. □

Definition Let V be a Hodge–Tate representation. The isomorphism

$$V \otimes_{\mathbf{Q}_p} \mathbf{C} \simeq \underset{i \in \mathbf{Z}}{\oplus} V\{i\}$$

is called the Hodge–Tate decomposition of V. If $V\{i\} \neq 0$, one says that the integer i is a Hodge–Tate weight of V, and that $d_i = \dim_{\mathbf{C}} V\{i\}$ is the multiplicity of i.

We will use the standard notation $\mathbf{C}(i) = \mathbf{C}(\chi_K^i)$ for the cyclotomic twists of $\mathbf{C}$. Then $V\{i\} = \mathbf{C}(i)^{d_i}$ as a Galois module. The Hodge–Tate decomposition of V can be written in the following form:

$$V \otimes_{\mathbf{Q}_p} \mathbf{C} = \underset{i \in \mathbf{Z}}{\oplus} \mathbf{C}(i)^{d_i}.$$

Example 10.1.6 (1) Let $\psi : G_K \to \mathbf{Z}_p^*$ be a continuous character. Then $\mathbf{Q}_p(\psi)$ is a Hodge–Tate of weight i if and only if

$$\psi|_{I'_K} = \chi_K^i|_{I'_K}$$

for some open subgroup I'_K of the inertia group I_K. This follows from Proposition 9.2.5.

(2) Assume that E is a subextension K such that $\tau E \subset K$ for each conjugate of E over $\mathbf{Q}_p$. Let $\psi : G_K \to O_E^*$ be a continuous character. Then $E(\psi)$ can be seen as a p-adic representation of dimension $[E : \mathbf{Q}_p]$ with coefficients in $\mathbf{Q}_p$ and

$$E(\psi)\otimes_{\mathbf{Q}_p}\mathbf{C}=\bigoplus_{\tau\in\mathrm{Hom}_{\mathbf{Q}_p}(E,K)}\mathbf{C}(\tau\circ\psi).$$

Therefore, $E(\psi)$ is of Hodge–Tate if and only if for each τ

$$\mathbf{C}(\tau\circ\psi)=\mathbf{C}(\chi_K^{n_\tau}),\qquad \text{for some } n_\tau\in\mathbf{Z}.$$

We come back to this example in Sect. 15.

10.2 Sen's Theory

10.2.1 Let V be a Galois representation of G_K. Then $V\otimes_{\mathbf{Q}_p}\mathbf{C}$ can be viewed as an object of the category $\mathbf{Rep}_{\mathbf{C}}(G_K)$ of finite-dimensional $\mathbf{C}$-vector spaces equipped with a *semi-linear* action of G_K. This category was first studied by Sen [136]. Let $K_\infty=K(\zeta_{p^\infty})$ denote the cyclotomic extension of K. Set $\Gamma_K=\mathrm{Gal}(K_\infty/K)$ and $H_K=\mathrm{Gal}(\overline{K}/K_\infty)$. Let $W\in\mathbf{Rep}_{\mathbf{C}}(G_K)$. Sen's method decomposes into 3 steps:

10.2.2 *Descent to* $\widehat{K}_\infty$. Set $\widehat{W}_\infty=W^{H_K}$. By Theorem 4.3.8 and the inflation-restriction exact sequence, one has

$$H^1(\Gamma_K,\mathrm{GL}_n(\widehat{K}_\infty))\simeq H^1(G_K,\mathrm{GL}_n(\mathbf{C})).$$

Therefore, the natural map

$$\widehat{W}_\infty\otimes_{\widehat{K}_\infty}\mathbf{C}\to W$$

is an isomorphism. Let $\mathbf{Rep}_{\widehat{K}_\infty}(\Gamma_K)$ be the category of finite-dimensional $\widehat{K}_\infty$-vector spaces equipped with a semi-linear action of Γ_K. Then the functor

$$\mathbf{Rep}_{\mathbf{C}}(G_K)\to\mathbf{Rep}_{\widehat{K}_\infty}(\Gamma_K),\qquad W\mapsto\widehat{W}_\infty$$

is an equivalence of categories. Its quasi-inverse is given by extension of scalars $X\mapsto X\otimes_{\widehat{K}_\infty}\mathbf{C}$.

10.2.3 *Undoing the completion*. For any $\widehat{K}_\infty$-representation X, let X_f denote the union of all finite-dimensional K-vector subspaces of X. Sen proves that the map

$$X_f\otimes_{K_\infty}\widehat{K}_\infty\to X$$

is an isomorphism. The key tool here is the canonical isomorphism

$$H^1(\Gamma_K,\mathrm{GL}_n(K_\infty))\simeq H^1(\Gamma_K,\mathrm{GL}_n(\widehat{K}_\infty))$$

(see [136, Proposition 6]). This implies that the functors $X\mapsto X_f$ and $U\to U\otimes_{K_\infty}\widehat{K}_\infty$ are mutually quasi-inverse equivalences between $\mathbf{Rep}_{\widehat{K}_\infty}(\Gamma_K)$ and $\mathbf{Rep}_{K_\infty}(\Gamma_K)$.

10.2.4 *Infinitesimal action of* Γ_K. Let U be a K_∞-representation of Γ_K. If $\gamma \in \Gamma_K$ is close to 1, the formal power series

$$\frac{\log(\gamma)}{\log(\chi_K(\gamma))} = \frac{1}{\log(\chi_K(\gamma))} \sum_{n=1}^{\infty} (-1)^{n-1} \frac{(\gamma - 1)^n}{n}$$

defines a K_∞-*linear* operator Θ on U, which does not depend on the choice of γ. There exists an open subgroup $\Gamma' \subset \Gamma_K$ such that

$$\gamma(x) = \exp\bigl(\log(\chi_K(\gamma))\,\Theta\bigr)\,(x) \qquad \forall \gamma \in \Gamma', x \in U.$$

Let $\mathbf{S}_{K_\infty}$ denote the category of finite dimensional K_∞-vector spaces equipped with a linear operator. The morphisms of $\mathbf{S}_{K_\infty}$ are defined as K_∞-linear maps which commute with the action of underlying operators. Using Hilbert's Theorem 90, it can be checked that the functor

$$\mathbf{Rep}_{K_\infty}(\Gamma_K) \to \mathbf{S}_{K_\infty}, \qquad U \mapsto (U, \Theta)$$

is exact and fully faithful.

10.2.5 Combining previous results, one can associate to any $\mathbf{C}$-representation W the K_∞-vector space $W_\infty = (\widehat{W}_\infty)_f$ equipped with the operator Θ. The main result of Sen's theory states as follows:

Theorem 10.2.6 (Sen) *The functor*

$$\mathbf{\Delta}_{\mathrm{Sen}} : \mathbf{Rep}_{\mathbf{C}}(G_K) \to \mathbf{S}_{K_\infty}, \qquad W \mapsto (W_\infty, \Theta)$$

is exact and fully faithful.

Proof See [136]. □

Remark 10.2.7 Let $\Theta_{\mathbf{C}} : W \to W$ denote the linear operator obtained from Θ by extension of scalars. The map

$$W^{G_K} \otimes_K \mathbf{C} \to W$$

is injective and identifies $W^{G_K} \otimes_K \mathbf{C}$ with $\ker(\Theta_{\mathbf{C}})$. In particular, W^{G_K} is a finite-dimensional K-vector space.

10.2.8 We discuss some applications of Sen's theory to p-adic representations. To any p-adic representation $\rho : G_K \to \mathrm{Aut}_{\mathbf{Q}_p} V$, we associate the $\mathbf{C}$-representation $W = V \otimes_{\mathbf{Q}_p} \mathbf{C}$ and set:

$$\mathbf{D}_{\mathrm{Sen}}(V) = \mathbf{\Delta}_{\mathrm{Sen}}(W).$$

Hodge-Tate representations have the following characterization in terms of the operator Θ :

Proposition 10.2.9 *V is a Hodge–Tate representation if and only if the operator $\Theta : \Delta_{\mathrm{Sen}}(V) \to \Delta_{\mathrm{Sen}}(V)$ is semi-simple and its eigenvalues belong to $\mathbf{Z}$.*

Proof See [136, Section 2.3]. □

10.2.10 We come back to general p-adic representations. The operator Θ allows to recover the Lie algebra of the image $\rho(I_K)$ of the inertia group:

Theorem 10.2.11 (Sen) *The Lie algebra $\mathfrak{g}$ of $\rho(I_K)$ is the smallest of the $\mathbf{Q}_p$-subspaces S of $\mathrm{End}_{\mathbf{Q}_p}(V)$ such that $\Theta \in S \otimes_{\mathbf{Q}_p} \mathbf{C}$.*

Proof See [136, Theorem 11]. □

The following corollary of this theorem generalizes Proposition 9.2.5.

Corollary 10.2.12 *$\rho(I_K)$ is finite if and only if $\Theta = 0$.*

Example 10.2.13 Let V be a two dimensional $\mathbf{Q}_p$-vector space with a fixed basis $\{e_1, e_2\}$. Let $\rho : G_K \to \mathrm{GL}(V)$ be the representation given by

$$\rho(g) = \begin{pmatrix} 1 & \log(\chi_K(g)) \\ 0 & 1 \end{pmatrix} \qquad \text{in the basis} \{e_1, e_2\}.$$

Prove that V is not Hodge–Tate. Let $\overline{e}_2 = e_2 \pmod{\mathbf{Q}_p e_1}$. Since V sits in the exact sequence

$$0 \to \mathbf{Q}_p e_1 \to V \to \mathbf{Q}_p \overline{e}_2 \to 0,$$

we have an exact sequence:

$$0 \to \mathbf{D}_{\mathrm{HT}}(\mathbf{Q}_p e_1) \to \mathbf{D}_{\mathrm{HT}}(V) \to \mathbf{D}_{\mathrm{HT}}(\mathbf{Q}_p \overline{e}_2).$$

Here $\mathbf{Q}_p e_1$ and $\mathbf{Q}_p \overline{e}_2$ are trivial p-adic representations, and

$$\mathbf{D}_{\mathrm{HT}}(\mathbf{Q}_p e_1) = K e_1, \qquad \mathbf{D}_{\mathrm{HT}}(\mathbf{Q}_p \overline{e}_2) = K \overline{e}_2.$$

Therefore, $\mathbf{D}_{\mathrm{HT}}(V)$ has dimension 2 if and only if $\overline{e}_2$ lifts to an element

$$x = e_2 + \lambda \otimes e_1 \in \mathbf{D}_{\mathrm{HT}}(V), \qquad \lambda \in \mathbf{B}_{\mathrm{HT}}.$$

The condition $x \in \mathbf{D}_{\mathrm{HT}}(V)$ reads:

$$g(\lambda) - \lambda = \log \chi_K(g), \qquad \forall g \in G_K.$$

Therefore, $\log \chi_K$ is a coboundary in $\mathbf{C}$, but this contradicts to Theorem 4.3.2. Hence, V is not Hodge–Tate. This example also shows that $\mathbf{Rep}_{\mathrm{HT}}(G_K)$ is not stable under extensions.

Note that in the same basis, the operator Θ reads:

$$\Theta = \begin{pmatrix} 0 & 1 \\ 0 & 0 \end{pmatrix}.$$

In particular, it is not semi-simple, and the above arguments agree with Proposition 10.2.9.

11 Rings of p-Adic Periods

11.1 The Field $\mathbf{B}_{\mathrm{dR}}$

11.1.1 In this section, we define Fontaine's rings of p-adic periods $\mathbf{B}_{\mathrm{dR}}$, $\mathbf{B}_{\mathrm{st}}$ and $\mathbf{B}_{\mathrm{cris}}$. For proofs and more detail, we refer the reader to [66, 68] and [70].

Let K be a local field of characteristic 0. Recall that the ring of integers of the tilt $\mathbf{C}^\flat$ of $\mathbf{C}$ was defined as the projective limit

$$O_{\mathbf{C}}^\flat = \varprojlim_{\varphi} O_{\mathbf{C}}/p\, O_{\mathbf{C}}, \qquad \varphi(x) = x^p$$

(see Sect. 5.2). By Propositions 5.2.3 and 5.2.4, $O_{\mathbf{C}}^\flat$ is a complete perfect valuation ring of characteristic p with residue field $\overline{k}_K$. The field $\mathbf{C}^\flat$ is a complete algebraically closed field of characteristic p.

11.1.2 We will denote by $\mathbf{A}_{\mathrm{inf}}$ the ring of Witt vectors

$$\mathbf{A}_{\mathrm{inf}}(\mathbf{C}) = W(O_{\mathbf{C}}^\flat).$$

Recall that $\mathbf{A}_{\mathrm{inf}}$ is equipped with the surjective ring homomorphism $\theta : \mathbf{A}_{\mathrm{inf}} \to O_{\mathbf{C}}$ (see Proposition 5.3.3, where it is denoted by θ_E). The kernel of θ is the principal ideal generated by any element $\xi = \sum_{n=0}^{\infty} [a_n] p^n \in \ker(\theta)$ such that a_1 is a unit in $O_{\mathbf{C}^\flat}$. Useful canonical choices are:

- $\xi = [\tilde{p}] - p$, where $\tilde{p} = (p^{1/p^n})_{n \geqslant 0}$;
- $\omega = \sum_{i=0}^{p-1} [\varepsilon]^{i/p}$, where $\varepsilon = (\zeta_{p^n})_{n \geqslant 0}$.

Let K_0 denote the maximal unramified subextension of K. Then $O_{K_0} = W(k_K) \subset \mathbf{A}_{\mathrm{inf}}$, and we set $\mathbf{A}_{\mathrm{inf},K} = \mathbf{A}_{\mathrm{inf}} \otimes_{O_{K_0}} K$. Then θ extends by linearity to a sujective homomorphism

$$\theta \otimes \mathrm{id}_K : \mathbf{A}_{\mathrm{inf}}(\mathbf{C}) \otimes_{O_{K_0}} K \to \mathbf{C}.$$

Again, the kernel $J_K := \ker(\theta \otimes \mathrm{id}_K)$ is a principal ideal. It is generated, for example, by $[\tilde{\pi}] - \pi$, where π is any uniformizer of K and $\tilde{\pi} = (\pi^{1/p^n})_{n \geqslant 0}$. The action of G_K extends naturally to $\mathbf{A}_{\mathrm{inf},K}$, and it's easy to see that J_K is stable under this action.

Let $\mathbf{B}^+_{\mathrm{dR},K}$ denote the completion of $\mathbf{A}_{\mathrm{inf},K}$ for the J_K-adic topology:

$$\mathbf{B}^+_{\mathrm{dR},K} = \varprojlim_n \mathbf{A}_{\mathrm{inf},K}/J_K^n.$$

The action of G_K extends to $\mathbf{B}^+_{\mathrm{dR},K}$. The main properties of $\mathbf{B}^+_{\mathrm{dR},K}$ are summarized in the following proposition:

Proposition 11.1.3 *(i)* $\mathbf{B}^+_{\mathrm{dR},K}$ *is a discrete valuation ring with maximal ideal*

$$\mathfrak{m}_{\mathrm{dR},K} = J_K \mathbf{B}^+_{\mathrm{dR},K}.$$

The residue field $\mathbf{B}^+_{\mathrm{dR},K}/\mathfrak{m}_{\mathrm{dR},K}$ *is isomorphic to* $\mathbf{C}$ *as a Galois module.*

(ii) The series

$$t = \log([\varepsilon]) = \sum_{n=1}^{\infty} (-1)^{n-1} \frac{([\varepsilon]-1)^n}{n}$$

converges in the J_K*-adic topology to a uniformizer of* $\mathbf{B}^+_{\mathrm{dR},K}$*, and the Galois group acts on t as follows:*

$$g(t) = \chi_K(g)t, \qquad g \in G_K.$$

(iii) If L/K *is a finite extension, then the natural map* $\mathbf{B}^+_{\mathrm{dR},K} \to \mathbf{B}^+_{\mathrm{dR},L}$ *is an isomorphism. In particular,* $\mathbf{B}^+_{\mathrm{dR},K}$ *depends only on the algebraic closure* $\overline{K}$ *of* K*.*

(iv) There exists a natural G_K*-equivariant embedding of* $\overline{K}$ *in* $\mathbf{B}^+_{\mathrm{dR},K}$*, and*

$$\left(\mathbf{B}^+_{\mathrm{dR},K}\right)^{G_K} = K.$$

11.1.4 We refer the reader to [66] and [70] for detailed proofs of these properties. Note that if L is a finite extension of K, then one checks first that $\mathbf{B}^+_{\mathrm{dR},K} \subset \mathbf{B}^+_{\mathrm{dR},L}$. From assertions i) and ii), it follows that this is an unramified extension of discrete valuation rings with the same residue field. This implies that $\mathbf{B}^+_{\mathrm{dR},K} = \mathbf{B}^+_{\mathrm{dR},L}$. Since $L \subset \mathbf{B}^+_{\mathrm{dR},L}$ for all L/K, this proves that $\overline{K} \subset \mathbf{B}^+_{\mathrm{dR},K}$.

11.1.5 The above proposition shows that $\mathbf{B}^+_{\mathrm{dR},K}$ depends only on the residual characteristic of the local field K. By this reason, we will omit K from notation and write $\mathbf{B}^+_{\mathrm{dR}} := \mathbf{B}^+_{\mathrm{dR},K}$.

Definition The field of p-adic periods $\mathbf{B}_{\mathrm{dR}}$ is defined to be the field of fractions of $\mathbf{B}^+_{\mathrm{dR}}$.

11.1.6 The field $\mathbf{B}_{\mathrm{dR}}$ is equipped with the canonical filtration induced by the discrete valuation, namely

$$\mathrm{Fil}^i \mathbf{B}_{\mathrm{dR}} = t^i \mathbf{B}^+_{\mathrm{dR}}, \qquad i \in \mathbf{Z}.$$

In particular, $\mathrm{Fil}^0 \mathbf{B}_{\mathrm{dR}} = \mathbf{B}^+_{\mathrm{dR}}$ and $\mathrm{Fil}^1 \mathbf{B}_{\mathrm{dR}} = \mathfrak{m}_{\mathrm{dR}}$. From Proposition 11.1.3, it follows that

$$\mathrm{Fil}^i\mathbf{B}_{\mathrm{dR}}/\mathrm{Fil}^{i+1}\mathbf{B}_{\mathrm{dR}} \simeq \mathbf{C}(i).$$

Therefore, for the associated graded module we have

$$\mathrm{gr}^\bullet(\mathbf{B}_{\mathrm{dR}}) \simeq \mathbf{B}_{\mathrm{HT}}.$$

Note that from this isomorphism it follows that $\mathbf{B}_{\mathrm{dR}}^{G_K} = K$ as claimed in Proposition 11.1.3, iii).

11.1.7 Recall that $\mathbf{A}_{\mathrm{inf}}$ is equipped with the canonical Frobenius operator φ. Set $X = [\varepsilon] - 1$. Then

$$\varphi(\omega) = \frac{\varphi(X)}{X} = \frac{(1+X)^p - 1}{X} = p + \binom{p}{2}X + \cdots + X^{p-1}.$$

From this formula it follows that $\ker(\theta)$ is *not* stable under the action of φ, and therefore φ can not be naturally extended to $\mathbf{B}_{\mathrm{dR}}$.

11.1.8 The field $\mathbf{B}_{\mathrm{dR}}$ is equipped with the topology induced by the discrete valuation. Now we equip it with a coarser topology, which is better adapted to the study of $\mathbf{B}_{\mathrm{dR}}$. Recall that the valuation topology on $\mathbf{C}^\flat$ induces a topology on $\mathbf{A}_{\mathrm{inf}}$, which we call the canonical topology (see Sect. 5.3). This topology induces a topology on $\mathbf{A}_{\mathrm{inf},K}$. The *canonical topology* on $\mathbf{B}_{\mathrm{dR}}^+ = \varprojlim_n \mathbf{A}_{\mathrm{inf},K}/J_K^n$ is defined as the topology of the inverse limit, where $\mathbf{A}_{\mathrm{inf},K}/J_K^n$ are equipped with the quotient topology. We refer the reader to [32, Exercise 4.5.3] for further detail.

11.2 *The Rings* $\mathbf{B}_{\mathrm{cris}}$ *and* $\mathbf{B}_{\mathrm{max}}$

11.2.1 We define the ring $\mathbf{B}_{\mathrm{cris}}$ of crystalline p-adic periods, which is a subring of $\mathbf{B}_{\mathrm{dR}}$ equipped with a natural Frobenius structure. The map $\theta : \mathbf{A}_{\mathrm{inf}} \to O_{\mathbf{C}}$ is the universal formal thickening of $O_{\mathbf{C}}$ in the sense of [70], and we denote by $\mathbf{A}_{\mathrm{inf}}^{\mathrm{PD}}$ the PD-envelop of $\ker(\theta)$ in $\mathbf{A}_{\mathrm{inf}}$ (see, for example, [22] for definition and basic properties of divided powers). Recall that

$$\xi = [\widetilde{p}] - p \in \mathbf{A}_{\mathrm{inf}}$$

is a generator of the $\ker(\theta)$. Then $\mathbf{A}_{\mathrm{inf}}^{\mathrm{PD}}$ can be seen as the submodule of $\mathbf{B}_{\mathrm{dR}}^+$ defined as

$$\mathbf{A}_{\mathrm{inf}}^{\mathrm{PD}} = \mathbf{A}_{\mathrm{inf}}\left[\frac{\xi^2}{2!}, \frac{\xi^3}{3!}, \ldots, \frac{\xi^n}{n!}, \ldots\right].$$

From the formula

$$\frac{\xi^n}{n!}\frac{\xi^m}{m!} = \binom{n+m}{n}\frac{\xi^{n+m}}{(n+m)!}$$

it follows that $\mathbf{A}_{\inf}^{\mathrm{PD}}$ is a subring of $\mathbf{B}_{\mathrm{dR}}$. Let

$$\mathbf{A}_{\mathrm{cris}}^{+} := \widehat{\mathbf{A}}_{\inf}^{\mathrm{PD}} = \varprojlim_{n} \mathbf{A}_{\inf}^{\mathrm{PD}} / p^n \mathbf{A}_{\inf}^{\mathrm{PD}}$$

denote its p-adic completion.

Proposition 11.2.2 *$\mathbf{A}_{\inf}^{\mathrm{PD}}$ is stable under the action of φ. Moreover, the action of φ extends to a continuous injective map $\varphi : \mathbf{A}_{\mathrm{cris}}^{+} \to \mathbf{A}_{\mathrm{cris}}^{+}$.*

Proof We have

$$\varphi(\xi) = [\widetilde{p}]^p - p = (\xi + p)^p - p = \xi^p + pz$$

for some $z \in \mathbf{A}_{\inf}$. Hence,

$$\frac{\varphi(\xi^n)}{n!} = \frac{p^n}{n!} \left(1 + (p-1)! \frac{\xi^p}{p!}\right)^n .$$

Since $\mathbf{A}_{\inf}^{\mathrm{PD}}$ is a ring, and $\frac{p^n}{n!} \in \mathbf{Z}_p$, this implies the proposition. □

11.2.3 It can be shown that the inclusion $\mathbf{A}_{\inf}^{\mathrm{PD}} \subset \mathbf{B}_{\mathrm{dR}}^{+}$ extends to a continuous embedding

$$\mathbf{A}_{\mathrm{cris}}^{+} \subset \mathbf{B}_{\mathrm{dR}}^{+},$$

where $\mathbf{A}_{\mathrm{cris}}^{+}$ and $\mathbf{B}_{\mathrm{dR}}^{+}$ are equipped with the p-adic and canonical topology respectively. In more explicit terms, $\mathbf{A}_{\mathrm{cris}}^{+}$ can be viewed as the subring

$$\mathbf{A}_{\mathrm{cris}}^{+} = \left\{ \sum_{n=0}^{\infty} a_n \frac{\xi^n}{n!} \mid a_n \in \mathbf{A}_{\inf}, \quad \lim_{n \to +\infty} a_n = 0 \right\} \subset \mathbf{B}_{\mathrm{dR}}^{+}.$$

The element $t = \log[\varepsilon]$ belongs to $\mathbf{A}_{\mathrm{cris}}^{+}$, and one has

$$\varphi(t) = pt.$$

Definition Set $\mathbf{B}_{\mathrm{cris}}^{+} = \mathbf{A}_{\mathrm{cris}}^{+}[1/p]$ and $\mathbf{B}_{\mathrm{cris}} = \mathbf{B}_{\mathrm{cris}}^{+}[1/t]$. The ring $\mathbf{B}_{\mathrm{cris}}$ is called the ring of crystalline periods.

It is easy to see that the rings $\mathbf{B}_{\mathrm{cris}}^{+}$ and $\mathbf{B}_{\mathrm{cris}}$ are stable under the action of G_K. The actions of G_K and φ on $\mathbf{B}_{\mathrm{cris}}$ commute to each other. The inclusion $\mathbf{B}_{\mathrm{cris}} \subset \mathbf{B}_{\mathrm{dR}}$ induces a filtration on $\mathbf{B}_{\mathrm{cris}}$ which we denote by $\mathrm{Fil}^i \mathbf{B}_{\mathrm{cris}}$. Note that $\mathbf{B}_{\mathrm{cris}}^{+} \subset \mathrm{Fil}^0 \mathbf{B}_{\mathrm{cris}}$ but the latter space is much bigger. Also the action of φ on $\mathbf{B}_{\mathrm{cris}}$ is not compatible with filtration i.e. $\varphi(\mathrm{Fil}^i \mathbf{B}_{\mathrm{cris}}) \not\subset \mathrm{Fil}^i \mathbf{B}_{\mathrm{cris}}$. We summarize some properties of $\mathbf{B}_{\mathrm{cris}}$ in the following proposition.

Proposition 11.2.4 *The following holds true:*

(i) The map

$$K \otimes_{K_0} \mathbf{B}_{\mathrm{cris}} \to \mathbf{B}_{\mathrm{dR}}, \qquad a \otimes x \to ax$$

is injective.
(ii) $\mathbf{B}_{\mathrm{cris}}^{G_K} = K_0$.
(iii) $\mathrm{Fil}^0 \mathbf{B}_{\mathrm{cris}}^{\varphi=1} = \mathbf{Q}_p$.
(iv) $\mathbf{B}_{\mathrm{cris}}$ *is* G_K*-regular.*

Proof See [70], especially Theorems 4.2.4 and 5.3.7. □

11.2.5 The main information about the relationship between the filtration on $\mathbf{B}_{\mathrm{cris}}$ and the Frobenius map is contained in the *fundamental exact sequence*:

$$0 \to \mathbf{Q}_p \to \mathbf{B}_{\mathrm{cris}}^{\varphi=1} \to \mathbf{B}_{\mathrm{dR}}/\mathrm{Fil}^0 \mathbf{B}_{\mathrm{dR}} \to 0. \tag{54}$$

The exactness in the middle term is equivalent to Proposition 11.2.4, iii) above. In addition, (54) says that and the projection $\mathbf{B}_{\mathrm{cris}}^{\varphi=1} \to \mathbf{B}_{\mathrm{dR}}/\mathbf{B}_{\mathrm{dR}}^+$ is surjective. We refer to [70] and [28] for proofs and related results.

11.2.6 The importance of the ring $\mathbf{B}_{\mathrm{cris}}$ relies on its connection to the crystalline cohomology [74]. On the other hand, the natural topology on $\mathbf{B}_{\mathrm{cris}}$ is quite ugly (see [40]). Sometimes, it is more convenient to work with the rings

$$\mathbf{A}_{\max}^+ = \left\{ \sum_{n=0}^{\infty} a_n \frac{\xi^n}{p^n} \mid a_n \in \mathbf{A}_{\mathrm{inf}}, \quad \lim_{n \to +\infty} a_n = 0 \right\},$$
$$\mathbf{B}_{\max}^+ = \mathbf{A}_{\max}^+ \otimes_{\mathbf{Z}_p} \mathbf{Q}_p,$$
$$\mathbf{B}_{\max} = \mathbf{B}_{\max}^+[1/t],$$

which are equipped with a natural action of φ and have better topological properties. One has

$$\varphi(\mathbf{B}_{\max}) \subset \mathbf{B}_{\mathrm{cris}} \subset \mathbf{B}_{\max}.$$

In particular, $\mathbf{B}_{\max}^{\varphi=1} = \mathbf{B}_{\mathrm{cris}}^{\varphi=1}$, and in the fundamental exact sequence $\mathbf{B}_{\mathrm{cris}}$ can be replaced by $\mathbf{B}_{\max}$. Note that the periods of crystalline representations (see Sect. 13) live in the ring

$$\widetilde{\mathbf{B}}_{\mathrm{rig}} = \bigcap_{i=0}^{\infty} \varphi^n(\mathbf{B}_{\mathrm{cris}}) = \bigcap_{i=0}^{\infty} \varphi^n(\mathbf{B}_{\max}).$$

We refer the reader to [40] for proofs and further results about these rings.

11.3 The Ring $\mathbf{B}_{\mathrm{st}}$

11.3.1 Morally $\mathbf{B}_{\mathrm{st}}$ is the ring of p-adic periods of varieties having semi-stable reduction modulo p. The simplest example of such a variety is provided by Tate elliptic curves E_q/K. Tate's original paper dated 1959 appeared only in [152], but an exposition of his theory can be found in [127]. See also [147] and [142]. For each $q \in K^*$ with $|q|_p < 1$, Tate constructs an elliptic curve E_q with modular invariant given by the usual formula

$$j(q) = \frac{1}{q} + 744 + 196884q + \dots$$

and having multiplicative split reduction modulo p. If E is an elliptic curve with modular invariant $j(E)$ such that $|j(E)|_p > 1$, then $j(E) = j(q)$ for some q, and E is isomorphic to E_q over a quadratic extension of K. The group of points $E_q(\overline{K})$ of E_q is isomorphic to $\overline{K}^*/q^{\mathbf{Z}}$, and the associated p-adic representation $V_p(E)$ is reducible and sits in an exact sequence

$$0 \to \mathbf{Q}_p(1) \to V_p(E) \to \mathbf{Q}_p \to 0.$$

There exists a basis $\{e_1, e_2\}$ of $V_p(E)$ such that the action of G_K is given by

$$g(e_1) = \chi_K(g)\, e_1, \quad g(e_2) = e_2 + \psi_q(g) e_1, \qquad g \in G_K,$$

where $\psi_q \,:\, G_K \to \mathbf{Z}_p$ is the cocycle defined by

$$g(\sqrt[p^m]{q}) = \zeta_{p^n}^{\psi_q(g)} \sqrt[p^m]{q}.$$

11.3.2 The ring $\mathbf{B}_{\mathrm{st}}$ is defined as the ring $\mathbf{B}_{\mathrm{cris}}[u]$ of polynomials with coefficients in $\mathbf{B}_{\mathrm{cris}}$. The Frobenius map extends to $\mathbf{B}_{\mathrm{st}}$ by $\varphi(u) = pu$. One equips $\mathbf{B}_{\mathrm{st}}$ by a *monodromy operator* N defined by $N = -\dfrac{d}{du}$. The operators φ and N are related by the formula:

$$N\varphi = p\varphi N.$$

This formula should be compared with the formulation of the ℓ-adic monodromy theorem (Theorem 7.2.3). One extends the Galois action on $\mathbf{B}_{\mathrm{st}}$ setting:

$$g(u) = u + \psi_p(g)t, \qquad g \in G_K,$$

where $\psi_p \,:\, G_K \to \mathbf{Z}_p$ is the cocycle defined by

$$g([\widetilde{p}]) = [\varepsilon]^{\psi_p(g)}[\widetilde{p}], \qquad g \in G_K.$$

There exists a G_K-equivariant embedding of $\mathbf{B}_{\mathrm{st}}$ in $\mathbf{B}_{\mathrm{dR}}$ which sends u onto the element

$$\log[\tilde{p}] = \log p + \sum_{n=1}^{\infty} \frac{(-1)^{n-1}}{n} \left(\frac{[\tilde{p}]}{p} - 1\right)^n.$$

We remark that this embedding is not canonical and depends on the choice of $\log p$. In particular, there is no canonical filtration on $\mathbf{B}_{\mathrm{st}}$. Note that it is customary to choose $\log p = 0$.

Finally we remark that sometimes it is more natural to work with the ring $\mathbf{B}_{\mathrm{mst}} = \mathbf{B}_{\max}[u]$ instead $\mathbf{B}_{\mathrm{st}}$, which is equipped with the same structures but has better topological properties.

12 Filtered (φ, N)-Modules

12.1 Filtered Vector Spaces

12.1.1 In this section, we review the theory of filtered Dieudonné modules. The main reference is [71]. We also refer the reader to [8] for the general formalism of slope filtrations. Let K be an arbitrary field.

Definition A filtered vector space over K is a finite dimensional K-vector space Δ equipped with an exhaustive separated decreasing filtration by K-subspaces $(\mathrm{Fil}^i \Delta)_{i\in\mathbf{Z}}$:

$$\ldots \supset \mathrm{Fil}^{i-1}\Delta \supset F^i\Delta \supset F^{i+1}\Delta \supset \ldots, \qquad \bigcap_{i\in\mathbf{Z}} \mathrm{Fil}^i\Delta = \{0\}, \quad \bigcup_{i\in\mathbf{Z}} \mathrm{Fil}^i\Delta = \Delta.$$

A morphism of filtered spaces is a linear map $f : \Delta' \to \Delta''$ which is compatible with filtrations:

$$f(\mathrm{Fil}^i\Delta') \subset \mathrm{Fil}^i\Delta'', \qquad \forall i \in \mathbf{Z}.$$

If Δ' and Δ'' are two filtered spaces, one defines the filtered space $\Delta' \otimes_K \Delta''$ as the tensor product of Δ' and Δ'' equipped with the filtration

$$\mathrm{Fil}^i(\Delta' \otimes_K \Delta'') = \sum_{i'+i''=i} \mathrm{Fil}^{i'}\Delta' \otimes_K \mathrm{Fil}^{i''}\Delta''.$$

The one-dimensional vector space $\mathbf{1}_K = K$ with the filtration

$$\mathrm{Fil}^i\mathbf{1}_K = \begin{cases} K, & \text{if } i \leqslant 0, \\ 0, & \text{if } i > 0 \end{cases}$$

is a unit object with respect to the tensor product defined above, namely

$$\Delta \otimes_K \mathbf{1}_K \simeq \Delta \qquad \text{for any filtered module } \Delta.$$

One defines the internal Hom in the category of filtered vector spaces as the vector space $\underline{\mathrm{Hom}}_K(\Delta', \Delta'')$ of K-linear maps $f : \Delta' \to \Delta''$ equipped with the filtration

$$\mathrm{Fil}^i\left(\underline{\mathrm{Hom}}_K(\Delta', \Delta'')\right) = \{f \in \underline{\mathrm{Hom}}_K(\Delta', \Delta'') \mid f(\mathrm{Fil}^j \Delta') \subset \mathrm{Fil}^{j+i}(\Delta'') \quad \forall j \in \mathbf{Z}\}.$$

In particular, we consider the dual space $\Delta^* = \underline{\mathrm{Hom}}_K(\Delta, \mathbf{1}_K)$ as a filtered vector space.

We denote by $\mathbf{MF}_K$ the category of filtered K-vector spaces. It is easy to check that the category $\mathbf{MF}_K$ is an additive tensor category with kernels and cokernels, but it is not abelian.

Example 12.1.2 Let W be a non-zero K-vector space. Let Δ' and Δ'' denote W equipped with the following filtrations:

$$\mathrm{Fil}^i \Delta' = \begin{cases} W, & \text{if } i \leqslant 0, \\ 0, & \text{if } i \geqslant 1, \end{cases} \qquad \mathrm{Fil}^i \Delta'' = \begin{cases} W, & \text{if } i \leqslant 1, \\ 0, & \text{if } i \geqslant 2. \end{cases}$$

The identity map $\mathrm{id}_W : W \to W$ defines a morphism $f : \Delta' \to \Delta''$ in $\mathbf{MF}_K$. It is easy to check that $\ker(f) = 0$ and $\mathrm{coker}(f) = 0$. Therefore, f is both a monomorphism and an epimorphism, but $\Delta' \not\simeq \Delta''$.

12.1.3 We adopt the following general definition:

Definition Let $\mathscr{C}$ be an additive category with kernels and cokernels. A sequence

$$0 \to X' \xrightarrow{f} X \xrightarrow{g} X'' \to 0$$

of objects in $\mathscr{C}$ is exact if $X' = \ker(g)$ and $X'' = \mathrm{coker}(f)$.

The following proposition describes short exact sequences in $\mathbf{MF}_K$:

Proposition 12.1.4 *(i) Let $f : \Delta' \to \Delta''$ be a morphism of filtered vector spaces. The canonical isomorphism*

$$\mathrm{coim}(f) = \Delta'/\ker(f) \to \mathrm{Im}(f)$$

is an isomorphism if and only if

$$f(\mathrm{Fil}^i \Delta') = f(\Delta') \cap \mathrm{Fil}^i \Delta'', \qquad \forall i \in \mathbf{Z}. \tag{55}$$

(ii) A short sequence of filtered spaces

$$0 \to \Delta' \to \Delta \to \Delta'' \to 0 \tag{56}$$

is exact if and only if for each $i \in \mathbb{Z}$ the sequence

$$0 \to \mathrm{Fil}^i \Delta' \to \mathrm{Fil}^i \Delta \to \mathrm{Fil}^i \Delta'' \to 0$$

is exact.

Proof The proof is left as an exercise. See also [50, Sect. 1]. □

12.1.5 For each filtered space, set:

$$t_{\mathrm{H}}(\Delta) = \sum_{i \in \mathbf{Z}} i \, \dim_K \left(\mathrm{gr}^i \Delta\right),$$

where $\mathrm{gr}^i \Delta = \mathrm{Fil}^i \Delta / \mathrm{Fil}^{i+1} \Delta$.

Proposition 12.1.6 *(i) The function t_{H} is additive, i.e. for any exact sequence of filtetred spaces (56) one has*

$$t_{\mathrm{H}}(\Delta) = t_{\mathrm{H}}(\Delta') + t_{\mathrm{H}}(\Delta'').$$

(ii) $t_{\mathrm{H}}(\Delta) = t_{\mathrm{H}}(\wedge^d \Delta)$, where $d = \dim_K \Delta$.

Proof (i) From the definition of an exact sequence it follows that the sequence

$$0 \to \mathrm{gr}^i \Delta' \to \mathrm{gr}^i \Delta \to \mathrm{gr}^i \Delta'' \to 0$$

is exact for all i. Therefore,

$$\dim_K(\mathrm{gr}^i \Delta) = \dim_K(\mathrm{gr}^i \Delta') + \dim_K(\mathrm{gr}^i \Delta'').$$

This implies (i).

(ii) For each i, choose a base $\{\overline{e}_{ij}\}_{j=1}^{d_i}$ of $\mathrm{gr}^i \Delta$ and denote by $\{e_{ij}\}_{j=1}^{d_i}$ its arbitrary lift in $\mathrm{Fil}^i \Delta$. Then $e = \underset{i,j}{\wedge} e_{ij}$ is a basis of $\wedge^d \Delta$. This description shows that $t_{\mathrm{H}}(\Delta)$ is the unique filtration break of $\wedge^d \Delta$. □

12.2 φ-Modules

12.2.1 In this section, we study in more detail the category of φ-modules over the field of fractions of Witt vectors, which was defined in Sect. 8.1. Here we change notation slightly and denote by k a perfect field of characteristic p and by K_0 the field $W(k)[1/p]$. This notation is consistent with the applications to the classification of p-adic representations of local fields of characteristic 0 which will be discussed in Sect. 13. As before, φ denotes the automorphism of Frobenius acting on K_0. Recall that a φ-module (or an φ-isocrystal) over K_0 is a finite dimensional K_0-vector space D equipped with a φ-semi-linear bijective map $\varphi : D \to D$. The category of φ-modules $\mathbf{M}_{K_0}^{\varphi}$ is a neutral tannakian category. In particular, it is abelian.

12.2.2 The structure of φ-modules is described by the theory of Dieudonné–Manin. Let v_p denote the valuation on K_0. First assume that D is a φ-module of dimension 1 over K_0. If d is a basis of D, then $\varphi(d) = \lambda d$ for some non-zero $\lambda \in K_0$, and we set $t_{\mathrm{N}}(D) = v_p(\lambda)$. Note that $v_p(\lambda)$ does not depend on the choice of d. Now, if D is a φ-module of arbitrary dimension n, its top exterior power $\wedge^n D$ is a one-dimensional vector space and we set

$$t_{\mathrm{N}}(D) = t_{\mathrm{N}}(\wedge^n D).$$

More explicitly, $t_{\mathrm{N}}(D) = v_p(A)$, where A is the matrix of φ with respect to any basis of M. The function t_{N} is additive on short exact sequences: if

$$0 \to D' \to D \to D'' \to 0$$

is exact, then $t_{\mathrm{N}}(D) = t_{\mathrm{N}}(D') + t_{\mathrm{N}}(D'')$.

Definition (i) The slope of a non-zero φ-module D is the rational

$$s(D) = \frac{t_{\mathrm{N}}(D)}{\dim_{K_0} D}.$$

(ii) A φ-module D is pure (or isoclinic) of slope λ if $s(D') = \lambda$ for any non-zero submodule $D' \subset D$.

If D is isoclinic, we will write its slope λ in the form:

$$\lambda = \frac{a}{b}, \qquad (a, b) = 1, \quad b > 0.$$

Theorem 12.2.3 (Dieudonné–Manin) *(i) D is isoclinic of slope $\lambda = a/b$ if and only if there exists an O_{K_0}-lattice $L \subset D$ such that $\varphi^b(L) = p^a L$.*

(ii) For all $a, b \in \mathbf{Z}$ such that $b > 0$ and $(a, b) = 1$, the φ-module

$$D_\lambda = K_0[\varphi]/(\varphi^b - p^a)$$

is isoclinic of slope $\lambda = a/b$. Moreover, if k is algebraically closed, then each isoclinic φ-module is isomorphic to a direct sum of copies of D_λ.

(iii) Each φ-module D over K_0 has a unique decomposition into a direct sum

$$D = \underset{\lambda \in \mathbf{Q}^*}{\oplus} D(\lambda),$$

where $D(\lambda)$ is isoclinic of slope λ.

Proof See [112, Section 2]. See also [56]. □

Corollary 12.2.4 *If k is algebraically closed, the category of φ-modules over K_0 is semi-simple. Its simple objects are Dieudonné modules which are isomorphic to D_λ.*

Remark 12.2.5 (1) A φ-module is étale in the sense of Sect. 8.1 if and only if it is isoclinic of slope 0.

(2) The theorem of Dieudonné–Manin allows to write $t_{\mathrm{N}}(D)$ in the form

$$t_{\mathrm{N}}(D) = \sum_{\lambda} \lambda \dim_{K_0} D(\lambda).$$

(3) Kedlaya [94] extended the theory of slopes to the category of φ-modules over the Robba ring.

12.3 Slope Filtration

12.3.1 Slope functions appear in several theories. Important examples are provided by the theory of vector bundles (Harder–Narasimhan theory [85]), differential modules [110, 155] and euclidian lattices [80, 148]. A unified axiomatic treatement of the theory of slopes was proposed by Y. André [8]. In this section, we discuss this formalism in relation with the examples seen in the previous sections. We work with additive categories and refer to [8] for the general treatement.

Definition Let $\mathscr{C}$ be an additive category with kernels and cokernels.

(i) A monomorphism $f : X \to Y$ is strict if there exists $g : Y \to Z$ such that $0 \to X \xrightarrow{f} Y \xrightarrow{g} Z \to 0$ is exact.

(ii) An epimorphism $g : Y \to Z$ is strict if there exists $f : X \to Y$ such that $0 \to X \xrightarrow{f} Y \xrightarrow{g} Z \to 0$ is exact.

(iii) $\mathscr{C}$ is quasi-abelian if every pull-back of a strict epimorphism is a strict epimorphism and every push-out of a strict monomorphism is a strict monomorphism.

Note that in the category $\mathbf{MF}_K$, a monomorphism (respectively epimorphism) $f : X \to Y$ is strict if and only if it satisfies the condition (55).

12.3.2 Let $\mathscr{C}$ be a quasi-abelian category. Assume that $\mathscr{C}$ is essentially small, i.e. that it is equivalent to a small category. A rank function on $\mathscr{C}$ is a function $\mathrm{rk} : \mathscr{C} \to \mathbf{N}$ such that:

(1) $\mathrm{rk}(X) = 0$ if and only if $X = 0$;
(2) rk is additive, i.e. for any exact sequence

$$0 \to X' \xrightarrow{f} X \xrightarrow{g} X'' \to 0$$

one has

$$\mathrm{rk}(X) = \mathrm{rk}(X') + \mathrm{rk}(X'').$$

We can now define the notion of a slope function.

Definition A slope function on $\mathscr{C}$ is a function $\mu : \mathscr{C} \setminus \{0\} \to \mathbf{R}$ such that:

(1) The associated degree function

$$\deg = \mathrm{rk} \cdot \mu \,:\, \mathscr{C} \to \mathbf{N}$$

(taking value 0 at the zero object) is additive on short exact sequences;

(2) For any morphism $f : X \to Y$ which is both a monomorphism and an epimorphism, one has

$$\mu(X) \leqslant \mu(Y).$$

An object $Y \in \mathscr{C}$ is called semi-stable if for any subobject X of Y, $\mu(X) \leqslant \mu(Y)$. We can now state the main theorem of this section.

Theorem 12.3.3 (Harder–Narasimhan, André) *For any $X \in \mathscr{C}$, there exists a unique filtration*

$$X = X_0 \supset X_1 \supset \ldots \supset X_k = \{0\}$$

such that:

(1) X_{i+1} is a strict subobject of X_i for all i;

(2) The quotients X_i/X_{i+1} are semi-stable, and the sequence $\mu(X_i/X_{i+1})$ is strictly increasing.

Proof The theorem was first proved for the category of vector bundles on a smooth projective curve over $\mathbf{C}$ [85]. André [8] extended the proof to the case of general quasi-abelian (and even proto-abelian) categories. □

We call the canonical filtration provided by Theorem 12.3.3 the Harder–Narasimhan filtration.

Example 12.3.4 (1) Let $\mathscr{C} = \mathbf{MF}_K$. Set $\mathrm{rk}(\Delta) = \dim_K \Delta$ and $\deg(\Delta) = t_{\mathrm{H}}(\Delta)$. Then

$$\mu_{\mathrm{H}}(\Delta) = \frac{t_{\mathrm{H}}(\Delta)}{\dim_K \Delta}$$

is a slope function. Semi-stable objects are filtered vector spaces with a unique filtration break. The Harder–Narasimhan filtration coincides (up to enumeration) with the canonical filtration on Δ.

(2) Let $\mathscr{C} = \mathbf{M}_{K_0}^{\varphi}$. Set $\mathrm{rk}(D) = \dim_{K_0} D$ and $\deg(D) = -t_{\mathrm{N}}(D)$. Then

$$\mu_{\mathrm{N}}(D) = s(D) = \frac{t_{\mathrm{N}}(D)}{\dim_{K_0} D}$$

is a slope function. Semi-simple objects are isoclinic φ-modules. On the other hand, it's easy to see that $-s(D)$ is also a slope function, which provides the opposite filtration on M and therefore its splitting in the direct sum of isoclinic components. This gives an interpretation of the decomposition of Dieudonné–Manin in terms of the slope filtration.

(3) Let $\mathscr{C} = \mathbf{Bun}(X)$ be the category of vector bundles on a smooth projective curve $X/\mathbf{C}$. To each object E of this category one associates its rank rk(E) and degree $\deg(E) := \deg(\wedge^{\mathrm{rk}(E)} E)$. Then

$$\mu_{\mathrm{HN}}(E) = \frac{\deg(E)}{\mathrm{rk}(E)}$$

is a slope function. This is the classical setting of the Harder–Narasimhan theory [85]. The semi-stable objects of $\mathscr{C}$ are described in [118]. The analog of this filtration in the setting of the curve of Fargues–Fontaine plays an important role in [60].

12.4 Filtered (φ, N)-Modules

12.4.1 Let K be a complete discrete valuation field of characteristic 0 with perfect residue field k of characteristic p, and let K_0 denote the maximal unramified subfield of K.

Definition (i) A filtered φ-module over K is a φ-module D over K_0 together with a structure of filtered K-vector space on $D_K = D \otimes_{K_0} K$.

(Ii) A filtered (φ, N)-module over K is a filtered φ-module D over K equipped with a K_0-linear operator $N : D \to D$ such that

$$N\varphi = p\varphi N.$$

Note that the relation $N\varphi = p\varphi N$ implies that $N : D \to D$ is *nilpotent*.

12.4.2 A morphism of filtered φ-modules (respectively, (φ, N)-modules) is a K_0-linear map $f : D' \to D''$ which is compatible with all additional structures. Filtered φ-modules (respectively (φ, N)-modules) form additive tensor categories which we denote by $\mathbf{MF}_K^{\varphi}$ and $\mathbf{MF}_K^{\varphi,N}$ respectively. Note that these categories are not abelian.

12.4.3 We define some subcategories of $\mathbf{MF}_K^{\varphi}$ and $\mathbf{MF}_K^{\varphi,N}$, which play an important role in the classification of p-adic representations. Equip $\mathbf{MF}_K^{\varphi}$ and $\mathbf{MF}_K^{\varphi,N}$ with the functions

$$\mathrm{rk}(D) := \dim_{K_0} K, \qquad \deg(D) := t_{\mathrm{H}}(D) - t_{\mathrm{N}}(D).$$

Proposition 12.4.4 $\mu(D) = \deg(D)/\mathrm{rk}(D)$ *is a slope function.*

Proof We only need to prove that if $f : D' \to D''$ is both a monomorphism and an epimorphism, then $\mu(D') \leqslant \mu(D'')$. We remark that such f is an isomorphism of φ-modules; hence $\mu_{\mathrm{N}}(D') = \mu_{\mathrm{N}}(D'')$. Set $d := \dim_{K_0} D' = \dim_{K_0} D''$. Then we have a monomorphism of one-dimensional filtered spaces $\wedge^d D' \to \wedge^d D''$, and therefore

$$t_{\mathrm{H}}(D') = t_{\mathrm{H}}\left(\wedge^d D'\right) \leqslant t_{\mathrm{H}}\left(\wedge^d D''\right) = t_{\mathrm{H}}(D'').$$

Hence, $\mu(D') \leqslant \mu(D'')$, and the proposition is proved. □

Definition A filtered φ-module (respectively, (φ, N)-module) is weakly admissible if it is semi-stable of slope 0.

More explicitly, D is weakly admissible if it satisfies the following conditions:

(1) $t_{\mathrm{H}}(D_K) = t_{\mathrm{N}}(D)$;
(2) $t_{\mathrm{H}}(D'_K) \leqslant t_{\mathrm{N}}(D')$ for any submodule D' of D.

This is the classical definition of the weak admissibility [65, 71]. We denote by $\mathbf{MF}_K^{\varphi,f}$ and $\mathbf{MF}_K^{\varphi,N,f}$ the resulting subcategories of $\mathbf{MF}_K^{\varphi}$ and $\mathbf{MF}_K^{\varphi,N}$.

Proposition 12.4.5 *(i) The categories $\mathbf{MF}_K^{\varphi,f}$ and $\mathbf{MF}_K^{\varphi,N,f}$ are abelian.*
(ii) If D is weakly admissible, then its dual D^* is weakly admissible.
(iii) If in a short exact sequence

$$0 \to D' \to D \to D'' \to 0$$

two of the three modules are weakly admissible, then so is the third.

Proof This is [65, Proposition 4.2.1]. See also [32, Proposition 8.2.10 &Theorem 8.2.11] for a detailed proof. □

Remark 12.4.6 The tensor product of two weakly admissible modules is weakly admissible. See [153] for a direct proof of this result. It also follows from the theorem “weakly admissible ⇒ admissible" of Colmez–Fontaine [48]. Therefore, the categories $\mathbf{MF}_K^{\varphi,f}$ and $\mathbf{MF}_K^{\varphi,N,f}$ are neutral tannakian.

13 The Hierarchy of p-Adic Representations

13.1 de Rham Representations

13.1.1 In this section, we come back to classification of p-adic representations. Let K be a local field. We apply the general formalism of Sect. 9.1 to the rings of p-adic periods constructed in Sect. 11.

13.1.2 Recall that $\mathbf{B}_{\mathrm{dR}}$ is a field with $\mathbf{B}_{\mathrm{dR}}^{G_K} = K$. In particular, it is G_K-regular. To any p-adic representation V of G_K we associate the finite-dimensional K-vector space

$$\mathbf{D}_{\mathrm{dR}}(V) = (V \otimes_{\mathbf{Q}_p} \mathbf{B}_{\mathrm{dR}})^{G_K}.$$

We equip it with the filtration induced from $\mathbf{B}_{\mathrm{dR}}$:

$$\mathrm{Fil}^i \mathbf{D}_{\mathrm{dR}}(V) = (V \otimes_{\mathbf{Q}_p} \mathrm{Fil}^i \mathbf{B}_{\mathrm{dR}})^{G_K}.$$

The mapping which assigns $\mathbf{D}_{\mathrm{dR}}(V)$ to each V defines a functor of tensor categories

$$\mathbf{D}_{\mathrm{dR}} \,:\, \mathbf{Rep}_{\mathbf{Q}_p}(G_K) \to \mathbf{MF}_K.$$

Definition A p-adic representation V is called de Rham if it is $\mathbf{B}_{\mathrm{dR}}$-admissible, i.e. if

$$\dim_K \mathbf{D}_{\mathrm{dR}}(V) = \dim_{\mathbf{Q}_p}(V).$$

We denote by $\mathbf{Rep}_{\mathrm{dR}}(G_K)$ the category of de Rham representations. By Proposition 9.1.7, it is tannakian and the restriction of $\mathbf{D}_{\mathrm{dR}}$ on $\mathbf{Rep}_{\mathrm{dR}}(G_K)$ is exact and faithful.

Proposition 13.1.3 *Each de Rham representation is Hodge–Tate.*

Proof Recall that we have exact sequences

$$0 \to \mathrm{Fil}^{i+1}\mathbf{B}_{\mathrm{dR}} \to \mathrm{Fil}^i\mathbf{B}_{\mathrm{dR}} \to \mathbf{C}t^i \to 0.$$

Tensoring with V and taking Galois invariants we have

$$\dim_K \left(\mathrm{gr}^i\mathbf{D}_{\mathrm{dR}}(V)\right) \leqslant \dim_K(V \otimes_{\mathbf{Q}_p} \mathbf{C}t^i).$$

From $\mathbf{B}_{\mathrm{HT}} = \bigoplus_{i\in\mathbf{Z}} \mathbf{C}t^i$ it follows that

$$\dim_K \mathbf{D}_{\mathrm{dR}}(V) = \sum_{i\in\mathbb{Z}} \dim_K \left(\mathrm{gr}^i\mathbf{D}_{\mathrm{dR}}(V)\right) \leqslant \dim_K \mathbf{D}_{\mathrm{HT}}(V) \leqslant \dim_{\mathbf{Q}_p}(V).$$

The proposition is proved. □

Remark 13.1.4 The functor $\mathbf{D}_{\mathrm{dR}}$ is not fully faithful. A p-adic representation cannot be recovered from its filtered module.

13.1.5 Using the fundamental exact sequence, one can construct Hodge–Tate representations which are not de Rham. Fix an integer $r \geqslant 1$ and consider an extension V of $\mathbf{Q}_p$ by $\mathbf{Q}_p(-r)$:

$$0 \to \mathbf{Q}_p(-r) \to V \to \mathbf{Q}_p \to 0.$$

Such extensions are classified by the first Galois cohomology group $H^1(G_K, \mathbf{Q}_p(-r))$, which is a one-dimensional K-vector space. Assume that V is a non-trivial extension. Since the Hodge–Tate weights of $\mathbf{Q}_p$ and $\mathbf{Q}_p(-r)$ are distinct, V is Hodge–Tate. However it is not de Rham (see [28, Section 4] for the proof).

13.2 *Crystalline and Semi-Stable Representations*

13.2.1 Recall that $\mathbf{B}_{\mathrm{cris}}$ is G_K-regular with $\mathbf{B}_{\mathrm{cris}}^{G_K} = K_0$. Therefore, for each p-adic representation V, the K_0-vector space

$$\mathbf{D}_{\mathrm{cris}}(V)=(V\otimes_{\mathbf{Q}_p}\mathbf{B}_{\mathrm{cris}})^{G_K}$$

is finite-dimensional with $\dim_{K_0}\mathbf{D}_{\mathrm{cris}}(V)\leqslant \dim_{\mathbf{Q}_p}(V)$. The action on φ on $\mathbf{B}_{\mathrm{cris}}$ induces a semi-linear operator on $\mathbf{D}_{\mathrm{cris}}(V)$, which we denote again by φ. Since φ is injective on $\mathbf{B}_{\mathrm{cris}}$, it is bijective on the finite-dimensional vector space $\mathbf{D}_{\mathrm{cris}}(V)$. The embedding $K\otimes_{K_0}\mathbf{B}_{\mathrm{cris}}\hookrightarrow \mathbf{B}_{\mathrm{dR}}$ induces an inclusion

$$K\otimes_{K_0}\mathbf{D}_{\mathrm{cris}}(V)\hookrightarrow \mathbf{D}_{\mathrm{dR}}(V).$$

This equips $\mathbf{D}_{\mathrm{cris}}(V)_K=K\otimes_{K_0}\mathbf{D}_{\mathrm{cris}}(V)$ with the induced filtration:

$$\mathrm{Fil}^i\mathbf{D}_{\mathrm{cris}}(V)_K=\mathbf{D}_{\mathrm{cris}}(V)_K\cap \mathrm{Fil}^i\mathbf{D}_{\mathrm{dR}}(V).$$

Thereore $\mathbf{D}_{\mathrm{cris}}$ can be viewed as a functor

$$\mathbf{D}_{\mathrm{cris}}\,:\,\mathbf{Rep}_{\mathbf{Q}_p}(G_K)\to \mathbf{MF}_K^{\varphi}.$$

Definition A p-adic representation V is crystalline if it is $\mathbf{B}_{\mathrm{cris}}$-admissible, i.e. if

$$\dim_{K_0}\mathbf{D}_{\mathrm{cris}}(V)=\dim_{\mathbf{Q}_p}V.$$

By Proposition 9.1.5, V is crystalline if and only if the map

$$\alpha_{\mathrm{cris}}\,:\,\mathbf{D}_{\mathrm{cris}}(V)\otimes_{K_0}\mathbf{B}_{\mathrm{cris}}\to V\otimes_{\mathbf{Q}_p}\mathbf{B}_{\mathrm{cris}} \tag{57}$$

is an isomorphism. We denote by $\mathbf{Rep}_{\mathrm{cris}}(G_K)$ the category of crystalline representations. From the general formalism of B-admissible representations it follows that $\mathbf{Rep}_{\mathrm{cris}}(G_K)$ is tannakian.

13.2.2 Similar arguments show that for each p-adic representation V the K_0-vector space

$$\mathbf{D}_{\mathrm{st}}(V)=(V\otimes_{\mathbf{Q}_p}\mathbf{B}_{\mathrm{st}})^{G_K}$$

is finite-dimensional and equipped with a natural structure of filtered (φ,N)-module. Since $\mathbf{B}_{\mathrm{st}}^{N=0}=\mathbf{B}_{\mathrm{cris}}$, we have:

$$\mathbf{D}_{\mathrm{cris}}(V)=\mathbf{D}_{\mathrm{st}}(V)^{N=0}.$$

Definition 13.2.3 A p-adic representation is called semi-stable if it is $\mathbf{B}_{\mathrm{st}}$-admissible, i.e. if $\dim_{K_0}\mathbf{D}_{\mathrm{st}}(V)=\dim_{\mathbf{Q}_p}V$.

By Proposition 9.1.5, V is semi-stable if and only if

$$\alpha_{\mathrm{st}}\,:\,\mathbf{D}_{\mathrm{st}}(V)\otimes_{K_0}\mathbf{B}_{\mathrm{st}}\to V\otimes_{\mathbf{Q}_p}\mathbf{B}_{\mathrm{st}} \tag{58}$$

is an isomorphism. We denote by $\mathbf{Rep}_{\mathrm{st}}(G_K)$ the tannakian category of semi-stable representations. The inclusions

$$K \otimes_{K_0} \mathbf{B}_{\mathrm{cris}} \hookrightarrow K \otimes_{K_0} \mathbf{B}_{\mathrm{st}} \hookrightarrow \mathbf{B}_{\mathrm{dR}}$$

show that

$$K \otimes_{K_0} \mathbf{D}_{\mathrm{cris}}(V) \hookrightarrow K \otimes_{K_0} \mathbf{D}_{\mathrm{st}}(V) \hookrightarrow \mathbf{D}_{\mathrm{dR}}(V).$$

Therefore, each crystalline representation is semi-stable, and each semi-stable representation is de Rham.

Example 13.2.4 The representation $V_p(E)$ constructed in Section 11.3 gives an example of semi-stable representation which is not crystalline.

Definition A filtered φ-module (respectively, (φ, N)-module) D is called admissible if it belongs to the essential image of $\mathbf{D}_{\mathrm{cris}}$ (respectively, $\mathbf{D}_{\mathrm{st}}$). In other words, D is admissible if $D \simeq \mathbf{D}_{\mathrm{cris}}(V)$ (respectively $D \simeq \mathbf{D}_{\mathrm{st}}(V)$) for some crystalline (respectively, semi-stable) representation V.

We denote by $\mathbf{MF}_K^{\varphi,a}$ and $\mathbf{MF}_K^{\varphi,N,a}$ the resulting subcategories. The following proposition shows that semi-stable representations can be recovered from their (φ, N)-modules.

Proposition 13.2.5 *The functors*

$$\mathbf{D}_{\mathrm{cris}} : \mathbf{Rep}_{\mathrm{cris}}(G_K) \to \mathbf{MF}_K^{\varphi,a}, \qquad \mathbf{D}_{\mathrm{st}} : \mathbf{Rep}_{\mathrm{st}}(G_K) \to \mathbf{MF}_K^{\varphi,N,a}$$

are equivalences of categories. The mappings

$$\mathbf{V}_{\mathrm{cris}} : D \to \mathrm{Fil}^0(D \otimes_{K_0} \mathbf{B}_{\mathrm{st}})^{\varphi=1}, \qquad \mathbf{V}_{\mathrm{st}} : D \to \mathrm{Fil}^0(D \otimes_{K_0} \mathbf{B}_{\mathrm{st}})^{N=0,\varphi=1}$$

define quasi-inverse functors of $\mathbf{D}_{\mathrm{cris}}$ *and* $\mathbf{D}_{\mathrm{st}}$.

Proof This follows from the equalities

$$\mathrm{Fil}^0(\mathbf{B}_{\mathrm{st}})^{N=0,\varphi=1} = \mathrm{Fil}^0(\mathbf{B}_{\mathrm{cris}})^{\varphi=1} = \mathbf{Q}_p.$$

Namely, assume that V is crystalline. Then using (58), we have

$$\mathbf{V}_{\mathrm{cris}}(\mathbf{D}_{\mathrm{cris}}(V)) = \mathrm{Fil}^0(\mathbf{D}_{\mathrm{cris}}(V) \otimes_{K_0} \mathbf{B}_{\mathrm{cris}})^{\varphi=1} = \mathrm{Fil}^0(V \otimes_{\mathbf{Q}_p} \mathbf{B}_{\mathrm{cris}})^{\varphi=1} = V.$$

The same argument applies in the semi-stable case. □

13.2.6 As in Sect. 9.1, one can also consider the contravariant functors

$$\mathbf{D}^*_{\mathrm{cris}} : \mathbf{Rep}_{\mathbf{Q}_p}(G_K) \to \mathbf{MF}_K^{\varphi}, \quad \mathbf{D}^*_{\mathrm{cris}}(V) = \mathrm{Hom}_{G_K}(V, \mathbf{B}_{\mathrm{cris}}),$$
$$\mathbf{D}^*_{\mathrm{st}} : \mathbf{Rep}_{\mathbf{Q}_p}(G_K) \to \mathbf{MF}_K^{\varphi,N}, \quad \mathbf{D}^*_{\mathrm{cris}}(V) = \mathrm{Hom}_{G_K}(V, \mathbf{B}_{\mathrm{st}}).$$

If V is crystalline (respectively, semi-stable), there is a canonical isomorphism

$$\mathbf{D}^*_{\mathrm{cris}}(V) \simeq \mathbf{D}_{\mathrm{cris}}(V)^*$$

(respectively, $\mathbf{D}^*_{\mathrm{st}}(V) \simeq \mathbf{D}_{\mathrm{st}}(V)^*$). The tautological map

$$V \otimes_{\mathbf{Q}_p} \mathbf{D}^*_\star(V) \to \mathbf{B}_\star, \qquad \star \in \{\mathrm{cris}, \mathrm{st}\}$$

can be viewed as an abstract p-adic integration pairing.

Proposition 13.2.7 *Each admissible (φ, N)-module is weakly admissible.*

Proof This is [65, Proposition 4.4.5]. We refer the reader to [32, Theorem 9.3.4] for a detailed proof. □

13.2.8 The converse statement is a fundamental theorem of the p-adic Hodge theory, which was first formulated as a conjecture in [65].

Theorem 13.2.9 (Colmez–Fontaine) *Each filtered weakly admissible module is admissible, i.e. we have equivalences of categories:*

$$\mathbf{MF}_K^{\varphi,a} \simeq \mathbf{MF}_K^{\varphi,f}, \qquad \mathbf{MF}_K^{\varphi,N,a} \simeq \mathbf{MF}_K^{\varphi,N,f}.$$

This theorem was first proved in [48]. Further development of ideas of this proof leads to the theory of p-adic Banach spaces [41] and almost $\mathbf{C}_p$-representations [72], [17]. Another proof, based on the theory of (φ, Γ)-modules was found by Berger [18]. A completely new insight on this theorem is provided by the theory of Fargues–Fontaine [60]. See [55] and [114] for an introduction to the work of Fargues and Fontaine.

Remark 13.2.10 The theorem of Colmez–Fontaine implies that the tensor product of two weakly admissible modules is weakly admissible. Recall that there exists a direct proof of this result [153].

13.3 The Hierarchy of p-Adic Representations

13.3.1 Let L be a finite extension of K. If $\rho : G_K \to \mathrm{Aut}_{\mathbf{Q}_p} V$ is a p-adic representation, one can consider its restriction on G_L and ask for the behavior of the functors $\mathbf{D}_{\mathrm{dR}}$, $\mathbf{D}_{\mathrm{st}}$ and $\mathbf{D}_{\mathrm{cris}}$ under restriction. Set:

$$\mathbf{D}_{\star/L}(V) = (V \otimes_{\mathbf{Q}_p} B_\star)^{G_L}, \qquad \star \in \{\,\mathrm{dR}, \mathrm{st}, \mathrm{cris}\}.$$

Applying Hilbert's theorem 90 (Theorem 1.6.3), we obtain that

$$\mathbf{D}_{\mathrm{dR}/L}(V) = \mathbf{D}_{\mathrm{dR}}(V) \otimes_K L.$$

In particular, V is a de Rham representation if and only if its restriction on G_L is a de Rham.

13.3.2 One says that a p-adic representation ρ is *potentially semi-stable* (respectively, *potentially crystalline*) if there exists a finite extension L/K such that the restriction of ρ on G_L is semi-stable (respectively, crystalline). Applying Hilbert's theorem 90 (Theorem 1.6.3), we obtain that in the case L/K is unramified, ρ is crystalline (respectively semi-stable) if and only if it's restriction on G_L is. The following proposition shows that ramified representations with finite image provide examples of potentially semi-stable representations that are not semi-stable.

Proposition 13.3.3 *A p-adic representation $\rho : G_K \to Aut_{\mathbf{Q}_p} V$ with finite image is semi-stable if and only if it is unramified.*

Proof Let ρ be a representation with a finite image. Let L/K be a finite extension such that $V^{G_L} = V$. Then

$$\mathbf{D}_{\mathrm{st}/L}(V) = V \otimes_{\mathbf{Q}_p} \mathbf{B}_{\mathrm{st}}^{G_L} = V \otimes_{\mathbf{Q}_p} L_0,$$

where L_0 is the maximal unramified subfield of L. One has

$$\mathbf{D}_{\mathrm{st}}(V) = (\mathbf{D}_{\mathrm{st}/L}(V))^{G_K} = (V \otimes_{\mathbf{Q}_p} L_0)^{\mathrm{Gal}(L/K)}.$$

Therefore, V is semi-stable if and only if it is L_0-admissible if and only if it is unramified (see Example 9.2). □

13.3.4 Set:

$$\mathbf{D}_{\mathrm{pst}}(V) = \varinjlim_{L/K} \mathbf{D}_{\mathrm{st}/L}(V),$$

where L runs all finite extensions of K. Then $\mathbf{D}_{\mathrm{pst}}(V)$ is a finite dimensional K_0^{ur}-vector space endowed with a natural structure of filtered (φ, N)-module. In addition, it is equipped with a discrete action of the Galois group G_K such that $\mathbf{D}_{\mathrm{st}}(V) = \mathbf{D}_{\mathrm{pst}}(V)^{G_K}$. This Galois action allows to define on $\mathbf{D}_{\mathrm{pst}}(V)$ the stucture of a Weil–Deligne representation. One can see $\mathbf{D}_{\mathrm{pst}}$ as a functor to the category of *filtered (φ, N, G_K)-modules*. One says that V is potentially semi-stable if and only if $\dim_{K_0^{\mathrm{ur}}} \mathbf{D}_{\mathrm{st}}(V) = \dim_{\mathbf{Q}_p}(V)$. The functor $\mathbf{D}_{\mathrm{pcris}}$ can be defined by the same way. See [71] for more detail.

The hierarchy of p-adic representations can be represented by the following diagram of full subcategories of $\mathbf{Rep}_{\mathbf{Q}_p}(G_K)$:

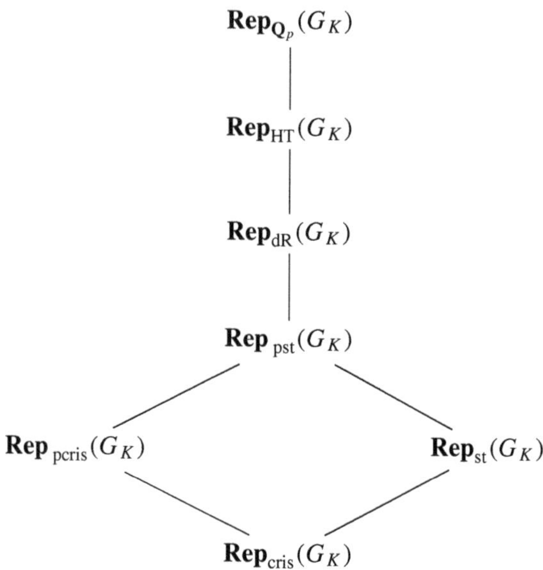

Finally, the categories $\mathbf{Rep}_{\mathrm{pst}}(G_K)$ and $\mathbf{Rep}_{\mathrm{dR}}(G_K)$ coincide as the following fundamental theorem shows:

Theorem 13.3.5 (p-adic monodromy conjecture) *Each de Rham representation is potentially semi-stable.*

This theorem was formulated as a conjecture by Fontaine. It can be seen as a highly non-trivial analog of Grothendieck's ℓ-adic monodromy theorem in the case $\ell = p$. The first proof, found by Berger [15], uses the theory of (φ, Γ_K)-modules (see below). Colmez [43] gave a completely different proof, based on the theory of p-adic Banach Spaces. See [60, Chap. 10] for the insight provided by the theory of Fargues–Fontaine.

13.3.6 Recall that Theorem 8.2.9 classifies *all* p-adic representations in terms of (φ, Γ_K)-modules. It is natural to ask how to recover $\mathbf{D}_{\mathrm{cris}}(V)$, $\mathbf{D}_{\mathrm{st}}(V)$ and $\mathbf{D}_{\mathrm{dR}}(V)$ from the étale (φ, Γ_K)-module $\mathbf{D}(V)$. This question is known as Fontaine's program. As a first step, Cherbonnier and Colmez [35] proved that each p-adic representation is overconvergent. As a second step, Berger [15] showed how to construct $\mathbf{D}_{\mathrm{cris}}(V)$, $\mathbf{D}_{\mathrm{st}}(V)$ and $\mathbf{D}_{\mathrm{dR}}(V)$ in terms of the overconvergent lattice $\mathbf{D}^{\dagger}(V)$ of $\mathbf{D}(V)$ using the Robba ring $\mathscr{R}_K$. Moreover, the infinitesimal action of Γ_K on $\mathbf{D}^{\dagger}(V) \otimes_{\mathbf{Q}_p} \mathscr{R}_K$ gives rise to a structure of a differential φ-module and associates to V a p-adic differential equation. This reduces the p-adic monodromy conjecture to a conjecture of Crew on p-adic differential equations. This last conjecture was proved by Kedlaya [94]. We refer the reader to [42] for a survey of these results. In another direction, the theory of (φ, Γ_K)-modules is closely related to the p-adic Langlands program for $\mathrm{GL}_2(\mathbf{Q}_p)$ [45–47].

13.4 Comparison Theorems

13.4.1 In [151], Tate considered the p-adic analog of the following situation. Let X be a smooth proper scheme over the field of complex numbers $\mathbb{C}$. To the analytic space $X(\mathbb{C})$ on can associate on the one hand, the singular cohomology $H^n(X(\mathbb{C}),\mathbf{Q})$ and on the other hand, the de Rham cohomology $H^n_{\mathrm{dR}}(X/\mathbb{C})$ defined as the hypercohomology of the complex $\Omega^\bullet_X$ of differential forms on X. The integration of differental forms against simplexes gives a non-degenerate pairing

$$H_n(X(\mathbb{C}),\mathbf{Q})\times H^n_{\mathrm{dR}}(X/\mathbb{C})\to\mathbb{C}, \tag{59}$$

which induces an isomorphism (comparison isomorphism):

$$H^n(X(\mathbb{C}),\mathbf{Q})\otimes_{\mathbf{Q}}\mathbb{C}\simeq H^n_{\mathrm{dR}}(X/\mathbb{C})$$

The spectral sequence

$$E_1^{ij}=H^j(X,\Omega^i_{X/\mathbb{C}})\Rightarrow H^{i+j}_{\mathrm{dR}}(X/\mathbb{C})$$

defines a decreasing exhaustive filtration $F^iH^n_{\mathrm{dR}}(X/\mathbb{C})$ on $H^n_{\mathrm{dR}}(X/\mathbb{C})$ such that

$$\mathrm{gr}^iH^n_{\mathrm{dR}}(X/\mathbb{C})=H^{n-i}(X,\Omega^i_X).$$

By Hodge theory, this filtration splits canonically and gives the decomposition of $H^n_{\mathrm{dR}}(X/\mathbb{C})$ into direct sum (Hodge decomposition):

$$H^n_{\mathrm{dR}}(X/\mathbb{C})=\bigoplus_{i+j=n}H^j(X,\Omega^i_X).$$

Therefore, one has the decomposition:

$$H^n(X(\mathbb{C}),\mathbf{Q})\otimes_{\mathbf{Q}}\mathbb{C}\simeq\bigoplus_{i+j=n}H^j(X,\Omega^i_X).$$

13.4.2 Now assume that X is a smooth proper scheme over a local field K of characteristic 0. The de Rham cohomologies $H^n_{\mathrm{dR}}(X/K)$ are still defined as the hypercohomology of $\Omega^\bullet_{X/K}$. Contrary to the complex case, the filtration $F^iH^n_{\mathrm{dR}}(X/K)$ has no canonical splitting [2]. One has

$$\mathrm{gr}^\bullet H^n_{\mathrm{dR}}(X/K)=\bigoplus_{i+j=n}H^j(X,\Omega^i_{X/K}).$$

In the p-adic situation, the singular cohomology is not defined, but it can be replaced by the p-adic étale cohomology $H^n_p(X)$, which has the additional structure of a

[2] However, see [162].

p-adic representation. The following result formulated by Tate as a conjecture was proved in full generality by Faltings [57].

Theorem 13.4.3 (Faltings) *There exists a functorial isomorphism*

$$H^n_p(X)\otimes_{\mathbf{Q}_p}\mathbf{C}\simeq \underset{i+j=n}{\oplus}\left(H^j(X,\Omega^i_{X/K})\otimes_K \mathbf{C}(-i)\right).$$

In particular, $H^n_p(X)$ is of Hodge–Tate, and

$$\mathbf{D}_{\mathrm{HT}}\left(H^n_p(X)\right)\simeq gr^{\bullet}H^n_{\mathrm{dR}}(X/K).$$

Tate proved this conjecture for abelian varieties having good reduction using his results about the continuous cohomology of G_K (see Sect. 4.3). Faltings' proof relies on the higher-dimensional generalization of Tate's method of almost étale extensions. The theory of almost étale extensions was systematically developped in [78]. See [130] for further generalization of Faltings' almost purity theorems.

13.4.4 Inspired by Grothendieck's problem of mysterious functor [83, 84], Fontaine [66, 71] formulated more precise conjectures, relating étale cohomology to other cohomology theories via the rings $\mathbf{B}_{\mathrm{cris}}$, $\mathbf{B}_{\mathrm{st}}$ and $\mathbf{B}_{\mathrm{dR}}$. These conjectures are actually theorems, which can be formulated as follows:

13.4.5 *Étale cohomology vs. de Rham cohomology*. Recall that the ring $\mathbf{B}_{\mathrm{dR}}$ is equipped with a canonical filtration and a continuous action of the Galois group G_K.

Theorem 13.4.6 (C_{dR}-conjecture) *Let X/K be a smooth proper scheme. There exists a functorial isomorphism*

$$H^i_p(X)\otimes_{\mathbf{Q}_p}\mathbf{B}_{\mathrm{dR}}\simeq H^i_{\mathrm{dR}}(X/K)\otimes_K\mathbf{B}_{\mathrm{dR}},\tag{60}$$

which is compatible with the filtration and the Galois action. In particular, $H^i_p(X)$ is de Rham, and

$$\mathbf{D}_{\mathrm{dR}}\left(H^n_p(X)\right)\simeq H^n_{\mathrm{dR}}(X/K).$$

Using the isomorphism $\mathrm{gr}^{\bullet}\mathbf{B}_{\mathrm{dR}}\simeq \underset{i\in\mathbf{Z}}{\oplus}\mathbf{C}(i)$ it is easy to see that this theorem implies Theorem 13.4.3.

13.4.7 *Étale cohomology vs. crystalline cohomology*. Let X/O_K be a smooth proper scheme having good reduction. The theory of crystalline cohomology [20] associates to the special fiber of X finite-dimensional K_0-vector spaces $H^i_{\mathrm{cris}}(X)$ equipped with a semi-linear Frobenius φ. By a theorem of Berhtelot–Ogus [22], there exists a canonical isomorphism

$$H^i_{\mathrm{dR}}(X/K)\simeq H^i_{\mathrm{cris}}(X)\otimes_{K_0}K,$$

which equips $H^i_{\mathrm{cris}}(X)\otimes_{K_0}K$ with a canonical filtration.

Theorem 13.4.8 (C_{cris}-conjecture) *Let X/O_K be a smooth proper scheme having good reduction.*

(i) There exists a functorial isomorphism

$$H_p^i(X) \otimes_{\mathbf{Q}_p} \mathbf{B}_{\mathrm{cris}} \simeq H_{\mathrm{cris}}^i(X) \otimes_{K_0} \mathbf{B}_{\mathrm{cris}}, \tag{61}$$

which is compatible with the Galois action and the action of φ. In particular, $H_p^i(X)$ is crystalline, and

$$\mathbf{D}_{\mathrm{cris}}\left(H_p^n(X)\right) \simeq H_{\mathrm{cris}}^n(X).$$

(ii) The isomorphism (60) can be obtained from (62) by the extension of scalars $\mathbf{B}_{\mathrm{cris}} \otimes_{K_0} K \subset \mathbf{B}_{\mathrm{dR}}$.

13.4.9 *Étale cohomology vs. log-crystalline cohomology*. Let X/O_K be a proper scheme having semi-stable reduction. The theory of log-crystalline cohomology [92] associates to X a finite-dimensional K_0-vector spaces $H_{\log-\mathrm{cris}}^i(X)$ equipped with a semi-linear Frobenius φ and a monodromy operator N such that $N\varphi = p\varphi N$. A theorem of Hyodo–Kato [87] shows the existence of an isomorphism

$$H_{\mathrm{dR}}^i(X/K) \simeq H_{\log-\mathrm{cris}}^i(X) \otimes_{K_0} K,$$

which equips $H_{\log-\mathrm{cris}}^i(X) \otimes_{K_0} K$ with the induced filtration. Note that if X has good reduction, then $N = 0$, and the log-crystalline cohomology coincides with the classical crystalline cohomology of X.

Theorem 13.4.10 (C_{st}-conjecture of Fontaine–Jannsen) *Let X/O_K be a proper scheme having semi-stable reduction.*

(i) There exists a functorial isomorphism

$$H_p^i(X) \otimes_{\mathbf{Q}_p} \mathbf{B}_{\mathrm{st}} \simeq H_{\log-\mathrm{cris}}^i(X) \otimes_{K_0} \mathbf{B}_{\mathrm{st}}, \tag{62}$$

which is compatible with the Galois action and the actions of φ and N. In particular, $H_p^i(X)$ is semi-stable, and

$$\mathbf{D}_{\mathrm{st}}\left(H_p^n(X)\right) \simeq H_{\log-\mathrm{cris}}^n(X).$$

13.4.11 These conjectures were first proved by two completely different methods:

- The method of almost étale extensions (Faltings [58, 59]);
- The method of syntomic cohomology of Fontaine–Messing (Fontaine–Messing, Hyodo–Kato, Tsuji [74, 154]).

Alternative proofs were found by Nizioł[120, 121] and Beilinson [26, 27]. The theory of perfectoids gave a new impetus to this subject [24, 25, 34, 49, 131]. The generalization of comparison theorems to cohomology with coefficients is intimately related to the theory of p-adic representations of affinoid algebras [9, 31, 95, 96, 115].

13.4.12 Over the field of complex numbers, the comparison isomorphism can be alternatively seen as the non-degenerate pairing of complex periods (59). In the p-adic case, such an interpretation exists for abelian varieties. Namely, if A is an abelian variety over K, then the p-adic analog of $H_1(A(\mathbb{C}), \mathbf{Q})$ is the p-adic representation $V_p(A) := T_p(A) \otimes_{\mathbf{Z}_p} \mathbf{Q}_p$. For the first p-adic cohomology of A, one has

$$H_p^1(A) \simeq V_p(A)^*.$$

The theory of p-adic integration [38, 39, 67] provides us with a non-degenerate pairing

$$H^1_{\mathrm{dR}}(A) \times T_p(A) \to \mathbf{B}_{\mathrm{dR}},$$

which gives an explicit approach to the comparison theorems for abelian varieties. The simplest case of p-divisible formal groups will be studied in the next section.

14 p-Divisible Groups

14.1 Formal Groups

14.1.1 In this section, we make first steps in studing p-adic representations arising from p-divisible groups. Such representations are crystalline and the associated filtered modules have an explicit description in geometric terms. We will focus our attention on formal groups because in this case many results can be proved by elementary methods, without using the theory of finite group schemes. We start with a short review of the theory of formal groups.

Definition Let A be an integral domain. A one-dimensional commutative formal group over A is a formal power series $F(X,Y) \in A[[X,Y]]$ satisfying the following conditions:

(i) $F(F(X,Y),Z) = F(X,F(Y,Z))$;
(ii) $F(X,Y) = F(Y,X)$;
(iii) $F(X,0) = X$ and $F(0,Y) = Y$;
(iv) There exists $i(X) \in XA[[X]]$ such that $F(X,i(X)) = 0$.

It can be proved that ii) and iv) follow from i) and iii) (see [109]). We will often write $X +_F Y$ instead $F(X,Y)$.

Example 14.1.2 (1) The additive formal group $\widehat{\mathbb{G}}_a(X,Y) = X + Y$. Here $i(X) = -X$.

(2) The multiplicative formal group $\widehat{\mathbb{G}}_m(X,Y)=X+Y+XY$. Note that $\widehat{\mathbb{G}}_m(X,Y)=(1+X)(1+Y)-1$. Here $i(X)=-\dfrac{X}{1+X}$.

(3) More generally, for each $a\in A$, the power series

$$F(X,Y)=X+Y+aXY$$

is a formal group over A. Here $i(X)=-\dfrac{X}{1+aX}$.

14.1.3 We introduce basic notions of the theory of formal groups. An homomorphism of formal groups $F\to G$ over A is a power series $f\in XA[[X]]$ such that $f\circ F(X,Y)=G(f(X),f(Y))$. The set $\mathrm{Hom}_A(F,G)$ of homomorphisms $F\to G$ is an abelian group with respect to the addition defined by the formula

$$f\oplus g=G(f(X),g(X)).$$

We set $\mathrm{End}_A(F)=\mathrm{Hom}_A(F,F)$. Then $\mathrm{End}_A(F)$ is a ring with respect to the addition defined above and the multiplication defined as the composition of power series:

$$f\circ g(X)=f(g(X)).$$

14.1.4 The module $\widehat{\Omega}^1_{A[[X]]}$ of formal Kähler differentials of $A[[X]]$ over A is the free $A[[X]]$-module generated by dX.

Definition We say that $\omega(X)=f(X)dX\in\widehat{\Omega}^1_{A[[X]]}$ is an invariant differential form on the formal group F if

$$\omega(X+_F Y)=\omega(X).$$

14.1.5 The next proposition describes invariant differential forms on one-dimensional formal groups. We will write $F_1'(X,Y)$ (respectively, $F_2'(X,Y)$) for the formal derivative of $F(X,Y)$ with respect to the first (respectively, second) variable.

Proposition 14.1.6 *The space of invariant differential forms on a one-dimensional formal group $F(X,Y)$ is the free A-module of rank one generated by*

$$\omega_F(X)=\frac{dX}{F_1'(0,X)}.$$

Proof See, for example, [88, Section 1.1].

(a) Since $F(Y,X)=Y+X+(\text{terms of degree}\geqslant 2)$, the series $F_1'(0,X)$ is invertible in $A[[X]]$, and one has

$$\omega(X):=\frac{dX}{F_1'(0,X)}\in A[[X]].$$

Differentiating the identity

$$F(Z, F(X, Y)) = F(F(Z, X), Y)$$

with respect ot Z, one has

$$F_1'(Z, F(X, Y)) = F_1'(F(Z, X), Y) \cdot F_1'(Z, X).$$

Setting $Z = 0$, we obtain that

$$\frac{F_1'(X, Y)}{F_1'(0, F(X, Y))} = \frac{1}{F_1'(0, X)},$$

or equivalently, that

$$\frac{dF(X, Y)}{F_1'(0, F(X, Y))} = \frac{dX}{F_1'(0, X)}.$$

This shows that $\omega(X)$ is invariant.

(b) Conversely, assume that $\omega(X) = f(X)dX$ is invariant. Then

$$f(F(X, Y))F_1'(X, Y) = f(X).$$

Setting $X = 0$, we obtain that $f(Y) = F_1'(0, Y)f(0)$. Therefore,

$$\omega(X) = f(0)\omega_F(X),$$

and the proposition is proved. □

Remark 14.1.7 We can write ω_F in the form:

$$\omega_F(X) = \left(\sum_{n=0}^{\infty} a_n X^n\right) dX, \qquad \text{where } a_n \in A \text{ and } a_0 = 1.$$

14.1.8 Let K denote the field of fractions of A. We say that a power series $\lambda(X) \in K[[X]]$ is a logarithm of F, if

$$\lambda(X +_F Y) = \lambda(X) + \lambda(Y).$$

Proposition 14.1.9 *Assume that* $\mathrm{char}(K) = 0$. *Then the map*

$$\omega \mapsto \lambda_\omega(X) := \int_0^X \omega$$

establishes an isomorphism between the one-dimensional K-vector space generated by ω_F and the K-vector space of logarithms of F.

Proof (a) Let $\omega(X) = g(X)dX$ be a non-zero invariant differential form on F. Set $g(X) = \sum\limits_{n=0}^{\infty} b_n X^n$. Since $\mathrm{char}(K) = 0$, the series $f(X)$ has the formal primitive

$$\lambda_\omega(X) := \int_0^X \omega = \sum_{n=1}^{\infty} \frac{b_{n-1}}{n} X^n \in K[[X]].$$

The invariance of ω reads

$$g(F(X, Y))F_1'(X, Y) = g(X),$$

and taking the primitives, we obtain:

$$\lambda_\omega(X +_F Y) = \lambda_\omega(X) + h(Y)$$

for some $h(Y) \in K[[Y]]$. Putting $X = 0$ in the last formula, we have $h(Y) = \lambda_\omega(Y)$, and $\lambda_\omega(X +_F Y) = \lambda_\omega(X) + \lambda_\omega(Y)$. Therefore, λ_ω is a logarithm of F.

(b) Conversely, let $\lambda(X)$ be a logarithm of F. Differentiating the identity $\lambda(Y +_F X) = \lambda(Y) + \lambda(X)$ with respect to Y and setting $Y = 0$, one has

$$\lambda'(X) = \frac{\lambda'(0)}{F_1(0, X)}.$$

Set $\omega = \lambda'(X)dX$. Then $\omega = \lambda'(0)\omega_F$, and the proposition is proved. □

Definition 14.1.10 Set

$$\lambda_F(X) = \int_0^X \omega_F.$$

Note that $\lambda_F(X)$ is the unique logarithm of F such that

$$\lambda_F(X) \equiv X \pmod{\deg 2}.$$

From Proposition 14.1.9, it follows that over a field of characteristic 0 all formal goups are isomorphic to the additive formal group. Indeed, λ_F is an isomorphism $F \simeq \widehat{\mathbb{G}}_a$.

Example 14.1.11 For the multiplicative group we have

$$\omega_{\mathbb{G}_m}(X) = \frac{dX}{1+X}, \qquad \lambda_{\mathbb{G}_m}(X) = \log(1+X) = \sum_{n=1}^{\infty} (-1)^{n-1} \frac{X^n}{n}.$$

14.1.12 We consider formal groups over the ring of integers of a local field K of characteristic 0 and residue caracteristic p.

For each $n \in \mathbf{Z}$, we denote by $[n]$ the formal multiplication by n:

$$[n] = \begin{cases} \underbrace{X +_F +X_F +_F \cdots + X}_{n}, & \text{if } n \geqslant 0, \\ i([-n]), & \text{if } n < 0. \end{cases}$$

This defines an injection

$$[\ ] : \mathbf{Z} \to \mathrm{End}_{O_K}(F), \qquad n \to [n](X) = nX + \cdots .$$

It can be easily checked that this map can be extended by continuity to an injective map

$$[\ ] : \mathbf{Z}_p \to \mathrm{End}_{O_K}(F), \qquad a \to [a](X) = aX + \cdots .$$

Proposition 14.1.13 *Let F be a formal group over O_K. Then either*

$$[p](X) \equiv 0 \pmod{\mathfrak{m}_K}$$

or there exists an integer $h \geqslant 1$ and a power series $g(X) = c_1X + \cdots$ such that $c_1 \not\equiv 0 \pmod{\mathfrak{m}_K}$ and

$$[p](X) \equiv g(X^{p^h}) \pmod{\mathfrak{m}_K}. \tag{63}$$

Proof The proof is not difficult. See, for example, [76, Chap. I, § 3, Theorem 2]. □

Definition 14.1.14 If $[p](X) \equiv 0 \pmod{\mathfrak{m}_K}$, we say that F has infinite height. Otherwise, we say that F is p-divisible and call the height of F the unique $h \geqslant 1$ satisfying condition (63).

14.1.15 Now we can explain the connection between formal groups and p-adic representations. Recall that we write $\mathbf{C}$ for the completion of $\overline{K}$. We denote by $O_{\mathbf{C}}$ the ring of integers of $\mathbf{C}$ and by $\mathfrak{m}_{\mathbf{C}}$ the maximal ideal of $O_{\mathbf{C}}$. Any formal group law $F(X,Y)$ over O_K defines a structure of $\mathbf{Z}_p$-module on $\mathfrak{m}_{\mathbf{C}}$ of $\overline{K}$:

$$\begin{aligned} &\alpha +_F \beta := F(\alpha, \beta), && \alpha, \beta \in \mathfrak{m}_{\mathbf{C}}, \\ &\mathbf{Z}_p \times \mathfrak{m}_{\mathbf{C}} \to \mathfrak{m}_{\mathbf{C}}, && (a, \alpha) \mapsto [a](\alpha). \end{aligned}$$

We will denote by $F(\mathfrak{m}_{\mathbf{C}})$ the ideal $\mathfrak{m}_{\mathbf{C}}$ equipped with this $\mathbf{Z}_p$-module structure. The analogous notation will be used for O_K-submodules of $\mathfrak{m}_{\mathbf{C}}$.

Proposition 14.1.16 *Assume that F is a formal group of finite height h. Then:*
(i) The map $[p] : F(\mathfrak{m}_{\mathbf{C}}) \to F(\mathfrak{m}_{\mathbf{C}})$ is surjecive.
(ii) The kernel $\ker([p])$ *is a free $\mathbf{F}_p$-module of rank h.*

Proof (i) Consider the equation

$$[p](X) = \alpha, \qquad \alpha \in F(\mathfrak{m}_{\mathbf{C}}).$$

A version of the Weierstrass preparation theorem (see, for example, the proof of [105, Theorem 4.2]) shows that this equation can be written in the form $f(X) = g(\alpha)$, where $f(X) \in O_K[X]$ is a polynomial of degree p^h such that $f(X) \equiv X^{p^h} \pmod{\mathfrak{m}_K}$, and $g \in O_K[[X]]$. Therefore, the roots of this equation are in $\mathfrak{m}_{\mathbf{C}}$.

(ii) To prove that $\ker([p])$ is a free $\mathbf{Z}/p\mathbf{Z}$-module of rank h, we only need to show that the roots of the equation $[p](X) = 0$ are all of multiplicity one. Differentiating the identity

$$[p](F(X,Y)) = F([p](X),[p](Y))$$

with respect to Y and setting $Y = 0$, we get:

$$[p]'(X) \cdot F_2'(X,0) = F_2'([p](X),0).$$

Let $[p](\xi) = 0$. Since $F_2'(X,0)$ is invertible in $O_K[[X]]$ and $\xi \in \mathfrak{m}_{\mathbf{C}}$, we have $F_2'(\xi,0) \neq 0$ and $[p]'(\xi) \neq 0$. Therefore, ξ is a simple root. □

14.1.17 For $n \geqslant 1$, let $T_{F,n}$ denote the p^n-torsion subgroup of $F(f m_{\mathbf{C}})$. From Proposition 14.1.16 it follows that as abelian group, it is not canonically isomorphic to $(\mathbf{Z}/p^n\mathbf{Z})^h$ and sits in the exact sequence

$$0 \to T_{F,n} \to F(\mathfrak{m}_{\mathbf{C}}) \xrightarrow{[p^n]} F(\mathfrak{m}_{\mathbf{C}}) \to 0.$$

As in the case of abelian varieties, the Tate module of F is defined as the projective limit

$$T(F) = \varprojlim_n T_{F,n}$$

with respect to the multiplication-by-p maps. Since the series $[p^n](X)$ have coefficients in O_K, the Galois group G_K acts on $E_{F,n}$, and this action gives rise to a $\mathbf{Z}_p$-adic representation:

$$\rho_F : G_K \to \mathrm{Aut}_{\mathbf{Z}_p}(T(F)) \simeq \mathrm{GL}_h(\mathbf{Z}_p).$$

We will denote by $V(F) = T(F) \otimes_{\mathbf{Z}_p} \mathbf{Q}_p$ the associated p-adic representation.

Example 14.1.18 (1) $F = \widehat{\mathbb{G}}_m$. One has $[p^n] = (1+X)^{p^n} - 1$. Therefore,

$$T_{\widehat{\mathbb{G}}_m,n} = \left\{\zeta - 1 \mid \zeta^{p^n} = 1\right\},$$

and the map

$$\mu_{p^n} \to T_{\widehat{\mathbb{G}}_m,n}, \qquad \zeta \mapsto \zeta - 1$$

is an isomorphism of G_K-modules. In particular, $T(\widehat{\mathbb{G}}_m) \simeq \mathbf{Z}_p(1)$.

(2) Let E/O_K be an elliptic curve having good reduction modulo $\mathfrak{m}_K$. Writing the group law on E in terms of a local parameter at 0, one obtains a formal power

series $F(X, Y)$, which is a formal group law over O_K. One can prove that F is of height 1 if E has ordinary reduction, and of height 2 if E has supersingular reduction. We have a canonical injection of $T(F)$ in the Tate module $T_p(E)$ of E, which is an isomorphism in the supersingular case. See [146, Chap. 4] for further detail and applications.

14.1.19 The notion of a formal group can be generalized to higher dimensions. Let $X = (X_1, \dots, X_d)$ and $Y = (Y_1, \dots, Y_d)$ be d-vectors of variables. A d-dimensional formal group over O_K is a d-tuple $F(X, Y) = (F_1(X, Y), \dots, F_d(X, Y))$ with

$$F_i(X, Y) \in O_K[[X, Y]], \qquad 1 \leqslant i \leqslant d,$$

which satisfies the direct analogs of conditions (i), (iii) and (iv) in the definition of a one-dimensional formal group. We remark that contrary to the one-dimensional case, there are non-commutative formal groups of dimension $\geqslant 2$. Non-commutative formal groups appear in Lie theory. Below, without special mentioning, we consider only commutative formal groups.

14.1.20 Propositions 14.1.6 and 14.1.9 generalize directly to the higher-dimensional case. Namely, let $I = (X_1, \dots, X_d) \subset O_K[[X]]$. We set:

$$t_F^*(O_K) = I/I^2$$

and call it the cotangent space of F over O_K. The module of invariant differential forms on F is canonically isomorphic to $t_F^*(O_K)$. Namely:

(1) For each $a_1X_1 + \cdots + a_dX_d \mod I^2 \in t_F^*(O_K)$, there exists a unique invariant differential form ω such that

$$\omega(0) = a_1 dX_1 + \cdots + a_d dX_d.$$

This correspondence gives an isomorphism:

$$t_F^*(O_K) \simeq \{\text{invariant differential forms on} F\}.$$

(2) Each invariant differential form ω is *closed*, i.e. there exists a unique $\lambda_\omega(X) \in K[X]$ such that $\lambda_\omega(0, \dots, 0) = 0$ and

$$d\lambda_\omega(X) = \omega.$$

(3) The map $\omega \mapsto \lambda_\omega$ establishes an isomorphism between the K-vector space Ω_F^1 generated by invariant differential forms on F and the K-vector space of logarithms of F.

The notion of the height of a formal group generalizes as follows:

Definition 14.1.21 A formal group F is p-divisible if the morphism

$$[p]^* : O_K[[X]] \to O_K[[X]], \qquad f(X) \mapsto f \circ [p](X)$$

makes $O_K[[X]]$ into a free module of finite rank over itself.

If F is p-divisible, then the degree of the map $[p]^*$ is of the form p^h for some $h \geqslant 1$. This follows from the fact that any finite connected group over k_K is of order p^h for some h (see, for example [64, Chapitre I, § 9]). We call h *the height* of F. A formal group of dimension d defines a structure of $\mathbf{Z}_p$-module on $\mathfrak{m}_{\mathbf{C}}^d$, which we will denote by $F(\mathfrak{m}_{\mathbf{C}})$. The definition of the Tate module $T(F)$ and the p-adic representation $V(F)$ generalizes directly to p-divisible formal groups.

14.2 p-Divisible Groups

14.2.1 The category of formal groups is too small to develop a satisfactory theory. In particular, it is not closed under taking duals. To remedy this problem, it is more convenient to work in the category of p-divisible groups, introduced by Tate [151].

Definition A p-divisible group of height h over O_K is a system $\mathscr{G} = (\mathscr{G}_n)_{n\in\mathbf{N}}$ of finite group schemes $\mathscr{G}_n$ of order p^{hn} equipped with injective maps $i_n : \mathscr{G}_n \to \mathscr{G}_{n+1}$ such that the sequences

$$0 \to \mathscr{G}_n \xrightarrow{i_n} \mathscr{G}_{n+1} \xrightarrow{p^n} \mathscr{G}_{n+1}, \qquad n \geqslant 1$$

are exact.

From the theory of finite group schemes, it is known that each $\mathscr{G}_n$ sits in an exact sequence

$$0 \to \mathscr{G}_n^0 \to \mathscr{G}_n \to \mathscr{G}_n^{\text{ét}} \to 0, \tag{64}$$

where $\mathscr{G}_n^0$ is a connected and $\mathscr{G}_n^{\text{ét}}$ is an étale group scheme. We will say that $\mathscr{G} = (\mathscr{G}_n)_{n\in\mathbf{N}}$ is connected (respectively, étale) if each $\mathscr{G}_n$ is. The exact sequences (64) give rise to an exact sequence of p-divisible groups

$$0 \to \mathscr{G}^0 \to \mathscr{G} \to \mathscr{G}^{\text{ét}} \to 0, \tag{65}$$

where $\mathscr{G}^0$ and $\mathscr{G}^{\text{ét}}$ are connected and étale respectively.

14.2.2 To each p-divisible group $\mathscr{G}$, one can naturally associate its Tate module, setting:

$$T(\mathscr{G}) = \varprojlim_n \mathscr{G}_n(O_{\mathbf{C}}).$$

Then $T(\mathscr{G})$ is a free $\mathbf{Z}_p$-module of rank h equipped with a natural action of G_K. We denote by $V(\mathscr{G}) := \mathbf{Q}_p \otimes_{\mathbf{Z}_p} T(\mathscr{G})$ the associated p-adic representation. From the exact sequence (65), one has an exact sequence of p-adic representations:

$$0 \to V(\mathscr{G}^0) \to V(\mathscr{G}) \to V(\mathscr{G}^{\text{ét}}) \to 0.$$

14.2.3 If $F(X, Y)$ is a p-divisible formal group, then the kernels $F[p^n]$ of the isogenies $[p^n] : F \to F$ form a system $F(p) = (F[p^n])_{n \in \mathbf{N}}$ of finite group schemes satisfying the above definition, and we have a functor $F \mapsto F(p)$ from the category of formal groups to the category of p-divisible groups.

Proposition 14.2.4 (Tate) *The functor $F \mapsto F(p)$ induces an equivalence between the category of p-divisible formal groups and the category of connected p-divisible groups.*

Proof See [151, Proposition 1] and the references in *op. cit.* □

14.2.5 If $\mathscr{G}$ is a p-divisible group, we call the dimension of $\mathscr{G}$ the dimension of the formal group F corresponding to its connected component. We also define the tangent space $t_{\mathscr{G}}(O_K)$ of $\mathscr{G}$ as the tangent space of F.

14.2.6 The Cartier duality for finite group schemes allows to associate to $\mathscr{G}$ a dual p-divisible group $\mathscr{G}^\vee$. We have fundamental relations between the heights and dimensions of $\mathscr{G}$ and $\mathscr{G}^\vee$:

$$\operatorname{ht}(\mathscr{G}) = \operatorname{ht}(\mathscr{G}^\vee), \qquad \dim(\mathscr{G}) + \dim(\mathscr{G}^\vee) = \operatorname{ht}(\mathscr{G})$$

([151, Proposition 3]). Moreover, the duality induces a non-degenerate pairing on Tate modules:

$$T(\mathscr{G}) \times T(\mathscr{G}^\vee) \to \mathbf{Z}_p(1).$$

Example 14.2.7 Let E/O_K be an elliptic curve having a good reduction modulo $\mathfrak{m}_K$. The kernel $E[p^n]$ of the multiplication-by-p^n map is a finite group scheme of order p^{2n}. The system $(E[p^n])_{n \in \mathbf{N}}$ is a p-divisible group of height 2. The connected component of this p-divisible group corresponds to the formal group associated by E in Example 14.1.18, 2).

14.3 Classification of p-Divisible Groups

14.3.1 In [64], Fontaine classified p-divisible groups over O_K up to isogeny in terms of filtered φ-modules. The idea of such classification goes back to Grothendieck [83, 84] and relies on the following principles:

(1) One associates to any p-divisible group $\mathscr{G}$ of dimension d and height h a φ-module $M(\mathscr{G})$ together with a d-dimensional subspace $L(\mathscr{G}) \subset M(\mathscr{G})_K$.
(2) The φ-module $M(\mathscr{G})$ is the Dieudonné module associated to the reduction $\overline{\mathscr{G}}$ of $\mathscr{G}$ modulo $\mathfrak{m}_K$ by the theory of formal group schemes in characteristic p (see, for example, [112]).

(3) The subspace $L(\mathscr{G}) \subset M(\mathscr{G})_K$ depends on the lift of $\overline{\mathscr{G}}$ in characteristic 0. The filtration on $M(\mathscr{G})_K$ is defined as follows:

$$\mathrm{Fil}^0 M(\mathscr{G})_K = M(\mathscr{G})_K, \qquad \mathrm{Fil}^1 M(F)_K = L(\mathscr{G}), \qquad \mathrm{Fil}^2 M(\mathscr{G})_K = \{0\}.$$

14.3.2 We give an interpretation of the module $(M(\mathscr{G}), L(\mathscr{G}))$ for formal p-divisible groups in terms of differential forms. This description is equivalent to Fontaine's general construction (see [64, Chap. V] for the proofs of the results stated below). Let F be a formal p-divisible group of dimension d and height h. Recall that a differential form

$$\omega = \sum_{i=1}^{d} a_i(X_1, \ldots, X_d) dX_i, \qquad a_i(X_1, \ldots, X_d) \in K[[X_1, \ldots, X_d]]$$

is closed if there exists a power series $\lambda_\omega \in K[[X_1, \ldots, X_d]]$ such that $\lambda_\omega(0, \ldots, 0) = 0$ and $d\lambda_\omega = \omega$. Note that if ω is an invariant form, then λ_ω is the associated logarithm of F. As before, we set $X = (X_1, \ldots, X_d)$ and $Y = (Y_1, \ldots, Y_d)$ to simplify notation.

Definition A closed differential form ω is
(i) of the second kind on F, if there exists $r \geqslant 0$ such that

$$\lambda_\omega(X +_F Y) - \lambda_\omega(X) - \lambda_\omega(Y) \in p^{-r} O_K[[X, Y]];$$

(ii) exact, if there exists $r \geqslant 0$ such that $\lambda_\omega \in p^{-r} O_K[[X]]$.

It is easy to see that each exact form is of the second kind. Consider the quotient:

$$H^1_{\mathrm{dR}}(F) = \frac{\{\text{differential forms of the second kind}\}}{\{\text{ exact forms}\}}.$$

Then $H^1_{\mathrm{dR}}(F)$ is a K-vector space of dimension h, which can be viewed as the first de Rham cohomology group of F. Let K_0 denote the maximal unramified subfield of K, and let $M(F)$ be the K_0-subspace of $H^1_{\mathrm{dR}}(F)$ generated by the forms with coefficients in K_0. Then $M(F)$ depends only on the reduction of F modulo $\mathfrak{m}_K$ and one has

$$H^1_{\mathrm{dR}}(F) = M(F)_K.$$

Moreover, $M(F)$ is equipped with the Frobenius operator φ which acts as the absolute Frobenius on the coefficients of power series and such that $\varphi(X_i) = X_i^p$:

$$\varphi\left(\sum_{i=1}^{d} a_i(X_1, \ldots, X_d) dX_i\right) = \sum_{i=1}^{d} a_i^{\varphi}(X_1^p, \ldots, X_d^p) dX_i^p.$$

Consider the K-vector space Ω^1_F generated by invariant forms on F. Recall that $\dim_K \Omega^1_F = d$. Each invariant form is clearly of the second kind, and Ω^1_F injects into $H^1_{\mathrm{dR}}(F)$. Set:

$$L(F) := \text{image of } \Omega^1_F \text{ in } H^1_{\mathrm{dR}}(F).$$

These data define a structure of filtered module on $M(F)$.

14.3.3 Assume that the local field K is absolutely unramified. In that case, formal groups over O_K were classified up isomorphism by Honda [88], purely in terms of their logarithms. In this section, we review Honda's classification. To simplify the exposition, we restrict our discussion to the one-dimensional case.

The ring of power series $K[[X]]$ is equipped with the Frobenius operator φ :

$$\varphi\left(\sum_{i=0}^{\infty} a_i X^i\right) = \sum_{i=0}^{\infty} \varphi(a_i) X^{ip}.$$

Assume that $\alpha_1, \dots, \alpha_{h-1}, \alpha_h \in O_K$ satisfy the following conditions:

$$\begin{aligned} &\alpha_1, \dots, \alpha_{h-1} \equiv 0 \pmod{p}, \\ &\alpha_h \in U_K. \end{aligned} \tag{66}$$

Set:

$$\mathscr{A}(\varphi) := \sum_{i=0}^{h} \alpha_i \varphi^i,$$

and consider the power series

$$\lambda(X) := \left(1 - \frac{\mathscr{A}(\varphi)}{p}\right)^{-1} (X) \in K[[X]].$$

For formal p-divisible groups of dimension one, the result of Honda states as follows:

Theorem 14.3.4 (Honda) *(i) Assume that $\alpha_1, \dots, \alpha_h$ satisfy conditions (66). Then $\lambda(X) = \lambda_G(X)$ for some one-dimensional formal group G of height h.*

(ii) Let F be a one-dimensional formal group over O_K of height h. Then there exists a unique system $\alpha_1, \dots, \alpha_h$ satisfying (66) such that

$$\left(1 - \frac{\mathscr{A}(\varphi)}{p}\right) \lambda_F(X) \in O_K[[X]].$$

Let G be the formal group associated to $\alpha_1, \dots, \alpha_h$ by part i). Then $F \simeq G$.

The relation between this theorem and Fontaine's classification is given by the following:

Proposition 14.3.5 *Assume that K is absolutely unramified. Let F be a one-dimensional formal group over O_K of height h. Denote by b_F the image of ω_F in $M(F)$. Then the following holds true:*

(i) The elements $b_F, \varphi(b_F), \dots, \varphi^{h-1}(b_F)$ form a basis of $M(F)$ over K.

(ii) Let $\alpha_1, \dots, \alpha_h$ be the parameters associated to F by Honda's theorem. Then

$$\alpha_1\varphi(b_F) + \alpha_2\varphi(b_F) + \cdots + \alpha_h\varphi^h(b_F) = pb_F.$$

(iii) One has an isomorphism of filtered φ-modules

$$M(F) \simeq K[\varphi]/(\mathscr{A}(\varphi) - p),$$

which sends $L(F) = K \cdot b_F$ to the one-dimensional K-vector space generated by 1.

Proof See [64, Chapitre V]. □

Remark 14.3.6 In fact, Fontaine's theory [64] gives more precise results that those that we have stated. Namely, if the absolute ramification index of K is $\leqslant p-1$, it allows to classify p-divisible groups up to isomorphism and not only up to isogeny. Using new ideas, Breuil [30] classified p-divisible groups up to isomorphism without any restriction on ramification. See [97] and [33] for further developments.

14.4 p-Adic Integration on Formal Groups

14.4.1 We maintain assumptions and conventions of the previous section. Let F be a formal p-divisible group of dimension d and height h. We denote by $T(F)$ the Tate module of F. Let $\xi = (\xi_n)_{n\geqslant 0} \in T(F)$, where $\xi_n \in T_{F,n}$ for each $n \geqslant 0$. Recall that we have the canonical map $\theta : \mathbf{A}_{\mathrm{inf}} \to O_{\mathbf{C}}$. For each n, choose $\widehat{\xi}_n \in \mathbf{A}_{\mathrm{inf}}^d$ such that $\theta(\widehat{\xi}_n) = \xi_n$.

Theorem 14.4.2 (Colmez, Fontaine) *(i) Let ω be a differential form of second kind. Then the sequence $(p^n\lambda_\omega(\widehat{\xi}_n))_{n\geqslant 0}$ converges in $\mathbf{B}^+_{\mathrm{cris},K} = K \otimes_{K_0} \mathbf{B}^+_{\mathrm{cris}}$. Its limit does not depend on the choice of $\widehat{\xi}_n$ and therefore defines the "p-adic integral":*

$$\int_\xi \omega := -\lim_{n\to+\infty} p^n\lambda_\omega(\widehat{\xi}_n). \tag{67}$$

If ω has coefficients in K_0, then $\int_\xi \omega \in \mathbf{B}^+_{\mathrm{cris}}$.

(ii) If ω is exact, then $\int_\xi \omega = 0$.

(iii) The p-adic integration (67) is compatible with the actions of the Galois group and the Frobenius φ. Namely, one has

$$\int_\xi \varphi(\omega) = \varphi\left(\int_\xi \omega\right),$$
$$\int_{g(\xi)} \omega = g\left(\int_\xi \omega\right), \qquad g \in G_K.$$

iv) The p-adic integration induces a non-degenerated pairing

$$M(F) \times T(F) \to \mathbf{B}_{\mathrm{cris}},$$

which is compatible with the Frobenius operator and the Galois action, and a non-degenerated pairing

$$H^1_{\mathrm{dR}}(F) \times T(F) \to \mathbf{B}^+_{\mathrm{dR}},$$

which is compatible with the Galois action and filtration.

Proof See [64, Chapitre V, §1], [66, Théorème 6.2] and [38, Proposition 3.1]. We remark that the delicate part here is the non-degeneracy of the constructed pairings. The proof of other points is straightforward. □

Example 14.4.3 Consider the case of the multiplicative formal group $\widehat{\mathbb{G}}_m$. Recall that $T(\widehat{\mathbb{G}}_m) \simeq \mathbf{Z}_p(1)$ is generated by any compatible system $(\xi_n)_{n\geqslant 0}$ such that $\xi_n = \zeta_{p^n} - 1$ and $\zeta_p \neq 1$. The space $H^1_{\mathrm{dR}}(\widehat{\mathbb{G}}_m)$ is generated over K by $\omega = \frac{dX}{1+X}$, and the formal primitive of ω is $\log(1+X)$. Take $\hat{\xi}_n = [\varepsilon]^{1/p^n} - 1$. One has

$$\int_\xi \omega = -\lim_{n\to+\infty} p^n \log[\varepsilon]^{1/p^n} = -t.$$

This formula can be seen as the p-adic analog of the following computation. Let C denote the unit circle on the complex plane parametrized by $e^{2\pi i x}$, $x \in [0,1]$. Then

$$\int_C \frac{dz}{z} = n\log(z)\Big|_0^{e^{\frac{2\pi i}{n}}} = 2\pi i.$$

Corollary 14.4.4 *The representation $V(F)$ is crystalline, and there exist canonical isomorphisms:*

$$\mathbf{D}^*_{\mathrm{cris}}(V(F)) \simeq M(F), \qquad \mathbf{D}^*_{\mathrm{dR}}(V(F)) \simeq H^1_{\mathrm{dR}}(F).$$

Corollary 14.4.5 (Tate) *The representation $V(F)$ is Hodge–Tate and there exists a canonical isomorphism*

$$V(F) \otimes_{\mathbf{Q}_p} \mathbf{C} \simeq \left(t^*_{F^\vee}(K) \otimes_K \mathbf{C}\right) \oplus \left(t_F(K) \otimes_K \mathbf{C}(1)\right). \tag{68}$$

Proof This follows from the previous corollary and the isomorphisms

$$t_F^*(K) \simeq \Omega_F^1, \qquad H_{\mathrm{dR}}^1(F)/\Omega_F^1 \simeq t_{F^\vee}(K)$$

(the second isomorphism is provided by duality).

Remark 14.4.6 (1) Corollary 14.4.4 holds for all p-divisible groups (see [66, Théorème 6.2]). Conversely, Breuil [30] proved that each crystalline representation with Hodge–Tate weights 0 and 1 arises from a p-divisible group.

(2) The Hodge–Tate decomposition (68) was first proved by Tate [151] for *all* p-divisible groups. Some constructions of this paper will be revewed in Sect. 16. The case of abelian variety with bad reduction follows from the semi-stable reduction theorem (Raynaud). A completely different proof was found by Fontaine [67].

(3) The construction of p-adic integration in Theorem 14.4.2 generalizes to the case of abelian varieties [38, 39].

15 Formal Complex Multiplication

15.1 Lubin–Tate Theory

15.1.1 In this section, we discuss the theory of complex multiplication in formal groups. We start with a brief overview of Lubin–Tate theory [111]. Let K is a local field of arbitrary characteristic. Set $q = |k_K| = p^f$. Fix an uniformizer π of K.

Theorem 15.1.2 *(i) Let $f(X) \in O_K[[X]]$ be a power series satisfying the following conditions:*

$$\begin{aligned} f(X) &\equiv \pi X \pmod{\deg 2}, \\ f(X) &\equiv X^q \pmod{\mathfrak{m}_K}. \end{aligned} \tag{69}$$

Then the following holds true:

(i) There exists a unique formal group $F_f(X,Y)$ over O_K such that $f(X) \in \mathrm{End}_{O_K}(F)$. Moreover, for each $a \in O_K$, there exists a unique endomorphism $[a](X) \in \mathrm{End}_{O_K}(F)$ such that $[a](X) \equiv aX \pmod{\deg 2}$.

(ii) Let $g(X)$ be another power series satisfying conditions (69) with the same uniformizer π. Then F_g and F_f are isomorphic over O_K. In the isomorphism class of F_f, there exists a formal group F_{LT} with the logarithm

$$\lambda_{\mathrm{LT}}(X) = X + \frac{X^q}{\pi} + \frac{X^{q^2}}{\pi^2} + \cdots.$$

(iii) Let π' be another uniformizer of O_K, and let $g(X)$ be a power series satisfying conditions (69) with π' in the place of π. Then F_f and F_g are isomorphic over the ring $\widehat{O}_K^{\mathrm{ur}}$.

Proof All these statements can be proved by successive approximation in the rings of formal power series. We refer the reader to [111] or to [140] for detailed proofs.

Definition F_f is called the Lubin–Tate formal group associated to f.

15.1.3 Let F_f be the Lubin–Tate formal group associated to $f(X) = \pi X + X^q$. The group of points $F_f(\mathfrak{m}_{\mathbf{C}})$ is an O_K-module with the action of O_K given by

$$(a, \alpha) \mapsto [a](\alpha), \qquad a \in O_K, \quad \alpha \in F_f(\mathfrak{m}_{\mathbf{C}}).$$

In particular, $[\pi](X) = f(X)$, and for any $n \geqslant 1$, one has

$$[\pi^n](X) = \underbrace{f \circ f \circ \cdots \circ f(X)}_{n}.$$

The polynomial

$$[\pi^n]/[\pi^{n-1}] = \pi + [\pi^{n-1}](X)^{q-1}, \tag{70}$$

is Eisenstein of degree $q^{n-1}(q-1)$. Let $T_{f,n}$ denote the group of π^n-torsion points of F_f. An easy induction together with the previous remark show that $T_{f,n}$ is an abelian group of order q^n. The endomorphism ring $\mathrm{End}_{O_K}(F_f) \simeq O_K$ acts on $T_{f,n}$ through the quotient $O_K/\pi^n O_K$, and $T_{f,n}$ is free of rank one over $O_K/\pi^n O_K$. The generators of $T_{f,n}$ are the roots of the polynomial (70). Let $K_{f,n}$ be the field generated over K by $T_{f,n}$. Then

$$K_{f,n} = K(\pi_n),$$

where π_n is any generator of $T_{f,n}$. In particular, $[K_{f,n} : K] = (q-1)q^{n-1}$, and π_n is a uniformizer of $K_{f,n}$.

15.1.4 Let g be another power series satisfying (69) with the same π. Then $F_g \simeq F_f$, $T_{g,n} \simeq T_{f,n}$, and $K_{f,n} = K_{g,n}$. Since the field generated by π^n-torsion points of a Lubin–Tate formal group depends only on the choice of the uniformizer π, we will write $K_{\pi,n}$ in the place of $K_{f,n}$. Set:

$$K_\pi = \bigcup_{n=1}^{\infty} K_{\pi,n}.$$

From the explicit form of Eisenstein polynomials (70), it follows that π is a universal norm in K_π/K.

The following theorem gives an explicit approach to local class field theory:

Theorem 15.1.5 (Lubin–Tate) *i) One has*

$$K^{\mathrm{ab}} = K^{\mathrm{ur}} \cdot K_\pi.$$

(ii) Let $\theta_K : K^* \to \mathrm{Gal}(K^{\mathrm{ab}}/K)$ denote the reciprocity map. For any $u \in U_K$, the automorphism $\theta_K(u)$ acts on the torsion points of F_f by the formula:

$$\theta_K(u)(\xi) = [u^{-1}](\xi), \qquad \forall \xi, \quad [\pi^n](\xi) = 0, \quad n \in \mathbf{N}.$$

Proof See [111] or [140]. □

Remark 15.1.6 (1) The torsion points of a one dimensional formal group are the roots of its logarithm (see Proposition 16.1.2 below). Therefore, K^{ab} is generated over K^{ur} by the roots of the power series $\lambda_{\mathrm{LT}}(X)$. This can be seen as a solution of Hilbert 12th problem for local fields. Theorem 15.1.5 is the local analog of the theory of complex multiplication.

(2) Let $K = \mathbf{Q}_p$. The multiplicative formal group $\widehat{\mathbb{G}}_m$ is the Lubin–Tate group associated to the series $f(X) = (X+1)^p - 1$. In that case, Theorem 15.1.5 says that $\mathbf{Q}_p^{\mathrm{ab}} = \overset{\infty}{\underset{n}{\cup}} \mathbf{Q}_p(\zeta_n)$ and that

$$\theta_{\mathbf{Q}_p}(u)(\zeta_{p^n}) = \zeta_{p^n}^{u^{-1}}, \qquad \forall u \in U_{\mathbf{Q}_p}.$$

This can be proved without using the theory of formal groups.

(3) Let π_n be a generator of the group of π^n-torsion points of F_f. Since π_n is a uniformizer of $K_{\pi,n}$, and Theorem 15.1.5 describes the action of $\mathrm{Gal}(K^{\mathrm{ab}}/K)$ on π_n, this allows to compute the ramification filtration on $\mathrm{Gal}(K^{\mathrm{ab}}/K)$. One has

$$\theta_K\left(U_K^{(v)}\right) = \mathrm{Gal}(K^{\mathrm{ab}}/K)^{(v)}, \qquad \forall v \geqslant 0.$$

See [140] for a detailed proof.

15.2 Hodge–Tate Decomposition for Lubin–Tate Formal Groups

15.2.1 In this section, we assume that K has characteristic 0. We fix a uniformizer π and write F for an unspecified Lubin–Tate formal group associated to π. Since $p = \pi^e u$ with $e = e(K/\mathbf{Q}_p)$, and $u \in U_K$, we see that F is a p-divisible group of height $h = ef = [K : \mathbf{Q}_p]$. Its Tate module $T(F)$ can be written as the projective limit of π^n-torsion subgroups with respect to the multiplication-by-π map. Since $T(F)$ is an O_K-module of rank one, the action of G_K on $T(F)$ is given by a character

$$\chi_\pi : G_K \to U_K.$$

The theory of Lubin–Tate (Theorem 15.1.5) says that $\chi_\pi^{-1} \circ \theta_K$ coincides with the projection of K^* onto U_K under the decomposition $K^* \simeq U_K \times \langle \pi \rangle$.

15.2.2 Let E be a finite extension of K containing all conjugates τK of K over $\mathbf{Q}_p$. By local class field theory, one has a commutative diagram

$$\begin{array}{ccc} E^* & \xrightarrow{\theta_E} & \mathrm{Gal}(E^{\mathrm{ab}}/E) \\ {\scriptstyle N_{E/K}}\downarrow & & \downarrow \\ K^* & \xrightarrow{\theta_K} & \mathrm{Gal}(K^{\mathrm{ab}}/K). \end{array}$$

Therefore, G_E acts on $T(F)$ via the character $\rho_E = \chi_\pi \circ N_{E/K}$. Consider the vector space $V(F) = T(F) \otimes_{O_K} K$ as a G_E-module. By the previous remark, $V(F) \simeq K(\rho_E)$, and one has

$$V(F) \otimes_{\mathbf{Q}_p} \mathbf{C} \simeq \bigoplus_{\tau \in \mathrm{Hom}(K,E)} \mathbf{C}(\tau \circ \rho_E).$$

Compare this decomposition with the Hodge–Tate decomposition:

$$V(F) \otimes_{\mathbf{Q}_p} \mathbf{C} \simeq t^*_{F^\vee}(\mathbf{C}) \oplus t_F(\mathbf{C})(1).$$

These decompositions are compatible with the K-module structures on the both sides. Since K acts on $t_F(E)$ via the embedding $K \hookrightarrow E$, one has

$$\mathbf{C}(\tau \circ \rho_E) \simeq \begin{cases} \mathbf{C}(1), & \text{if } \tau = \mathrm{id}, \\ \mathbf{C}, & \text{if } \tau \neq \mathrm{id}. \end{cases} \tag{71}$$

Proposition 15.2.3 *For any continuous character $\psi : G_E \to U_K$, the following conditions are equivalent:*

(a) ψ concides with $\prod\limits_{\tau \in \mathrm{Hom}(K,E)} \tau^{-1} \circ \rho_{\tau E}^{n_\tau}$ on some open subgroup of I_E;

(b) $\mathbf{C}(\tau \circ \psi) = \mathbf{C}(\chi_E^{n_\tau})$ for all $\tau \in \mathrm{Hom}(K, E)$.

Proof See [143, Section A5]. Recall that for two continuous characters ψ_1 and ψ_2 we write $\psi_1 \sim \psi_2$ if $\mathbf{C}(\psi_1)$ and $\mathbf{C}(\psi_2)$ are isomorphic as continuous Galois modules. From (71), one has

$$\begin{aligned} \tau \circ \sigma^{-1} \circ \rho_{\sigma K} &\sim \chi_E, \quad \text{if } \tau = \sigma, \\ \tau \circ \sigma^{-1} \circ \rho_{\sigma K} &\sim \mathrm{id}, \quad \text{if } \tau \neq \sigma. \end{aligned}$$

Set:

$$\psi_1 = \prod_{\tau \in \mathrm{Hom}(K,E)} \tau^{-1} \circ \rho_{\tau K}^{n_\tau}.$$

Then the previous formula gives:

$$\tau \circ \psi_1 \sim \chi_K^{n_\tau}, \qquad \forall \tau \in \mathrm{Hom}(E, K).$$

Now the proposition follows from Proposition 4.3.6. □

15.3 Formal Complex Multiplication for p-Divisible Groups

15.3.1 Using Proposition 15.2.3, we can prove a general result about formal complex multiplication for p-divisible groups.

Definition Let $\mathscr{G}$ be a p-divisible group over O_E of dimension d and height h. We say that $\mathscr{G}$ has a formal complex multiplication by a p-adic field $K \subset E$ if $[K : \mathbf{Q}_p] = h$ and there exists an injective ring map

$$K \to \mathrm{End}_{O_E}(\mathscr{G}) \otimes_{\mathbf{Z}_p} \mathbf{Q}_p.$$

If $\mathscr{G}$ has a complex multiplication by K, the p-adic representation $V(\mathscr{G})$ is a K-vector space of dimension 1, and G_E acts on $V(\mathscr{G})$ via a character $\psi_{\mathscr{G}} : G_E \to U_K$. On the other hand, the tangent space $t_{\mathscr{G}}(E)$ is a (E, K)-module, and the multiplication by E in $t_{\mathscr{G}}(E)$ gives rise to a map

$$\det_{\mathscr{G}} : E^* \to \mathrm{Aut}_K(t_{\mathscr{G}}(E)) \xrightarrow{\det} K^*.$$

Recall that $\theta_E : E^* \to \mathrm{Gal}(E^{\mathrm{ab}}/E)$ denotes the reciprocity map.

Theorem 15.3.2 *Let $\mathscr{G}$ be a p-divisible group having a formal complex multiplication by K. Assume that E contains all conjugates of K. Then one has*

$$\psi_{\mathscr{G}}(\theta_E(u)) = \det_{\mathscr{G}}(u)^{-1}, \qquad u \in U,$$

for some open subgroup U of U_E.

Proof Comparing the decomposition

$$V(\mathscr{G}) \otimes_{\mathbf{Q}_p} \mathbf{C} \simeq \bigoplus_{\tau \in \mathrm{Hom}(K,E)} \mathbf{C}(\tau \circ \psi_{\mathscr{G}})$$

and the Hodge–Tate decomposition of $V(\mathscr{G})$, we see that there exists a subset $S \subset \mathrm{Hom}(K, E)$ such that $t_{\mathscr{G}}(E) \simeq \bigoplus_{\tau \in S} \tau(K)$ as a K-module and that

$$\begin{aligned} \tau \circ \psi_{\mathscr{G}} &\sim \chi_E \quad \text{if } \tau \in S, \\ \tau \circ \psi_{\mathscr{G}} &\sim 1 \quad \text{if } \tau \notin S. \end{aligned}$$

Proposition 15.2.3 implies that $\psi_{\mathscr{G}}$ concides on an open subgroup of I_E with the character

$$\prod_{\tau \in \mathrm{Hom}(K,E)} \tau^{-1} \circ \rho_{\tau E}.$$

Now the theorem follows from the theory of Lubin–Tate together with the formula

$$\det_{\mathscr{G}}(u) = \prod_{\tau \in S} \tau^{-1} \circ N_{E/\tau(K)}(u).$$

□

Remark 15.3.3 Theorem 15.3.2 is mentioned in [144]. We remark that it implies the main theorem of complex multiplication of abelian varieties in the *global* setting.

16 The Exponential Map

16.1 The Group of Points of a Formal Group

16.1.1 In this section, we study the group of points of a formal group in more detail. Let F be a formal p-divisible group. We denote by $T_{F,\infty}$ the group of torsion points of F. Note that $T_{F,\infty} = \bigcup_{n=0}^{\infty} T_{F,n}$, and that there is a canonical isomorphism

$$T_{F,\infty} \simeq V(F)/T(F).$$

Proposition 16.1.2 *(i) For any invariant differential form ω on F, the logarithm $\lambda_\omega(X)$ converges on $\mathfrak{m}_{\mathbf{C}}$.*

(ii) The map

$$\log_F : F(\mathfrak{m}_{\mathbf{C}}) \to t_F(\mathbf{C}),$$
$$\log_F(\alpha)(\omega) = \lambda_\omega(\alpha), \qquad \forall \omega \in \Omega_F^1$$

is an homomorphism.

(iii) One has an exact sequence

$$0 \to T_{F,\infty} \to F(\mathfrak{m}_{\mathbf{C}}) \xrightarrow{\log_F} t_F(\mathbf{C}) \to 0. \qquad (72)$$

Moreover, $\log_F$ *is a local isomorphism.*

Proof (i) The space of invariant differential forms on F is generated by the forms $\omega_1, \dots, \omega_d$ such that $\omega_i(0) = dX_i$. Let $\lambda_1, \dots, \lambda_d$ denote the logarithms of these forms. Since ω_i have coefficients in O_K, the series λ_i can be written as

$$\lambda_i(X) = X_i + \sum_{n \geqslant 2} \left(\sum_{n_1 + \cdots n_d = n} a_{n_1,\dots,n_d} X_1^{n_1} \cdot \ldots \cdot X_d^{n_d} \right),$$

where

$$n \cdot a_{n_1,\dots,n_d} \in O_K, \qquad n = n_1 + \cdots + n_d. \qquad (73)$$

This implies that the series λ_i converge on $\mathfrak{m}_{\mathbf{C}}^d$. Moreover, any logarithm can be written as a linear combination of λ_i. Therefore, for any ω, the series λ_ω converges on $\mathfrak{m}_{\mathbf{C}}^d$. This proves that the map $\log_F$ is well defined.

(ii) Since $\lambda_\omega(X +_F Y) = \lambda_\omega(X) + \lambda_\omega(Y)$, we have

$$\log_F(\alpha +_F \beta) = \log_F(\alpha) + \log_F(\beta).$$

(iii) Fix $c \in O_K$ such that

$$v_K(c) > \frac{v_K(p)}{p-1}.$$

Then from (73) it follows that

$$c^{-1}\lambda_i(cX_1, \dots, cX_d) = X_i + \sum_{n\geqslant 2}\left(\sum_{n_1+\cdots+n_d=n} b_{n_1,\dots,n_d} X_1^{n_1} \cdot \ldots \cdot X_d^{n_d}\right),$$

where $b_{n_1,\dots,n_d} \in O_K$. Applying the p-adic version of the inverse function theorem to the function $\boldsymbol{\lambda}(X) = (\lambda_1, \dots, \lambda_n)$ (see, for example, [129, Chap. 1, Proposition 5.9]), we see that it establishes an analytic homeomorphism between $F(c\mathfrak{m}_{\mathbf{C}})$ and $(c\mathfrak{m}_{\mathbf{C}})^d$. This shows that $\log_F$ is a local analytic homeomorphism.

We show the exactness of the short exact sequence. Assume that $\alpha \in T_{F,\infty}$. Then there exists n such that $[p^n](\alpha) = 0$, and therefore for each invariant differential form ω one has $p^n\lambda_\omega(\alpha) = \lambda_\omega([p^n](\alpha)) = 0$. This shows that $\lambda_\omega(\alpha) = 0$ for all ω; hence $\alpha \in \ker(\log_F)$. Conversely, assume that $\alpha \in \ker(\log_F)$. Take n such that $[p^n](\alpha) \in F(c\mathfrak{m}_{\mathbf{C}})$. Then $\log_F([p^n](\alpha)) = p^n \log_F(\alpha) = 0$. Since $\log_F$ is an isomorphism on $F(c\mathfrak{m}_{\mathbf{C}})$, this shows that $\alpha \in T_{F,n}$. Thus $\ker(\log_F) = T_{F,\infty}$. Finally, since $\log_F$ is a local isomorphism and $F(\mathfrak{m}_{\mathbf{C}})$ is p-divisible, $\log_F$ is surjective. □

Corollary 16.1.3 *For each c such that $v_K(c) > \dfrac{v_K(p)}{p-1}$, the local inverse of* $\log_F$ *induces an isomorphism*

$$\exp_F \,:\, t_F(c\mathfrak{m}_{\mathbf{C}}) \simeq F(c\mathfrak{m}_{\mathbf{C}}).$$

Tensoring this local isomorphism with $\mathbf{Q}_p$, we obtain an isomorphism (which we denote again by $\exp_F$*):*

$$\exp_F \,:\, t_F(\mathbf{C}) \simeq F(\mathfrak{m}_{\mathbf{C}}) \otimes_{\mathbf{Z}_p} \mathbf{Q}_p. \tag{74}$$

Definition We call $\log_F$ and $\exp_F$ the logarithmic map and the exponential map respectively.

Example 16.1.4 For the multiplicative formal group, the exact sequence (72) reads:

$$0 \to \mu_{p^\infty} \to U_{\mathbf{C}}^{(1)} \xrightarrow{\log} \mathbf{C} \to 0, \tag{75}$$

where $U_{\mathbf{C}}^{(1)}=(1+\mathfrak{m}_{\mathbf{C}})^*$ is the multiplicative group of principal units of $\mathbf{C}$.

16.1.5 Following Tate [151], we give a description of the group of points $F(\mathfrak{m}_{\mathbf{C}})$ in terms of the Tate module of the dual p-divisible group $F^\vee$. Let $F(p)=(F[p^n])_{n\geqslant 1}$ be the p-divisible group associated to F. Then $F[p^n](O_{\mathbf{C}})=T_{F,n}$. Recall the injective maps $i_n\,:\,F[p^n]\to F[p^{n+1}]$. It's easy to see that for any s, one has

$$F(\mathfrak{m}_{\mathbf{C}}/p^s)=\varinjlim_{i_n} F[p^n](O_{\mathbf{C}}/p^s).$$

Therefore, $F(\mathfrak{m}_{\mathbf{C}})$ can be defined in terms of the p-divisible group $F(p)$:

$$F(\mathfrak{m}_{\mathbf{C}})=\varprojlim_{s} F(\mathfrak{m}_{\mathbf{C}}/p^s)=\varprojlim_{s}\varinjlim_{i_n} F[p^n](O_{\mathbf{C}}/p^s).$$

16.1.6 By Cartier duality, for any O_K-algebra R, we have a canonical isomorphism

$$F[p^n](R)\simeq \mathrm{Hom}_R(F^\vee[p^n],\mathbb{G}_m).$$

Taking $R=O_{\mathbf{C}}/p^s$ and passing to the limits on the both sides, we obtain a morphism

$$F(\mathfrak{m}_{\mathbf{C}})\to \mathrm{Hom}\left(T(F^\vee),U_{\mathbf{C}}^{(1)}\right). \tag{76}$$

Theorem 16.1.7 (Tate) *(i) We have a commutative diagram with exact rows*

$$\begin{array}{ccccccccc}
0 & \longrightarrow & V(F)/T(F) & \longrightarrow & F(\mathfrak{m}_{\mathbf{C}}) & \xrightarrow{\log_F} & t_F(\mathbf{C}) & \longrightarrow & 0\\
 & & \downarrow{=} & & \downarrow{f} & & \downarrow{g} & & \\
0 & \longrightarrow & V(F)/T(F) & \longrightarrow & \mathrm{Hom}\left(T(F^\vee),U_{\mathbf{C}}^{(1)}\right) & \longrightarrow & \mathrm{Hom}(T(F^\vee),\mathbf{C}) & \longrightarrow & 0,
\end{array}$$

where the morphisms are defined as follows:

- *the upper row is the short exact sequence (72);*
- *the bottom row is induced by the short exact sequence (75) and the isomorphism* $V(F)/T(F)\simeq \mathrm{Hom}(T(F^\vee),\mathbf{Q}_p/\mathbf{Z}_p(1))$*;*
- *the middle vertical map is (76).*

(ii) The maps f and g are injective.

(iii) The map g agrees with the Hodge–Tate decomposition of $V(F)$. Namely, the diagram

$$\begin{array}{ccc}
t_F(\mathbf{C}(1)) & \xrightarrow{g} & \mathrm{Hom}(T(F^\vee),\mathbf{C}(1))\\
 & \searrow^{\text{Hodge–Tate}} & \downarrow{\simeq\ \text{duality}}\\
 & & T(F)\otimes_{\mathbf{Q}_p}\mathbf{C}
\end{array}$$

commutes.

(iv) The middle vertical row of the diagram induces an isomorphism

$$F(\mathfrak{m}_K) \simeq \mathrm{Hom}_{G_K}\left(T(F^\vee), U_{\mathbf{C}}^{(1)}\right).$$

Proof (i) The commutativity of the diagram and the exactness of rows is clear from construction.

We omit the proof of (ii)–(iv), which are the key assertions of the proposition. We remark that assertions (ii) and (iv) are proved in [151, Proposition 11 and Theorem 3] without any referring to p-adic integration on formal groups. They imply immediately the Hodge–Tate decomposition for $V(F)$. Assertion (iii) says, roughly speaking, that the Hodge–Tate decomposition arising from p-adic integration agrees with Tate's one. See [64, Chap. V, §1]. □

Corollary 16.1.8 *The map f can be identified with the canonical injection*

$$F(\mathfrak{m}_{\mathbf{C}}) \hookrightarrow T(F) \otimes_{\mathbf{Z}_p} U_{\mathbf{C}}^{(1)}(-1)$$

which gives rise to an isomorphism

$$F(\mathfrak{m}_K) \simeq \left(T(F) \otimes_{\mathbf{Z}_p} U_{\mathbf{C}}^{(1)}(-1)\right)^{G_K}.$$

Proof This follows from Theorem 16.1.7 and the Cartier duality. □

16.2 The Universal Covering

16.2.1 In this section, we introduce the notion of the universal covering of a formal group, and relate it to the p-adic representation $V(F)$.

Definition We call the universal covering of $F(\mathfrak{m}_{\mathbf{C}})$ the projective limit

$$CF(\mathfrak{m}_{\mathbf{C}}) = \varprojlim_{[p]} F(\mathfrak{m}_{\mathbf{C}})$$

taken with respect to the multiplication-by-p map $[p] : F(\mathfrak{m}_{\mathbf{C}}) \to F(\mathfrak{m}_{\mathbf{C}})$.

We have an exact sequence

$$0 \to T(F) \to CF(\mathfrak{m}_{\mathbf{C}}) \xrightarrow{\mathrm{pr}_0} F(\mathfrak{m}_{\mathbf{C}}) \to 0, \tag{77}$$

where pr_0 denotes the projection map

$$\mathrm{pr}_0(\xi) = \xi_0, \qquad \forall \xi = (\xi_0, \xi_1, \ldots), \qquad [p](\xi_n) = \xi_{n-1}.$$

Combining this exact sequence with (72), we obtain an exact sequence

$$0 \to V(F) \to CF(\mathfrak{m}_{\mathbf{C}}) \xrightarrow{\log_F \circ \mathrm{pr}_0} t_F(\mathbf{C}) \to 0. \tag{78}$$

16.2.2 Let F_k denote the reduction of F modulo $\mathfrak{m}_K$, and let $S = \mathfrak{m}_{\mathbf{C}}/\mathfrak{m}_K$. Set:

$$CF_k(S) = \varprojlim_{[p]} F_k(S).$$

Proposition 16.2.3 *The canonical map $F(\mathfrak{m}_{\mathbf{C}}) \to F_k(S)$ induces an isomorphism*

$$CF(\mathfrak{m}_{\mathbf{C}}) \simeq CF_k(S).$$

In particular, $CF(\mathfrak{m}_{\mathbf{C}})$ depends only on the reduction of F.

Proof (a) The map $F(\mathfrak{m}_{\mathbf{C}}) \to F_k(S)$ is clearly an epimorphism. Let $y = (y_n)_{n \geqslant 0} \in CF_k(S)$. Let $\widehat{y}_n \in F(\mathfrak{m}_{\mathbf{C}})$ be any lift of y_n. It is easy to see that for each n, the sequence $[p^m](\widehat{y}_{n+m})$ converges to some $x_n \in F(\mathfrak{m}_{\mathbf{C}})$ and that $[p](x_{n+1}) = x_n$. This proves the surjectivity.

(b) The injectivity follows from the fact that for any non-zero $x = (x_n)_{n \geqslant 0} \in CF(\mathfrak{m}_{\mathbf{C}})$, there exists N such that $v_K(x_n) < 1$ for $n \geqslant N$. □

16.2.4 From Corollary 16.1.8, it follows that there exists a canonical isomorphism

$$CF(\mathfrak{m}_{\mathbf{C}}) \simeq T(F) \otimes_{\mathbf{Z}_p} CU_{\mathbf{C}}^{(1)}(-1). \tag{79}$$

Example 16.2.5 Consider the universal covering of $\widehat{\mathbb{G}}_m$. One has

$$\widehat{\mathbb{G}}_m(\mathfrak{m}_{\mathbf{C}}) \simeq U_{\mathbf{C}}^{(1)}, \qquad U_{\mathbf{C}}^{(1)} := (1 + \mathfrak{m}_{\mathbf{C}})^*,$$

and

$$C\widehat{\mathbb{G}}_m(\mathfrak{m}_{\mathbf{C}}) \simeq CU_{\mathbf{C}}^{(1)}, \qquad CU_{\mathbf{C}}^{(1)} := \varprojlim_{x^p \leftarrow x} U_{\mathbf{C}}^{(1)}.$$

The universal covering of the reduction of $\widehat{\mathbb{G}}_m$ is

$$C\widehat{\mathbb{G}}_{m,k}(S) = \varprojlim_{x^p \leftarrow x} (1 + S)^* \simeq (1 + \mathfrak{m}_{\mathbf{C}^\flat})^*,$$

and the isomorphism $C\mathbb{G}_m(\mathfrak{m}_{\mathbf{C}}) \simeq C\mathbb{G}_{m,k}(S)$ is induced by the isomorphism (32) for $E = \mathbf{C}$:

$$\varprojlim_{x^p \leftarrow x} O_{\mathbf{C}} \simeq O_{\mathbf{C}}^\flat.$$

The short exact sequence (77) reads:

$$0 \to \mathbf{Z}_p(1) \to CU_{\mathbf{C}}^{(1)} \to U_{\mathbf{C}}^{(1)} \to 0. \tag{80}$$

16.3 Application to Galois Cohomology

16.3.1 In this section, we consider the sequence

$$0 \to \mathbf{Q}_p(1) \to (\mathbf{B}^+_{\mathrm{cris}})^{\varphi=p} \xrightarrow{\theta} \mathbf{C} \to 0, \tag{81}$$

where the first map is the canonical identification of $\mathbf{Q}_p(1)$ with the submodule $\mathbf{Q}_p t$ of $(\mathbf{B}^+_{\mathrm{cris}})^{\varphi=p}$. The fundamental exact sequence (54) shows that the sequence (81) is also exact. Consider the diagram:

$$\begin{array}{ccccccccc} 0 & \longrightarrow & \mathbf{Z}_p(1) & \longrightarrow & CU^{(1)}_{\mathbf{C}} & \longrightarrow & U^{(1)}_{\mathbf{C}} & \longrightarrow & 0 \\ & & \big\downarrow = & & \big\downarrow \log[\cdot] & & \big\downarrow \log & & \\ 0 & \longrightarrow & \mathbf{Q}_p(1) & \longrightarrow & (\mathbf{B}^+_{\mathrm{cris}})^{\varphi=p} & \xrightarrow{\theta} & \mathbf{C} & \longrightarrow & 0. \end{array} \tag{82}$$

Here we use the isomorphism $CU^{(1)}_{\mathbf{C}} \simeq 1 + \mathfrak{m}_{\mathbf{C}^\flat}$ to define the middle vertical arrow as follows:

$$x \mapsto \log([x]) = \sum_{n=1}^{\infty} (-1)^{n+1} \frac{([x]-1)^n}{n}.$$

We omit the proof of convergence of this series in $\mathbf{B}^+_{\mathrm{cris}}$.

Proposition 16.3.2 *The diagram (82) commutes, and the middle vertical map is an isomorphism.*

Proof (a) The proof of commutativity is straightforward.

(b) The map $\log[\,\cdot\,]$ is surjective because the right vertical map log is surjective, and $CU^{(1)}_{\mathbf{C}}$ is a $\mathbf{Q}_p$-vector space. Since $\log[x] = 0$ implies that $[x]$ is a root of unity, and $CU^{(1)}_{\mathbf{C}}$ is torsion free, $\log[\,\cdot\,]$ is injective. □

16.3.3 The exact sequence (81) induces a long exact sequence of continuous Galois cohomology:

$$\begin{aligned} 0 \to H^0(G_K, \mathbf{Q}_p(1)) \to H^0(G_K, (\mathbf{B}^+_{\mathrm{cris}})^{\varphi=p}) \to H^0(G_K, \mathbf{C}) \xrightarrow{\partial_0} H^1(G_K, \mathbf{Q}_p(1)) \\ \to H^1(G_K, (\mathbf{B}^+_{\mathrm{cris}})^{\varphi=p}) \to H^1(G_K, \mathbf{C}) \xrightarrow{\partial_1} H^2(G_K, \mathbf{Q}_p(1)). \end{aligned}$$

We use Proposition 16.3.2 to compute the connecting homomorphisms ∂_0 and ∂_1.

16.3.4 Recall that μ_{p^n} denotes the group of p^nth roots of unity. For each n, the Kummer exact sequence

$$0 \to \mu_{p^n} \to \overline{K}^* \xrightarrow{p^n} \overline{K}^* \to 0$$

gives rise to the connecting map

$$\delta_n \,:\, K^* = H^0(G_K, \overline{K}^*) \to H^1(G_K, \mu_{p^n}).$$

Passing to the projective limit on n, we obtain a map

$$\delta \,:\, K^* \to H^1(G_K, \mathbf{Z}_p(1)).$$

The following proposition gives an interpretation of the Kummer map in terms of the fundamental exact sequence:

Proposition 16.3.5 *(i) The diagram*

$$\begin{array}{ccc} U_K^{(1)} & \xrightarrow{\delta} & H^1(G_K, \mathbf{Z}_p(1)) \\ \downarrow{\scriptstyle \log} & & \downarrow \\ K & \xrightarrow{\partial_1} & H^1(G_K, \mathbf{Q}_p(1)) \end{array}$$

is commutative.

(ii) The diagram

$$\begin{array}{ccc} H^1(G_K, \mathbf{C}) & \xrightarrow{\partial_2} & H^2(G_K, \mathbf{Q}_p(1)) \\ \uparrow{\scriptstyle \simeq} & & \downarrow{\scriptstyle \mathrm{inv}_K} \\ K & \xrightarrow{-\mathrm{Tr}_K} & \mathbf{Q}_p \end{array}$$

is commutative. Here the left vertical isomorphism is $a \mapsto a \log \chi_K$ (see Theorem 4.3.2), and the right vertical map is the canonical isomorphism of local class field theory [140, Theorem 3].

Proof (i) The commutative diagram (82) gives a commutative square:

$$\begin{array}{ccc} H^0(G_K, U_{\mathbf{C}}^{(1)}) & \longrightarrow & H^1(G_K, \mathbf{Z}_p(1)) \\ \downarrow & & \downarrow \\ H^0(G_K, \mathbf{C}) & \xrightarrow{\partial_i} & H^1(G_K, \mathbf{Q}_p(1)). \end{array}$$

Here $H^0(G_K, U_{\mathbf{C}}^{(1)}) = U_K^{(1)}$, and $H^0(G_K, \mathbf{C}) = K$ by Ax–Sen–Tate theorem. The explicit description of the connecting map shows that in this diagram, the upper row coincides with δ. This proves the first assertion.

(ii) Assertion (ii) is proved in [12, Proposition 1.7.2]. □

16.4 The Bloch–Kato Exponential Map

16.4.1 We maintain previous notation and conventions. Our first goal is to extend the definition of the Kummer map to the case of general p-divisible formal groups. Let $\mathfrak{m}_{\overline{K}}$ denote the maximal ideal of the ring of integers of $\overline{K}$.

For all $n \geqslant 1$, we have an exact sequence

$$0 \to T_{F,n} \to F(\mathfrak{m}_{\overline{K}}) \xrightarrow{[p^n]} F(\mathfrak{m}_{\overline{K}}) \to 0,$$

which can be seen as the analog of the Kummer exact sequence for formal groups. It induces a long exact sequence of Galois cohomology:

$$0 \to H^0(G_K, T_{F,n}) \to H^0(G_K, F(\mathfrak{m}_{\overline{K}})) \to H^0(G_K, F(\mathfrak{m}_{\overline{K}})) \xrightarrow{\delta_{F,n}} H^1(G_K, T_{F,n}) \to \dots.$$

Since $H^0(K, F(\mathfrak{m}_{\overline{K}})) = F(\mathfrak{m}_K)$, this exact sequence gives an injection

$$\delta_{F,n} \, : \, F(\mathfrak{m}_K)/p^n F(\mathfrak{m}_K) \to H^1(G_K, T_{F,n}).$$

Passing to the projective limit, we obtain a map

$$\delta_F \, : \, F(\mathfrak{m}_K) \to H^1(K, T(F)),$$

which is referred to as the Kummer map for F. This map plays an important role in the Iwasawa theory of elliptic curves (see, for example, [81] for an introduction to this topic).

16.4.2 Bloch and Kato [28] found a remarkable description of δ_F in terms of p-adic periods, which also allows to construct an analog of the Kummer map for a wide class of p-adic representations.

Definition Let V be a de Rham representation of G_K. The quotient

$$t_V(K) = \mathbf{D}_{\mathrm{dR}}(V)/\mathrm{Fil}^0\mathbf{D}_{\mathrm{dR}}(V)$$

is called the tangent space of V.

Using the isomorphisms $\mathrm{gr}_i(\mathbf{B}_{\mathrm{dR}}) \simeq \mathbf{C}(i)$, one can prove by devissage that the tautological exact sequence

$$0 \to \mathrm{Fil}^0\mathbf{B}_{\mathrm{dR}} \to \mathbf{B}_{\mathrm{dR}} \to \mathbf{B}_{\mathrm{dR}}/\mathrm{Fil}^0\mathbf{B}_{\mathrm{dR}} \to 0$$

induces an isomorphism

$$t_V(K) \simeq H^0(G_K, V \otimes_{\mathbf{Q}_p} \mathbf{B}_{\mathrm{dR}}/\mathrm{Fil}^0\mathbf{B}_{\mathrm{dR}}).$$

Consider the fundamental exact sequence (54):

$$0 \to \mathbf{Q}_p \to \mathbf{B}_{\mathrm{cris}}^{\varphi=1} \to \mathbf{B}_{\mathrm{dR}}/\mathrm{Fil}^0\mathbf{B}_{\mathrm{dR}} \to 0.$$

Tensoring this sequence with V and taking Galois cohomology, we obtain a long exact sequence

$$0 \to H^0(G_K, V) \to \mathbf{D}_{\mathrm{cris}}(V)^{\varphi=1} \to t_V(K) \xrightarrow{\exp_V} H^1(G_K, V).$$

Definition The connecting homomorphism

$$\exp_V : t_V(K) \to H^1(G_K, V)$$

is called the exponential map of Bloch and Kato.

16.4.3 We come back to representations arising from p-divisible formal groups. Since the Hodge–Tate weights of $V(F)$ are 0 and 1, we have

$$t_{V(F)}(K) \simeq H^0(G_K, V \otimes_{\mathbf{Q}_p} \mathbf{C}(-1)).$$

The Hodge–Tate decomposition of $V(F)$ provides us with a canonical isomorphism

$$t_F(K) \simeq t_{V(F)}(K). \tag{83}$$

In Proposition 16.1.2, we constructed the logarithmic map $\log_F : F(\mathfrak{m}_K) \to t_F(K)$. Taking the composition, we obtain a map $F(\mathfrak{m}_K) \to t_{V(F)}(K)$.

Theorem 16.4.4 (Bloch–Kato) *The diagram*

$$\begin{array}{ccc} F(\mathfrak{m}_K) & \xrightarrow{\delta_F} & H^1(G_K, T(F)) \\ \downarrow & & \downarrow \\ t_{V(F)}(K) & \xrightarrow{\exp_{V(F)}} & H^1(G_K, V(F)), \end{array}$$

where the left vertical map is the composition of the exponential map $\exp_F$ *with the isomorphism (83), is commutative.*

Proof This is [28, Example 3.10.1]. We first prove the following lemma, which gives an interpretation of the Kummer map in terms of universal coverings. □

Lemma 16.4.5 *(i) One has a commutative diagram with exact rows and injective vertical maps:*

$$\begin{array}{ccccccccc} 0 & \longrightarrow & T(F) & \longrightarrow & CF(\mathfrak{m}_{\mathbf{C}}) & \longrightarrow & F(\mathfrak{m}_{\mathbf{C}}) & \longrightarrow & 0 \\ & & \downarrow \simeq & & \downarrow & & \downarrow & & \\ 0 & \longrightarrow & T(F) & \longrightarrow & T(F) \otimes_{\mathbf{Z}_p} CU_{\mathbf{C}}^{(1)}(-1) & \longrightarrow & T(F) \otimes_{\mathbf{Z}_p} U_{\mathbf{C}}^{(1)}(-1) & \longrightarrow & 0. \end{array}$$

(ii) This diagram gives rise to a commutative diagram

$$\begin{array}{ccc} F(\mathfrak{m}_K) & \xrightarrow{\delta_F} & H^1(G_K, T(F)) \\ \downarrow \simeq & & \downarrow = \\ H^0\left(G_K, T(F) \otimes_{\mathbf{Z}_p} U_{\mathbf{C}}^{(1)}(-1)\right) & \longrightarrow & H^1(G_K, T(F)). \end{array}$$

Proof (i) The first statement follows from the exactness of the sequence (80) and Corollary 16.1.8.

(ii) Directly from construction, it follows that the upper connecting map is δ_F. Taking into account the isomorphism from Corollary 16.1.8, we obtain the lemma. □

16.4.6 *Proof of the theorem.* Consider the diagram

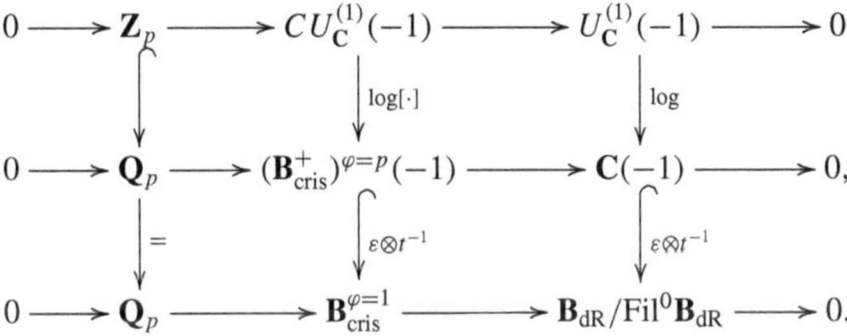

The upper part of the diagram is diagram (82) twisted by χ_K^{-1}. Therefore, the two upper squares commute. It is easy to check that the two lower squares commute too. Tensoring the diagram with $T(F)$ and taking Galois cohomology, we obtain a commutative diagram

$$\begin{array}{ccc} H^0\left(G_K, T(F) \otimes_{\mathbf{Z}_p} U_{\mathbf{C}}^{(1)}(-1)\right) & \longrightarrow & H^1(G_K, T(F)) \\ \downarrow & & \downarrow \\ t_{V(F)}(K) & \xrightarrow{\exp_{V(F)}} & H^1(G_K, V(F)). \end{array}$$

Combining this diagram with Lemma 16.4.5, we obtain the theorem.

16.5 Hilbert Symbols for Formal Groups

16.5.1 To illustrate the theory developed in previous sections, we sketch its application to an explicit description of Hilbert symbols on formal groups. Fix $n \geqslant 1$. Let

L/K be a finite extension containing the coordinates of all points of $T_{F,n}$. Recall that $\theta_L : L^* \to G_L^{\mathrm{ab}}$ denotes the reciprocity map.

Definition The Hilbert symbol on F is the pairing

$$(\ ,\ )_{F,n} : L^* \times F(\mathfrak{m}_L) \to T_{F,n} \tag{84}$$

defined by the formula

$$(\alpha, \beta)_{F,n} = x^{\theta_L(\alpha)} -_F x,$$

where $x \in F(\mathfrak{m}_{\overline{K}})$ is any solution of the equation $[p^n](x) = \beta$.

It is easy to see that this pairing is well defined, i.e. that $(\alpha, \beta)_{F,n}$ does not depend on the choice of x. If $F = \widehat{\mathbb{G}}_m$, and L contains the group μ_{p^n} of p^nth roots of unity, it reduces to the classical Hilbert symbol:

$$(\ ,\ )_{L,n} : L^* \times L^* \to \mu_{p^n},$$

$$(\alpha, \beta)_{L,n} = \left(\sqrt[p^n]{\beta}\right)^{\theta_L(\alpha)} / \sqrt[p^n]{\beta}.$$

16.5.2 By local class field theory, there exists a canonical isomorphism

$$H^2(G_L, \mu_{p^n}) \simeq \mathbf{Z}/p^n\mathbf{Z}$$

(see, for example, [142, Chap. VI]). Since $T_{F,n}$ is a trivial G_L-module, one has

$$H^2(G_L, \mu_{p^n} \otimes T_{F,n}) \simeq T_{F,n}.$$

Consider the cup product

$$H^1(G_L, \mu_{p^n}) \times H^1(G_L, T_{F,n}) \xrightarrow{\cup} H^2(G_L, \mu_{p^n} \otimes T_{F,n}) \simeq T_{F,n}.$$

Composing this pairing with the Kummer maps $\delta_{F,n} : F(\mathfrak{m}_L) \to H^1(G_L, T_{F,n})$ and $\delta_n : L^* \to H^1(G_L, \mu_{p^n})$, we obtain a pairing

$$L^* \times F(\mathfrak{m}_L) \to T_{F,n}.$$

From the cohomological description of the reciprocity map (see for example, [142, Chap. VI]), it follows that this pairing coincides with the Hilbert symbol (84).

16.5.3 Fix an uniformizer π_L of L. Let $f(X) \in O_K[X]$ denote the minimal polynomial of π_L over K. Writing O_L as $O_K[X]/(f(X))$ and taking into account that $\mathfrak{D}_{L/K} = (f'(\pi_L))$, we obtain an explicit description of the module of differentials $\Omega^1_{O_L/\mathbf{Z}_p}$ (see [142, Chap. III, §7]):

$$\Omega^1_{O_L/\mathbf{Z}_p} \simeq \left(O_L/\mathfrak{D}_{L/\mathbf{Q}_p}\right) d\pi_L$$

(recall that $\mathfrak{D}_{L/\mathbf{Q}_p}$ denotes the different of $L/\mathbf{Q}_p$). For any $\alpha \in O_L$ we write $\dfrac{d\alpha}{d\pi_L}$ for an element $a \in O_L$ such that $d\alpha = a \cdot d\pi_L$. Note that a is well defined modulo $\mathfrak{D}_{L/\mathbf{Q}_p}$. Set $d\log(\alpha) = \alpha^{-1}\dfrac{d\alpha}{d\pi_L}$.

16.5.4 Fix a base $(\xi_i)_{1\leqslant i\leqslant h}$ of $T_{F,n}$ over $\mathbf{Z}/p^{n\mathbf{Z}}$ and a basis $(\omega_j)_{1\leqslant j\leqslant h}$ of $H^1_{\mathrm{dR}}(F)$ in such a way that $(\omega_j)_{1\leqslant j\leqslant d}$ is a basis of Ω^1_F. Set:

$$\Theta_{L,n} = p^n \begin{pmatrix} \lambda'_{\omega_1}(\xi_1)\dfrac{d\xi_1}{d\pi_L} & \lambda'_{\omega_1}(\xi_2)\dfrac{d\xi_2}{d\pi_L} & \cdots & \lambda'_{\omega_1}(\xi_h)\dfrac{d\xi_h}{d\pi_L} \\ \vdots & \vdots & \ddots & \vdots \\ \lambda'_{\omega_d}(\xi_1)\dfrac{d\xi_1}{d\pi_L} & \lambda'_{\omega_d}(\xi_2)\dfrac{d\xi_2}{d\pi_L} & \cdots & \lambda'_{\omega_d}(\xi_h)\dfrac{d\xi_h}{d\pi_L} \\ \lambda_{\omega_{d+1}}(\xi_1) & \lambda_{\omega_{d+1}}(\xi_2) & \cdots & \lambda_{\omega_{d+1}}(\xi_h) \\ \vdots & \vdots & \ddots & \vdots \\ \lambda_{\omega_h}(\xi_1) & \lambda_{\omega_h}(\xi_2) & \cdots & \lambda_{\omega_h}(\xi_h) \end{pmatrix},$$

where we adopt the notation:

$$\lambda'_{\omega_j}(\xi_i)\frac{d\xi_i}{d\pi_L} := \sum_{k=1}^{d} \frac{d\lambda_{\omega_j}(\xi_i)}{dX_k}\frac{d\xi_i^{(k)}}{d\pi_L}, \qquad \text{if } \xi_i = \left(\xi_i^{(1)}, \ldots, \xi_i^{(d)}\right).$$

Let $X = (X_{ij})_{1\leqslant i,j\leqslant h}$ denote the inverse matrix of $\Theta_{L,n}$. The theory of p-adic integration together with Bloch–Kato's interpretation of the Kummer map allow to give the following explicit formula for this pairing:

Theorem 16.5.5 *For all $\alpha \in L^*$ and $\beta \in F(\mathfrak{m}_L)$ such that $v_p(\beta) > \dfrac{2}{p-1}$, one has*

$$(\alpha,\beta)_{F,n} = \sum_{i=1}^{h}\sum_{j=1}^{d}\left[\mathrm{Tr}_{L/\mathbf{Q}_p}\left(X_{ij}d\log(\alpha)\lambda_{\omega_j}(\beta)\right)\right](\xi_j).$$

Corollary 16.5.6 *Applying this formula to the multiplicative formal group we obtain the explicit formula of Sen [137] for the classical Hilbert symbol:*

$$(\alpha,\beta)_{L,n} = \zeta_{p^n}^{[\alpha,\beta]_n}, \qquad \textit{where } [\alpha,\beta]_n := \frac{1}{p^n}\mathrm{Tr}_{L/\mathbf{Q}_p}\left(\frac{d\log(\alpha)}{d\log(\zeta_{p^n})}\log(\beta)\right).$$

For Lubin–Tate formal groups, this formula impoves the explicit reciprocity law of Wiles [158].

Comments on the proof (a) This formula was proved in [12] assuming that $v_p(\beta) > c$ for some constant c independent of n. In [77], it was noticed that one can take $c = \frac{2}{p-1}$.

(b) Let $\widehat{\pi}_L \in \mathbf{A}_{\rm inf}$ be any lift of π_L under the map $\theta\,:\,\mathbf{A}_{\rm inf} \to O_{\mathbf{C}}$. Note that $\pi_L - \widehat{\pi}_L \in \mathrm{Fil}^1\mathbf{B}_{\rm dR}$. Take $u = (u_i)_{i\geqslant 0} \in T(F)$, where $[p](u_{i+1}) = u_i$. Let ω be a differential form of the second kind. From the definition of the p-adic integration in Theorem 14.4.2, it follows that the p-adic period $\int_u \omega$ can be approximated as follows:

$$\int_u \omega \approx \begin{cases} p^n\lambda'_\omega(u_n)\dfrac{du_n}{d\pi_L}(\pi_L - \widehat{\pi}_L) \pmod{\mathrm{Fil}^2\mathbf{B}_{\rm dR}}, & \text{if } \omega \in \Omega^1_F, \\ -p^n\lambda_\omega(u_n) \pmod{\mathrm{Fil}^1\mathbf{B}_{\rm dR}}, & \text{otherwise} \end{cases}$$

(see [12, Sect. 2.4] for precise statements). Therefore, the matrix $\Theta_{L,n}$ can be seen as "the matrix of p-adic periods of F modulo p^n".

(c) The Hodge–Tate decomposition gives an isomorphism

$$t_F(L) \simeq H^0(G_L, T(F)\otimes_{\mathbf{Z}_p}\mathbf{C}(-1)),$$

which can be described in terms of the matrix of p-adic periods. We consider an integral $\bmod p^n$ version of this isomorphism. Namely, set:

$$\mathfrak{m}'_{\mathbf{C}} = \left\{x \in \mathbf{C} \mid v_K(x) > \frac{v_K(p)}{p-1}\right\},$$

and $\mathfrak{m}'_L = \mathfrak{m}'_{\mathbf{C}} \cap \mathfrak{m}_L$. Since $T_{F,n}$ is a trivial G_L-module, we have a map

$$\eta_n\,:\, t_F(\mathfrak{m}'_L) \to H^0(G_L, T_{F,n}\otimes_{\mathbf{Z}_p}\mathfrak{m}'_{\mathbf{C}}(-1)) \simeq H^0(G_L, \mathfrak{m}'_{\mathbf{C}}(-1))\otimes_{\mathbf{Z}_p} T_{F,n},$$

which has an explicit description in terms of the matrix $\Theta_{L,n}$.

(c) The plan of the proof is the following. Using a $\bmod p^n$ version $\exp_{F,n}$ of the Bloch–Kato exponential map, we construct a commutative diagram

$$\begin{array}{ccccc} L^*\times t_F(\mathfrak{m}'_L) & \xrightarrow{\delta_{p^n}\cup\eta_n} & H^1(G_L, \mathfrak{m}'_{\mathbf{C}}/p^n)\otimes_{\mathbf{Z}_p} T_{F,n} & & \\ \downarrow{\scriptstyle(\delta_{p^n},\exp_{F,n})} & & \downarrow{\scriptstyle(\tau_n,\mathrm{id})} & & \\ H^1(G_L,\mu_{p^n})\times H^1(G_L,T_{F,n}) & \xrightarrow{\cup} & H^2(G_L,\mu_{p^n})\otimes T_{F,n} & \xrightarrow{\sim} & T_{F,n}. \end{array}$$

From the cohomological interpretation of the Hilbert symbol and Theorem 16.4.4, it follows that the Hilbert symbol $(\alpha,\beta)_{F,n}$ can be computed as the image of $(\alpha, \log_F)$ under the map $\delta_{p^n}\cup\exp_{F,n}$. We compute it using the above diagram, as the image of $(\alpha,\log_F)$ under the composition $(\tau_n,\mathrm{id})\circ(\delta_{p^n}\cup\eta_n)$. From construction, τ_n is the integral $\bmod p^n$ version of the connecting map $\partial_2\,:\,H^1(G_L,\mathbf{C}) \to H^2(G_L,\mathbf{Q}_p(1))$ associated to the exact sequence

$$0 \to \mathbf{Q}_p(1) \to (\mathbf{B}^+_{\rm cris})^{\varphi=p} \to \mathbf{C} \to 0.$$

Therefore, it can be computed in terms of the trace map using Proposition 16.3.5. The computation of the cup product $\delta_{p^n} \cup \eta_n$ is more subtle, and we refer the reader to [12] for further details.

Remark 16.5.7 (1) Explicit formulas of other types are proved in [3] and [150]. They generalize the explicit reciprocity law of Vostokov [156] and also use information about the matrix of p-adic periods.

(2) The exponential map of Bloch–Kato is closely related to special values of L-functions and Iwasawa theory [28, 125]. For further reading, see [13, 14, 16, 40, 116, 117, 124].

17 The Weak Admissibility: The Case of Dimension One

17.1 Formal Groups of Dimension One

17.1.1 In this section, we assume that K is a finite totally ramified extension of $K_0 = \widehat{\mathbf{Q}}_p^{\mathrm{ur}}$. Assume that M is an irreducible filtered φ-module over K of rank h satisfying the following conditions:

(1) $M = M_{1/h}$.
(2) $\mathrm{Fil}^0 M_K = M_K$, $\mathrm{Fil}^2 M_K = \{0\}$, and $\dim_K \mathrm{Fil}^1 M_K = 1$.

The first condition means that $M \simeq K_0[\varphi]/(\varphi^h - p)$, and by the theory of Dieudonné–Manin, M is the unique irreducible φ-module with $\mu_{\mathrm{N}}(M) = 1/h$. Since $t_{\mathrm{H}}(M) = 1/h$, we see that M is weakly admissible.

17.1.2 Let F_{LT} denote the Lubin–Tate formal group with the logarithm

$$\lambda_{\mathrm{LT}}(X) = \left(1 - \frac{\varphi^h}{p}\right)^{-1}(X) = X + \frac{X^{p^h}}{p} + \frac{X^{p^{2h}}}{p^2} + \cdots .$$

Extending scalars, we consider F_{LT} as a formal group over K. The filtered φ-module $M(F_{\mathrm{LT}})$ has the following description. The class b_{LT} of the canonical differential $\omega_{\mathrm{LT}} = d\lambda_{\mathrm{LT}}$ in $M(F_{\mathrm{LT}})$ satisfies the relation

$$\varphi^h(b_{\mathrm{LT}}) = p b_{\mathrm{LT}},$$

and the vectors $b_{\mathrm{LT}}, \varphi(b_{\mathrm{LT}}), \ldots, \varphi^{h-1}(b_{\mathrm{LT}})$ form a basis of $M(F_{\mathrm{LT}})$ over K_0. The filtration on $M(F_{\mathrm{LT}})_K$ is given by

$$\mathrm{Fil}^1 M(F_{\mathrm{LT}})_K = K \cdot b_{\mathrm{LT}}.$$

In particular, $M(F_{\mathrm{LT}})$ and M are isomorphic as φ-modules. Let v_p denote the valuation normalized as $v_p(p) = 1$.

Theorem 17.1.3 (Laffaille) *Assume that M is a filtered φ-module satisfying the conditions (1)–(2) above. The following holds true:*

(i) There exists $b \in M$ such that:

(a) b is a generator of M as a φ-module, and $\varphi^h(b) = pb$;
(b) There exist $c_0 = 1, c_1, \dots, c_{h-1} \in K$ such that

$$v_p(c_i) \geqslant -i/h \quad \textit{for all} \quad 1 \leqslant i \leqslant h-1, \tag{85}$$

and

$$\ell := \sum_{i=0}^{h-1} c_i \varphi^i(b) \in \mathrm{Fil}^1 M_K.$$

(ii) For all $c_0 = 1, c_1, \dots, c_{h-1} \in K$ satisfying condition (85), the series

$$\lambda(X) = \sum_{i=0}^{h-1} c_i \lambda_{\mathrm{LT}}(X^{p^{ih}})$$

is the logarithm of some formal p-divisible group over O_K of height h.

(iii) M is admissible. More precisely, there exists a formal group F of dimension one over O_K such that $M(F) \simeq M$ as filtered φ-modules.

Proof This theorem is proved in [102].

(i) By the discussion preceding the theorem, there exists a generator b' of M such that $\varphi^h(b') = pb'$ and $b', \varphi(b'), \dots, \varphi^{h-1}(b')$ is a K_0-basis of M. Then for any non-zero $\ell \in \mathrm{Fil}^1 M_K$, one has

$$\ell' = \sum_{i=0}^{h-1} c_i' \varphi^i(b'), \qquad \text{for some } c_i' \in K. \tag{86}$$

Note that $c_i' \neq 0$ for some i. Replacing, if necessary, b' by $\varphi^i(b')$ and dividing ℓ by c_i', we can assume that in (86), $c_0' = 1$. Let j be such that

$$v_p(c_j') + j/h \leqslant v_p(c_i') + i/h, \qquad \forall i = 0, \dots, h-1.$$

If $v_p(c_j') + j/h \geqslant 0$, then $v_p(c_i') \geqslant -i/h$ for all i, and we can take

$$c_i = c_i', \qquad \ell = \ell'.$$

Otherwise $c_j' \neq 0$. In that case, set:

$$b = \varphi^j(b'), \qquad \ell = \ell'/c_j'.$$

Then

$$\ell = \sum_{i=0}^{h-1} c_i \varphi^i(b),$$

where the coefficients c_i are given by

$$c_i = \begin{cases} c'_{i+j}/c'_j, & \text{if } 0 \leqslant i \leqslant h-j-1 \\ c'_{i+j-h}/pc'_j, & \text{if } h-j \leqslant i \leqslant h-1. \end{cases}$$

For $0 \leqslant i \leqslant h-j-1$, one has

$$v_p(c_i) + i/h = v_p(c'_{i+j}) - v_p(c'_j) + i/h = \left(v_p(c'_{i+j}) + (i+j)/h\right) - \left(v_p(c'_j) + j/h\right) \geqslant 0.$$

For $h-j \leqslant i \leqslant h-1$, one has

$$\begin{aligned} v_p(c_i) + i/h &= v_p(c'_{i+j-h}) - v_p(c'_j) - 1 + i/h \\ &= \left(v_p(c'_{i+j-h}) + (i+j-h)/h\right) - \left(v_p(c'_j) + j/h\right) \geqslant 0. \end{aligned}$$

This shows that $c_0, c_1, \dots, c_{h-1}$ satisfy (85).

(ii) By [86, §15.2], a power series of the form $\sum_{n=0}^{\infty} a_n X^{p^n}$ with $a_0 = 1$ is the logarithm of a formal group if and only if the sums

$$\begin{aligned} A_1 &:= pa_1, \\ A_2 &:= pa_2 - a_1 A_1^p, \\ &\dots\dots\dots\dots\dots\dots \\ A_n &:= pa_n - \sum_{i=0}^{n-1} a_i A^{p^i}, \\ &\dots\dots\dots\dots\dots\dots\dots\dots\dots\dots\dots\dots \end{aligned}$$

are in O_K. The verification of these conditions for the series $\lambda(X)$ is quite technical and is omitted here. See [102, proof of Proposition 2.4].

(iii) Let M be a filtered φ-module satisfying conditions (1)–(2). By part (i), there exists a generator b of M such that conditions a-b) hold for some $c_1, \dots, c_{h-1}$. By part (ii), the formal power series $\lambda(X) = \sum_{i=0}^{h-1} c_i \lambda_{\mathrm{LT}}(X^{p^{ih}})$ is the logarithm of some formal group F of height h. Then $M(F) \simeq M$ as filtered φ-modules. By Theorem 14.4.2, one has $M(F) \simeq \mathbf{D}^*_{\mathrm{cris}}(V(F))$. Hence, M is admissible. □

Remark 17.1.4 This theorem implies the surjectivity of the Gross–Hopkins period map [82]. See also [103] for the case of Drinfeld spaces.

17.2 Geometric Interpretation of $(\mathbf{B}_{\mathrm{cris}}^+)^{\varphi^h=p}$

17.2.1 We maintain previous notation and consider the Lubin–Tate formal group F_{LT} of height h with the logarithm $\lambda_{\mathrm{LT}}(X)$. Note that F_{LT} is defined over $\mathbf{Z}_p$. Let $F_{\mathrm{LT},k}$ denote the reduction of F_{LT} modulo p. We have the following interpretation of the universal covering of F_{LT}, which generalizes Example 16.2.5:

Proposition 17.2.2 *There is a canonical isomorphism*

$$CF_{\mathrm{LT}}(\mathfrak{m}_{\mathbf{C}}) \simeq F_{\mathrm{LT},k}(\mathfrak{m}_{\mathbf{C}^\flat}).$$

Proof Since $[p](X) \equiv X^q \pmod{p}$, the multiplication by p in $F_{\mathrm{LT},k}$ is given by φ^h. Set $S = \mathfrak{m}_{\overline{K}}/(p)$. Then

$$CF_{\mathrm{LT},k}(S) \simeq \varprojlim_{\varphi^h} F_{\mathrm{LT},k}(S) \simeq F_{\mathrm{LT},k}(\varprojlim_{\varphi^h} S) \simeq F_{\mathrm{LT},k}(\mathfrak{m}_{\mathbf{C}^\flat}).$$

Now the proposition follows from Proposition 16.2.3. □

17.2.3 Since $\mathbf{A}_{\mathrm{inf}}/(p) \simeq O_{\mathbf{C}}^\flat$, we have a well defined composition

$$\kappa : F_{\mathrm{LT},k}(\mathfrak{m}_{\mathbf{C}^\flat}) \xrightarrow{\sim} F_{\mathrm{LT},k}(\mathbf{A}_{\mathrm{inf}}/(p)) \xrightarrow{\sim} CF_{\mathrm{LT}}(\mathbf{A}_{\mathrm{inf}}) \xrightarrow{\mathrm{pr}_0} F_{\mathrm{LT}}(\mathbf{A}_{\mathrm{inf}}).$$

Here, $CF_{\mathrm{LT}}(\mathbf{A}_{\mathrm{inf}}) := \varprojlim_{[p]} F_{\mathrm{LT}}(\mathbf{A}_{\mathrm{inf}})$, and pr_0 denotes the projection on the ground level.

Theorem 17.2.4 (Fargues–Fontaine) *The map*

$$\mathrm{Log}(x) := \lambda_{\mathrm{LT}}(\kappa(x))$$

establishes an isomorphism $F_{\mathrm{LT},k}(\mathfrak{m}_{\mathbf{C}^\flat}) \simeq (\mathbf{B}_{\mathrm{cris}}^+)^{\varphi^h=p}$.

Proof *(Sketch of the proof)* The proof of the convergence of the series $\lambda_{\mathrm{LT}}(y)$ in $\mathbf{B}_{\mathrm{cris}}^+$ for $y \in F_{\mathrm{LT}}(\mathbf{A}_{\mathrm{inf}})$ is routine, and we omit it. Since , $F_{\mathrm{LT}}(\mathbf{A}_{\mathrm{inf}})$ does not contain torsion points of F_{LT}, the map Log is injective.

The series $F_{\mathrm{LT}}(X, Y)$ has coefficients in $\mathbf{Z}_p$. Hence, the formal group law commutes with φ, and one has

$$\varphi^h \lambda_{\mathrm{LT}}(\kappa(x)) = \lambda_{\mathrm{LT}}(\varphi^h(\kappa(x))) = \lambda_{\mathrm{LT}}(\kappa(\varphi^h(x))).$$

On the other hand, $\varphi^h(x) = [p](x)$ in $F_{\mathrm{LT},k}(\mathfrak{m}_{\mathbf{C}^\flat})$, and therefore

$$\lambda_{\mathrm{LT}}(\kappa(\varphi^h(x))) = \lambda_{\mathrm{LT}}([p](\kappa(x))) = p\lambda_{\mathrm{LT}}(\kappa(x)).$$

This proves that $\mathrm{Log}(x) \in (\mathbf{B}_{\mathrm{cris}}^+)^{\varphi^h=p}$.

The proof of the surjectivity is more subtle and we refer the reader to [60, Chap. 4], where this map is studied in all detail and in a more general setting. □

Example 17.2.5 If $h = 1$, then F_{LT} is isomorphic to $\widehat{\mathbb{G}}_m$. Therefore, $F_{\mathrm{LT},k}(\mathfrak{m}_{\mathbf{C}^\flat}) \simeq (1+\mathfrak{m}_{\mathbf{C}^\flat})^*$, and the map κ can be identified with the map $\log[\cdot]$ introduced in Proposition 16.3.2.

17.2.6 The next theorem furnishes further information about the structure of $(\mathbf{B}_{\mathrm{cris}}^+)^{\varphi^h=p}$.

Theorem 17.2.7 (Fargues–Fontaine) *For any family of elements*

$$\alpha_0, \alpha_1, \alpha_2, \dots, \alpha_{h-1} \in \mathbf{C},$$

not all zero, consider the map:

$$f : (\mathbf{B}_{\mathrm{cris}}^+)^{\varphi^h=p} \to \mathbf{C}, \qquad f(x) = \sum_{i=1}^{h-1} \alpha_i \theta(\varphi^i(x)).$$

Then f is surjective, and $\ker(f)$ *is a $\mathbf{Q}_p$-vector space of dimension h.*

Proof See [60, Théorème 8.1.2]. Without loss of generality, we can assume that $v_p(\alpha_i) \geqslant 0$ and $\alpha_0 = 1$. The arguments used in the proof of Theorem 17.1.3 apply and show that there exists a formal group F over $O_{\mathbf{C}}$ such that

$$\lambda_F = \sum_{i=0}^{h-1} \alpha_i \varphi^i(\lambda_{\mathrm{LT}}).$$

Consider the diagram

$$\begin{array}{ccccccccc} 0 & \longrightarrow & V(F) & \longrightarrow & CF(\mathfrak{m}_{\mathbf{C}}) & \xrightarrow{\lambda_F \circ \mathrm{pr}_0} & \mathbf{C} & \longrightarrow & 0, \\ & & & & \uparrow \simeq & \nearrow f & & & \\ & & & & (\mathbf{B}_{\mathrm{cris}}^+)^{\varphi^h=p} & & & & \end{array}$$

where the first line is the exact sequence (78) for F, and the vertical isomorphism is provided by Theorem 17.2.4. Since $\dim_{\mathbf{Q}_p} V(F) = h$, the theorem is proved. □

We refer the reader to [41] for the interpretation of this result in terms of the theory of Banach Spaces, and to [60] and [55] for applications to the theory of Fargues–Fontaine.

Acknowledgements The author is very grateful to Nicola Mazzari and the anonymous referee for pointing out several inaccuracies in the first version of this text. This work was partially supported by the Agence National de Recherche (grant ANR-18-CE40-0029) in the framework of the ANR-FNR project "Galois representations, automorphic forms and their L-functions."

References

1. V.A. Abrashkin, A ramification filtration of the Galois group of a local field, Proc. St. Petersburg Math. Soc. III, Amer. Math. Soc. Transl., Ser.2, **166** (Amer. Math. Soc., Providence 1996) pp. 35-100
2. V.A. Abrashkin, Ramification filtration of the Galois group of a local field. Proc. Steklov Math. Inst. **208**, 15–62 (1995)
3. V.A. Abrashkin, Explicit formulas for the Hilbert symbol of a formal group over Witt vectors, Izvestiya: Mathematics **61:3**, 463-615 (1997)
4. V.A. Abrashkin,The ramification filtration of the Galois group of a local field III, Izvestiya: Mathematics **62**(5), 857–900 (1998)
5. V.A. Abrashkin, A local analogue of the Grothendieck conjecture. Int. J. Math. **11**, 3–43 (2000)
6. V.A. Abrashkin, An analogue of the field-of-norms functor and the Grothendieck conjecture, Journal of Algebraic. Geometry **16**, 671–730 (2007)
7. V.A. Abrashkin, Galois groups of local fields, Lie algebras and ramification, in *Arithmetic and Geometry*, ed. by L. Dieulefait, G. Faltings, D.R. Heath-Brown, Yu. I. Manin, B.Z. Moroz, J.-P. Wintenberger. London Mathematical Society Lecture Note Series, vol. 420, pp. 1–23 (2015)
8. Y. André, Slope filtrations. Confluentes Mathematici **1**, 1–85 (2009)
9. F. Andreatta and A. Iovita, Comparison Isomorphisms for Smooth Formal Schemes. J. de l'Inst. de Math. de Jussieu **12**, 77–151 (2013)
10. E. Artin, *Algebraic Numbers and Algebraic Functions* (Gordon and Breach, New York, 1967)
11. J. Ax, Zeros of polynomials over local fields - The Galois action. J. Algebra **15**, 417–428 (1970)
12. D. Benois, Périodes p-adiques et lois de réciprocité explicites. J. Reine Angew. Math. **493**, 115–151 (1997)
13. D. Benois, On Iwasawa theory of crystalline representations. Duke Math. J. **104**, 211–267 (2000)
14. D. Benois, L. Berger, Théorie d'Iwasawa des représentations cristallines. Comm. Math. Helvetici **83**, 603–677 (2008)
15. L. Berger, Représentations p-adiques et équations différentielles. Invent. Math. **48**, 219–284 (2002)
16. L. Berger, Bloch and Kato's exponential map: three explicit formulas, Doc. Math., Extra Vol.: Kazuya Kato's Fiftieth Birthday, 99–129 (2003)
17. L. Berger, Construction de (φ, Γ) -modules: représentations p-adiques et B-paires. Algebra Num. Theo. **2**, 91–120 (2008)
18. L. Berger, Équations différentielles p-adiques et (φ, N) -modules filtrés. Astérisque **319**, 13–38 (2008)
19. L. Berger, Presque $\mathbf{C}_p$ -représentations et (φ, Γ) -modules. J. Inst. Math. Jussieu **8**, 653–668 (2009)
20. P. Berthelot, *Cohomologie cristalline des schémas de caractéristique* $p{>}0$, Lecture Notes in Math, vol. 407 (Springer, 1974)
21. P. Berhtelot, A. Ogus, F-isocrystals and de Rham cohomology, I. Invent. Math. **72**, 159–199 (1983)
22. P. Berthelot, A. Ogus, *Notes on Crystalline Cohomology*, Mathematical Notes, vol. 21 (Princeton University Press, 1978)
23. B. Bhatt, Lecture Notes for a Class on Perfectoid Spaces http://www-personal.umich.edu/~bhattb/teaching/mat679w17/lectures.pdf
24. B. Bhatt, M. Morrow, P. Scholze, Integral p-adic Hodge theory. Publ. Math. de l'IHÉS **128**, 219–397 (2018)
25. B. Bhatt, M. Morrow, P. Scholze, Topological Hochschild homology and integral p-adic Hodge theory. Publ. Math. de l'IHÉS **129**, 199–310 (2019)

26. A. Beilinson, p-adic periods and derived de Rham cohomology. J. Am. Math. Soc. **25**, 715–738 (2012)
27. A. Beilinson, On the crystalline period map. Cambridge J. Math. **1**, 1–51 (2013)
28. S. Bloch, K. Kato, L-functions and Tamagawa numbers of motives, in *The Grothendieck Festschrift*, ed. by P. Cartier, L. Illusie, N.M. Katz, G. Laumon, Y.I. Manin, K.A. Ribet, vol. I, Progress in Math, vol. 86, (Birkhäuser, Boston 1990), pp. 333–400
29. C. Breuil, Une application de corps des normes. Compositio Math. **117**, 189–203 (1999)
30. C. Breuil, Groupes p-divisibles, groupes finis et modules filtrés. Ann. Math. **152**, 489–549 (2000)
31. O. Brinon, Représentations p-adiques cristallines et de de Rham dans le cas relatif. Mém. Soc. Math. Fr. (N.S.), **112** (2008), vi+159 pages
32. O. Brinon, B. Conrad, CMI summer school notes on p-adic Hodge theory, http://math.stanford.edu/~conrad/papers/notes.pdf
33. X. Caruso, Représentations galoisiennes p-adiques et (φ, τ) -modules. Duke Math. J. **162**, 2525–2607 (2013)
34. K. Česnavičius, T. Koshikawa, The $\mathbf{A}_{\text{inf}}$ -cohomology in the semistable case. Compositio Math. **155**, 2039–2128 (2019)
35. F. Cherbonnier, P. Colmez, Représentations p-adiques surconvergentes. Invent. Math. **133**, 581–611 (1998)
36. F. Cherbonnier, P. Colmez, Théorie d'Iwasawa des représentations p-adiques d'un corps local. J. Am. Math. Soc. **12**, 241–268 (1999)
37. J. Coates, R. Greenberg, Kummer theory for abelian varieties over local fields. Invent. Math. **124**, 129–174 (1996)
38. P. Colmez, Périodes p-adiques des variétés abéliennes. Math. Annalen **292**, 629–644 (1992)
39. P. Colmez, Intégration sur les variétés p-adiques, Astérisque **248** (1998)
40. P. Colmez, Théorie d'Iwasawa des représentations de de Rham d'un corps local. Ann. Math. **148**, 485–571 (1998)
41. P. Colmez, Espaces de Banach de dimension finie. J. Inst. Math. Jussieu **1**, 331–439 (2002)
42. P. Colmez, Les conjectures de monodromie p-adiques, Séminaire Bourbaki 2001/02. Astérisque **290**, 53–101 (2003)
43. P. Colmez, Espaces Vectoriels de dimension finie et représentations de de Rham. Astérisque **319**, 117–186 (2008)
44. P. Colmez, Représentations triangulines de dimension 2. Astérisque **319**, 213–258 (2008)
45. P. Colmez, (φ, Γ)-modules et représentations du mirabolique de $GL_2(\mathbf{Q}_p)$. Astérisque **330**, 61–153 (2010)
46. P. Colmez, La série principale unitaire de $GL_2(\mathbf{Q}_p)$. Astérisque **330**, 213–262 (2010)
47. P. Colmez, Représentations de $GL_2(\mathbf{Q}_p)$ et (φ, Γ)-modules. Astérisque **330**, 281–509 (2010)
48. P. Colmez, J.-M. Fontaine, Construction des représentations p-adiques semi-stables. Invent. Math. **140**, 1–43 (2000)
49. P. Colmez, W. Nizioł, Syntomic complexes and p-adic nearby cycles. Invent. Math. **208**, 1–108 (2017)
50. P. Deligne, Théorie de Hodge, II, Publ. math. de l'I.H.É.S. **40**, 5–57 (1971)
51. P. Deligne, J. Milne, Tannakian categories, in *Hodge Cycles, Motives, and Shimura Varieties*, ed. by P. Deligne, J. Milne, A. Ogus, K.-Y. Shih. Lecture Notes in Math, vol. 900, pp. 101–228 (Springer 1982)
52. P. Deligne, Les corps locaux de caractéristique p, limites de corps locaux de caractéristique 0, In: *Représentations des groupes réductifs sur un corps local* (Hermann, Paris, 1984) pp. 120–157
53. M. Demazure, P. Gabriel, *Groupes algébriques. Tome I: Géométrie algébrique, généralités, groupes commutatifs* (Masson Cie, Éditeur, Paris; North-Holland Publishing Co., Amsterdam, 1970) Avec un appendice *Corps de classes local* par Michiel Hazewinkel
54. S. Demushkin, On the maximal p-extension of a local field. Izv. Akad. Nauk, USSR. Math. Ser., **25**, 329–346 (1961)
55. E. de Shalit, *The Fargues-Fontaine curve and p-adic Hodge theory*, 2020 (this volume)

56. Y. Ding, Y. Ouyang, A simple proof of Dieudonné-Manin classification theorem. Acta Math. Sinica, English Series **28**, 1553–1558 (2012)
57. G. Faltings, p-adic Hodge theory. J. Amer. Math. Soc. **1**, 255–299 (1988)
58. G. Faltings, Crystalline cohomology and p-adic Galois representations, in *Algebraic Analysis, Geometry, and Number Theory* (Johns Hopkins University Press, Baltimore, 1989), pp. 25–80
59. G. Faltings, Almost étale extensions, in *Cohomologies p-adiques et applications arithmétiques (II)*, Astérisque, **279**, 185–270 (2002)
60. L. Fargues, J.-M. Fontaine, Courbes et fibrés vectoriels en théorie de Hodge p-adique, Astérisque **406** (2018)
61. I. Fesenko, On deeply ramified extensions. J. London Math. Soc. **57**, 325–335 (1998)
62. I. Fesenko, On just infinite pro-p-groups and arithmetically profinite extensions of local fields. J. für die reine angew. Math. **517**, 61–80 (1999)
63. I. Fesenko and M. Kurihara, *Invitation to Higher Local Fields. Conference on higher local fields,* Münster, Germany, August 29-September 5, 1999, Geometry and Topology Monographs 3 (Mathematical Sciences Publishers, 2000)
64. J.-M. Fontaine, Groupes p-divisibles sur les corps locaux, Astérisque **47–48** (1977)
65. J.-M. Fontaine, Modules galoisiens, modules filtrés et anneaux de Barsotti-Tate. Astérisque **65**, 3–80 (1979)
66. J.-M. Fontaine, Sur certaines types de représentations p-adiques du groupe de Galois d'un corps local; construction d'un anneau de Barsotti-Tate. Ann. of Math. **115**, 529–577 (1982)
67. J.-M. Fontaine, Formes Différentielles et Modules de Tate des variétés Abéliennes sur les Corps Locaux. Invent. Math. **65**, 379–409 (1982)
68. J.-M. Fontaine, Cohomologie de de Rham, cohomologie cristalline et représentations p-adiques, in: *Algebraic Geometry Tokyo-Kyoto*. Lecture Notes in Math. **1016** (1983), pp. 86–108
69. J.-M. Fontaine, Représentations p-adiques des corps locaux, in *The Grothendieck Festschrift*, ed. by P. Cartier, L. Illusie, N.M. Katz, G. Laumon Y.I. Manin, K.A. Ribet, vol. II, Progress in Math. **87** (Birkhäuser, Boston 1991) pp. 249–309
70. J.-M. Fontaine, Le corps des périodes p-adiques. Astérisque **223**, 59–102 (1994)
71. J.-M. Fontaine, Représentations p-adiques semistables. Astérisque **223**, 113–184 (1994)
72. J.-M. Fontaine, Presque $\mathbf{C}_p$-représentations, Doc. Math. Extra Volume: Kazuya Kato's Fiftieth Birthday, 285–385 (2003)
73. J.-M. Fontaine, Arithmétique des représentations galoisiennes p-adiques. Astérisque **295**, 1–115 (2004)
74. J.-M. Fontaine, W. Messing, p-adic periods and p-adic étale cohomology, In: Current trends in arithmetical algebraic geometry, Proc. Summer Res. Conf., Arcata/Calif. 1985, Contemp. Math. **67** (1987), pp. 179–207
75. J. Fresnel, M. Matignon, Produit tensoriel topologique de corps valués. Can. J. Math. **35**(2), 218–273 (1983)
76. A. Fröhlich, Formal groups, Lecture Notes in Math. **74** (Springer 1968)
77. T. Fukaya, Explicit reciprocity laws for p-divisible groups over higher dimensional local fields. J. für die reine und angew. Math. **531**, 61–119 (2001)
78. O. Gabber, L. Ramero, Almost Ring Theory, Lecture Notes in Math. **1800** (Springer 2003)
79. N.L. Gordeev, Infinitude of the number of relations in the Galois group of the maximal p-extension of a local field with restricted ramification. Math. USSR Izv. **18**, 513–524 (1982)
80. D. Grayson, Higher algebraic K-theory II (after Daniel Quillen), in *Algebraic K-theory, Proc. Conf. Nothwestern Univ.* 1976, Lect. Notes in Math. **551** (Springer, 1976), pp. 217–240
81. R. Greenberg, Iwasawa theory for elliptic curves, in *Arithmetic Theory of Elliptic Curves*, ed. by C. Viola. Lecture Notes in Math. **1716** (Springer 1999), pp. 51–144
82. B. H. Gross, M. J. Hopkins, Equivariant vector bundles on the Lubin–Tate moduli space, in: "Topology and representation theory (Evanston, IL, 1992)" Contemp. Math. **158**, Amer. Math. Soc., Providence, RI, 1994, pp. 23–88
83. A. Grothendieck, Groupes de Barsotti-Tate et cristaux, Actes Congrès Int. Math. Nice, I, Gautiers-Villars. Paris **1971**, 431–436 (1970)

84. A. Grothendieck, *Groupes de Barsotti-Tate et cristaux de Dieudonné*, Séminaire de mathématiques supérieures **45**, Presses de l'Université de Montréal, 1974
85. G. Harder, M. Narasimhan, On the Cohomology groups of moduli spaces of vector bundles on curves. Math. Annalen **212**, 215–248 (1974)
86. M. Hazewinkel, *Formal Groups and Applications* (Academic Press, New York, 1978)
87. O. Hyodo, K. Kato, Semi-stable reduction and crystalline cohomology with logarithmic poles. Astérisque **223**, 221–268 (1994)
88. T. Honda, On the theory of commutative formal groups. J. Math. Soc. Japan **22**, 213–246 (1970)
89. K. Iwasawa, On Galois groups of local fields. Trans. AMS **80**, 448–469 (1955)
90. U. Jannsen, K. Wingberg, Die Struktur der absoluten Galoisgruppe p-adischer Zahlkörper. Invent. Math. **70**, 71–98 (1982)
91. K. Kato, A generalization of local class field theory by using K-groups. II. J. Fac. Sci. Univ. Tokyo **27**, 603–683 (1980)
92. K. Kato, Logarithmic structures of Fontaine–Illusie, in *Algebraic Analysis, Geometry, and Number Theory* (Johns Hopkins University Press, Baltimore, 1989), pp. 191–224
93. M. Katz, Review of ℓ-adic cohomology, in Motives, Proc. Symp. in Pure Math. **55**, part 1 (1994), pp. 21–30
94. K. Kedlaya, A p-adic local monodromy theorem. Ann. Math. **160**, 93–184 (2004)
95. K. Kedlaya, R. Liu, Relative p-adic Hodge theory: foundations, Astérisque **371** (2015)
96. K. Kedlaya, R. Liu, Relative p-adic Hodge theory, II: Imperfect period rings, https://arxiv.org/abs/1602.06899
97. M. Kisin, Crystalline representations and F-crystals, in *Algebraic geometry and number theory*, Progr. Math. **253** (Birkhäuser, Boston, 2006) pp. 459–496
98. H. Koch, Über die Galoissche Gruppe der algebraischen Abschliessung eines Potenzreihenkörpers mit endlichem Konstantenkörper. Math. Nachr. **35**, 323–327 (1967)
99. H. Koch, *Galois theory of p-extensions* (Springer Monographs in Mathematics, Springer, 2002)
100. M. Krasner, Approximation des corps valués complets de caractéristique p par ceux de caractéristique 0, in *Colloque d'algèbre supérieure, Centre Belge de Recherches Mathématiques*, (Gautier-Villars, Paris 1957) pp. 129–206
101. J. Labute, Classification of Demushkin groups. Canadian J. Math. **19**, 106–132 (1967)
102. G. Laffaille, Construction de groupes p-divisibles. Astérisque **65**, 103–123 (1979)
103. G. Laffaille, Groupes p-divisibles et corps gauches. Compositio Math. **56**, 221–232 (1985)
104. S. Lang, *Algebra*, Graduate Texts in Mathematics **211** (2002)
105. S. Lang, *Introduction to Modular forms*, Grundlehren der mathematischen Wissenschaften **222** (1976)
106. S. Lang, Algebraic Number Theory, Graduate Texts in Mathematics **110**, 1986
107. F. Laubie, Extensions de Lie et groupes d'automorphismes de corps locaux. Compositio Math. **67**, 165–189 (1988)
108. F. Laubie, M. Saine, Ramification of automorphisms of $k((t))$. J. Number Theory **63**, 143–145 (1997)
109. M. Lazard, *Groupes de Lie formels à un paramètre*, Séminaire Dubreil. Algèbre et théorie des nombres, tome 8 (1954–1955), exp. no. 5 et 7, pp. 1–31
110. A. Levelt, Jordan decomposition for a class of singular differential operators. Arkiv för Mat. **13**, 1–27 (1975)
111. J. Lubin, J. Tate, Formal Complex Multiplication in Local Fields. Annals Math. **81**, 380–387 (1965)
112. Yu. Manin, The theory of commutative formal groups over fields of finite characteristic. Russian Math. Surv. **18**, 1–80 (1963)
113. Sh. Mochizuki, A version of the Grothendieck conjecture for p-adic local fields. Int. J. Math. **8**, 499–506 (1997)
114. M. Morrow, The Fargues-Fontaine curve and diamonds, Séminaire Bourbaki 2017/18. Astérisque **414**, 533–572 (2019)

115. M. Morrow, T. Tsuji, Generalised representations as q-connections in integral p-adic Hodge theory, https://arxiv.org/pdf/2010.04059.pdf
116. K. Nakamura, Iwasawa theory of de Rham (φ, Γ) -modules over the Robba ring. J. Inst. Math. Jussieu **13**, 65–118 (2014)
117. K. Nakamura, A generalization of Kato's local epsilon-conjecture for (φ, Γ)-modules over the Robba ring. Algebra Num. Theo. **11**, 319–404 (2017)
118. M. Narasimhan, C. Seshadri, Stable and unitary vector bundles on a compact Riemann surface. Ann. Math. **82**, 540–567 (1965)
119. J. Neukirch, A. Schmidt, K. Wingberg, *Cohomology of Number Fields*, Grundlehren der mathematischen Wissenschaften Bd. **323** (Springer, 2013)
120. W. Nizioł, Crystalline conjecture via K-theory. Annales Sci. de l'ENS **31**, 659–681 (1998)
121. W. Nizioł, Semistable conjecture via K-theory. Duke Math. J. **141**, 151–178 (2008)
122. A. Parshin, Local class field theory. Proc. Steklov Inst. Math. (3), 157–185 (1985)
123. A. Parshin, Galois cohomology and Brauer group of local fields. Proc. Steklov Inst. Math. (4), 191–201 (1991)
124. B. Perrin-Riou, Théorie d'Iwasawa des représentations p-adiques sur un corps local. Invent. Math. **115**, 81–149 (1994)
125. B. Perrin-Riou, *Fonctions L p-adiques des représentations p-adiques*, Astérisque **229** (1995)
126. J. Rodrigues Jacinto, (φ, Γ)-modules de de Rham et fonctions L p-adiques. Algebra Num. Theo. **12**, 885–934 (2018)
127. P. Roquette, *Analytic theory of Elliptic functions over local fields*, Hamburger Mathematische Einzelschriften (N.F.), Heft 1 (Göttingen: Vandenhoeck and Ruprecht, 1970)
128. A.J. Scholl, Higher fields of norms and (φ, Γ)-modules, Documenta Math., Extra Volume: John H. Coates' Sixtieth Birthday 682–709 (2006)
129. P. Schneider, *p-adic Lie Groups*, Grundlehren der mathematischen Wissenschaften **344** (Springer 2011)
130. P. Scholze, Perfectoid spaces. Publ. math. de l'IHÉS **116**(1), 245–313 (2012)
131. P. Scholze, p-adic Hodge theory for rigid-analytic varieties. Forum Math. Pi **1**, e1 (2013)
132. P. Scholze, J. Weinstein, Berkeley lectures on p-adic geometry. Annals Math. Stud. **207** (2020)
133. S. Sen, On automorphisms of local fields. Ann. Math. **90**, 33–46 (1969)
134. S. Sen, Ramification in p-adic Lie extensions. Invent. Math. **17**, 44–50 (1972)
135. S. Sen, Lie algebras of Galois groups arising from Hodge-Tate modules. Ann. Math. **97**, 160–170 (1973)
136. S. Sen, Continuous Cohomology and p-adic Galois Representations. Invent. Math. **62**, 89–116 (1980)
137. S. Sen, On explicit reciprocity laws. J. reine angew. Math. **313**, 1–26 (1980)
138. J.-P. Serre, Sur les corps locaux à corps résiduel algébriquement clos. Bull. Soc. Math. France **89**, 105–154 (1961)
139. J.-P. Serre, Structure de certains pro-p-groupes (d'après Demuškin). Séminaire Bourbaki, Vol. 8, Exposé no. 252. Paris, Société Mathématique de France 1995, 145–155 (1964)
140. J.-P. Serre, Local class field theory, in *Algebraic Number Theory*, ed. by J.W.S. Cassels, A. Fröhlich (Academic Press, 1967)
141. J.-P. Serre, Sur les groupes de Galois attachées aux groupes p-divisibles, in *Proc. Conf. Local Fields,* Driebergen, Springer T.A. ed. (Springer 1967), pp. 118–131
142. J.-P. Serre, *Corps locaux* (Hermann, Paris, 1968)
143. J.P. Serre, *Abelian ℓ-adic representations and elliptic curves*, McGill University Lecture Notes (Benjamin, W.A, 1968)
144. J.-P. Serre, J. Tate, Good reduction of abelian varieties. Ann. Math. **88**, 492–517 (1968)
145. I. Shafarevich, On p-extensions. Mat. Sbornik, N. Ser. **20**(62), 351–363 (1947)
146. J. Silverman, *The Arithmetic of Elliptic Curves*. Graduate Texts in Mathematics **106**, (Springer 1986)
147. J. Silverman, *Advanced Topics in the Arithmetic of Elliptic Curves*, Graduate Texts in Mathematics, **151** (Springer, 1994)

148. U. Stuhler, Eine Bemerkung zur Reduktionstheorie quadratischen Formen. Archiv der Math. **27**, 604–610 (1976)
149. T. Suzuki, M. Yoshida, A refinement of the local class field theory of Serre and Hazewinkel, *Algebraic number theory and related topics, RIMS Kôkyûroku Bessatsu*, B32, Res. Inst. Math. Sci. (RIMS) (Kyoto 2012), pp. 163–191
150. F. Tavares Ribeiro, An explicit formula for the Hilbert symbol of a formal group. Annales de l'Institut Fourier, **61**, 261–318 (2011)
151. J. Tate, p-divisible groups, in Proc. Conf. Local Fields, Driebergen, Springer T.A. ed. (Springer 1967), pp. 158–183
152. J. Tate, A review of non-Archimedian elliptic functions in, *Elliptic curves, modular forms and Fermat's last theorem (Hong-Kong 1993)*, Series in Number Theory I (Int. Press Cambridge MA, 1995) pp. 162–184
153. B. Totaro, Tensor products in p-adic Hodge Theory. Duke Math. J. **83**, 79–104 (1996)
154. T. Tsuji, p-adic étale cohomology and crystalline cohomology in the semi-stable reduction case. Invent. Math **137**, 233–411 (1999)
155. H. Turrittin, Convergent solutions of ordinary differential equation in the neighborhood of an irregular point. Acta Math. **93**, 27–66 (1955)
156. S. V. Vostokov, *Explicit form of the law of reciprocity*, Izvestia: Mathematics **13:3** 557–646 (1979)
157. J. Weinstein, The Galois group of $\mathbf{Q}_p$ as a geometric fundamental group. Int. Math. Res. Notices **10**, 2964–2997 (2017)
158. A. Wiles, Higher explicit reciprocity laws. Ann. Math. **107**, 235–254 (1978)
159. J.-P. Wintenberger, Extensions de Lie et groupes d'automorphismes des corps locaux de caractéristique p, C. R. Acad. Sc. **288**, série A (1979), pp. 477–479
160. J.-P. Wintenberger, Extensions abéliennes et groupes d'automorphismes des corps locaux, C. R. Acad. Sc. **290**, série A (1980), pp. 201–203
161. J.-P. Wintenberger, Le corps des normes de certaines extensions infinies de corps locaux; applications. Ann. Sc. ENS **16**, 59–89 (1983)
162. J.-P. Wintenberger, Un scindage de la filtration de Hodge pour certaines variétés algébriques sur les corps locaux. Ann. Math. **119**, 511–548 (1984)
163. A.V. Yakovlev, The Galois group of the algebraic closure of a local field. Math. USSR Izvestiya **2**, 1231–1269 (1968)
164. I.G. Zel'venski, On the algebraic closure of a local field for p=2. Math. USSR-Izvestiya **6**, 925–937 (1972)
165. I.G. Zel'venski, Maximal extension without simple ramification of a local field. Math. USSR-Izvestiya **13**, 647–661 (1979)

Perfectoid Spaces: An Annotated Bibliography

Kiran S. Kedlaya

This annotated bibliography was prepared as part of a five-lecture series at the summer school on perfectoid spaces held at the International Centre for Theoretical Sciences (ICTS), Bengaluru, September 9–13, 2019. It is not intended to be a freestanding reference, although we do include a few short proofs and some sketches of longer proofs; instead, I have attempted to give some complements to my Arizona Winter School 2017 lecture notes [28], which provide a far more complete version of the story.

Throughout, fix a prime number p.

1 Perfectoid Fields

Primary references: [26, §1], [44, §3], [32, §3.5].

Proposition 1.1 (Fontaine–Wintenberger theorem) *The Galois groups of the fields* $\mathbb{Q}_p(\mu_{p^\infty})$ *and* $\mathbb{F}_p((t))$ *are isomorphic. More precisely, this isomorphism arises from an explicit isomorphism of Galois categories.*

Proof This is a consequence of results announced in [16, 17] and proved in detail in [51]. It is also a special case of Proposition 1.16 via Krasner's lemma (Remark 1.10).

We expand briefly on how Proposition 1.1 is embedded in the aforementioned papers of Fontaine–Wintenberger. By a theorem of Sen [43], the field $\mathbb{Q}_p(\mu_{p^\infty})$ is *strictly arithmetically profinite* in the sense of Fontaine–Wintenberger (we do not need the exact definition here). This then implies that its *norm field* is a local field of

K. S. Kedlaya (✉)
Department of Mathematics, University of California San Diego, 9500 Gilman Drive #0112, La Jolla, CA 92093, USA
e-mail: kedlaya@ucsd.edu

© The Author(s), under exclusive license to Springer Nature Singapore Pte Ltd. 2022
D. Banerjee et al. (eds.), *Perfectoid Spaces*, Infosys Science Foundation Series in Mathematical Sciences https://doi.org/10.1007/978-981-16-7121-0_6

characteristic p with residue field $\mathbb{F}_p$ (see [16, Théorème 2.4], [51, Théorème 2.1.3]), and so may be identified with $\mathbb{F}_p((t))$ via the Cohen structure theorem. The norm field construction then defines a bijection between finite extensions of $\mathbb{Q}_p(\mu_{p^\infty})$ and finite separable extensions of its norm field [51, Théorème 3.2.2]. □

Remark 1.2 The results of Fontaine–Wintenberger cited above also imply that for any strictly arithmetically profinite algebraic extension K of $\mathbb{Q}_p$, the Galois group of K and its norm field are isomorphic. This more general statement can also be recovered from Proposition 1.16, by showing that the completion of K is perfectoid with tilt isomorphic to the completed perfect closure of the norm field.

Before relating the Fontaine–Wintenberger theorem to perfectoid fields, we introduce some background on nonarchimedean fields.

Definition 1.3 A *nonarchimedean field* is a topological field whose topology is defined by some *nontrivial* nonarchimedean absolute value, with respect to which the field is complete. For K a nonarchimedean field, write $|K^\times|$ for the value group, $\mathfrak{o}_K$ for the valuation ring, $\mathfrak{m}_K$ for the maximal ideal, and κ_K for the residue field.

Proposition 1.4 *Let K be a nonarchimedean field (a field complete with respect to a nontrivial nonarchimedean absolute value). Let L/K be a finite extension.*

(i) The absolute value on K extends uniquely to a nonarchimedean absolute value on L.
(ii) There is a unique maximal subextension U of K which is unramified *over K: $|U^\times| = |K^\times|$ and κ_U/κ_K is separable of degree $[U : K]$. In particular, $\mathfrak{o}_U$ can be written as $\mathfrak{o}_K[\lambda]$ where λ maps to a primitive element of the residue field extension; in particular, if we write an element x of U as $\sum_{i=0}^{d-1} a_i\lambda^i$ with $a_i \in K$, then $x \in \mathfrak{o}_U$ if and only if $a_i \in \mathfrak{o}_U$ (or equivalently $a_i \in \mathfrak{o}_K$) for all i.*
(iii) If κ_K has characteristic p, then there is a unique maximal subextension T of K containing U which is totally tamely ramified *over U: $\kappa_T = \kappa_U$ and $[|T^\times| : |U^\times|] = [T : U]$ is coprime to p. That is, $[T : U]$ is coprime to p, the residue fields of T and U coincide, and the value group extension of T/U has index $[T : U]$. Moreover, T can be written as $U(\lambda^{1/d})$ in such a way that $\lambda^{1/d}$ generates the quotient of the value groups; in particular, if we write an element x of T as $\sum_{i=0}^{d-1} a_i\lambda^{i/d}$ with $a_i \in U$, then $x \in \mathfrak{o}_T$ if and only if $a_i\lambda^{i/d} \in \mathfrak{o}_T$ for all i.*
(iv) With notation as in (iii), the degree $[L : U]$ is a power of p. In particular, if L/K is Galois with group G, then Gal(L/U) *admits a subnormal series in which each successive quotient is cyclic of order p.*

Proof See, for example, [6, Chapter XIII]. □

Remark 1.5 If you are used to thinking about local fields as examples of nonarchimedean fields, a warning is in order: for L/K a finite extension of nonarchimedean fields, the inequality

$$[L : K] \geq [|L^\times| : |K^\times|][\kappa_L : \kappa_K],$$

which is always an equality when K is a local field, and can be strict in general. See [36] for a detailed discussion of this phenomenon.

Definition 1.6 For $P(T) = P_n T^n + \cdots + P_0 \in K[T]$ a polynomial over a nonarchimedean field K, the *Newton polygon* of P is the open polygon which forms the lower boundary of the convex hull of the set

$$\bigcup_{i=0}^{n} \{i\} \times [-\log |P_{n-i}|, \infty) \subset \mathbb{R}^2.$$

For $i = 1, \ldots, n$, the section of this polygon with x-coordinates in the range $[i-1, i]$ is a line segment. The slopes of these n line segments form the *slope multiset* of P.

Proposition 1.7 (Properties of Newton polygons) *Let $P(T)$ be a polynomial over a nonarchimedean field K.*

(i) Choose an extension L of K over which $P(T)$ factors as $(T-\alpha_1)\cdots(T-\alpha_n)$. Then the slope multiset of P consists of $-\log|\alpha_1|, \ldots, -\log|\alpha_n|$ in some order. In particular, the slope multiset of a product of two polynomials is the union of the slope multisets of the two polynomials.
(ii) If P is irreducible, then the Newton polygon is a straight line segment.

Proof There are many references for this material, see, for example, [23, Chapter 2].

Proposition 1.8 (Krasner's lemma) *Let L/K be a (not necessarily finite) extension of nonarchimedean fields. Let $P(T) \in K[T]$ be a polynomial which factors completely over L as $(T-\alpha_1)\cdots(T-\alpha_n)$. Then for any $\beta \in L$ such that*

$$|\beta - \alpha_1| < |\alpha_i - \alpha_1| \qquad (i = 2, \ldots, n),$$

we have $K(\alpha_1) \subseteq K(\beta)$.

Proof For $i = 2, \ldots, n$, $|\beta - \alpha_i| = |\alpha_i - \alpha_1| > |\beta - \alpha_1|$ by the non-Archimedean triangle inequality. By Proposition 1.7, the Newton polygon of $P(T-\beta)$ includes a segment of length 1, which must correspond to an irreducible factor. Alternatively, see [42, Theorem III.1.5.1], it is assumed therein that K is a discretely valued field, but the proof remains unchanged in the general case. □

Proposition 1.9 *Let K be a nonarchimedean field and let x be a nonzero element of K of positive valuation. Then K is algebraically closed if and only if:*

(a) the value group of K is not discrete and
(b) every polynomial over $\mathfrak{o}_K/(x)$ has a root in $\mathfrak{o}_K/(x)$.

Proof (The following argument is extracted from [26, Lemma 1.5.4], see also [44, Proposition 3.8], [32, Lemma 3.5.5].) It is clear that both conditions are necessary. To check sufficiency, note first that (a) and (b) together imply that the value group of K is in fact divisible. With this in mind, we show that every polynomial $P(T) \in \mathfrak{o}_K[T]$ has a root in $\mathfrak{o}_K$, by induction on the degree n of P. (This implies the same with $\mathfrak{o}_K$ replaced by K, by rescaling in T.)

To this end, we construct a sequence $z_0, z_1, \dots$ as follows. Start with $z_0 = 0$. Given z_i, if $P(z_i) = 0$ there is nothing more to check. If $P(z_i) \neq 0$ but the polynomial $P(T + z_i)$ has more than one distinct slope in its slope multiset, Proposition 1.7 allows us to factor it nontrivially and proceed by induction. Otherwise, because K has divisible value group, we can find a nonzero value $u_i \in K$ for which $P(u_i T + z_i)$ has all slopes equal to 0. By hypothesis (b), there exists $y_i \in \mathfrak{o}_K$ such that $P(u_i y_i + z_i) \in x\mathfrak{o}_K$, put $z_{i+1} = z_i + u_i y_i$.

To conclude the argument, it will suffice to check that if the construction of the sequence continues infinitely, then the sequence converges to a limit z which is a root of P. Since $P(T + z_i)$ has only one slope in its slope multiset, we must have $|u_i| = |P(z_i)|^{1/n}$. Since $|P(z_{i+1})| \leq |x|\,|P(z_i)|$, it follows that $u_i \to 0$ as $i \to \infty$, so the z_i do converge to a limit z satisfying $|P(z)| = 0$. □

Remark 1.10 From the previous discussion, we deduce that an algebraic extension L of K is algebraically closed if and only if its completion is algebraically closed: the "if" assertion follows from Krasner's lemma (Proposition 1.8) while the "only if" assertion follows from Proposition 1.9.

One consequence of this observation for the Fontaine–Wintenberger theorem is that, on one hand, $\mathbb{Q}_p(\mu_{p^\infty})$ and its completion have the same Galois group; on the other hand, $\mathbb{F}_p((t))$, its perfect closure, and the completion of its perfect closure all have the same Galois group.

Definition 1.11 A *perfectoid field* is a nonarchimedean field K with residue field of characteristic p and nondiscrete value group, for which the Frobenius map $x \mapsto x^p$ on $\mathfrak{o}_K/(p)$ is surjective. We allow the possibility that K is of characteristic p, in which case K is forced to be perfect.

Remark 1.12 Any algebraically closed nonarchimedean field with residue field of characteristic p is perfectoid. The completion K of $\mathbb{Q}_p(p^{p^{-\infty}})$ is perfectoid:

$$\mathfrak{o}_K/(p) \cong \mathbb{F}_p[\overline{T}_1, \overline{T}_2, \dots]/(\overline{T}_1^p, \overline{T}_2^p - \overline{T}_1, \dots).$$

The completion of $\mathbb{Q}_p(\mu_{p^\infty})$ is perfectoid, and we have

$$\mathfrak{o}_K/(p) \cong \mathbb{F}_p[\overline{T}_1, \overline{T}_2, \dots]/(\overline{T}_1^{p-1} + \cdots + \overline{T}_1 + 1. \overline{T}_2^p - \overline{T}_1, \dots).$$

In both cases, the tilt is isomorphic to the completion of $\mathbb{F}_p((T))[T^{1/p^\infty}]$. In particular, one cannot recover K from $K^\flat$ alone; some extra data is needed which we describe in the next lecture.

Remark 1.13 As noted in [28, Remark 2.1.8], the definition of a perfectoid field first appeared in [39] in 1984 under the terminology *hyperperfect field* (in French, *corps hyperparfait*), but the significance of this went unnoticed at the time.

Proposition 1.14 (after Fontaine) *Let K be a perfectoid field.*

(i) The natural map

$$\varprojlim_{x \mapsto x^p} \mathfrak{o}_K \to \varprojlim_{x \mapsto x^p} \mathfrak{o}_K/(p)$$

is an isomorphism of multiplicative monoids.

(ii) Using (i) to upgrade $\mathfrak{o}_{K^\flat} := \varprojlim_{x \mapsto x^p} \mathfrak{o}_K$ to a ring, it becomes a perfect valuation ring of characteristic p with fraction field $K^\flat := \varprojlim_{x \mapsto x^p} K$. The valuation on $K^\flat$ is the restriction along the final projection $\sharp : K^\flat \to K$.

(iii) The map $\sharp$ induces an isomorphism $|K^\times| \cong |K^{\flat\times}|$, in particular, both value groups are p-divisible.

(iv) The fields κ_K and $\kappa_{K^\flat}$ are isomorphic, in particular, both residue fields are perfect. Moreover, for $\overline{x} \in K^\flat$ such that $\sharp(\overline{x})/p \in \mathfrak{o}_K^\times$ (which exists by (iii)), the rings $\mathfrak{o}_K/(p)$ and $\mathfrak{o}_{K^\flat}/(\overline{x})$ are isomorphic.

We call $K^\flat$ the tilt *of K.*

Proof See [26, Lemma 1.3.3] or [44, Lemma 3.4]. (While the basic construction described here was known to Fontaine, the term *tilt*, and the notations $\flat$ and $\sharp$, were introduced by Scholze in [44].) □

Proposition 1.15 *Let K be a perfectoid field. Then K is algebraically closed if and only if $K^\flat$ is algebraically closed.*

Proof This follows from Propositions 1.9 and 1.14(iii, iv). □

We are not yet able to prove the following result; we state the proof modulo a key construction which we will introduce in the next lecture.

Proposition 1.16 (Generalized Fontaine–Wintenberger theorem) *Let K be a perfectoid field with tilt $K^\flat$.*

(i) Every finite extension of K is perfectoid.

(ii) The functor $L \mapsto L^\flat$ defines an equivalence of Galois categories between finite extensions of K and $K^\flat$, and hence an isomorphism between the absolute Galois groups of K and $K^\flat$.

Proof We follow the proof of [26, Theorem 1.5.6]. See [44, Theorem 3.7] for a somewhat different approach (using almost ring theory in place of Witt vectors). We may omit the case where K is of characteristic p, as in this case $K = K^\flat$ and the claim is trivial.

We will show in the next lecture (see Proposition 2.16) that there exists a surjective homomorphism $\theta : W(\mathfrak{o}_{K^\flat}) \to \mathfrak{o}_K$ with the property that for each finite extension E of $K^\flat$, there exists a perfectoid field L with

$$W(\mathfrak{o}_E) \otimes_{W(\mathfrak{o}_{K^\flat}),\theta} \mathfrak{o}_K \cong \mathfrak{o}_L;$$

this isomorphism will induce an isomorphism $L^\flat \cong E$. By inverting p in the previous isomorphism, we will also have an identification

$$W(\mathfrak{o}_E)[p^{-1}] \otimes_{W(\mathfrak{o}_{K^\flat}),\theta} \mathfrak{o}_K \cong L;$$

in case $E/K^\flat$ is a Galois extension with group G, it will follow that G acts on L with invariant subring K, so by Artin's lemma L/K is a finite Galois extension with Galois group G. If $E/K^\flat$ is not necessarily Galois, we may first go up to a Galois closure of $E/K^\flat$ and then come back down to deduce that $[L : K] = [E : K^\flat]$.

In this way, we will obtain a functor from finite extensions of $K^\flat$ to finite perfectoid extensions of K which, when followed by the tilting functor, yields an equivalence of categories (by the degree preservation property from the previous paragraph). In particular, this functor is fully faithful, and it remains only to check that it is essentially surjective. For this, let E be a completed algebraic closure of $K^\flat$. By Proposition 2.14 again, we may realize E as the tilt of some extension L of K; by Proposition 1.15, L is algebraically closed. By Remark 1.10(ii), the union of the finite extensions of K arising from finite extensions of $K^\flat$, or equivalently finite Galois extensions of $K^\flat$, is also algebraically closed. Hence, every finite extension L of K is contained in a finite Galois extension of K arising from a finite Galois extension of $K^\flat$; as in the previous paragraph, we deduce that L is itself perfectoid. This proves the claim. □

Remark 1.17 Let $\mathbb{C}$ be a completed algebraic closure of $\mathbb{Q}_p$. By Proposition 1.16; we can identify $\mathbb{C}^\flat$ with a completed algebraic closure of $\mathbb{F}_p((t))$ in various ways; for example, the two calculations from Remark 1.12 give rise to two distinct isomorphisms of this sort.

Suppose now that K is an arbitrary untilt of $\mathbb{C}^\flat$ of characteristic 0. Since K is algebraically closed and contains $\mathbb{Q}_p$, the completed algebraic closure of $\mathbb{Q}_p$ within K is isomorphic to $\mathbb{C}$. However, the resulting inclusion $\mathbb{C} \subseteq K$ can be strict, see [35] for examples.

2 Tilting, Untilting, and Witt Vectors

In the previous lecture, the proof of Proposition 1.16 hinged on being able to find perfectoid fields with a specified tilt using Witt vectors. In order to better understand the relationship between perfectoid fields and their tilts, we use Witt vectors to describe all possible fields with a given tilt.

Definition 2.1 A ring R of characteristic p is *perfect* if the Frobenius homomorphism $x \mapsto x^p$ is an isomorphism; note that injectivity of this map is equivalent to R being reduced. When R is a field, this is equivalent to the Galois-theoretic condition that every finite extension of R is separable.

Proposition 2.2 *Let R be a perfect ring of characteristic p.*

(a) *There exists a p-adically separated and complete ring $W(R)$ with $W(R)/(p) \cong R$ (the* ring of p-typical Witt vectors *with coefficients in R).*
(b) *The reduction map $W(R) \to R$ admits a unique multiplicative section $\overline{x} \mapsto [\overline{x}]$ (called the* Teichmüller map*).*
(c) *The construction of $W(R)$ is functorial in R. In particular, $W(R)$ itself is unique up to unique isomorphism.*

Proof See, for example, [26, §1.1]. Note that the Teichmüller map can be characterized by the formula $x = \lim_{n\to\infty} x_n^{p^n}$ where $x_n \in W(R)$ is any element satisfying $x_n^{p^n} \equiv x \pmod{p}$; the limit exists because $x_n^{p^n} \equiv x_{n+1}^{p^{n+1}} \pmod{p^{n+1}}$. □

Remark 2.3 The Witt vector construction was first introduced in the context where R is a perfect field. In this case, $W(R)$ is a complete discrete valuation ring with maximal ideal p and residue field R. For example, $W(\mathbb{F}_p) \cong \mathbb{Z}_p$.

Remark 2.4 One may fancifully think of $W(R)$ as $R[\![p]\!]$ except with some "carries" in the arithmetic. More precisely, every element $x \in W(R)$ has a unique representation as a convergent series $\sum_{n=0}^{\infty}[\overline{x}_n]p^n$ with $\overline{x}_n \in R$, but the arithmetic operations are somewhat complicated to express in terms of these coordinates. (Note that $\overline{x}_n$ is not the n-th Witt vector coefficient, but rather its p^n-th root.)

Since $W(R)$ is functorial in R, it admits a unique lift φ of the Frobenius map on R. This map has the property that $\varphi([\overline{x}]) = [\overline{x}^p]$ for $\overline{x} \in R$, that is, the elements $[\overline{x}]$ form the kernel of the associated p-derivation

$$\delta(x) := \frac{\varphi(x) - x^p}{p},$$

which occurs prominently in the context of *prismatic cohomology* [5, 31].

Definition 2.5 For the remainder of this lecture, let F denote a perfect nonarchimedean field of characteristic p, and define the ring $\mathbf{A}_{\mathrm{inf}}(F) := W(\mathfrak{o}_F)$. This is a local ring with residue field equal to that of F.

Remark 2.6 Since $\mathbf{A}_{\mathrm{inf}}(F)$ is to be interpreted as $\mathfrak{o}_F[\![p]\!]$, one can form a tenuous analogy between $\mathbf{A}_{\mathrm{inf}}(F)$ and a two-dimensional complete local ring such as $\mathbb{F}_p[\![x, y]\!]$. On one hand, the ring $\mathbf{A}_{\mathrm{inf}}(F)$ does not have any reasonable finiteness properties. For starters, it is certainly not noetherian: for any $\overline{x} \in \mathfrak{o}_F$ of positive valuation, the ideal

$$([\overline{x}], [\overline{x}^{1/p}], [\overline{x}^{1/p^2}], \dots)$$

is not finitely generated. In fact, $\mathbf{A}_{\mathrm{inf}}(F)$ has infinite [37] and even uncountable [13] global dimension, and, in general, is not even coherent [30].

On the other hand, it is true that every vector bundle on the punctured spectrum of $\mathbf{A}_{\mathrm{inf}}(F)$ extends uniquely over the puncture. See [30].

Proposition 2.7 (after Fontaine) *Let K be a perfectoid field.*

(i) *There is a unique homomorphism* $\theta : \mathbf{A}_{\mathrm{inf}}(K^\flat) \to \mathfrak{o}_K$ *whose restriction along the Teichmüller map is the map* $\sharp$.
(ii) *The map* θ *is surjective.*

Proof Part (i) is a formal consequence of the basic properties of p-typical Witt vectors, see [26, §1.1]. Part (ii) follows from Proposition 1.14(iv). □

To further analyze the kernel of θ, we make a key definition.

Definition 2.8 An element $z = \sum_{n=0}^{\infty} [\overline{z}_n] p^n \in \mathbf{A}_{\mathrm{inf}}(F)$ is *primitive* if $\overline{z}_0 \in \mathfrak{m}_F$ and $\overline{z}_1 \in \mathfrak{o}_F^\times$. An ideal of $\mathbf{A}_{\mathrm{inf}}(F)$ is *primitive* if it is principal generated by some primitive element. (It will follow from the following remark that every generator is then primitive.)

Remark 2.9 In the definition of a primitive element, the condition that $\overline{z}_1 \in \mathfrak{o}_F^\times$ may be replaced by the condition that $(z - [\overline{z}_0])/p \in \mathbf{A}_{\mathrm{inf}}(F)^\times$ or the condition that $\delta(z) \in \mathbf{A}_{\mathrm{inf}}(F)^\times$ (because $\delta(z) \equiv [\overline{z}_1^p] \pmod{p}$). From the latter formulation and the identity

$$\delta(yz) = y^p \delta(z) + z^p \delta(y) + p\delta(y)\delta(z),$$

we see that the product of a primitive element with a unit is a primitive element.

Remark 2.10 In the analogy between $\mathbf{A}_{\mathrm{inf}}(F)$ and $\mathbb{F}_p[\![x, y]\!]$, primitive elements correspond to power series $\sum_{m,n=0}^{\infty} a_{m,n} x^m y^n$ with $a_{0,0} = 0, a_{0,1} \neq 0$. By the Weierstrass preparation theorem, any such power series can be written as a unit of $\mathbb{F}_p[\![x, y]\!]$ times $y - cx$ for some $c \in \mathbb{F}_p$; consequently, the quotient by the ideal generated by such a power series is isomorphic to $\mathbb{F}_p[\![x]\!]$.

Proposition 2.11 *For* I *a primitive ideal, every class in* $\mathbf{A}_{\mathrm{inf}}(F)/I$ *can be represented by some element of* $\mathbf{A}_{\mathrm{inf}}(F)$ *which is a unit times a Teichmüller lift.*

Proof See [26, Lemma 1.4.7]. To summarize, let z be a generator of I. Given $x = \sum_{n=0}^{\infty} [\overline{x}_n] p^n \in \mathbf{A}_{\mathrm{inf}}(F)$, x generates the same class in $\mathbf{A}_{\mathrm{inf}}(F)/(z)$ as

$$x - \frac{(x - [\overline{x}_0])/p}{(z - [\overline{z}_0])/p} z = [\overline{x}_0] + [\overline{z}_0](x - [\overline{x}_0])/p.$$

By repeating this construction, we either produce a representative of the desired form, or verify that $x \in (z)$ (in which case we take the representative 0). We will take a more detailed look at what is going on here in the third lecture. □

Remark 2.12 A more "prismatic" version of the construction from Proposition 2.11 would be to replace x with

$$x - \varphi^{-1}\left(\frac{\delta(x)}{\delta(z)}\right) z.$$

However, we have not checked that this has the same convergence property as the construction given above.

Proposition 2.13 *For K a perfectoid field, the kernel of $\theta : \mathbf{A}_{\text{inf}}(K^\flat) \to \mathfrak{o}_K$ is a primitive ideal.*

Proof We reproduce here [26, Corollary 1.4.14]. By Proposition 1.14(iii), there exists $\overline{x} \in \mathfrak{o}_{K^\flat}$ such that $y := \theta(\overline{x})/p$ is a unit in $\mathfrak{o}_K$. By Proposition 2.7, there exists $w \in \mathbf{A}_{\text{inf}}(K^\flat)$ with $\theta(w) = y$. Since w must be a unit in $\mathbf{A}_{\text{inf}}(K^\flat)$, the element $z := pw - [\overline{x}]$ is a primitive element in $\ker(\theta)$.

By Proposition 2.11, every nonzero class in $\mathbf{A}_{\text{inf}}(K^\flat)/(z)$ can be represented by an element of $\mathbf{A}_{\text{inf}}(K^\flat)$ which is a unit times a Teichmüller lift; any such element has nonzero image in θ. It follows that $\ker(\theta) = (z)$, as claimed. □

Proposition 2.14 (Tilting correspondence) *For every primitive ideal I of $\mathbf{A}_{\text{inf}}(F)$, the quotient $\mathbf{A}_{\text{inf}}(F)/I$ can be identified with $\mathfrak{o}_K$ for some perfectoid field K. We then have an isomorphism $K^\flat \cong F$ for which I occurs as the kernel of $\theta : \mathbf{A}_{\text{inf}}(F) \cong \mathbf{A}_{\text{inf}}(K^\flat) \to \mathfrak{o}_K$. (Any such K is called an* untilt *of F.)*

Proof See [26, Theorem 1.4.13]. To summarize, by Proposition 2.11, we can represent each class in the quotient by a unit times a Teichmüller lift, and use the latter to define the valuation (modulo showing that this does not depend on the choice of representative). □

Remark 2.15 Note that $z = p$ is a primitive element, and consistently F is an untilt of itself. Any other untilt of F is of characteristic 0.

As noted earlier, the following result completes the proof of Proposition 1.16.

Proposition 2.16 (Untilting of extensions) *Let K be a perfectoid field. For any nonarchimedean field E containing $K^\flat$, the ring*

$$\mathfrak{o}_L := \mathbf{A}_{\text{inf}}(E) \otimes_{\mathbf{A}_{\text{inf}}(K^\flat),\theta} \mathfrak{o}_K$$

is the valuation ring of a perfectoid field L with $L^\flat \cong E$.

Proof By Proposition 2.7, the map $\theta : \mathbf{A}_{\text{inf}}(K^\flat) \to \mathfrak{o}_K$ is surjective and its kernel I is a primitive ideal. The ideal $I\mathbf{A}_{\text{inf}}(E)$ is again primitive, so by Proposition 2.14, there is a perfectoid field L for which $L^\flat \cong E$ and $\theta : \mathbf{A}_{\text{inf}}(E) \to \mathfrak{o}_L$ has kernel $I\mathbf{A}_{\text{inf}}(E)$. This field has the desired property. □

Remark 2.17 Now that we have a reasonable way to describe the untilts of F, one can try to construct a moduli space of these untilts. Before doing so, we must observe that for any primitive ideal I of $\mathbf{A}_{\text{inf}}(F)$, $\varphi(I)$ is also a primitive ideal and φ induces an isomorphism $\mathbf{A}_{\text{inf}}(F)/I \cong \mathbf{A}_{\text{inf}}(F)/\varphi(I)$, that is, I and $\varphi(I)$ define "the same" untilt of F.

In order to construct the desired moduli space, we must therefore find a way to define a space associated to $\mathbf{A}_{\text{inf}}(F)$ and then quotient by the action of φ. Since φ is of infinite order, there is no hope of doing this within the category of schemes, at least not directly. We will compare two different constructions of this form in the fourth lecture.

Remark 2.18 The ring $\mathbf{A}_{\text{inf}}(K)$ plays a central role in Fontaine's construction of p-adic period rings. We recommend [3] for a development of this point in modern language.

3 Perfectoid Rings and Spaces

In this lecture, we describe how the tilting equivalence can be extended to certain rings and spaces. Some detailed historical remarks, including many original references for the following statements, can be found in [28, Remarks 2.1.8, 2.3.18, 2.4.11, 2.5.13]; we do not attempt to reproduce these here.

Remark 3.1 These lectures will not include any review of Huber's theory of adic spaces, as these are covered in other lectures. For the reader reading this document in isolation, some introductory sources for the theory are [11] (in the context of rigid analytic geometry), [28, Lecture 1], [47, Lectures I–V], and [50].

One caution is in order: we will only consider Huber rings A, and Huber pairs (A, A^+), in which A is Hausdorff, complete, and contains a topologically nilpotent unit (also called a *pseudouniformizer*); this last condition is usually called *Tate*. In [28, Lecture 1], the Tate condition is relaxed to the condition that the topologically nilpotent elements of A generate the unit ideal; this condition is called *analytic*.

Definition 3.2 A Huber ring A is *perfectoid* if the following conditions hold:

(a) The ring A is *uniform*: its subring A° of power-bounded elements is bounded (and hence a ring of definition).
(b) There exists a pseudouniformizer ϖ with $p \in \varpi^p A^\circ$ such that the Frobenius map $\varphi : A^\circ/(\varpi) \to A^\circ/(\varpi^p)$ is surjective.

A Huber pair (A, A^+) is *perfectoid* if A is perfectoid. This implies an analogue of (b) with A° replaced by A^+, see [28, Corollary 2.3.10].

Remark 3.3 Beware that different sources use the term *perfectoid* at different levels of generality. In [44], the only rings considered are perfectoid K-algebras where K is itself a perfectoid field. In [32], only perfectoid $\mathbb{Q}_p$-algebras are considered. The definition we give above was introduced by Fontaine [15] and adopted by Kedlaya–Liu in [33] and Scholze in [47]. Even more general definitions are also possible, as in [4].

Remark 3.4 Given a perfectoid ring A, there is not much wiggle room left in the choice of A^+; it is a subring of A° and the quotient A°/A^+ is an *almost zero* A^+-module, meaning that it is annihilated by every topologically nilpotent element of A^+.

The notion of an almost zero module is the starting point of *almost commutative algebra* as introduced by Faltings and developed by Gabber–Ramero [18], in which one systematically defines *almost* versions of various ring-theoretic and module-theoretic concepts consistent with the previous definition.

A crucial first example is given by the perfectoid analogues of Tate algebras.

Definition 3.5 For K a perfectoid field of characteristic 0, the rings

$$K\langle T^{p^{-\infty}}\rangle := (\mathfrak{o}_K[T^{p^{-\infty}}])_p^\wedge[p^{-1}], \qquad K\langle T^{\pm p^{-\infty}}\rangle := (\mathfrak{o}_K[T^{\pm p^{-\infty}}])_p^\wedge[p^{-1}]$$

are perfectoid rings for the p-adic topologies. More generally, if (A, A^+) is a perfectoid Huber pair, we may similarly define perfectoid rings $A\langle T^{p^{-\infty}}\rangle$, $A\langle T^{\pm p^{-\infty}}\rangle$.

The following is true but not straight forward to prove.

Proposition 3.6 *A perfectoid ring which is a field is a perfectoid field. That is, if the underlying ring is a field, then the topology is induced by some nonarchimedean absolute value.*

Proof See [29, Theorem 4.2]. □

Remark 3.7 A related statement is that for A a perfectoid ring, the residue field of any maximal ideal of A is a perfectoid field. See [28, Corollary 2.9.14].

As for perfectoid fields, there is a tilting construction that plays a pivotal role in the theory.

Proposition 3.8 *Let (A, A^+) be a perfectoid Huber pair.*

(i) The natural map

$$\varprojlim_{x\mapsto x^p} A^+ \to \varprojlim_{x\mapsto x^p} A^+/(p)$$

is an isomorphism of multiplicative monoids.

(ii) Using (ii) to upgrade $\varprojlim_{x\mapsto x^p} A^+$ to a ring $A^{\flat+}$, this ring occurs in a perfectoid Huber pair $(A^\flat, A^{\flat+})$ of characteristic p in which the underlying multiplicative monoid of $A^\flat$ is $\varprojlim_{x\mapsto x^p} A$. (Moreover, $A^\flat$ depends only on A, not on A^+.)

(iii) Let $\sharp : A \to A^\flat$ be the final projection. Then there exists a pseudouniformer $\overline{\varpi}$ of $A^\flat$ such that $\sharp(\overline{\varpi})/\varpi$ is a unit in A^+.

(iv) With notation as in (iii), the rings $A^+/(\varpi)$ and $A^{\flat+}/(\overline{\varpi})$ are isomorphic.

Proof See [28, Theorem 2.3.9]. □

Remark 3.9 The construction of perfectoid Tate algebras (Definition 3.5) commutes with tilting.

Definition 3.10 Let (R, R^+) be a perfectoid Huber pair of characteristic p. An element $z = \sum_{n=0}^{\infty}[\overline{z}_n]p^n \in W(R^+)$ is *primitive* if $\overline{z}_0$ is topologically nilpotent and $\overline{z}_1$ is a unit. Any associate of a primitive element is again primitive; we thus say that an ideal of $W(R^+)$ is *primitive* if it is principal and some (hence any) generator is primitive.

Proposition 3.11 *Let (A, A^+) be a perfectoid Huber pair.*

(i) There is a unique homomorphism $\theta : W(A^{\flat +}) \to A^+$ *whose restriction along the Teichmüller map is the map* $\sharp$.
(ii) The map θ *is surjective.*
(iii) The kernel of θ *is primitive.*

Proof See [28, Theorem 2.3.9]. □

Proposition 3.12 *Let* (R, R^+) *be a perfectoid Huber pair of characteristic* p. *For every primitive ideal* I *of* $W(R^+)$, *there exist a perfectoid Huber pair* (A, A^+) *and an identification* $(A^\flat, A^{\flat +}) \cong (R, R^+)$ *under which* I *corresponds to the kernel of* θ.

Proof See [28, Theorem 2.3.9]. □

Remark 3.13 For A a perfectoid ring, the categories of perfectoid rings over A and $A^\flat$ are equivalent via tilting, using the primitive ideal coming from A to untilt extensions of $A^\flat$. The case where A is a perfectoid field is the form of the tilting equivalence stated in [44].

The compatibility of tilting with finite extensions of fields has the following analogue for rings.

Proposition 3.14 *Let* (A, A^+) *be a perfectoid Huber pair.*

(i) Let $A \to B$ *be a finite étale morphism and let* B^+ *be the integral closure of* A^+ *in* B. *Then* (B, B^+) *is again a perfectoid Huber pair.*
(ii) The categories of finite étale algebras over A *and over* $A^\flat$ *are equivalent via tilting.*

Proof See [28, Theorem 2.5.9]. □

A new feature in the ring case is that we also have a compatibility of tilting with localization.

Proposition 3.15 *Let* (A, A^+) *be a perfectoid Huber pair.*

(i) Let $(A, A^+) \to (B, B^+)$ *be a rational localization. Then* (B, B^+) *is again a perfectoid Huber pair. (In particular,* B *is again uniform.)*
(ii) The categories of rational localizations of (A, A^+) *and of* $(A^\flat, A^{\flat +})$ *are equivalent via tilting.*

Proof See [28, Theorem 2.5.3]. □

This allows to construct adic spaces using the following criterion for sheafiness of Huber rings.

Definition 3.16 A Huber pair (A, A^+) is *stably uniform* if for every rational localization $(A, A^+) \to (B, B^+)$, the Huber ring B is uniform. This depends only on A, not on A^+ [28, Definition 1.2.12].

Proposition 3.17 (Buzzard–Verberkmoes, Mihara) *Any stably uniform Huber pair is sheafy.*

Proof This is due independently to Buzzard–Verberkmoes [7, Theorem 7] and Mihara [40, Theorem 4.9]. See also [32, Theorem 2.8.10] and [28, Theorem 1.2.13] (which also covers the case where A is analytic but not Tate). □

Proposition 3.18 *Every perfectoid Huber pair is sheafy.*

Proof This follows by combining Proposition 3.15 with Proposition 3.17. □

Remark 3.19 In the previous statement, it is crucial that we have a criterion for sheafiness without a noetherian hypothesis: a perfectoid ring cannot be noetherian unless it is a finite product of perfectoid fields. See [28, Corollary 2.9.3].

Definition 3.20 For (A, A^+) a perfectoid Huber pair, by Proposition 3.18 the structure presheaf on $\mathrm{Spa}(A, A^+)$ is a sheaf. We may thus define a *perfectoid space* to be a locally v-ringed space which is locally of this form.

Remark 3.21 As a first example, one can use the perfectoid Tate algebra to define analogues of projective spaces in the category of perfectoid algebras; these play an important role in the application to the weight-monodromy conjecture given in [44], in which one exploits the fact that the conjecture is known in the equal-characteristic setting to deduce certain cases of it in mixed characteristic. One can also extend both the construction and the application to toric varieties; we leave this to the interested reader.

Proposition 3.22 *For (A, A^+) a sheafy Huber pair, the structure sheaf on (A, A^+) is acyclic. In particular, by Proposition 3.18, this holds when (A, A^+) is perfectoid.*

Proof See [28, Theorem 1.4.16]. □

Remark 3.23 There are some further compatibilities of tilting with other algebraic operations or properties.

- Tilting commutes with taking completed tensor products [28, Theorem 2.4.1]. This implies the existence of fiber products in the category of perfectoid spaces.
- Certain properties of morphisms of perfectoid rings are compatible with tilting, including injectivity [28, Corollary 2.9.13], strict injectivity [28, Theorem 2.4.2], surjectivity [28, Theorem 2.4.4], or having dense image [28, Theorem 2.4.3].

4 Fargues–Fontaine Curves

We now show how to construct "moduli spaces of untilts" in the spirit of Remark 2.17, leading to the schematic and adic Fargues–Fargues curves.

Throughout this lecture, let F be a perfect(oid) nonarchimedean field of characteristic p.

Definition 4.1 For any element ϖ of the maximal ideal of $\mathfrak{o}_F$, the ring $\mathbf{A}_{\mathrm{inf}}(F)$ is complete for the $(p, [\varpi])$-adic topology; we may thus view it as a Huber ring using itself as the ring of definition, and then form the adic spectrum $\mathrm{Spa}(\mathbf{A}_{\mathrm{inf}}(F), \mathbf{A}_{\mathrm{inf}}(F))$. From this space, remove the zero locus of $p[\varpi]$; we denote the resulting space by Y_F.

Proposition 4.2 *The action of φ on Y_F is without fixed points, and moreover is properly discontinuous: every point admits a neighborhood whose images under the various powers of φ are pairwise disjoint.*

Proof For $n \in \mathbb{Z}$, define the rational subsets U_n, V_n of Y_F by the formulas

$$U_n := \{v \in Y_F : v([\varpi])^{p^n+p^{n-1}} \leq v(p) \leq v([\varpi])^{p^n}\}$$
$$V_n := \{v \in Y_F : v([\varpi])^{p^{n+1}} \leq v(p) \leq v([\varpi])^{p^n+p^{n-1}}\}.$$

Then the U_n are pairwise disjoint and $\varphi(U_n) = U_{n+1}$; the V_n are pairwise disjoint and $\varphi(V_n) = V_{n+1}$; and the union of all of the U_n and V_n is all of Y_F. □

Definition 4.3 By Proposition 4.2, we may form the quotient $X_F^{\mathrm{an}} := Y_F/\varphi$. This quotient is the *adic Fargues–Fontaine curve* associated to F. (We will define later a *schematic Fargues–Fontaine curve* which has X_F as its "analytification".)

In order to say anything more, we must analyze the rings that arise in the construction.

Definition 4.4 Fix a normalization of the absolute value on F. For $\rho \in (0, 1)$, we define the *ρ-Gauss norm* on $\mathbf{A}_{\mathrm{inf}}(F)$ as the function $|\bullet|_\rho : \mathbf{A}_{\mathrm{inf}}(F) \to [0, +\infty)$ defined by

$$x = \sum_{n=0}^{\infty} [\overline{x}_n] p^n \mapsto \max_n \{\rho^n \, |\overline{x}_n|\}.$$

Remark 4.5 Recalling that we think of $W(\mathfrak{o}_F)$ as an interpretation of the nonsensical expression $\mathfrak{o}_F[\![p]\!]$, we keep in mind that the following facts about the ρ-Gauss norm on $\mathbf{A}_{\mathrm{inf}}(F)$ parallel more elementary facts about the ρ-Gauss norm on $\mathfrak{o}_F[\![T]\!]$:

$$x = \sum_{n=0}^{\infty} \overline{x}_n T^n \mapsto \max_n \{\rho^n \, |\overline{x}_n|\}.$$

For any closed interval $I \subset (0, 1)$, define also

$$|x|_I = \sup\{|x|_\rho : \rho \in I\}.$$

Proposition 4.6 (Hadamard three circles property) *For any fixed $x \in \mathbf{A}_{\mathrm{inf}}(F)$, the function $\rho \mapsto |x|_\rho$ is continuous and log-convex. The latter means that for $\rho_1, \rho_2 \in (0, 1)$ and $t \in [0, 1]$, for $\rho := \rho_1^t \rho_2^{1-t}$ we have*

$$\left|x\right|_\rho \leq \left|x\right|_{\rho_1}^t \left|x\right|_{\rho_2}^{1-t}.$$

In particular, for any closed interval $I = [\rho_1, \rho_2] \subset (0, 1)$, *we have*

$$\left|x\right|_I = \max\{\left|x\right|_{\rho_1}, \left|x\right|_{\rho_2}\}.$$

Proof The log-convexity inequality is an equality in case $x = [\overline{x}_n]p^n$, and hence a valid inequality, in general. This in turn implies continuity. □

Proposition 4.7 *For* $\rho \in (0, 1)$, *the function* $\left|\bullet\right|_\rho$ *is a nonarchimedean absolute value on* $\mathbf{A}_{\text{inf}}(F)$.

Proof Modulo changes of notation, this can be found in any of [14, Lemme 1.4.2], [22, Lemma 2.1.7], [24, Lemma 2.2], [25, Lemma 4.1], [32, Proposition 5.1.2]. To summarize, the strong triangle inequality follows from the homogeneity of Witt vector arithmetic, we have

$$\sum_{n=0}^{\infty}[\overline{x}_n]p^n + \sum_{n=0}^{\infty}[\overline{y}_n]p^n = \sum_{n=0}^{\infty}[\overline{z}_n]p^n, \qquad \overline{z}_n = \overline{x}_n + \overline{y}_n + P(\overline{x}_0, \dots, \overline{x}_{n-1}, \overline{y}_0, \dots, \overline{y}_{n-1}),$$

where P is homogeneous of degree 1 with coefficients in $\mathbb{Z}$. The multiplicative property is easiest to derive in an indirect way. For any given x and y, the multiplicativity is clear for those values of ρ for which both maxima are achieved by a unique index; this omits a discrete set of values of ρ, which we can fill in by continuity (Proposition 4.6). □

Definition 4.8 The *Newton polygon* associated to an arbitrary element $x = \sum_{n=0}^{\infty}[\overline{x}_n]p^n$ of $\mathbf{A}_{\text{inf}}(F)$ is the lower boundary of the convex hull of the set

$$\bigcup_{n=0}^{\infty}[n, \infty) \times [-\log\left|\overline{x}_n\right|, \infty) \subset \mathbb{R}^2.$$

The multiplicativity of the Gauss norms implies that this Newton polygon has the usual property: the slope multiset of a product xy is the multiset union of the slope multisets of x and y. (See [14, Définition 1.6.18].)

Definition 4.9 For $I \subseteq (0, 1)$ a closed interval, let B_I be the completion of $\mathbf{A}_{\text{inf}}(F)[p^{-1}, [\varpi]^{-1}]$ with respect to the norm $\left|\bullet\right|_I = \sup\{\left|\bullet\right|_\rho : \rho \in I\}$ (extending $\left|\bullet\right|_\rho$ to $\mathbf{A}_{\text{inf}}(F)[p^{-1}, [\varpi]^{-1}]$ by multiplicativity). This norm is power multiplicative (for all x, $\left|x\right|_I^2 = \left|x^2\right|_I$); consequently, B_I is a uniform Huber ring.

In case $I = [\rho_1, \rho_2]$ where $\rho_i = \left|\varpi\right|^{s_i}$ for some $s_i \in \mathbb{Q}$, the ring B_I is the ring associated to the rational subspace

$$\{v \in Y_F : v([\varpi])^{s_2} \leq v(p) \leq v([\varpi])^{s_1}\}$$

of Y_F. In the analogy between $\mathbf{A}_{\text{inf}}(F)$ and $\mathfrak{o}_F[\![p]\!]$, B_I corresponds to the expression

$$F\left\langle \frac{p}{\varpi^{s_1}}, \frac{\varpi^{s_2}}{p} \right\rangle.$$

Remark 4.10 Beware that one cannot express an arbitrary element of B_I as a sum $\sum_{n\in\mathbb{Z}}[\overline{x}_n]p^n$ (see the published erratum to [22]). However, for any $x \in B_I$ and any $\epsilon > 0$, one can find a finite sum $y = \sum_{n\in\mathbb{Z}}[\overline{y}_n]p^n$ such that $|x - y|_I < \epsilon$.

Proposition 4.11 *For $I \subseteq (0, 1)$ a closed interval, the ring B_I is a principal ideal domain.*

Proof See [14, Théorème 2.5.1], [22, Proposition 2.6.8]. The key point is that the Banach ring B_I has the property that its associated graded ring is a Laurent polynomial ring (generated by the image of p) over the associated graded ring of F. □

Proposition 4.12 *The ring B_I is* strongly noetherian *(every Tate algebra over it is noetherian) and sheafy. Consequently, the structure presheaf on Y_F is a sheaf, so we may view Y_F and X_F^{an} as "honest" noetherian adic spaces, and consider coherent sheaves on them.*

Proof The strongly noetherian property is proved in [27], using similar ideas as in the proof of Proposition 4.11. This implies the sheafy property by a result of Huber [21, Theorem 2]. □

Remark 4.13 The rings B_I share other properties with the usual affinoid algebras appearing in rigid analytic geometry, in particular, they are known to be excellent [49].

Remark 4.14 One can also define the ring B_I when I is a half-open or open interval, but not as a Banach ring. Rather, one takes the *Fréchet completion* of $\mathbf{A}_{\text{inf}}(F)[p^{-1}, [\varpi]^{-1}]$ with respect to the family of norms $|\bullet|_\rho$ for $\rho \in I$, that is, one declares a sequence to be Cauchy (and thus to have a limit) if it is Cauchy for each Gauss norm individually, but with no uniform control on the rate of convergence. (One can also use this definition when I is closed, by the last part of Proposition 4.6, it gives the same definition as before.)

The rings B_I correspond to the *extended Robba rings* of [32].

Definition 4.15 Since $Y_F \to X_F^{\text{an}}$ is a free quotient by the action of φ, we can specify sheaves on X_F^{an} by specifying φ-equivariant sheaves on Y_F. For example, for $n \in \mathbb{Z}$, we can define a line bundle $\mathcal{O}(n)$ on X_F^{an} by taking the trivial line bundle on Y_F on a generator $\mathbf{v}$, then specifying that the action of φ takes $\mathbf{v}$ to $p^{-n}\mathbf{v}$.

Define the graded ring

$$P_F := \bigoplus_{n=0}^{\infty} P_{F,n}, \qquad P_{F,n} = \Gamma(X_F^{\text{an}}, \mathcal{O}(n)) = \Gamma(Y_F, \mathcal{O})^{\varphi=p^n}.$$

The scheme $X_F := \operatorname{Proj} P_F$ is the *schematic Fargues–Fontaine curve* associated to F. It is a scheme over $\operatorname{Spec} \mathbb{Q}_p$ but not over $\operatorname{Spec} F$ (because P_F is not an F-algebra).

Proposition 4.16 *The scheme X_F has the following properties:*

(a) *It is connected, separated, noetherian, and regular of dimension* 1.
(b) *For each closed point $x \in X_F$, the residue field of x is a perfectoid field whose tilt may be naturally viewed as a finite extension of F; we write $\deg(x)$ for the degree of this extension. (In particular, if F is algebraically closed, then $\deg(x) = 1$ always.)*
(c) *The degree map on divisors induces a morphism $\deg : \mathrm{Pic}(X_F) \to \mathbb{Z}$ taking $\mathcal{O}(n)$ to n. Moreover, if F is algebraically closed, then $\mathrm{Pic}(X_F) \cong \mathbb{Z}$.*

Proof See [14, Théorème 6.5.2] for the case where F is algebraically closed, and [14, Théorème 7.3.3] for the general case. □

Remark 4.17 Proposition 4.16 states that X_F, together with the degree function on closed points, constitutes an *abstract complete curve* in the sense of [14, §5].

Definition 4.18 By construction, there is a morphism $X_F^{\mathrm{an}} \to X_F$ of locally ringed spaces, along which the canonical ample line bundle $\mathcal{O}(1)$ on X_F pulls back to the prescribed $\mathcal{O}(1)$ on X_F^{an}. This morphism should be thought of as a form of "analytification", analogous to the morphism $X^{\mathrm{an}} \to X$ where X is a scheme locally of finite type over $\mathbb{C}$ and X^{an} is its associated complex analytic space [20, Expose XII], or similarly with $\mathbb{C}$ replaced by a nonarchimedean field, using rigid analytic geometry in place of complex analytic geometry [10, Appendix].

Proposition 4.19 (GAGA for X_F)

(a) *The line bundle $\mathcal{O}(1)$ on X_F^{an} is ample. More precisely, for every coherent sheaf $\mathcal{F}$ on X_F^{an}, there exists a positive integer N such that for each integer $n \geq N$, $\mathcal{F}(n)$ is generated by global sections and $H^i(X_F^{\mathrm{an}}, \mathcal{F}(n)) = 0$ for all $i > 0$. (Note that this vanishing only has content for $i = 1$, because X_F^{an} admits a covering by two affinoids.)*
(b) *Pullback from X_F to X_F^{an} defines an equivalence of categories between coherent sheaves on the two spaces. Moreover, the sheaf cohomology of a coherent sheaf is preserved by pullback from X_F to X_F^{an}.*

Proof See [14, Théoréme 11.3.1]. □

Remark 4.20 In general, the cohomology groups of a coherent sheaf on X_F are Banach spaces over $\mathbb{Q}_p$ which are typically not finite dimensional. However, they do have a somewhat weaker finiteness property: they are *Banach–Colmez spaces* [9]. In fact, the derived categories of coherent sheaves on X_F and Banach–Colmez spaces are equivalent [38].

Proposition 4.21 *One consequence of Proposition 4.19 is that the category of vector bundles on X_F is equivalent to the category of φ-equivariant vector bundles on Y_F. These can themselves be described algebraically: the space Y_F is a quasi-Stein space, so vector bundles correspond to finite projective modules over $\Gamma(Y_F, \mathcal{O}) = B_{(0,\infty)}$, and moreover the ring $B_{(0,\infty)}$ is a* Bézout domain *(every finitely generated ideal* *is principal), which implies that finite projective modules are free. Consequently, vector bundles on X_F can be equated with φ-modules over $B_{(0,\infty)}$; this is the basis for the description of (φ, Γ)-modules in the sense of Berger using vector bundles on a Fargues–Fontaine curve.*

Another description of vector bundles can be given using the Beauville–Laszlo theorem [1] *to glue them from their restriction to the completed local ring at some point and to the complement of that point. In the case where we have a specified untilt K of F in mind, that defines a degree-1 point of X_F and the completion of the local ring is Fontaine's period ring $\mathbf{B}^+_{\mathrm{dR}}$ associated to K. This then leads to the description of Berger's (φ, Γ)-modules in terms of B-pairs* [2].

5 Vector Bundles on Fargues–Fontaine Curves

We give the classification of vector bundles on Fargues–Fontaine curves, then briefly introduce the relative version of the construction. See [28, Lecture 3] for a more detailed discussion.

As in the previous lecture, let F be a perfect nonarchimedean field of characteristic p.

Definition 5.1 Let V be a vector bundle on either X_F or X_F^{an} (by Proposition 4.19 these are interchangeable). Since X_F is connected, the *rank* of V is a well-defined nonnegative integer. The *degree* of V is the degree of the top exterior power $\wedge^{\mathrm{rank}(V)} V$ via the map $\deg : \mathrm{Pic}(X_F) \to \mathbb{Z}$. For V nonzero, the *slope* of V is the ratio $\mu(V) := \deg(V)/\mathrm{rank}(V)$. We say that V is *semistable* (resp., *stable*) if every proper nonzero subbundle W of V satisfies $\mu(W) \leq \mu(V)$ (resp., $\mu(W) < \mu(V)$).

Remark 5.2 The definitions in Definition 5.1 are copied verbatim from the theory of vector bundles on curves in algebraic geometry. In particular, the term *semistable*, having its origins in geometric invariant theory, is quite entrenched within that subject. This creates a terminological issue in p-adic Hodge theory, where we also consider *semistable* Galois representations. This may be unfortunate but is in no way an accident; this second use of the word can be traced back to the notion of *semistable reduction* of families of curves, which is named as such again because it relates to the same phenomenon in geometric invariant theory.

Proposition 5.3 *Let V, V' be semistable vector bundles on X_F. If $\mu(V) > \mu(V')$, then* $\mathrm{Hom}(V, V') = 0$.

Proof As per [28, Lemma 3.4.5], this reduces to the fact that rank-1 bundles are stable, which in turn reduces to the case of $\mathcal{O}$. This case follows by calculating that $H^0(X_F, \mathcal{O}) = \mathbb{Q}_p$. □

Proposition 5.4 *Every vector bundle V on X_F admits a unique filtration*

$$0 = V_0 \subset \cdots \subset V_l = V$$

in which each quotient V_i / V_{i-1} *is a vector bundle which is semistable of some slope* μ_i, *and* $\mu_1 > \cdots > \mu_l$. *This is called the* Harder–Narasimhan filtration *of* V.

Proof This is essentially a formal consequence of Proposition 4.16 and Proposition 5.3. See [14, Théorème 5.5.2] or [28, Lemma 3.4.9]. □

Definition 5.5 For V a vector bundle on X_F, the *Harder–Narasimhan polygon* (or *HN polygon*) of V is the Newton polygon associated to the Harder–Narasimhan filtration. It has length equal to the rank of V, and for $i = 1, \ldots, l$, the slope μ_i occurs with multiplicity $\text{rank}(V_i / V_{i-1})$.

When F is algebraically closed, one can give a complete classification of vector bundles on X_F.

Definition 5.6 Let r/s be a rational number written in lowest terms, that is, r and s are integers with $\gcd(r, s) = 1$ and $s > 0$. Let $\mathcal{O}(r/s)$ be the vector bundle of rank s on X_F corresponding (via Proposition 4.19) to the trivial vector bundle generated by $\mathbf{v}_1, \ldots, \mathbf{v}_s$ on Y_F equipped with the φ-action defined by

$$\varphi(\mathbf{v}_1) = \mathbf{v}_2, \quad \cdots \quad \varphi(\mathbf{v}_{s-1}) = \mathbf{v}_s, \quad \varphi(\mathbf{v}_s) = p^{-r}\mathbf{v}_1.$$

In case $s = 1$, this reproduces the definition of $\mathcal{O}(r)$.

Proposition 5.7 (Classification of vector bundles) *Suppose that* F *is algebraically closed.*

(ii) A vector bundle V *on* F *of slope* μ *is stable if and only if it is isomorphic to* $\mathcal{O}(\mu)$.

(ii) Every vector bundle V *on* F *can be expressed (nonuniquely) as a direct sum of stable subbundles (of various slopes). In particular, the HN filtration of* V *splits (nonuniquely).*

Proof This result has a slightly complicated history. As formulated, it is due to Fargues–Fontaine [14, Théorème 8.2.10], who give two distinct proofs: one using periods of p-divisible groups, and another using the theory of Banach–Colmez spaces (see Remark 4.20). However, using Proposition 4.19 it can also be deduced from earlier results of Kedlaya, see [28, Theorem 3.6.13] for more discussion of this point (and a high-level sketch of the proof). The key point is to show that any V which sits in a nonsplit short exact sequence

$$0 \to \mathcal{O}(-1/n) \to V \to \mathcal{O}(1) \to 0$$

is trivial; the space of such extensions is essentially the Scholze–Weinstein moduli space of p-divisible groups [46]. □

Remark 5.8 Proposition 5.7 is formally similar to the classification of vector bundles on the projective line over a field, in which every vector bundle splits as a direct sum of various $\mathcal{O}(n)$. A more apt analogy is the classification of rational Dieudonné modules over an algebraically closed field (Dieudonné–Manin classification) in which some higher rank objects with fractional slopes also appear; indeed, some of the precursor statements to Proposition 5.7 mentioned above are formulated as Dieudonné–Manin classifications for φ-modules over the ring $B_{(0,1)}$ or other related rings.

Proposition 5.9 (Analogue of Narasimhan–Seshadri) *The functor*

$$V \mapsto \Gamma(X_{\widehat{\overline{F}}}, V)$$

defines an equivalence of categories between semistable vector bundles of slope 0 *on* X_F *and continuous representations of the absolute Galois group* G_F *of* F *on finite-dimensional* $\mathbb{Q}_p$*-vector spaces.*

Proof This follows from Proposition 5.7 and the equality $\Gamma(X_F, \mathcal{O}) = \mathbb{Q}_p$. □

Remark 5.10 For line bundles, Proposition 5.9 gives rise to a canonical isomorphism

$$\mathrm{Pic}(X_F) \cong \mathbb{Z} \oplus \mathrm{Hom}_{\mathrm{cont}}(G_F, \mathbb{Q}_p^\times).$$

Remark 5.11 Proposition 5.9 is meant to evoke the Narasimhan–Seshadri theorem [41]: for X a compact Riemann surface, and there is a canonical equivalence of categories between stable vector bundles of slope 0 on X and irreducible finite-dimensional unitary representations of the fundamental group of X.

In the theory of vector bundles on curves in algebraic geometry, the Narasimhan–Seshadri theorem implies that the tensor product of two semistable vector bundles on a curve is semistable *provided* that the base field is of characteristic 0. The fact that this is a highly nonformal statement can be seen by its failure to carry over to positive characteristic, which was first observed by Gieseker [19]. Correspondingly, Proposition 5.9 implies that the tensor product of two semistable vector bundles on X_F is semistable.

Many applications of the theory of vector bundles on curves involve moduli spaces of these bundles. In order to study these for Fargues–Fontaine curves, we need to introduce the relative form of the construction, in which the base field is replaced by a perfectoid ring (or a space, or...).

Definition 5.12 Let (R, R^+) be a perfectoid Huber pair of characteristic p. Let Y_R be the complement of the zero locus of $p[\varpi]$ in $\mathrm{Spa}(W(R^+), W(R^+))$, where $\varpi \in R$ is any pseudouniformizer (the answer does not depend on the choice).

Now fix a power-multiplicative Banach norm on R. For $\rho \in (0, 1)$, we may define the ρ-Gauss norm on $W(R^+)$ by the same formula as before. For I a closed interval in $(0, 1)$, we may then define a ring $B_{I,R}$ by completing $W(R^+)[p^{-1}, [\varpi]^{-1}]$ for the supremum of the ρ-Gauss norms for all $\rho \in I$, and again use the spectra of these rings to cover Y_R.

One cannot hope for the ring $B_{I,R}$ to be noetherian, in general, nor is it perfectoid (because this was already not true when R was a field). However, it is close enough to being perfectoid to inherit the sheafy property.

Proposition 5.13 *The Huber ring $B_{I,R}$ is stably uniform, and hence sheafy.*

Proof While $B_{I,R}$ is not a perfectoid ring, it turns out that it becomes perfectoid after taking the completed tensor product over $\mathbb{Q}_p$ with any perfectoid field. This can be used to recover the stably uniform property by a splitting construction, see [28, Lemma 3.1.3]. □

Definition 5.14 By Proposition 5.13, Y_R is an adic space. Following the previous model, we form the quotient $X_R^{\mathrm{an}} := Y_R/\varphi$ by the totally discontinuous action of φ; we define the line bundles $\mathcal{O}(n)$ on X_R^{an} in terms of φ-equivariant line bundles on Y_R; we define the graded ring $P_R := \bigoplus_{n=0}^{\infty} P_{R,n}$ by taking $P_{R,n}$ to be the sections of $\mathcal{O}(n)$; we define the scheme $X_R := \operatorname{Proj} P_R$; and we obtain a morphism $X_R^{\mathrm{an}} \to X_R$ of locally ringed spaces.

Remark 5.15 There is a natural continuous map $X_R^{\mathrm{an}} \to \operatorname{Spa}(R, R^+)$ of topological spaces; however, this morphism does not promote to a morphism of locally ringed spaces due to the mismatch of characteristics (namely, p is invertible on the source and zero on the target). That said, any untilt (A, A^+) of (R, R^+) over $\mathbb{Q}_p$ gives rise to a section of this map which does promote to a morphism of adic spaces.

Remark 5.16 Since neither X_R nor X_R^{an} is noetherian, we cannot easily handle coherent sheaves on these spaces. In [33] and [28, Lecture 1] one finds a theory of *pseudocoherent* sheaves, which obey a stronger finiteness condition; we omit this here and instead restrict attention to vector bundles in what follows. Before doing so, we point out that the following discussion implicitly uses the analogue of Kiehl's theorem for vector bundles on affinoid adic spaces: for (A, A^+) a sheafy Huber pair, the global sections functor defines an equivalence of categories between vector bundles on $\operatorname{Spa}(A, A^+)$ and finite projective A-modules [28, Theorem 1.4.2].

Proposition 5.17 (GAGA revisited)

(a) *For every vector bundle V on X_R^{an}, there exists a positive integer N such that for each integer $n \geq N$, $V(n)$ is generated by global sections and $H^i(X_R^{\mathrm{an}}, V(n)) = 0$ for all $i > 0$.*
(b) *Pullback from X_R to X_R^{an} defines an equivalence of categories between vector bundles on the two spaces. Moreover, the sheaf cohomology of a vector bundle is preserved by pullback from X_R to X_R^{an}.*

Proof See [32, Theorem 8.7.7]. □

The following is analogue of the usual semicontinuity for families of vector bundles on a curve, or more generally on a family of varieties [48].

Proposition 5.18 (Kedlaya–Liu semicontinuity theorem) *Let V be a vector bundle on X_R.*

(*i*) *The Harder–Narasimhan polygons of the fibers of V form a lower semicontinuous function on* $\mathrm{Spa}(R, R^+)$.

(*ii*) *If this function is constant, then the Harder–Narasimhan filtrations of the fibers of V arise by specialization from a filtration of V.*

Proof See [32, Theorem 4.7.5, Corollary 7.4.10]. Additional discussion found in [28, Theorem 3.7.2]. □

There is also a relative form of the Narasimhan–Seshadri theorem.

Proposition 5.19 *There is an equivalence of categories between étale $\mathbb{Q}_p$-local systems on* $\mathrm{Spa}(R, R^+)$ *(see below) and vector bundles on X_R which are fiberwise semistable of degree* 0.

Proof See [32, Theorem 9.3.13]. Additional discussion found in [28, Theorem 3.7.5] □

Remark 5.20 In Proposition 5.19, one must be careful about the meaning of the phrase "étale $\mathbb{Q}_p$-local system". One way to interpret this correctly is via de Jong's theory of étale fundamental groups [12]; this amounts to saying that an étale $\mathbb{Q}_p$-local system is étale-locally the isogeny object associated to a $\mathbb{Z}_p$-local system. Another correct interpretation can be obtained by replacing the étale topology with a certain *pro-étale topology*; this is the approach taken in [32] based on a construction of Scholze [45].

Remark 5.21 The preceding discussion lies at the heart of the construction of moduli spaces of vector bundles on Fargues–Fontaine curves. This of course requires a globalization of the definition of the relative Fargues–Fontaine curve, first to perfectoid spaces, and second to certain stacks on the category of perfectoid spaces (in particular to what Scholze calls *diamonds*). See [47] for further discussion of these stacks and their role in the study of moduli spaces of vector bundles.

Another application of relative Fargues–Fontaine curves is to the study of cohomology of $\mathbb{Q}_p$-local systems on rigid analytic spaces over p-adic fields. See [34].

References

1. A. Beauville, Y. Laszlo, Un lemme de descente. C.R. Acad. Sci. Paris **320**, 335–340 (1995)
2. L. Berger, Construction de (φ, Γ)-modules: représentations p-adiques et B-paires. Alg. Num. Theory **2**, 91–120 (2008)
3. B. Bhatt, The Hodge-Tate decomposition via perfectoid spaces, in [8]
4. B. Bhatt, M. Morrow, P. Scholze, Integral p-adic Hodge theory. Publ. Math. IHÉS **128**, 219–397 (2018)
5. B. Bhatt and P. Scholze, Prisms and prismatic cohomology. arXiv:1905.08229v2 (2019)
6. N. Bourbaki, *Commutative Algebra* (Springer, Berlin, 1989)
7. K. Buzzard, A. Verberkmoes, Stably uniform affinoids are sheafy. J. Reine Angew. Math. **740**, 25–39 (2018)

8. B. Cais et al., *Perfectoid Spaces: Lectures from the 2017 Arizona Winter School*, Math. Surveys and Monographs 242, Amer. Math. Soc., 2019
9. P. Colmez, Espaces de Banach de dimension finie. J. Inst. Math. Jussieu **1**, 331–439 (2002)
10. B. Conrad, Relative ampleness in rigid geometry. Ann. Inst. Fourier **56**, 1049–1126 (2006)
11. B. Conrad, Several approaches to non-archimedean geometry, in p-adic Geometry: Lectures from the, Arizona Winter School, University Lecture Series 45, Amer. Math. Soc. Providence **2008**, 24–79 (2007)
12. A.J. de Jong, Étale fundamental groups of non-archimedean analytic spaces. Compos. Math. **97**, 89–118 (1995)
13. H. Du, $\mathbf{A}_{\text{inf}}$ has uncountable Krull dimension, arXiv:2002.10358v1 (2020)
14. L. Fargues and J.-M. Fontaine, Courbes et fibrés vectoriels en théorie de Hodge p-adique, *Astérisque* **406** (2018)
15. J.-M. Fontaine, Perfectoïdes, presque pureté et monodromie-poids (d'après Peter Scholze), Séminaire Bourbaki, volume 2011/2012, *Astérisque* **352** (2013)
16. J.-M. Fontaine and J.-P. Wintenberger, Le "corps des normes" de certaines extensions algébriques de corps locaux, *C.R. Acad. Sci. Paris* **288** (1979), 367–370
17. J.-M. Fontaine and J.-P. Wintenberger, Extensions algébriques et corps de normes des extensions APF des corps locaux, *C.R. Acad. Sci. Paris* **288** (1979), 441–444
18. O. Gabber, L. Ramero, *Almost Ring Theory*, Lecture Notes in Math, vol. 1800. (Springer-Verlag, Berlin, 2003)
19. D. Gieseker, Stable vector bundles and the Frobenius morphism. Ann. Scient. Éc. Norm. Sup. **6**, 95–101 (1973)
20. A. Grothendieck, *Revêtements Étales et Groupe Fondamental (SGA 1)*, Lecture Notes in Math, vol. 224. (Springer-Verlag, Berlin, 1971)
21. R. Huber, A generalization of formal schemes and rigid analytic varieties. Math. Z. **217**, 513–551 (1994)
22. K.S. Kedlaya, Slope filtrations revisited, *Doc. Math.* **10** (2005), 447–525; errata, *ibid.* **12** (2007), 361–362. Additional errata at https://kskedlaya.org/papers
23. K.S. Kedlaya, *p-adic Differential Equations*, Cambridge Univ. Press, 2010. Errata at https://kskedlaya.org/papers
24. K.S. Kedlaya, Relative p-adic Hodge theory and Rapoport-Zink period domains, in *Proceedings of the International Congress of Mathematicians (Hyderabad, 2010)*, Volume II, Hindustan Book Agency/World Scientific, 2010, 258–279
25. K.S. Kedlaya, Nonarchimedean geometry of Witt vectors. Nagoya Math. J. **209**, 111–165 (2013)
26. K.S. Kedlaya, *New methods for* (φ, Γ)*-modules* (Res. Math, Sci, 2015)
27. K.S. Kedlaya, *Noetherian properties of Fargues-Fontaine curves* (Int. Math. Res, Notices, 2015)
28. K.S. Kedlaya, Sheaves, stacks, and shtukas, in [8]
29. K.S. Kedlaya, On commutative nonarchimedean Banach fields. Doc. Math. **23**, 171–188 (2018)
30. K.S. Kedlaya, *Some ring-theoretic properties of* $\mathbf{A}_{\text{inf}}$, *in p-adic Hodge Theory* (Springer, Simons Symposia, 2020), pp. 129–141
31. K.S. Kedlaya, Notes on prismatic cohomology, https://kskedlaya.org/prismatic/ (2021)
32. K.S. Kedlaya and R. Liu, Relative p-adic Hodge theory: Foundations, *Astérisque* **371** (2015)
33. K.S. Kedlaya and R. Liu, Relative p-adic Hodge theory, II: Imperfect period rings, arXiv:1602.06899v3 (2019)
34. K.S. Kedlaya and R. Liu, Finiteness of cohomology of local systems on rigid analytic spaces, arXiv:1611.06930v2 (2019)
35. K.S. Kedlaya, M. Temkin, Endomorphisms of power series fields and residue fields of Fargues-Fontaine curves. Proc. Amer. Math. Soc. **146**, 489–495 (2018)
36. F.-V. Kuhlmann, *The defect, in Commutative Algebra-Noetherian and Non-Noetherian Perspectives* (Springer, New York, 2011), pp. 277–318
37. J. Lang and J. Ludwig, $\mathbf{A}_{\text{inf}}$ is infinite dimensional, *J. Inst. Math. Jussieu*, to appear (published online 2020)

38. A.-C. Le Bras, Espaces de Banach-Colmez et faisceaux cohérents sur la courbe de Fargues-Fontaine. Duke Math. J. **167**, 3455–3532 (2018)
39. M. Matignon, M. Reversat, Sous-corps fermés d'un corps valué. J. Algebra **90**, 491–515 (1984)
40. T. Mihara, On Tate acyclicity and uniformity of Berkovich spectra and adic spectra. Israel J. Math. **216**, 61–105 (2016)
41. M.S. Narasimhan, C.S. Seshadri, Stable and unitary vector bundles on a compact Riemann surface. Ann. of Math. **82**, 540–567 (1965)
42. A. Robert, *A Course in p-adic Analysis*, Graduate Texts in Math. 198, Springer, New York, 2000
43. S. Sen, Ramification in p-adic Lie extensions. Invent. Math. **17**, 44–50 (1972)
44. P. Scholze, Perfectoid spaces. Publ. Math. IHÉS **116**, 245–313 (2012)
45. P. Scholze, p-adic Hodge theory for rigid analytic varieties, *Forum of Math. Pi* **1** (2013), https://doi.org/10.1017/fmp.2013.1; errata available at http://www.math.uni-bonn.de/people/scholze/pAdicHodgeErratum.pdf
46. P. Scholze, J. Weinstein, Moduli of p-divisible groups. Cambridge J. Math. **1**, 145–237 (2013)
47. P. Scholze and J. Weinstein, *Berkeley Lectures on p-adic Geometry*, Annals of Math. Studies 207, Princeton Univ. Press, Princeton, 2020
48. S. Shatz, The decomposition and specialization of algebraic families of vector bundles. Compos. Math. **35**, 163–187 (1977)
49. P. Wear, Properties of extended Robba rings, arXiv:1709.06221v1 (2017)
50. J. Weinstein, Adic spaces, in [8]
51. J.-P. Wintenberger, Le corps des normes de certaines extensions infinies de corps locaux; applications. Ann. Scient. Éc. Norm. Sup. **16**, 59–89 (1983)

The Fargues-Fontaine Curve and p-Adic Hodge Theory

Ehud de Shalit

Abstract This survey paper is based, in part, on the author's lectures at the summer school that was held at Tata Institute's ICTS in Bangalore in August 2019, and in part on a web-seminar that was held at the Hebrew University in the Spring Semester of 2020. I would like to thank the organizers of the summer school and the participants of both events for their contribution. Special thanks go to David Kazhdan for leading the seminar at the Hebrew University.
The goal of this survey is to explain the main results of [5]. To be able to do so in a reasonable amount of space we have sacrificed generality and omitted many interesting topics, but we did choose to give some background, whenever we felt it was necessary. The reader should always refer to the book by Fargues and Fontaine for details, missing explanations, and other developments. We have also benefitted from the excellent Bourbaki seminar by Morrow [16], which we recommend as a starting point for anybody encountering the topic for the first time.
The informal style of the lectures, especially in their later sections, where they became increasingly sketchy, was also kept in the printed version. Needless to say, none of the results surveyed here are due to the author, but errors, as much as they have escaped my attention, are all original errors.

1 Introduction

Let p be a prime. The Fargues-Fontaine curve is a fundamental geometric object associated with p, introduced in [5]. It serves as the arena "where p-adic Hodge theory takes place". Historically, it was discovered rather late in the development of the subject, but its discovery offered a new point of view on Fontaine's rings of periods, yielded simpler more conceptual proofs to several of the principal theorems in the field, and gave hope for a geometrization of the classical local Langlands correspondence (work in progress of Fargues and Scholze). As the fundamental group of the curve is just $Gal(\bar{\mathbb{Q}}_p/\mathbb{Q}_p)$, from the point of view of local Galois

E. de Shalit (✉)
Einstein Institute of Mathematics, The Hebrew University,Jerusalem, Israel
e-mail: ehud.deshalit@mail.huji.ac.il

© The Author(s), under exclusive license to Springer Nature Singapore Pte Ltd. 2022
D. Banerjee et al. (eds.), *Perfectoid Spaces*, Infosys Science Foundation Series in Mathematical Sciences https://doi.org/10.1007/978-981-16-7121-0_7

representations, it may be considered as a richer substitute for $Spec(\mathbb{Q}_p)$, over which such representations should be studied. One cannot avoid a distant dream, that one day, when the analogy between primes and knots will be made more precise, a global (three-dimensional?) object will be found, into which the Fargues-Fontaine curves for the various primes will naturally embed, linked in a way that reflects the relative position of the various decomposition groups in $Gal(\bar{\mathbb{Q}}/\mathbb{Q})$.

This survey will focus on the construction of the Fargues-Fontaine curve, its main properties, and the classification of vector bundles over the curve. With the exception of "weakly admissible equals admissible" we shall not go into any of its beautiful applications. The reader may find several of them well explained in the original work.

1.1 The Fargues-Fontaine Curve

1.1.1 Curves

Let X be a separated noetherian scheme.

Definition 1 (i) X is a *curve* if it is regular, *one*-dimensional, and connected. (ii) A curve X is *complete* if

$$\sum_{x\in|X|} ord_x(f) = 0 \tag{1.1}$$

for every $f \in \mathcal{O}_{X,\eta}^{\times}$.

Here $|X|$ denotes the set of closed (i.e., codimension 1) points of X and η the unique non-closed (i.e., generic) point; $\mathcal{O}_{X,\eta}$ is the function field of the scheme X. Thus X can be covered by a finite number of affine sets, each of which is of the form $Spec(A)$ for a Dedekind domain A. The local rings of X at $x \in |X|$ are DVRs, and ord_x are the respective normalized valuations.

Example 2 (i) A smooth connected curve over an algebraically closed field k is a curve in this definition. It is complete if it is projective. If k were not algebraically closed, we should have allowed a more general condition instead of (1.1), multiplying each term by the positive integer $deg(x) = [k(x) : k]$.

(ii) $Spec(\mathbb{Z})$ is a non-complete curve.

(iii) The Fargues-Fontaine curve X^{FF}, to be discussed in these lectures.

It would be extremely interesting to discover any new class of examples. Surprisingly, they are not easy to come by.

1.1.2 Some Well-Known Facts About $\mathbb{P}^1_{\mathbb{C}}$

The Fargues-Fontaine curve will resemble the simplest curve over $\mathbb{C}$, the projective line $\mathbb{P}^1_{\mathbb{C}}$. To make this comparison, we recall some well-known facts.

(1) The projective line can be constructed by either (a) dividing $\mathbb{A}^2_{\mathbb{C}} - \{0\}$ by the equivalence relation of "generating the same line", or (b) gluing $\mathbb{A}^1_{\mathbb{C}} = Spec(\mathbb{C}[z])$ and the formal disk $Spec(\mathbb{C}[[z^{-1}]])$ along the punctured formal disc $Spec(\mathbb{C}((z^{-1})))$.

(2) There is a *fundamental exact sequence* of vector spaces

$$0 \to \mathbb{C} \to \mathbb{C}[z] \to \mathbb{C}((z^{-1}))/\mathbb{C}[[z^{-1}]] \to 0. \tag{1.2}$$

(3) For every $f \in \mathbb{C}(z)$

$$\sum_{\zeta \in \mathbb{C}} ord_{\zeta}(f) + ord_{\infty}(f) = 0.$$

(4) The function $deg(f) = -ord_{\infty}(f)$ makes $\mathbb{C}[z]$ into a Euclidean domain. Recall the definition.

Definition 3 A *Euclidean domain* is a commutative ring R, equipped with a function $deg : R \to \mathbb{N} \cup \{-\infty\}$ satisfying:

- (E1) $deg(f) = -\infty$ iff $f = 0$; $deg(1) = 0$.
- (E2) If $g \neq 0$ then $deg(f) \leq deg(fg)$
- (E3) If $g \neq 0$ then for any f there are q and r with $deg(r) < deg(g)$ such that

$$f = qg + r.$$

It follows from the axioms that R is a domain and (E3$'$) $deg(g) = 0$ if and only if $g \in R^{\times}$. We let (E3$''$) be the same condition as in (E3) where we only demand $deg(r) \leq deg(g)$. Anticipating later developments, we make the following definition, unmotivated at present. A *semi-Euclidean* domain is a pair (R, deg) satisfying (E1), (E2), and (E3$'$) and (E3$''$) instead of (E3).

(5) There are line bundles $\mathcal{O}(n) = \mathcal{O}(n\infty)$ $(n \in \mathbb{Z})$, and every vector bundle is isomorphic to a unique direct sum

$$\bigoplus_{i=1}^{k} \mathcal{O}(n_i)$$

(up to permutation of the factors).

(6) We have

$$H^0(\mathbb{P}^1_{\mathbb{C}}, \mathcal{O}(n)) = 0 \;\; (n < 0)$$

$$H^1(\mathbb{P}^1_{\mathbb{C}}, \mathcal{O}(n)) = 0 \;\; (n \geq -1).$$

The last vanishing of H^1 comes from Serre duality, $H^1(\mathbb{P}^1_{\mathbb{C}}, \mathcal{O}(n))^{\vee} \simeq H^0(\mathbb{P}^1_{\mathbb{C}}, \mathcal{O}(-n-2))$ and reflects the fact that $\mathbb{P}^1_{\mathbb{C}}$ has genus 0.

Point 5 is a theorem of Grothendieck. Let us recall its proof. Cover $\mathbb{P}^1_{\mathbb{C}}$ with $U = Spec(\mathbb{C}[z])$ and $V = Spec(\mathbb{C}[z^{-1}])$. Then $U \cap V = Spec(\mathbb{C}[z, z^{-1}])$. Since any rank k vector bundle is trivial when restricted to U or V, which are spectra of PIDs, we may trivialize it by specifying global bases on each of the two affine pieces. What completely determines the vector bundle is then the transition matrix between the bases over $U \cap V$, which is a matrix from $GL_k(\mathbb{C}[z, z^{-1}])$. Taking into account the freedom to change bases in each piece separately, we see that isomorphism classes of rank k vector bundles are classified by the double coset space

$$GL_k(\mathbb{C}[z^{-1}])\backslash GL_k(\mathbb{C}[z, z^{-1}])/GL_k(\mathbb{C}[z]).$$

The claim that we have to prove is that any double coset in this space is represented by a unique diagonal matrix with entries z^{n_i} where $n_1 \geq n_2 \geq \cdots \geq n_k$. This is done by induction on k, where the key ingredient is the Euclidean algorithm in $\mathbb{C}[z]$ and $\mathbb{C}[z^{-1}]$, applied in column and row operations to reduce the given matrix to its standard form. We leave the details to the reader as an exercise, but emphasize the close connection between Point 4 and Point 5: it is *not enough* to know that $\mathbb{C}[z]$ and $\mathbb{C}[z^{-1}]$ are PIDs to deduce Grothendieck's theorem. Essential use is made of them being Euclidean. As we shall see later, Point 6 is also closely connected to the Euclidean property.

Finally, we caution the reader that the category of vector bundles over $\mathbb{P}^1_{\mathbb{C}}$, as a full subcategory of the abelian category of coherent sheaves, is not abelian. Point 5 sheds only little light on its structure. Even if we restrict our attention to short exact sequences, where the morphisms are morphisms of vector bundles (the image is locally a direct summand), one encounters non-split short exact sequences such as

$$0 \to \mathcal{O}(-1) \to \mathcal{O}^2 \to \mathcal{O}(1) \to 0.$$

Here we may identify $\mathcal{O}(1)$ with the line bundle $\mathcal{O}(\infty)$; the first morphism is $f \mapsto (zf, f)$ and the second is $(g_1, g_2) \mapsto g_1 - zg_2$. On the other hand, there does not exist a short exact sequence as above with the roles of $\mathcal{O}(1)$ and $\mathcal{O}(-1)$ interchanged (why?).

1.1.3 Comparison with the Fargues-Fontaine Curve

The Fargues-Fontaine curve $X = X^{FF}$ (in its simplest form, there are relative versions nowadays) will be a complete curve over $Spec(\mathbb{Q}_p)$. Nevertheless, it will not be of finite type over $\mathbb{Q}_p$. To stress how much *not* of finite type it will be, we mention that at one of its closed points, denoted ∞, the residue field will be $\mathbb{C}_p$, the completion of a fixed algebraic closure of $\mathbb{Q}_p$. In contrast, residue fields at closed points of curves of finite type over $\mathbb{Q}_p$ are finite extensions of $\mathbb{Q}_p$!

The curve X will have a distinguished point ∞ whose complement will be affine, $Spec(B_e)$ for a huge ring B_e that will be constructed explicitly. The completion of the local ring at ∞ will be one of Fontaine's rings B_{dR}^{+}. As expected, this is a complete DVR, whose field of fractions is denoted by B_{dR}. The associated valuation (analogue of ord_{∞}) is denoted by v_{dR}.

Points 1–3 will have good analogues. We shall construct X (at least the set $|X|$ of its closed points with an appropriate analytic topology) as the set of Frobenius equivalence classes of $|Y|$, "the space of untilts" of a certain characteristic p field F. We will also have, in retrospect, a second construction, by gluing $Spec(B_e)$ and $Spec(B_{dR}^{+})$ along $Spec(B_{dR})$.

The analogue of point 2 will become the *fundamental exact sequence of p-adic Hodge theory*

$$0 \to \mathbb{Q}_p \to B_{cris}^{\varphi=1} \to B_{dR}/B_{dR}^{+} \to 0. \tag{1.3}$$

Here we encounter for the first time another one of Fontaine's rings, the ring B_{cris}. Both the field B_{dR} and its subring B_{cris} carry an action of $G_{\mathbb{Q}_p} = Gal(\bar{\mathbb{Q}}_p/\mathbb{Q}_p)$. The field B_{dR} is a discrete valuation field, hence it carries a natural decreasing filtration $Fil^{\bullet}$. The subring B_{cris} carries a Frobenius endomorphism φ (that does not extend to B_{dR}) and

$$B_e = H^0(X - \{\infty\}, \mathcal{O}_X) = B_{cris}^{\varphi=1}.$$

The exact sequence (1.3), tensored with a p-adic Galois representation V of $G_{\mathbb{Q}_p}$, and the long exact sequence in Galois cohomology that ensues, is the starting point in any application of p-adic Hodge theory to the study of local Galois representations. It is of utmost importance. The Fargues-Fontaine curve allows us to give it a geometric interpretation.

The analogue of point 3 is (1.1); it rephrases the fact that X is complete.

Points 4–6 exhibit subtle differences between X and the projective line that make the theory of vector bundles over X much richer.

As already hinted above, the ring B_e, while a PID, will only be an almost Euclidean domain with respect to $deg = -v_{dR}$ (the negative of the valuation at ∞). Given the important role played by the Euclidean property in the proof of Grothendieck's theorem on vector bundles on $\mathbb{P}^1_{\mathbb{C}}$, it is not surprising that being only almost Euclidean results in a different classification theorem. For each rational number $\lambda = d/h$ (in reduced terms) there will be an non-decomposable vector bundle $\mathcal{O}(\lambda)$ of rank h and degree d (thus slope λ), and every vector bundle on X will be a direct sum of such. The Harder-Narasimhan formalism of slopes will apply, and the only semi-stable vector bundles will be, as in the case of $\mathbb{P}^1_{\mathbb{C}}$, the isoclinic ones, vector bundles of the form $\mathcal{O}(\lambda)^n$ for some n.

This structure theorem for vector bundles over X is the most difficult part of [5]. The whole theory of p-divisible groups, the crystals associated with them, and period maps enters in its proof. Is there a simpler proof that avoids this, and instead only uses Point 2 (the fundamental exact sequence) and the almost Euclidean structure of B_e, as in the proof of Grothendieck's theorem sketched above?

A closely related difference between X and $\mathbb{P}^1_{\mathbb{C}}$ is in Point 6. The H^1 cohomologies of the $\mathcal{O}(\lambda)$ will only vanish for $\lambda \geq 0$, not for $\lambda \geq -1$. While $H^1(X, \mathcal{O}) = 0$ is often regarded as evidence for “genus 0”, this deviation is sometimes summarized by saying that the Fargues-Fontaine curve has “genus $0+\varepsilon$”.

1.2 Applications

The construction of X and the study of its properties intrinsically belong to p-adic Hodge theory. They provide a geometric set-up for Fontaine’s rings of periods and the relations between them.

The classification of vector bundles over X is related to the classification of isocrystals over $\bar{\mathbb{F}}_p$, or, almost equivalently, to the classification of p-divisible groups up to isogeny (Manin-Dieudonné theory). The finer study of *modifications* of vector bundles at the point ∞ (injective maps between vector bundles whose cokernels are skyscraper sheaves supported at ∞) is related to *deformations* of p-divisible groups to p-adically complete rings such as $\mathcal{O}_{\mathbb{C}_p}$.

There are several deep theorems for which the Fargues-Fontaine curve supplied new more transparent proofs. The list below is far from complete.

(a) Fontaine’s conjecture that weakly admissible (i.e., semi-stable of slope 0) filtered φ-modules are admissible (i.e., are of the form $D_{cris}(V)$ for a crystalline Galois representation V). This was proved by Colmez and Fontaine, and later a different proof was found by Berger, but the proof using the Fargues-Fontaine curve is remarkably short (once all the prerequisites are in place)! The direction “admissible implies weakly admissible” is an old, easier, result of Fontaine.

(b) Faltings’ theorem on the “isomorphism between the Lubin-Tate and Drinfeld towers” (towers of generic fibers of certain moduli spaces of p-divisible groups with level structures added). Understanding this theorem was the subject of a full-size book by Fargues (Genestier and V. Lafforgue contributing to the function-field case). Again, the Fargues-Fontaine curve presents a new perspective and a much shorter proof, as well as a far-reaching generalization, due to Weinstein and Scholze (last section of [18]).

(c) Geometrization of the local Langlands correspondence—ongoing work of Fargues and Scholze.

2 Construction of X^{FF}

2.1 Tilting and the Space $|Y|$

2.1.1 Perfectoid Fields and Tilts

Let L be a *complete non-Archimedean* field of residual characteristic p, and denote ν_L the associated valuation normalized by[1]

$$|x| = p^{-\nu_L(x)}.$$

Definition 4 The field L is called *perfectoid* if ν_L is non-discrete, and the p-power map $\phi : \mathcal{O}_L/p\mathcal{O}_L \to \mathcal{O}_L/p\mathcal{O}_L$ is surjective.

The non-discreteness of ν_L is imposed to exclude "small" fields such as $\mathbb{Q}_p$. If L has characteristic p then it is perfectoid if and only if it is perfect. In characteristic 0 the field $\mathbb{C}_p$ (the completion of a fixed algebraic closure of $\mathbb{Q}_p$) is perfectoid, but there are much smaller examples, e.g., $\mathbb{Q}_p^{cycl} = \mathbb{Q}_p(\mu_{p^\infty})^\wedge$ (exercise!).

If L is perfectoid we define an $\mathbb{F}_p$-algebra

$$\mathcal{O}_L^\flat = \varprojlim(\mathcal{O}_L/p\mathcal{O}_L)$$

where the inverse limit is taken with respect to iterations of the homomorphism ϕ. Thus an element of $\mathcal{O}_L^\flat$ is a sequence $x = (x_0, x_1, \dots)$ with $x_i \in \mathcal{O}_L/p\mathcal{O}_L$ and $x_{i+1}^p = x_i$. If the characteristic of L is p, projection to x_0 is an isomorphism $\mathcal{O}_L^\flat \simeq \mathcal{O}_L$, so from now on we assume that L has characteristic 0. Let $\tilde{x}_i$ be a representative of x_i in $\mathcal{O}_L$. It is easy to check that

$$x^{(i)} = \lim_{j\to\infty} \tilde{x}_{i+j}^{p^j} \in \mathcal{O}_L$$

exist, depending only on x (and not on the chosen representatives) and satisfy $x^{(i+1)p} = x^{(i)}$. Conversely, starting with such a sequence and defining $x_i = x^{(i)}$ mod p we get a point $x \in \mathcal{O}_L^\flat$. We may therefore identify $\mathcal{O}_L^\flat$, as a set, with the sequences $\xi = (\xi_0, \xi_1, \dots)$ of elements of $\mathcal{O}_L$ satisfying $\xi_{i+1}^p = \xi_i$. With this identification, the ring operations of $\mathcal{O}_L^\flat$ have the following description. Multiplication is done component-wise. For the addition, $\xi + \eta = \zeta$ where

$$\zeta_i = \lim_{j\to\infty} (\xi_{i+j} + \eta_{i+j})^{p^j}.$$

---

[1] By definition, ν_L is non-trivial.

For $x \in \mathcal{O}_L^\flat$ let

$$x^\sharp = x^{(0)}, \quad \nu_{L^\flat}(x) = \nu_L(x^\sharp), \quad |x|_{L^\flat} = |x^\sharp|.$$

It is easily checked that $x \mapsto x^\sharp$ is a multiplicative map and that $\nu_{L^\flat}$ is a complete valuation on $\mathcal{O}_L^\flat$. The residue fields of $\mathcal{O}_L^\flat$ and $\mathcal{O}_L$ are canonically identified.

We let $L^\flat$ be the fraction field of $\mathcal{O}_L^\flat$ and we extend the valuation $\nu_{L^\flat}$ and the map $x \mapsto x^\sharp$ to $L^\flat$. Then $F = L^\flat$ is a perfectoid field in characteristic p. We remark that the association $L \rightsquigarrow L^\flat$ is functorial: a field homomorphism $\alpha : L_1 \to L_2$ yields a homomorphism $\alpha^\flat : L_1^\flat \to L_2^\flat$. The field $L^\flat$ is called the *tilt* of L.

Example 5 Let $L = \mathbb{Q}_p^{cycl}$. Then, fixing a basis ε of the Tate module $T_p\mu = \lim_{\leftarrow} \mu_{p^n}(L)$, there is a unique identification of $L^\flat$ with the completed perfect closure of $\mathbb{F}_p((\varpi))$, taking $\varepsilon - 1$ to ϖ. Starting, instead, with $M = \mathbb{Q}_p^{ab}$ (the completion of the maximal abelian extension of $\mathbb{Q}_p$) and making the same choice, $M^\flat$ is identified with the completed perfect closure of $\bar{\mathbb{F}}_p((\varpi))$. An element of this completed perfect closure is a formal power series

$$\sum_{i \in \mathbb{Z}[1/p]} a_i \varpi^i$$

with the provision that for any T, only finitely many $i < T$ have $a_i \neq 0$. The denominators of the i's in the support of such an element need not be bounded.

2.1.2 Untilts

Definition 6 (i) Starting with a perfectoid field F in characteristic p we define an *untilt* of F to be a pair (L, ι) where L is a perfectoid field of characteristic 0, and

$$\iota : L^\flat \simeq F$$

is an isomorphism of valued fields.

(ii) A homomorphism of untilts is a map $\alpha : L_1 \to L_2$ of valued fields, satisfying $\iota_2 \circ \alpha^\flat = \iota_1$. It can be shown that α must then be an isomorphism, in which case the two untilts are called (surprise!) *isomorphic*. (If L_2 is algebraic over $\alpha(L_1)$ this follows from Theorem 7 below.)

(iii) If (L, ι) is an untilt of F, any untilt which is isomorphic to $(L, \phi^n \circ \iota)$ for some $n \in \mathbb{Z}$ is said to be *Frobenius equivalent* to (L, ι).

Theorem 7 *Fix an algebraic closure L^{alg} of L. Every finite extension $L \subset M \subset L^{alg}$ is perfectoid and*

$$\lim_{\to} M^\flat$$

is an algebraic (separable) closure of $L^\flat$. *We have* $[M^\flat : L^\flat] = [M : L]$, M/L *is Galois if and only if* $M^\flat/L^\flat$ *is Galois, and in this case their Galois groups are canonically identified.*

The proof is not difficult, and we refer to the literature. What is not evident, perhaps, is that starting with F, there are many non-isomorphic[2] untilts (L, ι). But this will become evident soon.

We are now able to make our first shot at X^{FF}.

Definition 8 Denote by $|Y_F|$ the set of isomorphism classes of untilts of F, and by

$$|X_F| = |Y_F|/\phi^{\mathbb{Z}}$$

the set of Frobenius equivalence classes of such untilts.

Exercise 9 *Show that* $(L, \iota) \in |Y_F|$ *and* $(L, \phi^n \circ \iota) \in |Y_F|$ *are distinct, unless* $n = 0$.

It turns out that there are *adic spaces* (a p-adic analytic notion not defined in these lecture notes) $\mathscr{Y}_F$ and $\mathscr{X}_F$, of which $|X_F|$ and $|Y_F|$ are the sets of closed points, ϕ acts discretely on $\mathscr{Y}_F$ and $\mathscr{X}_F = \mathscr{Y}_F/\phi^{\mathbb{Z}}$. Moreover, there will be a curve X_F (in the sense of the definition given in the introduction) whose associated adic space is $X_F^{ad} = \mathscr{X}_F$. We can then identify $|X_F|$ with the set of closed points of X_F. However, there does not exist a curve "Y_F". The space $\mathscr{Y}_F$ and the action of ϕ exist only in the analytic category. This situation is vaguely analogous to the construction of the Tate elliptic curve as a quotient, in the analytic category, of the analytic multiplicative group by an infinite cyclic group acting discretely.

Although it is possible to give the definition of X_F as a scheme, independently of the adic space $\mathscr{X}_F$, for the "relative theory", making the same construction over a large base, rather than over the point $Spec(\mathbb{Q}_p)$, it is essential to work in the analytic category. We shall not touch upon the relative theory at all in our lectures.

2.2 *Rings of Functions on* $|Y_F|$

Let F be a perfectoid field in characteristic p. Our goal is to introduce certain "rings of functions" on the set $|Y_F|$ and a topology in which they will be continuous. The Witt vector construction, which we review briefly, becomes an indispensible tool.

2.2.1 Witt Vectors

We recall the main facts about Witt vectors. Fix a prime p and let, for $n \geq 0$

[2] In fact, even the isomorphism type of L, disregarding ι, might be different, but this is more difficult and not needed below.

$$\mathcal{W}_n(x_0, x_1, ...) = x_0^{p^n} + p x_1^{p^{n-1}} + \cdots + p^n x_n.$$

If R is any $\mathbb{Z}_{(p)}$-algebra and $x_i, y_i \in R$ satisfy $x_i \equiv y_i \mod p^s R$ for $s \geq 1$ then $\mathcal{W}_n(x_0, x_1, \ldots) \equiv \mathcal{W}_n(y_0, y_1 \ldots) \mod p^{s+n} R$.

Proposition 10 *There exists a unique affine ring scheme W over $\mathbb{Z}_{(p)}$, whose underlying scheme is $Spec(\mathbb{Z}_{(p)}[X_0, X_1, \ldots])$ such that*

$$\mathcal{W} = (\mathcal{W}_0, \mathcal{W}_1, ...) : W \to \mathbb{A}^{\mathbb{N}}$$

is a ring homomorphism.

The ring structure in $\mathbb{A}^{\mathbb{N}}$ is by component-wise addition and multiplication. What this means is that there are polynomials $S_0(X, Y), S_1(X, Y), ...$ and $P_0(X, Y), P_1(X, Y), ...$ with coefficients in $\mathbb{Z}_{(p)}$ such that

$$\mathcal{W}_n(S_0, S_1, ...) = \mathcal{W}_n(X_0, X_1, ...) + \mathcal{W}_n(Y_0, Y_1, ...)$$

$$\mathcal{W}_n(P_0, P_1, ...) = \mathcal{W}_n(X_0, X_1, ...) \cdot \mathcal{W}_n(Y_0, Y_1, ...).$$

The polynomials S_n and P_n will involve, in fact, only $X_0, \ldots, X_n, Y_0, \ldots, Y_n$. This has the following consequence: if we denote by W_n the truncated Witt vectors of length $n+1$ they also form an affine ring scheme and

$$W = \lim_{\leftarrow} W_n.$$

For the proofs, see [20], II.6. Check that

$$S_0 = X_0 + Y_0, \quad S_1 = X_1 + Y_1 + \frac{1}{p}\left(X_0^p + Y_0^p - (X_0 + Y_0)^p\right)$$

$$P_0 = X_0 Y_0, \quad P_1 = X_1 Y_0^p + X_0^p Y_1 + p X_1 Y_1.$$

The polynomials $\mathcal{W}_n$ are called the *ghost components* of the *Witt vector* $(x_0, x_1, \ldots)$. If $p = 0$ in R then $\mathcal{W}_n(x) = x_0^{p^n}$ and the higher x_i do not show up. Although we are primarily interested in $W(R)$ where R is an $\mathbb{F}_p$-algebra, to *prove* the proposition one must work over $\mathbb{Z}_{(p)}$.

The map $V : W(R) \to W(R)$, $V(x_0, x_1, ...) = (0, x_0, x_1, ...)$ is an additive group homomorphism.

Over an $\mathbb{F}_p$*-algebra* we also have the Frobenius[3] $F : W(R) \to W(R)$, raising the coordinates to power p, which is a *ring* endomorphism (automorphism if R is perfect). We have $F \circ V = V \circ F = p$ (multiplication by p). We shall sometimes denote F also by φ. It should be said that F always exists (even if R is not an $\mathbb{F}_p$-algebra), but it is in general not given by raising the coordinates to pth powers,

[3] This notation is standard, and there should be no confusion with the field F.

and while $F \circ V = p$, instead of $V \circ F = p$ we only get the "projection formula" $V(Fx \cdot y) = x \cdot Vy$.

The construction of Witt vectors $W(R)$ is clearly functorial in R. The previous remark about F is a special case of this functoriality.

Example: $W(\mathbb{F}_p) = \mathbb{Z}_p$, $W(\bar{\mathbb{F}}_p) = \mathbb{Z}_p^{nr}$.

A *strict p-ring* is a ring A which is complete and separated in the p-adic topology, without p-torsion. Recall that the first assumption means that the natural map induces an isomorphism

$$A \simeq \lim_{\leftarrow} A/p^n A.$$

If R is a *perfect* $\mathbb{F}_p$-algebra (i.e., $\phi(x) = x^p$ is bijective) then $W(R)$ is, up to an isomorphism, the unique strict p-ring A with $A/pA = R$. ([20], II, Theorems 5 and 8). In this case we define the Teichmüller representative of $a \in R$ to be

$$[a] = (a, 0, 0, ...) \in W(R).$$

If $\widetilde{a}$ denotes *any* lifting of a to $W(R)$, then

$$[a] = \lim \widetilde{(a^{p^{-n}})}^{p^n}.$$

We clearly have $[ab] = [a][b]$. Any element $x \in W(R)$ has a unique representation

$$x = (x_0, x_1^p, x_2^{p^2}, \ldots) = \sum_{n=0}^{\infty} p^n [x_n].$$

We caution the reader that these results all fail if R is non-perfect. While Witt rings of non-perfect $\mathbb{F}_p$-algebras do appear in p-adic Hodge theory, they are not as well-behaved.

One concludes that the functor

$$W : \{\text{perfect } \mathbb{F}_p \text{ algebras}\} \to \{\text{strict } p \text{ rings}\}$$

is the *left adjoint* of the functor "reduction modulo p".

In the computations below we shall need the following lemma.

Lemma 11 *Let R be a perfect $\mathbb{F}_p$-algebra equipped with a non-Archimedean absolute value . Assume that $|x_n| \le \rho$ and $|y_n| \le \rho$. If*

$$\sum_{n=0}^{\infty} p^n [x_n] + \sum_{n=0}^{\infty} p^n [y_n] = \sum_{n=0}^{\infty} p^n [z_n]$$

then $|z_n| \le \rho$ too. If the assumption holds for $n \le k$ only, the conclusion also holds for $n \le k$.

Proof If we give the variables X_i and Y_i weight p^i then the universal polynomials $S_n(X_0, X_1, ..., Y_0, Y_1, ...)$ describing the addition law in the Witt group scheme become isobaric-homogeneous of weight p^n. This means that S_n is a polynomial, all of whose monomials have weight p^n. Its coefficients lie in $\mathbb{Z}_{(p)}$. Since

$$z_n^{p^n} = S_n(x_0, x_1^p, x_2^{p^2}, ..., y_0, y_1^p, y_2^{p^2}, ...)$$

the lemma follows at once. □

2.2.2 The Ring A_{inf}

Recall that F is a perfectoid field of characteristic p. The ring $\mathcal{O}_F$ is perfect ($\phi(x) = x^p$ is an automorphism) and we define

$$A_{inf} = W(\mathcal{O}_F).$$

As we remarked above, this ring is characterized by the unique[4] p-adically complete and torsion-free ring A with $A/pA \simeq \mathcal{O}_F$. It is local with maximal ideal $pW(\mathcal{O}_F) + W(\mathfrak{m}_F)$, but of course it is not noetherian. An element of A_{inf} has a unique "*Teichmüller expansion*"

$$a = (\alpha_0, \alpha_1^p, \alpha_2^{p^2}, \ldots) = \sum_{n=0}^{\infty} p^n[\alpha_n] \tag{2.1}$$

where

$$[\alpha] = (\alpha, 0, \ldots) = \lim \widetilde{(\alpha^{p^{-j}})}^{p^j}$$

is the Teichmüller representative of α. Here $\widetilde{x}$ denotes any lift of $x \in \mathcal{O}_F$ to $W(\mathcal{O}_F)$.

The *Frobenius* φ is the automorphism

$$\varphi(a) = \sum_{n=0}^{\infty} p^n[\alpha_n^p].$$

Remark 12 Suppose that $F = (\mathbb{Q}_p^{cycl})^\flat$. We have noted that this field is isomorphic to the field of all power series

$$\sum_{-\infty<<m\in\mathbb{Z}[1/p]} a_m \varpi^m$$

[4] up to unique isomorphism

where $\varpi = \varepsilon - 1$, $\varepsilon = (1, \zeta_1, \zeta_2, \dots)$, ζ_i a primitive p^ith root of unity. Here the $a_m \in \mathbb{F}_p$ and for any T, $a_m \neq 0$ for only finitely many $m < T$. The power series in $\mathcal{O}_F$ are those with $a_m = 0$ for $m < 0$. It follows that $W(\mathcal{O}_F)$ is isomorphic to the ring $\mathscr{A}$ of all power series

$$\sum_{0 \leq m \in \mathbb{Z}[1/p]} a_m X^m$$

with $a_m \in \mathbb{Z}_p$ such that for any T there are only finitely many $m < T$ with $v_p(a_m) < T$. To prove this claim, note that this $\mathscr{A}$ is a strict p-ring and that the map $\mathscr{A} \to \mathcal{O}_F$ sending a_m to $a_m \mod p$ and X to ϖ is a surjective homomorphism.

In the earlier days of p-adic Hodge theory, this presentation of $W(\mathcal{O}_F)$ was being used. The power series expansion in p used in (2.1) goes in the opposite direction. If $\mathbb{Q}_p$ were replaced by $k((\pi))$, this is similar to considering $(k[[\varpi]])[[\pi]]$ instead of $(k[[\pi]])[[\varpi]]$. In this purely characteristic p analogue, the symmetry between π and ϖ can be exploited further.

2.2.3 The Rings $B^{b,+} \subset B^b$ and Their Gauss Norms

Elements of A_{inf} will be viewed, in analogy with the ring $\mathbb{Z}_p[[X]]$, as "analytic functions of the variable p with coefficients in $\mathcal{O}_F$, convergent in the punctured unit disk" (we say "punctured" because we will soon want to invert p). This point of view works very well, and allows to use tools such as the Newton polygon, Weierstrass preparation, and Weierstrass division, much as they are used in the classical function theory of $\mathbb{Z}_p[[X]]$. Unlike $\mathbb{Z}_p[[X]]$, arithmetic in A_{inf} involves "carrying" between the coefficients of the p^n's, but thanks to Lemma 11, the analogues of the above-mentioned tools are still valid.

Let ϖ be any element of $\mathcal{O}_F$ with $0 < |\varpi| < 1$. Define

$$B^{b,+} = A_{inf}[\frac{1}{p}], \quad B^b = B^{b,+}[\frac{1}{[\varpi]}].$$

Both are subrings of the very large field $\mathscr{E} = W(F)[1/p] \supset \mathcal{O}_{\mathscr{E}} = W(F)$. (The ring $\mathcal{O}_{\mathscr{E}}$ plays a fundamental role in the theory of (φ, Γ)-modules.) We have

$$B^b = \left\{ x = \sum_{n >> -\infty} p^n [x_n] | \ x_n \in F \text{ bounded} \right\}$$

and $B^{b,+}$ is the subring where $x_n \in \mathcal{O}_F$.

The Frobenius φ extends to an automorphism of these rings.

If $x = \sum p^n [x_n] \in B^b$ and $0 < \rho = p^{-r} \leq 1$ (thus $0 \leq r < \infty$) we let

$$|x|_\rho = \sup_n |x_n| \rho^n, \quad v_r(x) = \inf_n (v(x_n) + nr).$$

You should think of p as a variable and x as an analytic function on the open unit disk, and then $|x|_\rho$ is like the sup norm on the open disk of radius ρ. This analogy will become precise once we interpret elements of B^b as functions on $|Y_F|$, the set of untilts of F.

Proposition 13 *The v_r is a non-Archimedean valuation and $|.|_\rho$ is the associated absolute value.*

Proof It suffices to work in A_{inf} instead of B^b. Let $x \in A_{inf}$. Define $N_k(x) = \sup_{0\le n\le k} |x_n|$. Since $\rho \le 1$,

$$|x|_\rho = \sup_{k\ge 0} N_k(x)\rho^k.$$

If $\alpha \in \mathcal{O}_F$, then $|N_k(x)| \le |\alpha|$ if and only if $x \in [\alpha]A_{inf} + p^{k+1}A_{inf}$. Using this it is easy to see that

$$N_k(x+y) \le \max(N_k(x), N_k(y)),$$

the essential point being that if $\alpha, \beta \in \mathfrak{a} \subset \mathcal{O}_F$ (an ideal) and $[\alpha] + [\beta] = \sum_{n\ge 0} p^n[\gamma_n]$ then all the $\gamma_n \in \mathfrak{a}$. See Lemma 11. The two displayed formulae imply

$$|x+y|_\rho \le \max\{|x|_\rho, |y|_\rho\}.$$

To show $|xy|_\rho = |x|_\rho |y|_\rho$ we may assume (by the continuity of the norm in ρ) that $\rho < 1$. In this case

$$|x|_\rho = \max_n |x_n|\rho^n.$$

If the $|x_n|\rho^n$, $|y_n|\rho^n$ different from 0 are all distinct, then the multiplicativity of $|.|_\rho$ follows from the strong triangle inequality, since in each of the expressions

$$x = \sum_{n\ge 0} p^n[x_n],\quad y = \sum_{m\ge 0} p^m[x_m],\quad xy = \sum_{n,m\ge 0} p^{n+m}[x_n y_m]$$

there will be precisely one term of maximal ρ-norm. For given x, y the set of ρ's for which this happens is dense in $(0, 1)$, so by the continuity of the norm in ρ we get the multiplicativity everywhere. □

Note that v_0 is the "Gauss norm" $\inf_n v(x_n)$. Note also that if $r > 0$ the equivalent valuation $r^{-1}v_r(x) = \inf_n(r^{-1}v(x_n) + n)$ is such that its limit, as $r \to \infty$ is

$$v_\infty(x) = \inf\{n|\, x_n \ne 0\}.$$

This is the valuation inherited from the p-adic valuation of $\mathscr{E}$. The corresponding absolute value is $|x|_0 = p^{-v_\infty(x)}$.

The *weak topology* of A_{inf} is the $(p, [\varpi])$-adic topology. The following proposition is easy and left as an exercise.

Proposition 14 *Let $0 \leq \rho \leq 1$. Then A_{inf} is complete for $|.|_\rho$. The resulting topology on A_{inf} is the p-adic topology if $\rho = 0$, the weak topology if $0 < \rho < 1$, and the $[\varpi]$-adic topology if $\rho = 1$. If $\rho < 1$ then $\mathbb{Z}_p \subset A_{inf}$ inherits the p-adic topology, but if $\rho = 1$ then $\mathbb{Z}_p$ is discrete.*

Fix $x \in B^b$. Since $r \mapsto \nu_r(x)$ is the infimum of the linear functions $\nu(x_n) + nr$ it is *concave* for $0 \leq r < \infty$ (analogue of Hadamard's three circles theorem in complex variables). The *maximum principle* is easily seen to be valid: if $0 < \rho_1 \leq \rho \leq \rho_2 \leq 1$, then

$$|x|_\rho \leq \max\{|x|_{\rho_1}, |x|_{\rho_2}\}.$$

2.2.4 An Interlude: The Legendre Transform

Let $\varphi(x) : \mathbb{R} \to \mathbb{R} \cup \{+\infty\}$ be any function. We define its Legendre transform $\psi = \mathscr{L}(\varphi)$

$$\psi(r) : \mathbb{R} \to \mathbb{R} \cup \{-\infty\}, \quad \psi(r) = \inf_x \{\varphi(x) + xr\}.$$

The inverse Legendre transform is $\varphi = \mathscr{L}^\vee(\psi)$ where

$$\varphi(x) = \sup_r \{\psi(r) - xr\}.$$

These two transforms are "tropical" analogues of the Fourier transform and its inverse. The function $\mathscr{L}(\varphi)$ is always concave, $\mathscr{L}^\vee(\psi)$ is always convex, and if φ were convex to begin with, $\mathscr{L}^\vee(\mathscr{L}(\varphi)) = \varphi$. In general, we obtain the *convex hull* of the function φ. If φ is piecewise linear and convex, then we can talk about its *slopes*. To conform with the conventions used in the theory of Newton polygons, these are the *negatives* of the slopes of the graph of φ. The *break points* of the graph of φ are the points where left and right derivatives aren't the same.

A *left shift* of φ by x_0 results in the *subtraction* of the linear function $x_0 r$ from its Legendre transform. *Adding* a linear function xr_0 to $\varphi(x)$ results in a *left shift* of its Legendre transform. Dual statements hold for the inverse Legendre transform.

If $\varphi(x)$ is convex decreasing and asymptotic to a horizontal line when $x \to \infty$ then its Legendre transform is finite for $r \geq 0$ and $-\infty$ elsewhere.

In general, the Legendre transform of a piecewise linear function interchanges slopes and break points. The slopes (resp. break points) of $\mathscr{L}(\varphi)$ are the break points (resp. slopes) of φ. Note that for a convex (dually, concave) piecewise linear function, the slopes and the break points determine the function up to an additive constant.

Although not necessary for our purposes, it is clear that these notions have analogues where the two copies of $\mathbb{R}$ (with coordinates x, r) are replaced by a finite-dimensional real vector space and its dual.

In the discussion of the previous section, given $f \in B^b$, $f = \sum_{n>>-\infty} p^n[x_n]$, its Newton polygon $\mathcal{N}_f(x)$ is the largest non-increasing convex function lying on or below the points $(n, \nu_F(x_n))$. Since x_n are bounded elements of F, $\mathcal{N}_f$ is asymptotic to the horizontal line at height $\nu_0(f) = \inf \nu_F(x_n)$. Its Legendre transform is the function $r \mapsto v_r(f)$ if $r \geq 0$, $r \mapsto -\infty$ if $r < 0$. (The fact that the Legendre transform lives only in the first quadrant reflects the fact that $\mathcal{N}_f$ was defined to be non-increasing.)

2.2.5 The Rings B_I

Let $I \subset [0, 1]$ be a non-empty interval (closed, open, or half-closed). Pushing the analogy with $\mathbb{Z}_p[[X]]$ we now define rings that represent functions "converging in the annulus of radii in I" (if $0 \in I$ this is a disk).

Definition 15 Let B_I be the completion of B^b in the family of norms $|.|_\rho$, $\rho \in I$.

Here are some properties of these rings.

- $A_{inf} \subset B_I$ is closed (since it is complete in any $|.|_\rho$).
- If $J \subset I$ there is a continuous map $B_I \to B_J$ (it is injective, but this is non-trivial: [5], 1.6.15).
- $B_{[0,1]} = B^b$ (i.e., B^b is already complete in the family of *all* norms).
- If $I = [\rho_1, \rho_2]$ where $0 < \rho_1 \leq \rho_2 \leq 1$ then B_I is a Banach algebra in the norm $||.||_I = \max\{|.|_{\rho_1}, |.|_{\rho_2}\}$. If $\rho_2 = 1$ this norm is trivial on $\mathbb{Z}_p$ (gives 1 to any non-zero element). If $\rho_2 < 1$ it induces on $\mathbb{Z}_p$ the p-adic topology.
- If $I = [\rho_1, \rho_2]$ where $0 < \rho_1 = |a| \leq |b| = \rho_2 < 1$ for some $a, b \in \mathcal{O}_F$ then

$$B_I = \widehat{A_{inf}[\frac{[a]}{p}, \frac{p}{[b]}]}[\frac{1}{p}]$$

  where the completion is w.r.t. the p-adic topology. Note $|p^{-1}[a]|_{\rho_2} \leq |p^{-1}[a]|_{\rho_1} = 1$ and $|p[b]^{-1}|_{\rho_1} \leq |p[b]^{-1}|_{\rho_2} = 1$.
- $B_{\{0\}} = \mathscr{E} = W(F)[\frac{1}{p}]$.
- In general, if $I = \bigcup[\rho_n, \rho_n']$ is an increasing union then letting $I_n = [\rho_n, \rho_n']$, B_I is Fréchet (its topology is defined by a countable family of norms) and in fact is equal to $\lim_{\leftarrow} B_{I_n}$, an inverse limit of Banach algebras.
- If $0 \in I$ but $1 \notin I$ then

$$B_I = \left\{ \sum_{n>>-\infty} p^n[x_n] \in \mathscr{E} | \lim_{n\to\infty} |x_n|\rho^n \to 0 \ \forall \rho \in I\} \right\} \subset B_{\{0\}} = \mathscr{E}.$$

Definition 16 Let $B_F = B_{(0,1)}$. If the reference to F is clear, we shall write simply B for this ring.

Remark 17 Consider formal expressions $\sum_{n=-\infty}^{\infty} p^n[x_n]$ where $x_n \in F$ and for any $0 < \rho < 1$ we have $|x_n|\rho^n \to 0$ when $n \to \pm\infty$. They converge in B_F. Surprisingly, it is not known (?) if any element of B_F admits such a "Laurent expansion" in p, and if distinct Laurent expansions represent distinct elements of B_F. The same question arises with the B_I for any interval I such that $0 \notin I$.

For any ρ let $\varphi(\rho) = \rho^p$. Since $|\varphi(x)|_{\varphi(\rho)} = |x|_\rho^p$, it is immediate that $\varphi : B^b \simeq B^b$ extends to an isomorphism $\varphi : B_I \simeq B_{\varphi(I)}$. In particular, φ is a continuous automorphism of B_F.

The ring

$$\bigoplus_{k=0}^{\infty} B_F^{\varphi=p^k}$$

is a graded ring. Its structure is so far unclear. But we can at least give the scheme-theoretic definition of the Fargues-Fontaine curve

$$X_F = Proj\left(\bigoplus_{k=0}^{\infty} B_F^{\varphi=p^k}\right).$$

At this point this is still a useless definition.

2.2.6 The Robba Ring

Using the rings B_I we can define two of the most important rings of p-adic Hodge theory:

$$\mathscr{E}^\dagger = \lim_{\rightarrow} B_{[0,\rho]} = \left\{ \sum_{n>>-\infty} p^n[x_n] \in \mathscr{E} \,|\, \exists \rho > 0,\ \sup |x_n|\rho^n < \infty \right\} \subset B_{\{0\}} = \mathscr{E}.$$

(the limit over $\rho > 0$). This is the ring of "overconvergent" elements of $\mathscr{E}$.

Proposition 18 *$\mathscr{E}^\dagger$ with the p-adic valuation v_∞ is a henselian discretely valued field, whose completion is $\mathscr{E}$.*

Proof [5], 1.8.2. □

The Robba ring is

$$\mathscr{R} = \lim_{\rightarrow} B_{(0,\rho]}.$$

This is a much larger ring, but nevertheless any unit of $\mathscr{R}$ appears already in $\mathscr{E}^\dagger$.

Proposition 19 *The inclusion $\mathscr{E}^\dagger \subset \mathscr{R}$ induces $(\mathscr{E}^\dagger)^\times \simeq \mathscr{R}^\times$.*

Proof [5] 1.8.6. □

We shall not use these rings, but we remark that they play an important role in the theory of p-adic representations and (φ, Γ)-modules. Here we take (in the simplest example) $F = (\mathbb{Q}_p^{cycl})^\flat$. In addition to the Frobenius φ the perfect field F has a commuting action of $\Gamma = Gal(\mathbb{Q}_p(\mu_{p^\infty})/\mathbb{Q}_p) \simeq \mathbb{Z}_p^\times$. These actions induce actions on $\mathscr{E}$ and $\mathscr{E}^\dagger$. A (φ, Γ)-module is a finite-dimensional vector space M over $\mathscr{E}$, with semi-linear commuting actions of φ and Γ. The matrix of φ in any basis should, furthermore, be invertible. M is "overconvergent" if it comes, via base change, from a similar object $M^\dagger$ over $\mathscr{E}^\dagger$. Overconvergent (φ, Γ)-modules can be base-changed to $\mathscr{R}$. The big advantage of the Robba ring is that one can perform hard p-adic analysis (p-adic differential equations) over it, a tool that is missing over the "formal" $\mathscr{E}$.

Basic theorems of Katz and Fontaine connect p-adic local Galois representations to *étale* (φ, Γ)-modules (albeit over a smaller $\mathscr{E}$...). Here a (φ, Γ)-module M is étale if with respect to a suitable basis the matrix of φ is in $GL_n(\mathcal{O}_\mathscr{E})$. Luckily, by a theorem of Cherbonnier and Colmez, étale (φ, Γ)-modules are overconvergent, so the procedure described above applies to them.

We make one final remark. The reader might have seen, when $F = (\mathbb{Q}_p^{cycl})^\flat$, a description of the Robba ring as a ring of germs of functions converging in some annulus $R < |X| < 1$, when $R \to 1$. Here we look at germs of functions in the punctured disk $0 < |p| < \rho$ (regarding p as a variable), when $\rho \to 0$. This is not a mistake. It is related to the dual points of view discussed in Remark 12.

2.2.7 The Map θ_y

From now on we assume, to simplify things, that *F is algebraically closed*. By Theorem 7, any untilt of F would then be algebraically closed too. Let $y = (C_y, \iota_y) \in |Y_F|$ be an untilt of F. The multiplicative map sending $x \in \mathcal{O}_F$ to $x^\sharp \in \mathcal{O}_{C_y}$ extends to $x \in W(\mathcal{O}_F) = A_{inf}$ as follows:

$$\theta_y(\sum_{n=0}^{\infty} p^n[x_n]) = \sum_{n=0}^{\infty} p^n x_n^\sharp.$$

Lemma 20 *This map is a surjective homomorphism $\theta_y : A_{inf} \to \mathcal{O}_{C_y}$. Its kernel is principal*

$$\ker(\theta_y) = (\xi_y) = (p - [\varpi_y]),$$

where $\varpi_y = \iota_y(p, p^{1/p}, p^{1/p^2}, \dots) = \iota_y(p^\flat)$ ($p^\flat$ is unique only up to multiplication by an element of $T_p\mu$).

Proof Write $C = C_y$. The "ghost component" homomorphism

$$\mathcal{W}_n : W_n(\mathcal{O}_C/p^n) \to \mathcal{O}_C/p^n$$

$$\mathcal{W}_n(a_0, a_1, \dots, a_{n-1}) = a_0^{p^{n-1}} + pa_1^{p^{n-2}} + \dots + p^{n-1}a_{n-1}$$

depends only on $a_i \mod p$, so factors through a homomorphism $\theta_n : W_n(\mathcal{O}_C/p) \to \mathcal{O}_C/p^n$. We have the commutative diagram

$$\begin{array}{ccc} W_{n+1}(\mathcal{O}_C/p) & \stackrel{\theta_{n+1}}{\to} & \mathcal{O}_C/p^{n+1} \\ \downarrow \phi & & \downarrow \\ W_n(\mathcal{O}_C/p) & \stackrel{\theta_n}{\to} & \mathcal{O}_C/p^n \end{array}$$

where $\phi(a_0, \dots, a_n) = (a_0^p, \dots, a_{n-1}^p)$. Taking the inverse limit gives a homomorphism $\theta : W(\mathcal{O}_F) \to \mathcal{O}_C$. Since it sends $[x]$ to $x^\sharp$ it agrees with the map defined above. To check the surjectivity, it is enough to check it modulo p, but then we recover the surjectivity of $x \mapsto x^\sharp$. Finally, the element $p - [\varpi_y]$ is evidently in the kernel. Any element of A_{inf} may be written as $a = [x] + (p - [\varpi_y])b$ ("division with remainder", see Appendix 7). We then have $x^\sharp = \theta(a) = 0$ if and only if $x = 0$, if and only if $a \in (p - [\varpi_y])$. □

2.2.8 Primitive Elements of A_{inf} and the Perfectoid Correspondence for Algebraically Closed Fields

The element $a = \sum p^n[\alpha_n] \in A_{inf}$ is called *primitive of degree n* if $\alpha_0 \neq 0$, $\alpha_n \in \mathcal{O}_F^\times$ and $\alpha_i \in \mathfrak{m}_F$ for $i < n$. For example, "primitive of degree 0" is equivalent to being a unit. An element is primitive of degree 1 if it is of the form

$$[\alpha] - pu,$$

where $0 \neq \alpha \in \mathfrak{m}_F$ and $u \in A_{inf}^\times$. The element $p - [\varpi_y]$ encountered before is such an element.

It turns out that a primitive element of degree n is a product of n primitive elements of degree 1, much as a polynomial over an algebraically closed field factors into a product of linear terms.

We claim that *ideals generated by primitive elements of degree 1* are in a bijection with the untilts $(L, \iota) \in |Y_F|$ of F. We have already seen one direction, starting with an untilt and finding that the kernel of θ_y is such an ideal. In the converse direction, let ξ be a primitive element of degree 1. Let $\mathcal{O}_C = A_{inf}/(\xi)$ and let $\theta : A_{inf} \to \mathcal{O}_C$ be the canonical projection. By "division with remainder", since ξ is primitive of degree 1, any element of A_{inf} is of the form $[x] + \xi b$, so every element of $\mathcal{O}_C$ is of the form $\theta([x])$. One proves that if $\theta([x]) = \theta([y])$ then $v_F(x) = v_F(y)$ and that if we set $v_C(\theta([x])) = v_F(x)$ the ring $\mathcal{O}_C$ becomes a complete valuation ring untilting $\mathcal{O}_F$. Admitting these unchecked details, we have proved the following theorem.

Theorem 21 (Perfectoid correspondence for algebraically closed fields) *The maps described above establish a bijection between ideals in $A_{inf} = W(\mathcal{O}_F)$ generated by primitive elements of degree 1, and untilts $(C, \iota) \in |Y_F|$ of F.*

Corollary 22 *Every primitive element of degree 1 is of the form $u(p - [\varpi])$ for some $\varpi \in \mathfrak{m}_F$ and $u \in A_{inf}^\times$.*

Proof If ξ is a primitive element of degree 1 consider the corresponding untilt $\mathcal{O}_C = A_{inf}/(\xi)$ and the corresponding map θ. We have seen that $\ker(\theta) = (p - [\varpi])$ for a suitable ϖ. It follows that $\xi = u(p - [\varpi])$ for $u \in A_{inf}^{\times}$. □

Remark 23 In Lemma 20 we seemed to have answered the question: when do $(p - [\varpi])$ and $(p - [\varpi'])$ generate the same ideal of A_{inf}, i.e., when do they give the same untilt? Indeed, if the untilt is (C_y, ι_y) then $\varpi_y = \iota_y(p^\flat)$ where $p^\flat = (p, p^{1/p}, \dots) \in C_y^\flat$ is unique precisely up to multiplication by a generator of $T\mu(C_y)$, the Tate module of p-power roots of unity in C_y. This answer is not satisfactory, as it depends on knowing the untilt (C_y, ι_y). One may wonder if, *intrinsically in* F, we can find a criterion telling us when ϖ and ϖ' give rise to the same untilt. This is possible via a different approach to primitive ideals of degree 1, which we do not describe here. See [5], 2.3.2, and Sect. 2.2.12.

2.2.9 The Ring B_F as a Ring of Functions on $|Y_F|$

We keep our assumption that F is an algebraically closed complete valued field in characteristic p. The perfectoid correspondence, which identifies primitive ideals of degree 1 in A_{inf} with untilts of F, has two consequences. It allows us to interpret A_{inf}, B^b, and its completion $B_F = B_{(0,1)}$, as rings of functions on $|Y_F|$. These functions take values in changing fields, but we are accustomed to this already in algebraic geometry, or (say, Berkovich) analytic geometry. It also allows us to put a topology on $|Y_F|$ in which these functions become continuous.

To achieve the first goal, if $y = (C_y, \iota_y) \in |Y_F|$ and $f \in A_{inf}$ define

$$f(y) = \theta_y(f) \in C_y.$$

Since p and $[\varpi]$ map under θ_y to p and $\varpi^\sharp$, which are both non-zero, this extends to $f \in B^b$. We shall write

$$|f(y)| = |f(y)|_{C_y}.$$

If $f \equiv [x] \mod \xi_y$, $x \in \mathcal{O}_F$, then $|f(y)| = |x^\sharp|_{C_y} = |x|_F$. In particular, $|[x](y)| = |x|_F$ is constant on $|Y_F|$. In contrast

$$|p(y)| = |\theta_y(p)|_{C_y} = |\theta_y([\varpi_y])|_{C_y} = |\varpi_y^\sharp|_{C_y} = |\varpi_y|_F$$

can get any value in $(0, 1)$. Thus elements of the form $[x]$ should be regarded as "constants" and p as a "variable". This explains our previous remark that the Teichmüller expansion $\sum_{n>>-\infty} p^n[x_n]$ of an element from B^b should be regarded as a Laurent power series in the variable p with coefficients from $\mathcal{O}_F$.

For $\rho \in (0, 1)$ let $\mathfrak{a}_\rho = \{x = \sum_{n=0}^{\infty} p^n[x_n] \in A_{inf} | \forall n \ |x_n|_F \le \rho\}$. By Lemma 11, this is an ideal. If $I, J \subset A_{inf}$ are two ideals define

$$d(I, J) = \inf\{\rho |\, I + \mathfrak{a}_\rho = J + \mathfrak{a}_\rho\}.$$

If $y_1, y_2 \in |Y_F|$ define

$$d(y_1, y_2) = d((\xi_{y_1}), (\xi_{y_2})).$$

Warning: As we have seen, it may well happen that $\varpi_1 \neq \varpi_2$, but nevertheless $(p - [\varpi_1]) = (p - [\varpi_2])$. However, with our standard notation, it can be easily shown that

$$d(y_1, y_2) = |\xi_{y_1}(y_2)| = |\xi_{y_2}(y_1)|.$$

This is intuitively pleasing: the function ξ_{y_1} is a local parameter at y_1 and the absolute value at y_2 (an element of C_{y_2}) measures the distance of y_2 from y_1.

We also warn the reader that Fargues and Fontaine denote by d the logarithmic distance function, i.e., $-\log_p$ of our $d(-,-)$.

The metric $d(-,-)$ is an ultrametric distance function on $|Y_F|$. Via this distance function $|Y_F|$ resembles the *punctured unit disk*. If we define "distance to the origin"

$$\mathfrak{r}(y) = d((\xi_y), (p)) = |\varpi_y|_F = |p(y)|$$

then $\mathfrak{r} : |Y_F| \to (0, 1)$ is continuous and $\mathfrak{r}^{-1}([\rho, 1))$ is *complete* for any $\rho > 0$ (the proof of completeness is straightforward, see [5], Proposition 2.3.4).

For a fixed y, the function

$$\theta_y : B^b \to C_y$$

is continuous with respect to the Gauss norm $|.|_{\mathfrak{r}(y)}$ on B^b, as can be seen from $(\rho = \mathfrak{r}(y))$

$$|f(y)| = |\sum p^n x_n^\sharp|_{C_y} \leq \sup |p^n x_n^\sharp|_{C_y} = \sup \rho^n |x_n|_F = |f|_\rho,$$

so extends to the completion of B^b with respect to this norm. This way we can consider B_F as a ring of continuous functions on $|Y_F|$. Continuity means that for any $\varepsilon > 0$ and any $f \in B_F$ the set

$$\{y \in |Y_F| \mid |f(y)| < \varepsilon\}$$

is open in $|Y_F|$, where we remember that $|f(y)| = |f(y)|_{C_y}$. More generally, if $I \subset (0, 1)$ is an interval then B_I is a ring of continuous functions on the "annulus" $\mathfrak{r}^{-1}(I)$.

Recall that ϕ acted on $|Y_F|$ and it is straightforward to check that

$$\mathfrak{r}(\phi(y)) = \mathfrak{r}(y)^{1/p}.$$

The action of ϕ is therefore discrete.

At this point it is natural to ask whether $|Y_F|$ is the set of closed points of a certain (Huber, adic, or Berkovich)-analytic space $\mathscr{Y}_F$ (justifying, in retrospect, the cumbersome notation $|Y_F|$ that we have been carrying all along). This is indeed so, and the rings B_I play an important role in defining this analytic structure. To construct $\mathcal{Y}_F$ as an adic space we would have to review adic geometry first. The adic approach is indispensible when one develops the Fargues-Fontaine curve in the relative set-up, but as we do it here, it can be postponed.

2.2.10 Weierstrass Factorization for A_{inf}

Theorem 24 *Let $f \in A_{inf}$ be a primitive element of degree $k \geq 1$. Then there are $\varpi_i \in \mathfrak{m}_F$ $(1 \leq i \leq k)$ and $u \in A_{inf}^{\times}$ such that*

$$f = u(p - [\varpi_1]) \cdots (p - [\varpi_k]).$$

Proof In view of Corollary 22 it is enough to prove that $f = \xi_1 \cdots \xi_k$ with ξ_i primitive of degree 1. We do it by induction on k. The key step is the next lemma, asserting that f has a zero at some $y \in |Y_F|$. Applying division with remainder (see Appendix), $f = [x] + \xi_y g$ and $f(y) = 0$ implies $x = 0$. It is easy to see that g is primitive of degree $k-1$, and we may apply the induction hypothesis. □

Lemma 25 *Let f be as above. Then $f(y) = 0$ for some $y \in |Y_F|$.*

Proof If $f = \sum_{n=0}^{\infty} p^n[x_n] \in A_{inf}$ its Newton polygon $\mathcal{N}_f(x)$ is the largest *non-increasing* convex function $\mathbb{R} \to \mathbb{R} \cup \{+\infty\}$ lying below (or on) the points $(n, v_F(x_n))$. We have $\mathcal{N}_f(x) = +\infty$ for $x < v_\infty(f)$ (the p-adic valuation of f, or the first n with $x_n \neq 0$) and

$$\lim_{x \to \infty} \mathcal{N}_f(x) = v_0(f) = \inf v_F(x_n).$$

The *slopes* of $\mathcal{N}_f$ are by definition the *negatives* of the slopes of the graph. They are non-negative. The definitions extend naturally to $f \in B^b$.

If $f \in A_{inf}$ is primitive of degree k then $v_F(x_0) < \infty$ and $v_F(x_k) = 0$, so $\mathcal{N}_f(x) = 0$ for $x \geq k$ and $\mathcal{N}_f$ has k positive slopes. If f is in the form of the theorem then its positive slopes are the $v_F(\varpi_i)$, $1 \leq i \leq k$ (with multiplicities). This would be well-known if p were replaced by a variable X, but holds true also in A_{inf}, in view of what might be called the "arithmetic of carrying". For example, suppose that $k = 2$, $u = 1$, and $|\varpi_1| \leq |\varpi_2|$. Then

$$f = [\varpi_1 \varpi_2] - ([\varpi_1] + [\varpi_2])p + p^2.$$

Let

$$[\varpi_1] + [\varpi_2] = [\varpi_1 + \varpi_2] + p[s_1] + p^2[s_2] + \cdots$$

where by Lemma 11 $v_F(s_i) \geq \min\{v_F(\varpi_1), v_F(\varpi_2)\} > 0$. It follows that in the expression for f

$$v_F(x_0) = v_F(\varpi_1) + v_F(\varpi_2), \quad v_F(x_1) \geq \min\{v_F(\varpi_1), v_F(\varpi_2)\},$$

with equality if $\nu_F(\varpi_1) > \nu_F(\varpi_2)$, $\nu_F(x_2) = 0$, and of course $\nu_F(x_n) \geq 0$ for $n \geq 3$. Thus the claim follows in this case, and the proof of the general case is similar.

To prove the lemma, let λ be the *smallest* positive slope of $\mathcal{N}_f$. We have to show that f has a zero $y \in |Y_F|$ with $\mathfrak{r}(y) = p^{-\lambda}$ (this will be the largest zero, i.e., "the one farthest from the origin"). In other words, f will be divisible by

$$\xi_y = p - [\varpi_y]$$

with $\nu_F(\varpi_y) = \lambda$. This is done by successive approximations. Start with $z \in \mathcal{O}_F$ with $\nu_F(z) = \lambda$ solving

$$\sum_{n=0}^{k} z^n x_n = 0$$

(which exists since F is algebraically closed). Since λ was the smallest slope of $\mathcal{N}_f$, it is the smallest valuation of a root of the polynomial $\sum_{n=0}^{k} Z^n x_n$, so $\nu_F(x_n) \geq (k-n)\lambda$. It follows that we can write (for $0 \leq n \leq k$)

$$z^n x_n = z^k w_n$$

with $w_n \in \mathcal{O}_F$, and $\sum_{n=0}^{k} w_n = 0$. Letting y_1 correspond to the primitive element $\xi = (p - [z])$ we have $\theta_{y_1}([z]) = p$, so ($\theta = \theta_{y_1}$)

$$f(y_1) = \theta(\sum_{n=0}^{\infty} p^n[x_n]) \equiv \theta(\sum_{n=0}^{k}[z^n x_n]) \equiv p^k \sum_{n=0}^{k} \theta([w_n]) \equiv 0 \mod p^{k+1}\mathcal{O}_{C_y}.$$

The last congruence follows from the fact that $\sum_{n=0}^{k}[w_n] \in pA_{inf}$.

We have already remarked that applying "division with remainder" we can write $f = [a_0] + f_1\xi$. Applying this remark inductively we obtain an expression

$$f = \sum_{n=0}^{\infty}[a_n]\xi^n,$$

expressing f as a power series in ξ instead of p. As before, $a_n \in \mathfrak{m}_F$ for $0 \leq n < k$, and $a_k \in \mathcal{O}_F^\times$. Furthermore, $\theta_{y_1}([a_0]) = f(y_1) \equiv 0 \mod p^{k+1}$, or $\nu_{C_{y_1}}(f(y_1)) = \nu_F(a_0) \geq (k+1)\lambda$.

We now *improve* y_1. Let y_2 be the point corresponding to $\xi - [u]$, where u is the smallest root (i.e., root of *largest* valuation) solving

$$\sum_{n=0}^{k} u^n a_n = 0.$$

Note that since $\nu_F(a_0)$ is the sum of the valuations of the k roots, $\nu_F(u) \geq \nu_F(a_0)/k \geq (k+1)\lambda/k$. This $\nu_F(u)$ gives us the bound

$$d(y_1, y_2) = p^{-\nu_F(u)} \leq p^{-(k+1)\lambda/k},$$

so in passing we see that $\mathfrak{r}(y_2) = \mathfrak{r}(y_1) = p^{-\lambda}$. We now compute $f(y_2)$ and show that it is smaller (in absolute value; it lies in a different field!) than $f(y_1)$. Since u was a root of largest valuation, $\nu_F(a_n) \geq \nu_F(a_0) - n\nu_F(u)$, so there are $b_n \in \mathcal{O}_F$ such that $u^n a_n = b_n a_0$. As before, $\sum_{n=0}^k b_n = 0$. We get ($\theta = \theta_{y_2}$)

$$\theta(f) = \sum_{n=0}^{\infty} \theta([a_n][u^n]) = \theta(\sum_{n=0}^{k}[b_n])\theta([a_0]) + \theta(\sum_{n=k+1}^{\infty}[a_n][u^n]).$$

The first sum has valuation at least $(k+2)\lambda$, $(k+1)\lambda$ coming from a_0 and one extra λ coming from $\sum_{n=0}^k [b_n] \equiv 0 \mod p$. The second term has valuation at least $(k+1)\nu_F(u) \geq (k+1)^2\lambda/k \geq (k+2)\lambda$. Thus $\nu_{C_{y_2}}(f(y_2)) \geq (k+2)\lambda$.

Repeating this procedure we improve y_1 successively to obtain a sequence of points $y_i \in |Y_F|$ such that $f(y_i) \to 0$, and $\{y_i\}$ is a Cauchy sequence. By the completeness of $|Y_F|$ "away from 0" (all the y_i have the same $\mathfrak{r}(y_i) = p^{-\lambda}$), the y_i converge to a zero y of f. For more details, see [5], Theorem 2.4.1. □

Let $I \subset [0, 1)$ be an interval not containing 1. It can be open, closed, or half-open. To extend the Newton polygon from $f \in B^b$ to $f \in B_I$, we must have a version of the Newton polygon that "sees" only the slopes of f that belong to the interval I. Start with $f \in B^b$. Using the Legendre transform Fargues and Fontaine define a "partial" Newton polygon $\mathcal{N}_f^I(x)$ which is "the part of $\mathcal{N}_f$ with slopes in I". Its domain is therefore limited to an interval. If there are no slopes of $\mathcal{N}_f$ in I we write $\mathcal{N}_f^I = \emptyset$. (We say, by abuse of language, that a slope $\lambda > 0$ belongs to I if $p^{-\lambda} \in I$.) By continuity, $\mathcal{N}_f^I$ can be defined now for $f \in B_I$. Note however that if I is not compact and $f \notin B^b$ $\mathcal{N}_f^I$ may have infinitely many positive slopes (accumulating at the missing end-points of I).

Corollary 26 *Let $I \subset [0, 1)$ be a compact interval. If $I = \{\rho\}$ with $\rho = 0$ or $\rho \notin |F|$ then $B_{\{\rho\}}$ is a field. In any other case, B_I is a PID, and its maximal ideals are in bijection with $|Y_I| = \mathfrak{r}^{-1}(I)$.*

Proof (*Sketch*). We have already seen that $B_{\{0\}} = \mathcal{E}$ is a field. Assume that $I = [\rho_1, \rho_2]$ with $0 \leq \rho_1 < \rho_2 < 1$ or $\rho_1 = \rho_2 \in |F^\times|$. Since I is compact, for each non-zero $f \in B_I$ its Newton polygon $\mathcal{N}_f^I$ contains only finitely many slopes λ with $p^{-\lambda} \in I$. If this set of slopes (with multiplicities), call it $Slopes_I(f)$, is empty, one proves that $f \in B_I^\times$. The proof of this fact is essentially *identical* to the proof of the statement that a power series in $\mathbb{Z}_p[[X]]$ not having λ as a slope is invertible on the annulus of radius $p^{-\lambda}$. See [5], 1.6.25. On the other hand, any time a slope $\lambda \in Slopes_I(f)$ occurs, we can, by the previous theorem (extended from A_{inf} to B_I—this needs justification, and of course relies on $p^{-\lambda} \in I$), find a factor $\xi_y =$

$p - [\varpi_y]$ of f , $\mathfrak{r}(y) = p^{-\lambda}$, the quotient by which has one less I-slope. Here the assumption that $\rho_2 < 1$ is being used to apply a change of variable $p \mapsto p[z]$ for some $z \in \mathfrak{m}_F$ (multiplying the radii by $|z|_F$), after which we may assume that f is primitive (extracting a certain $p^n[\varpi]$ from all the coefficients).

It follows that any element of B_I can be written as $u\xi_1 \ldots \xi_n$, with u a unit, and the ξ_i primitive of degree 1, hence irreducible (since they generate a maximal ideal in B_I). This factorization is unique up to order and multiplication by a unit. The corollary follows from the well-known fact that a unique factorization domain is a PID if and only if the ideals (ξ) generated by its irreducible elements are maximal.

The remaining case where $I = \{\rho\}$ but $\rho \notin |F|$ is treated similarly, noting that in this case $Slopes_I(f) = \emptyset$, so every non-zero f is invertible and $B_{\{\rho\}}$ is a field. $\square$

If $J \subset I \subset (0, 1)$ are two compact intervals the map $B_I \to B_J$ is a "localization and completion". The irreducibles ξ_y for $y \in I - J$ are inverted. Since $B = B_F$ is the projective limit of the B_I where $I \subset (0, 1)$ is a compact interval, we deduce that the maximal ideals of B are in bijection with $|Y_F|$.

The ring B is not principal, nor is B_I if I is not compact. However, similar to the classical (complex or p-adic) situation, a function $0 \neq f \in B$ admits an infinite "Weierstrass product" decomposition "near 0". See [5], Theorem 2.6.1.

Passing from maximal ideals to arbitrary closed ideals allows to identify them with $Div^+(Y_F)$, the monoid of formal expressions $\sum_{y\in|Y_F|} n_y[y]$, $n_y \geq 0$, where for any compact $I \subset (0, 1)$, only finitely many $y \in |Y_I|$ have $n_y \neq 0$. This follows formally from the fact that if $\mathfrak{a}$ is a closed ideal in $B - \lim_{\leftarrow} B_I$ (inverse limit over compact intervals in $(0, 1)$) then

$$\mathfrak{a} = \lim_{\leftarrow} \mathfrak{a}B_I,$$

while

$$Div^+(Y_F) = \lim_{\leftarrow} Div^+(Y_I).$$

Corollary 27 *The correspondence*

$$D = \sum_{y\in|Y_F|} n_y[y] \mapsto \mathfrak{a}_D = \{f \in B |\, ord_y(f) \geq n_y \ \forall y \in |Y_F|\}$$

is an isomorphism of monoids between $Div^+(Y_F)$ and the monoid of closed ideals of B.

There is an obvious notion of a divisor $div(f)$ associated to $f \in B$ (or B_I in general), and $f|g$ if and only if $div(g) \geq div(f)$. The closed ideal $\mathfrak{a}_D$ is principal if and only if D is a principal divisor.

The nature of the rings B_I for non-compact I is not easily determined. If $I = (0, \rho]$ with $\rho < 1$ then it can be proved that B_I is Bézout, and the principal ideals, the finitely generated ideals, and the closed ideals are all the same. Equivalently, every $D \in Div^+(Y_I)$ is principal, $D = div(f)$ for some $f \in B_I$. As a result, the Robba ring

$$\mathscr{R} = \lim_{\to} B_{(0,\rho]}$$

is also Bézout.

To summarize what we have seen so far, $|Y_F|$ is exhausted by an increasing union of the $|Y_I|$, for compact intervals I. These indeed look like curves–$|Y_I|$ is identified with the maximal spectrum of a PID B_I.

2.2.11 Dividing by the Action of Frobenius

It is now time to divide $|Y| = |Y_F|$ by the action of ϕ. As usual, the ring isomorphism $\varphi : B_I \simeq B_{\phi(I)}$ induces a map in the opposite direction on maximal spectra $\phi = \varphi^* : |Y_{\phi(I)}| \to |Y_I|$, so if $y \leftrightarrow (\xi_y) = (p - [\varpi_y])$ we have $\varpi_{\phi(y)} = \varpi_y^{1/p}$ and $\mathfrak{r}(\phi(y)) = \mathfrak{r}(y)^{1/p}$. Similarly, $d(\phi(y_1), \phi(y_2)) = d(y_1, y_2)^{1/p}$. As already noted, the action of ϕ is discrete, and we define a topological space

$$|X| = |X_F| = |Y|/\phi^{\mathbb{Z}}.$$

We identify

$$Div(X) = Div(Y)^{\phi=1}$$

and similarly for the monoid of effective divisors $Div^+(X)$.

There is a homomorphism

$$deg : Div(X) \to \mathbb{Z}$$

taking $\sum n_y[y]$ to $\sum_{y \bmod \phi} n_y$.

The ring $B^{\varphi=1}$ is too small; in fact, it will be shown to coincide with $\mathbb{Q}_p$. However, if $f \in B^{\varphi=p^k}$ then $div(f)$ is ϕ-invariant and effective (these functions should be regarded as analogues of theta functions in the classical theory of elliptic curves).

Denote by $\coprod_{k\geq 0}(B_F - \{0\})^{\varphi=p^k}$ the monoid which is the disjoint union of $(B_F - \{0\})^{\varphi=p^k}$, with multiplication as the monoid operation.

Theorem 28 *([5], Theorem 6.2.1.) The homomorphism of monoids*

$$div : \left(\coprod_{k\geq 0}(B_F - \{0\})^{\varphi=p^k}\right)/\mathbb{Q}_p^\times \to Div^+(X_F)$$

is an isomorphism. It respects degrees: if $\varphi h = p^k h$ then $deg(div(h)) = k$. In particular, $B_F^{\varphi=1} = \mathbb{Q}_p$.

Proof *(sketch)* Before we prove the theorem, let us make some remarks on Newton polygons. Let $h \in B$. Its Newton polygon, $\mathcal{N}_h^{(0,1)}$ will be denoted for simplicity by

$\mathcal{N}_h$. If $h \in B^b$, this is the Newton polygon of h, with the slope 0 part removed (if slope 0 occurs, corresponding to $\rho = 1$). In general, for I a compact sub-interval of $(0, 1)$ and $h \in B_I$ the Newton polygon $\mathcal{N}_h^I$ is the limit of $\mathcal{N}_{h_n}^I$ for a sequence h_n converging to h in the Banach norm of B_I. These Newton polygons are the same for $n >> 0$, so $\mathcal{N}_h$ has finitely many slopes. For I non-compact, like $(0, 1)$, $\mathcal{N}_h^I$ is the union of $\mathcal{N}_h^J$ for $J \subset I$ a compact sub-interval, and may have infinitely many slopes. Going back to $h \in B$ and $\mathcal{N}_h = \mathcal{N}_h^{(0,1)}$, it follows from the definitions that it is defined for all x sufficiently small and that $\lim_{x \to -\infty} \mathcal{N}_h(x) = +\infty$. For $f \notin B^b$, the Newton polygon may or may not attain the value $+\infty$ for some x.

Finally, $\mathcal{N}_{\varphi(h)}(x) = p\mathcal{N}_h(x)$ and $\mathcal{N}_{p^m h}(x) = \mathcal{N}_h(x - m)$ (shift to the left if $m < 0$). This is readily proved for $h \in B^b$ and extends by continuity to $h \in B$.

Injectivity: Suppose two non-zero functions $f \in B^{\varphi=p^k}$ and $g \in B^{\varphi=p^\ell}$ with $k \geq \ell$ have the same divisor. Then $h = gf^{-1} \in B_I^\times$ for every compact I (B_I being a PID). It therefore belongs to $B = \lim_\leftarrow B_I$. It even belongs to $B^{\varphi=p^m}$ where $m = \ell - k$. It suffices to show that $B^{\varphi=p^m} = 0$ if $m < 0$ and $\mathbb{Q}_p$ if $m = 0$. This is again a Newton polygon argument.

If $m < 0$ and $h \neq 0$ $p\mathcal{N}_h(x) = \mathcal{N}_h(x - m)$ implies that $\mathcal{N}_h$ must be increasing, while by definition it is non-increasing, a contradiction (note that it cannot get only the two values $0, +\infty$ in this case). If $m = 0$ $\mathcal{N}_h(x)$ can get only the values 0 or ∞ and for $x << 0$ $\mathcal{N}_h(x) = +\infty$. This implies ([5], Proposition 1.9.1, the proof seems to contain a gap, but can be fixed) that h lies in $B_{[0,1)} \subset B_{(0,1)} = B$. But $B_{[0,1)} \subset B_{\{0\}} = \mathscr{E} = W(F)[1/p]$ and

$$W(F)[1/p]^{\varphi=1} = W(\mathbb{F}_p)[1/p] = \mathbb{Q}_p.$$

Surjectivity: It is enough to show that if $y \in |Y|$ then there exists a function $t_y \in B^{\varphi=p}$ with

$$div(t_y) = \sum_{n \in \mathbb{Z}} [\phi^n(y)].$$

Let $\xi_y = p - [\varpi_y]$ be a primitive element of degree 1 corresponding to y. Consider the product

$$t_y^+ = \prod_{n=0}^{\infty} \varphi^n(\xi_y/p) = \prod_{n=0}^{\infty} (1 - [\varpi^{p^n}]/p).$$

The product converges in B and satisfies $\xi_y \varphi(t_y^+) = pt_y^+$, and $div(t_y^+) = \sum_{n \leq 0} [\phi^n(y)]$. On the other hand, the equation

$$\varphi(T) = \xi_y T$$

can be shown to have a solution $t_y^- \in A_{inf}$. This is easy and done by successive approximations modulo p^n, using the fact that F, being algebraically closed, admits $p - 1$ roots and solutions of Artin-Schreier equations, see [5], Proposition 6.2.10. Its divisor satisfies

$$\phi^{-1}(div(t_y^-)) = [y] + div(t_y^-),$$

so $div(t_y^-) = \sum_{n>0}[\phi^n(y)]$, and is supported on $Y_{[\rho^{1/p},1)}$ where $\rho = \mathfrak{r}(y)$.

It follows that $t_y = t_y^+ t_y^-$ is the desired element. Note that this element cannot belong to B^b as its Newton polygon satisfies $\mathcal{N}_{t_y}(x+1) = \mathcal{N}_{t_y}(x)/p$, so it has infinitely many slopes (in both directions). □

Remark 29 The space $B^{\varphi^h = p^d}$ is a $\mathbb{Q}_p$-vector space and is in fact a closed Banach subspace of $B_{[\rho^p,\rho]}$ if we choose any $\rho \in (0,1)$. This is because if a sequence of functions $f_n \in B^{\varphi^h = p^d}$ converges in $B_{[\rho^p,\rho]}$ one may use the action of Frobenius to show that they converge in any $B_{[\rho^{p^N},\rho^{p^{-N}}]}$ and the limit function clearly also satisfies $\varphi^h(f) = p^d h$. When $N \to \infty$ these compact intervals exhaust $(0,1)$. While the norm induced on this space depends on the choice of ρ, the Banach topology does not. We shall always consider it with this Banach space topology.

2.2.12 An Alternative Parametrization of $|Y|$ via the Multiplicative Formal Group

Let $\varepsilon \in \mathfrak{m}_F - \{0\}$ and

$$u_\varepsilon = \frac{[1+\varepsilon]-1}{[1+\varepsilon^{1/p}]-1} = \sum_{i=0}^{p-1}[1+\varepsilon^{1/p}]^i \in A_{inf}.$$

Modulo p, $u_\varepsilon \equiv \varepsilon^{(p-1)/p} \neq 0$, and under $W(\mathcal{O}_F) \to W(\kappa_F)$ $(\kappa_F = \mathcal{O}_F/\mathfrak{m}_F)$ the second expression shows that it maps to p. Thus u_ε is a primitive element of degree 1. Let y_ε be the point of $|Y|$ corresponding to $(u_\varepsilon) \subset A_{inf}$. If $a \in \mathbb{Z}_p$ we can look at[5]

$$\{a\}(\varepsilon) = (1+\varepsilon)^a - 1 = \sum_{n=1}^{\infty}\binom{a}{n}\varepsilon^n \in \mathfrak{m}_F$$

and then

$$u_{\{a\}(\varepsilon)} = \frac{[(1+\varepsilon)^a]-1}{[(1+\varepsilon^{1/p})^a]-1}.$$

If $a \in \mathbb{Z}_p^\times$ then $[(1+\varepsilon)^a]-1$ and $[1+\varepsilon]-1$ divide each other in A_{inf} so $u_{\{a\}(\varepsilon)}$ and u_ε differ by a unit, and $y_\varepsilon = y_{\{a\}(\varepsilon)}$. If $a = p$ the two points differ by Frobenius ϕ. Since F is perfect the notation $\{a\}(\varepsilon)$ can be extended to $a \in \mathbb{Q}_p$.

Proposition 30 *The association $\varepsilon \mapsto y_\varepsilon$ gives bijections*

$$(\mathfrak{m}_F - \{0\})/\mathbb{Z}_p^\times \simeq |Y_F|, \quad (\mathfrak{m}_F - \{0\})/\mathbb{Q}_p^\times \simeq |X_F|.$$

---

[5] Endomorphisms of formal groups are usually denoted by $[a]$. In order not to conflict with the notation for Teichmüller representatives, we use $\{a\}$.

Proof It is enough to prove the first claim. To prove injectivity, let $C = C_{y_\varepsilon}$, and suppose that $y = y_\varepsilon = y_{\varepsilon'}$. First note that since $\theta_y(u_\varepsilon) = 0$ we must have $(1+\varepsilon)^\sharp = 1$ in C, so if $1 + \varepsilon = \zeta \in F = C^\flat$ then

$$\zeta = (1, \zeta_1, \zeta_2, ...)$$

where ζ_i is a p^i root of unity in C. In fact, $(1+\varepsilon^{1/p})^\sharp \neq 1$, or else $\theta_y(u_\varepsilon) = p$, so ζ_i should be a *primitive* p^i root of unity. The same holds for ε', so $\zeta' = \zeta^a$ for some $a \in \mathbb{Z}_p^\times$ and $\varepsilon' = \{a\}(\varepsilon)$.

To prove surjectivity, given $y \in |Y_F|$ let $C = C_y$ and choose $\varepsilon \in C^\flat = F$ in such a way that $\zeta = 1 + \varepsilon$ is a basis of the Tate module of $T\mu(C)$. Then working the above arguments backward we see that u_ε is a primitive element of degree 1, and is in the kernel of θ_y, so must generate it, and $y = y_\varepsilon$. □

Corollary 31 *Up to a $\mathbb{Q}_p$-multiple, the element t_y constructed before by means of Weierstrass products is also given by*

$$t_y = \log([1+\varepsilon]) = \sum_{n=1}^{\infty} (-1)^{n-1} \frac{([1+\varepsilon]-1)^n}{n}$$

for ε such that $y = y_\varepsilon$. (We denote this element also by t_ε.)

Proof It is easy to check that the power series converges in B_F, so it represents an element there (that does not belong, of course, to B^b!). Clearly $\varphi(t_y) \in B^{\varphi=p}$, but the elements in this space vanishing at y form a one-dimensional space over $\mathbb{Q}_p$, as we have seen. □

Remark 32 We shall see later that the spaces $B^{\varphi=1}$ and $B^{\varphi=p}$ are related to periods of the p-divisible groups $\mathbb{Q}_p/\mathbb{Z}_p$ and μ_{p^∞}. More generally, the "Banach-Colmez" space $B^{\varphi^h=p^d}$ for $0 \le d \le h$ relatively prime will be related to the periods of p-divisible groups of dimension d and height h. Here we are only considering, for the construction of X_F, the case $h = 1$, so the relation to p-divisible groups occurs only for $d = 0, 1$. For the study of vector bundles over X_F we shall have to consider all values of h.

Banach-Colmez spaces is one of the topics completely absent from these notes, for reasons of space. They are intimately woven into the fabric; however, the reader can learn more about them from the original book by Fargues and Fontaine, or from Colmez' original paper.

2.3 The Schematic Fargues-Fontaine Curve and Its Main Properties

2.3.1 The Definition of the Curve

Having studied the ring $B = B_{(0,1)}$ and its completions B_I for $I \subset (0, 1)$ a compact interval (PIDs !), we can finally define the curve X and get its main properties.

Let

$$P = \bigoplus_{k=0}^{\infty} B^{\varphi=p^k}.$$

This is a graded ring and we let

$$X_F = Proj(P).$$

A graded ring $P = \bigoplus_{k=0}^{\infty} P_k$ is called *graded factorial with irreducible elements in degree 1* if P_0 is a field E and the multiplicative monoid

$$\coprod_{k=0}^{\infty} (P_k - \{0\})/E^{\times}$$

is free and generated by $(P_1 - \{0\})/E^{\times}$. Theorem 28 yields the following:

Corollary 33 *The ring P is graded factorial with irreducible elements in degree 1, and $P_0 = \mathbb{Q}_p$.*

2.3.2 The Fundamental Exact Sequence

Besides the last corollary, there is another important ingredient needed to prove that X_F is a "curve". It is an exact sequence which yields the "fundamental exact sequence of p-adic Hodge theory". In this sub-section we explain what it is and how to derive it.

Lemma 34 *Let $y \in |Y|$. Let $\xi_y \in A_{inf}$ be a corresponding primitive element of degree 1, and $\mathfrak{m}_y = \xi_y B = \ker(\theta_y)$ the corresponding maximal ideal of B. The homomorphism $\theta_y : B \to B/\mathfrak{m}_y = C_y$ is surjective when restricted to $B^{\varphi=p}$.*

Proof We have seen that every element $h \in B^{\varphi=p}$ is associated with $div(h) = \sum_{n\in\mathbb{Z}}[\phi^n y']$ for some y', and that $div(h)$ determines h up to a $\mathbb{Q}_p^{\times}$-multiple. Since every element of $|Y|$ is of the form y_ε, every element of $B^{\varphi=p}$ is $t_\varepsilon = \log([1+\varepsilon])$ for a unique $\varepsilon \in \mathfrak{m}_F$. Replacing ε by $\{a\}(\varepsilon)$ for $a \in \mathbb{Q}_p^{\times}$ (the freedom allowed keeping the Frobenius orbit of y unchanged) results in multiplying t_ε by a. But $\theta_y(t_\varepsilon) = \log((1+\varepsilon)^\sharp)$ and $(1+\varepsilon)^\sharp$ can be an arbitrary element of $1 + \mathfrak{m}_{C_y}$. The lemma follows from the fact that $\log : 1 + \mathfrak{m}_C \to C$ is surjective. Recall that by the convergence of exp on $p\mathcal{O}_C$ ($4\mathcal{O}_C$ if $p = 2$) its image contains a neighborhood of 0. But since $1 + \mathfrak{m}_C$ is p-divisible, so is the image of log; hence it is all of C. □

Proposition 35 *The sequence*

$$0 \to \mathbb{Q}_p t_y \to B^{\varphi=p} \xrightarrow{\theta_y} B/\mathfrak{m}_y = C_y \to 0$$

is an exact sequence of $\mathbb{Q}_p$-Banach spaces.

Proof We have checked that θ_y is onto, and that $\ker(\theta_y) = \mathbb{Q}_p t_y$. □

Theorem 36 *Let $y_1, \ldots, y_r$ be pairwise ϕ-inequivalent and $e_i \geq 1$. Let $t_i = t_{y_i}$. Let $d = \sum e_i$. Then the sequence*

$$0 \to \mathbb{Q}_p t_1^{e_1} \cdots t_r^{e_r} \to B^{\varphi=p^d} \to B/\mathfrak{m}_{y_1}^{e_1} \cdots \mathfrak{m}_{y_r}^{e_r} \to 0$$

is exact.

Proof The sequence is clearly a 0-sequence. If $h \in B^{\varphi=p^d}$ maps to 0 then $div(h) \geq \sum e_i[y_i]$. Since it is ϕ-invariant and the y_i are Frobenius inequivalent,

$$div(h) \geq div(t_1^{e_1} \cdots t_r^{e_r}).$$

Since both divisors are of degree d, they are equal. This means that $h \in \mathbb{Q}_p t_1^{e_1} \cdots t_r^{e_r}$. The surjectivity of the map from $B^{\varphi=p^d}$ is proved by induction on d. The case $d = 1$ was already done. The map $B^{\varphi=p^d} \to B/\mathfrak{m}_{y_1} = C_{y_1}$ is surjective since we can multiply any element from $B^{\varphi=p}$ by $t_{y'}^{d-1}$ where $y' \neq y_1$ and $\theta_{y_1}(t_{y'}) \neq 0$. It remains therefore to show that we can get any element in $\mathfrak{m}_{y_1}/\mathfrak{m}_{y_1}^{e_1} \cdots \mathfrak{m}_{y_r}^{e_r}$. By induction we can get any element in $B/\mathfrak{m}_{y_1}^{e_1-1} \cdots \mathfrak{m}_{y_r}^{e_r}$ as the image of an element from $B^{\varphi=p^{d-1}}$. Multiplying by t_{y_1} gives the desired result. □

Remark 37 The B-module $B/\mathfrak{m}_y^2$ has a non-split filtration

$$0 \to \mathfrak{m}_y/\mathfrak{m}_y^2 \to B/\mathfrak{m}_y^2 \to B/\mathfrak{m}_y \to 0$$

where the two terms at the extremes are isomorphic to C_y. This suffices to show that multiplication by $t_{y'}$ for $y' \notin \phi^{\mathbb{Z}} y$ is bijective on $B/\mathfrak{m}_y^2$. Similar claims hold for any $B/\mathfrak{m}_{y_1}^{e_1} \cdots \mathfrak{m}_{y_r}^{e_r}$.

Corollary 38 *Let $0 \neq t \in P_1 = B^{\varphi=p}$ and choose $y \in |Y|$ so that $div(t) = \sum_{n \in \mathbb{Z}} [\phi^n y]$. Then there is a canonical isomorphism of graded algebras*

$$P/tP = P_0 \oplus \bigoplus_{k=1}^{\infty} P_k/tP_{k-1} \simeq \{f \in C_y[T] \mid f(0) \in \mathbb{Q}_p\}.$$

Proof Send $a \in P_k/tP_{k-1}$ to $\theta_y(a)T^k$ (here $P_{-1} = 0$). □

2.3.3 The Ring $B^+_{dR,y}$ and the Field $B_{dR,y}$

Let $y \in |Y|$. We denote the completion of B^b at the maximal ideal $\mathfrak{m}_y = (\xi_y) = \ker(\theta_y)$ by $B^+_{dR,y}$. Thus

$$B^+_{dR,y} = \lim_{\leftarrow} B^b/(\xi_y^n).$$

By abuse of language we denote by $\mathfrak{m}_y$ the kernel of θ_y in *any* of the rings B^b, B_I, and B.

Proposition 39 *$B^+_{dR,y}$ is a complete DVR, and ξ_y is a uniformizer. The homomorphisms*

$$B^b \hookrightarrow B \twoheadrightarrow B_I$$

(I a compact interval in $(0, 1)$ *containing* $\mathfrak{r}(y)$*) induce isomorphisms on the completions of their localizations at y.*

Proof It is enough to prove the second claim because B_I is a PID and ξ_y is a generator of $\mathfrak{m}_y \subset B_I$. Recall that B and B_I were obtained as certain topological completions of B^b in the family of norms $|.|_\rho$, while $B^+_{dR,y}$ is the formal completion at $\mathfrak{m}_y$. The proposition follows from the fact that the kernel of θ_y in B or B_I is still principal and generated by ξ_y, in itself proven via approximations, and from the identity

$$B_I/\xi_y B_I = B/\xi_y B = B^b/\xi_y B^b = C_y.$$

The proof should be compared to the proof that the formal completion of the ring of germs of holomorphic functions at 0 is the same as the formal completion of $\mathbb{C}[z]$ at 0. In this classical example one argues with Taylor expansions. In our case, such expansions do not exist, as $B^+_{dR,y}$ is not a vector space over C_y. Instead, one has to argue with filtrations and graded objects. Except for this, the proof is the same. □

Viewing $B^+_{dR,y}$ as a completion of the localization of B at y we can also consider (the image of) t_y as a uniformizer. We denote by $B_{dR,y}$ the field of fractions of $B^+_{dR,y}$, i.e., $B^+_{dR,y}[1/t_y]$.

Remark 40 We have started from an arbitrary perfectoid field F in characteristic p and looked at all its untilts, parametrized by $|Y_F|$. Had we started with $\mathbb{C}_p$ and tilted it to get $F := \mathbb{C}_p^b$, we would have a distinguished untilt (namely $\mathbb{C}_p$ with the canonical identification of $\mathbb{C}_p^b$ with F), i.e., a point $\infty \in |Y_F|$ and a corresponding point in $|X_F|$. Originally, Fontaine called the resulting $B_{dR,\infty}$ simply B_{dR}. It was only later realized that such a field exists for every untilt y.

We make another remark about notation.

Remark 41 The use of the superscript + might be occasionally confusing. In the context of B_{dR} it is used to denote the valuation ring before t is inverted. However, in the context of the ring B^b it was used to denote an intermediate ring

$$A_{inf} \subset B^{b,+} \subset B^b = A_{inf}[1/p, 1/[\varpi]]$$

in which p was inverted, but $[\varpi]$ not yet. The convergent power series

$$\frac{1}{[\varpi]} = \frac{1}{p}\sum_{n=0}^{\infty}(\frac{\xi}{p})^n$$

($\xi = p - [\varpi]$) shows that

$$B^{b,+}/(\xi^n) = B^b/(\xi^n),$$

hence $B_{dR}^+ = \lim_{\leftarrow} B^{b,+}/(\xi^n)$ as well. But beware: B_{dR}^+ contains B^b and B, not only $B^{b,+}$.

There is also the ring $B^+ \subset B$, which is the closure of $B^{b,+}$ in B. It can be shown that it coincides with the $f \in B = B_{(0,1)}$ for which $\mathcal{N}_f(x) \geq 0$ for all x. Note that in passing from A_{inf} to $B^b = A_{inf}(\frac{1}{p}, \frac{1}{[\varpi]})$ division by p moves Newton polygons horizontally to the left, invading the second quadrant, while further division by $[\varpi]$ moves them vertically downwards, thus invading the third and fourth quadrants. Using this characterization it is easy to see that for $d, h \geq 0$

$$B^{\varphi^h = p^d} = (B^+)^{\varphi^h = p^d}.$$

We shall later discuss also divided power completions of A_{inf} and the ring

$$B_{cris}^+ = A_{inf}[\widehat{\xi^n/n!;\ n \geq 1}][1/p]$$

(the big hat signifying p-adic completion). It will then be true also that

$$B^{+\varphi^h = p^d} = B_{cris}^{+\varphi^h = p^d}.$$

The fundamental exact sequence (associated with the point ∞) will take the more familiar form

$$0 \to \mathbb{Q}_p t^d \to (B_{cris}^+)^{\varphi = p^d} \to B_{dR}^+/t^d B_{dR}^+ \to 0.$$

2.3.4 The Main Theorem ([5], Théorème 6.5.2)

Theorem 42 *(i) The scheme X is a complete curve, whose field of definition $H^0(X, \mathcal{O}_X) = \mathbb{Q}_p$. All the closed points have degree 1.*

(ii) For $t \in P_1 = B^{\varphi = p}$ the locus $V^+(t) \subset Proj(P) = X$ consists of a single point ∞_t. The residue field C_t at ∞_t is a complete valued field, algebraically closed, whose tilt is canonically identified with F.

(iii) The map $\mathbb{Q}_p^\times t \mapsto \infty_t$ is a bijection $(P_1 - \{0\})/\mathbb{Q}_p^\times \leftrightarrow |X|$.

(iv) The ring $\mathcal{O}_{X,\infty_t}$ *is a DVR, whose completion is canonically identified with* $B^+_{dR,y}$ *where* $y \in |Y|$ *is any point such that* $div(t) = \sum_{n\in\mathbb{Z}} \phi^n([y])$.

(v) Let $B_e = B[1/t]^{\varphi=1} = B^+[1/t]^{\varphi=1}$. *Then* B_e *is a PID and* $Spec(B_e) = D^+(t) = X - \{\infty_t\}$. *Furthermore, if* ord_{∞_t} *is the valuation on* $Frac(B_e) \subset B_{dR,\infty_t}$, *then the couple* $(B_e, -ord_{\infty_t})$ *is an almost Euclidean domain.*

(vi) The degree homomorphism induces $Pic(X) \simeq \mathbb{Z}$.

(vii) We have $H^1(X, \mathcal{O}_X) = 0$.

The proof of the theorem rests on a general construction of complete curves. The next section is motivated by the application we have in mind, but is set in a general axiomatic framework.

2.4 Construction of Curves ([5], Chap. 5)

2.4.1 Curves

We recall the definition from the introduction. A separated noetherian scheme X is called a *curve* if it is regular, one-dimensional, and connected. We denote by η its generic point, by $E(X) = \mathcal{O}_{X,\eta}$ its function field, and by $ord_x : E(X) \to \mathbb{Z} \cup \{\infty\}$ the normalized valuation associated with a closed point $x \in |X|$.

Let $deg(x) \in \mathbb{N}$ be given for every $x \in |X|$. For a divisor $D \in Div(X)$ we define $deg(D)$ as usual. The group $Div(X)$ is identified with the group of Cartier divisors, hence with the group of pairs $(\mathcal{L}, s)$ where $\mathcal{L}$ is a line bundle and s is a rational section. The sequence

$$0 \to \Gamma(X, \mathcal{O}_X)^\times \to E(X)^\times \stackrel{div}{\to} Div(X) \to Pic(X) \to 0$$

is exact.

A curve X, equipped with a degree function, is called *complete*, if

$$deg(div(f)) = \sum ord_x(f)deg(x) = 0$$

for any $f \in E(X)^\times$. If X is complete and $0 \neq f \in \Gamma(X, \mathcal{O}_X)$ then $div(f) \geq 0$, but since $deg(div(f)) = 0$, $div(f) = 0$ and f is invertible. Thus $E = \Gamma(X, \mathcal{O}_X)$ is a field, called the *field of definition* of X.

2.4.2 Almost Euclidean Rings

A ring B equipped with a degree function $deg : B \to \mathbb{N} \cup \{-\infty\}$ satisfying

(1) $deg(a) = -\infty$ iff $a = 0$, $deg(1) = 0$

(2) $deg(a) \leq deg(ab)$ for $b \neq 0$
is called *almost Euclidean* if in addition
(3) $deg(a) = 0$ iff $a \in B^\times$
(4) For any $x, y \neq 0$ there exist a, b with $x = ay + b$ and $deg(b) \leq deg(y)$.

Note that (B, deg) is Euclidean if in the last property we have a strict inequality.

Suppose B is an integral domain with fraction field K. Let $ord_\infty : K \to \mathbb{Z} \cup \{\infty\}$ be a normalized discrete valuation with valuation ring A. Suppose that

- $ord_\infty(b) \leq 0$ for every $0 \neq b \in B$ and $ord_\infty(b) = 0$ iff $b \in B^\times$ (i.e., $A \cap B = B^\times \cup \{0\}$).

Note that in this case $E = A \cap B$ is a field, and the function $deg = -ord_\infty : B \to \{-\infty\} \cup \mathbb{N}$ satisfies 1. and 2. above. It even satisfies

$$deg(ab) = deg(a) + deg(b).$$

Let $Fil_i B = \{b \in B | \, deg(b) \leq i\}$. This is an increasing filtration on B, $Fil_0 B = E$.

Proposition 43 *Suppose that for* $i \geq 1$ *the map* $Fil_i B / Fil_{i-1} B \to \mathfrak{m}_A^{-i} / \mathfrak{m}_A^{-i+1}$ *is surjective. Then* (B, deg) *is almost Euclidean.*

Proof Point 3. is satisfied by assumption. Let us prove "weak division with remainder" by induction on $deg(x) - deg(y)$. If $deg(x) \leq deg(y)$ let $a = 0$, $b = x$. Assume that $deg(x) = i > j = deg(y)$. Then $xy^{-1} \in \mathfrak{m}_A^{j-i}$ and there exists an $\alpha \in Fil_{i-j} B$ mapping to it modulo $\mathfrak{m}_A^{j-i+1}$. It follows that $\beta = x - \alpha y = y(xy^{-1} - \alpha) \in B$ satisfies $deg(\beta) \leq j + (i - j - 1) < i$. Thus $deg(\beta) - deg(y) < deg(x) - deg(y)$ and by the induction hypothesis we can write $\beta = ay + b$, $deg(b) \leq deg(y)$. But then

$$x = (\alpha + a)y + b$$

as desired. □

Example 44 Fix $t \in P_1$ where $P_k = B_F^{\varphi = p^k}$. Let $\infty_t \in |X|$ be the corresponding point and $y \in |Y|$ a point mapping to it. Let $B_{dR} = B_{dR,y}$, and B_{dR}^+ its valuation ring. Observe that $B_{dR}^+ = \lim_{\leftarrow} B_F/(\xi_y^n)$ contains B_F. In the role of B of the Proposition we take

$$B_e = B_F[1/t]^{\varphi=1} \subset K = Frac(B_e).$$

Clearly $B_e = \bigcup_{k=1}^{\infty} t^{-k} B_F^{\varphi=p^k}$ and since $t^{-k} b = t^{-k-1}(tb)$ this is an increasing union. The discrete valuation on B_{dR} induces a discrete valuation on K. Its valuation ring is

$$A = B_{dR}^+ \cap K,$$

and $A \cap B_e = B_{dR}^+ \cap B_e = \mathbb{Q}_p$. In fact

$$Fil_k B_e = t^{-k} B_{dR}^+ \cap B_e = t^{-k} B_F^{\varphi=p^k}$$

precisely. The fundamental exact sequence implies that for $k \geq 1$

$$t^{-k} B_F^{\varphi=p^k} / t^{-k+1} B_F^{\varphi=p^{k-1}} \simeq t^{-k} C_y \simeq \mathfrak{m}_A^{-k} / \mathfrak{m}_A^{-k+1}$$

so the condition of the Proposition is satisfied. We conclude that $(B_e, -ord_\infty)$ is an almost Euclidean ring. We shall see later that it is even a PID.

2.4.3 Construction of Complete Curves

Let $P = \bigoplus_{k \geq 0} P_k$ be a graded integral domain in which $P_0 = E$ is a field. Assume $\dim_E P_1 \geq 2$. Let

$$X = Proj(P),$$

a scheme over E.

Theorem 45 *([5], Théorème 5.2.7) Assume*

(1) The multiplicative monoid $\coprod_{k \geq 0} (P_k - \{0\})/E^\times$ is free on $(P_1 - \{0\})/E^\times$ as generators.
(2) For every $t \in P_1 - \{0\}$ there exists a field $E \subset C$ such that

$$P/Pt \simeq D = \{f \in C[T] | \, f(0) \in E\}$$

as a graded E-algebra.

Then:

(a) For every $t \in P_1 - \{0\}$, the locus $V^+(t) = \{\infty_t\}$ is a single (necessarily closed) point.

(b) The association $t \mapsto \infty_t$ induces a bijection

$$(P_1 - \{0\})/E^\times \simeq |X|$$

with the closed points of X.

(c) Letting $deg(x) = 1$ for every $x \in |X|$, X is a complete curve.

(d) For any $\infty \in |X|$, $X - \{\infty\}$ is affine open of the form $Spec(B)$ for a PID B, i.e., $Pic(X - \{\infty\}) = 0$. Moreover, $(B, -ord_\infty)$ is almost Euclidean.

Proof (1) Let $t \in P_1 - \{0\}$. Using 2. for the structure of P/Pt we get that

$$V^+(t) = Proj(P/Pt)$$

is a closed point ∞_t, as any non-zero homogenous prime ideal of D contains $TC[T]$ (*exercise*: note that $C \not\subseteq D$, so a little argument is needed!). This proves (a).

(2) We have

$$X - \{\infty_t\} = Spec(B), \quad B = P[1/t]_0.$$

Every non-zero element of B is of the form xt^{-k} with $x \in P_k$. By 1. it can be written uniquely up to an $E^\times$-multiple as $\prod_{i=1}^k (s_i/t)$ with $s_i \in P_1 - Et$. Thus B is a unique factorization domain with irreducible elements of the form s/t for $s \in P_1 - Et$. To show that it is a PID it is enough to verify that for any such s, (s/t) is a maximal ideal. But

$$B/(s/t) = (P/sP)[1/t]_0.$$

By assumption 2., now for s, P/sP maps isomorphically onto another ring D' constructed from another field extension C' of E. The element t is homogenous of degree 1 so must map to cT for some $c \in C'$. But $D'[1/cT]_0 \simeq C'$ is a field, so (s/t) is a maximal ideal of B. This proves the first part of (d). Furthermore, the closed points of $X - \{\infty_t\}$ are in bijection with the irreducible s as above (up to $E^\times$), and this proves (b). If s, t are in $P_1 - \{0\}$ as above then $div(s/t) = \{\infty_s\} - \{\infty_t\}$, so is of degree 0. As any element of the function field of X is a finite product of such s/t, X is a complete curve and (c) is proved.

(3) It remains to check that[6] $B = P[1/t]_0$ is almost Euclidean with respect to $deg = -ord_\infty$ ($\infty = \infty_t$). Here $deg(f)$ for $f \in B$ is the usual degree of the divisor $div(f)$ (recall B is a PID). It is checked immediately that ord_∞ is a discrete valuation on $K = Frac(B)$ (the function field of X) and that t is a uniformizer. For "almost Euclidean" we use the criterion from Proposition 43. Since every element of B is

$$f = u \prod_{i=1}^{k} \frac{s_i}{t}$$

($s_i \in P_1 - Et, \quad u \in E^\times$), $ord_\infty(f) = -k \leq 0$ and if it is 0, we must have $k = 0$ so $f = u \in E^\times$ is invertible. Let $i \geq 1$. An element of $\mathfrak{m}_K^{-i}$ is of the form

$$u \frac{s_1 \cdots s_{i+j}}{s'_1 \cdots s'_j t^i}$$

where $u \in E^\times$, the $s'_\ell \in P_1 - Et$ and $s_\ell \in P_1 - \{0\}$. $Fil_i B$ consists of the same elements with $j = 0$. We claim that

$$Fil_i B / Fil_{i-1} B \simeq \mathfrak{m}_K^{-i} / \mathfrak{m}_K^{1-i}$$

(an isomorphism of modules over P/tP). Let S be the multiplicative subset of P generated by $P_1 - Et$. Then $K = (S^{-1}P[1/t])_0$ and as $Fil_i B = t^{-i} P_i$ after multiplication by t^i the claim becomes the claim that for $i \geq 1$

[6] In the application to the Fargues-Fontane curve, this will be the ring B_e, not the much larger B_F!

$$P_i/tP_{i-1} \simeq [S^{-1}(P/tP)]_i.$$

But identifying $P/tP = E \oplus \bigoplus_{i=1}^{\infty} CT^i$ we get $S^{-1}(P/tP) = \bigoplus_{i\in\mathbb{Z}} CT^i$ (Laurent polynomials), hence the desired isomorphism. □

2.5 *Vector Bundles on Curves*

2.5.1 Beauville-Laszlo Gluing

Let X be a complete curve with a field of definition $E = H^0(X, \mathcal{O}_X)$. Let $K = E(X) = \mathcal{O}_{X,\eta}$ be the function field of X. Let $\infty \in |X|$ be a closed point and $U = X \setminus \{\infty\}$. Consider the category $\mathscr{C}$ of triples $(\mathcal{E}, N, u)$ consisting of a vector bundle $\mathcal{E}$ on U, a free finite rank module N over $\mathcal{O}_{X,\infty}$ and an isomorphism

$$u : N \otimes_{\mathcal{O}_{X,\infty}} K \simeq \mathcal{E}_\eta.$$

Similarly, let $\widehat{\mathscr{C}}$ be the category of triples $(\mathcal{E}, \widehat{N}, \widehat{u})$ where $\mathcal{E}$ is as above, $\widehat{N}$ is a free finite rank module over $\widehat{\mathcal{O}}_{X,\infty}$ and

$$\widehat{u} : \widehat{N} \otimes_{\widehat{\mathcal{O}}_{X,\infty}} \widehat{K}_\infty \simeq \widehat{\mathcal{E}}_\eta := \mathcal{E}_\eta \otimes_K \widehat{K}_\infty.$$

Let $\mathscr{V}\mathscr{B}_X$ be the category of vector bundles of finite rank on X.

Proposition 46 *The functors $\mathscr{V}\mathscr{B}_X \to \mathscr{C}$ and $\mathscr{V}\mathscr{B}_X \to \widehat{\mathscr{C}}$ given by*

$$\mathcal{E} \mapsto (\mathcal{E}|_U, \mathcal{E}_\infty, can), \quad (\mathcal{E}|_U, \widehat{\mathcal{E}}_\infty, can)$$

are both equivalences of categories.

Proof Call these functors α_∞ and $\widehat{\alpha}_\infty$. For any open V containing ∞ there is a similar functor α_V built from the cover $\{U, V\}$ of X, which is a usual Zariski gluing, hence an equivalence of categories. Thus $\alpha_\infty = \lim_\to \alpha_V$ is also an equivalence. To show that $\widehat{\alpha}_\infty$ is an equivalence one uses an approximation argument, based on the fact that

$$GL_n(K)/GL_n(\mathcal{O}_{X,\infty}) \simeq GL_n(\widehat{K}_\infty)/GL_n(\widehat{\mathcal{O}}_{X,\infty}).$$

□

Remark. (i) If $U = Spec(B)$ is affine and B is a PID, then the isomorphism classes of rank n vector bundles over X are given by

$$GL_n(B) \setminus GL_n(\widehat{K}_\infty)/GL_n(\widehat{\mathcal{O}}_{X,\infty}).$$

(ii) There is an obvious generalization to the situation where $X \setminus U$ is a finite number of points.

2.5.2 Cohomology and Twisting

Assume now that $U = X \setminus \{\infty\}$ is affine, $U = Spec(B)$ where B is a PID (equivalently $Pic(U) = 0$) and let $\mathcal{E}|_U$ correspond to the B-module M, $N = \mathcal{E}_\infty$ and $\widehat{N} = \widehat{\mathcal{E}}_\infty$. Let t be a uniformizer at ∞. We want to describe the cohomology of $\mathcal{E}$ in terms of (M, N, u) where $u : M \otimes_B K \simeq N \otimes_{\mathcal{O}_{X,\infty}} K$ (we follow the notation of [5] which for some reason switch in the middle between u and u^{-1}) and similarly in terms of $(M, \widehat{N}, \widehat{u})$.

Proposition 47 *The complex of abelian groups $R\Gamma(X, \mathcal{E})$ is canonically isomorphic to the complex*

$$M \oplus N \to N \otimes_{\mathcal{O}_{X,\infty}} K$$

$$(x, y) \mapsto u(x) - y$$

and similarly to the same complex where N, u are replaced by $\widehat{N}, \widehat{u}$ and K by $\widehat{K}_\infty$. In particular

$$H^0(X, \mathcal{E}) = u(M) \cap N = \widehat{u}(M) \cap \widehat{N}$$

$$H^1(X, \mathcal{E}) = N \otimes_{\mathcal{O}_{X,\infty}} K/(u(M) + N) = \widehat{N} \otimes_{\widehat{\mathcal{O}}_{X,\infty}} \widehat{K}_\infty/(\widehat{u}(M) + \widehat{N}).$$

Proof Once again, if X is covered by $U = Spec(B)$ and $V = Spec(A)$ with both rings PIDs, the analogous statement is the familiar comparison between derived functor cohomology and Čech cohomology. The Proposition follows from this case in the same way as the previous proposition. □

We let $\mathcal{E}(k\infty)$ denote the twist of the vector bundle $\mathcal{E}$ by the line bundle $\mathcal{O}(k\infty)$ associated with the divisor $k[\infty]$. We leave the proof of the next proposition to the reader.

Proposition 48 *Suppose that $\mathcal{E}$ is represented by a triple (M, N, u) (or $(M, \widehat{N}, \widehat{u})$). Then $\mathcal{E}(k\infty)$ is represented by*

$$(M, t^{-k}N, u), \quad (M, t^{-k}\widehat{N}, \widehat{u}).$$

2.5.3 The Relation with the (Almost) Euclidean Property

We assume that X is a complete curve such that $deg(\infty) = 1$ and $X \setminus \{\infty\} = U = Spec(B)$ with B a PID, i.e., $Pic(U) = 0$. It is then easily verified that via deg, $Pic(X) \simeq \mathbb{Z}$. We let $E = H^0(X, \mathcal{O}_X)$ be the field of definition and $K = E(X) = \mathcal{O}_{X,\eta}$ the functions field, $A = \mathcal{O}_{X,\infty}$ and $\widehat{K} = \widehat{K}_\infty$. We let $t \in K$ be a uniformizer at

∞. Note $A \cap B = E$. We write $deg = -ord_\infty : B \to \mathbb{N} \cup \{-\infty\}$. The line bundle $\mathcal{O}(1)$ is endowed with a canonical section "1″ with a simple zero at ∞ and nowhere else. Tensoring with it gives

$$E = H^0(X, \mathcal{O}) \subset H^0(X, \mathcal{O}(1)) \subset \cdots$$

which corresponds to the filtration by deg on B

$$E = B^{deg \leq 0} \subset B^{deg \leq 1} \subset \cdots$$

(via Proposition 47). For $k < 0$ $H^0(X, \mathcal{O}(k)) = 0$. For $k \in \mathbb{Z}$ cupping with "1″ is a surjection

$$H^1(X, \mathcal{O}(k)) \twoheadrightarrow H^1(X, \mathcal{O}(k+1))$$

$$K/(B + t^{-k}A) \twoheadrightarrow K/(B + t^{-k-1}A).$$

It follows that if $H^1(X, \mathcal{O}(k)) = 0$ then H^1 vanishes for all larger ks.

Let $i : \{\infty\} \hookrightarrow X$. Then there is an exact sequence of sheaves

$$0 \to \mathcal{O}(k-1) \to \mathcal{O}(k) \to i_*(\mathfrak{m}_A^{-k}/\mathfrak{m}_A^{-k+1}) \to 0.$$

It follows that if $H^1(X, \mathcal{O}) = 0$ then for all $k \geq 1$ the map

$$B^{deg \leq k}/B^{deg \leq k-1} \to \mathfrak{m}_A^{-k}/\mathfrak{m}_A^{-k+1}$$

is surjective (in fact an isomorphism). Recall that this was our criterion for (B, deg) to be almost Euclidean. If P is as in theorem 45 and $t \in P_1^\times$ an element such that $\infty = \infty_t$ then for $k \geq 0$ we have

$$P_k \simeq B^{deg \leq k} = H^0(X, \mathcal{O}(k))$$

$$b \mapsto \frac{b}{t^k}$$

so $X = Proj\left(\oplus_{k=0}^\infty \Gamma(X, \mathcal{O}(k))\right)$. The sequence of inclusions of the $H^0(X, \mathcal{O}(k))$'s becomes the sequence

$$P_0 \overset{\times t}{\to} P_1 \overset{\times t}{\to} P_2 \to \cdots$$

Proposition 49 *The following equivalences hold:*
(i) (B, deg) *is almost Euclidean* $\Leftrightarrow H^1(X, \mathcal{O}) = 0$
(ii) (B, deg) *is Euclidean* $\Leftrightarrow H^1(X, \mathcal{O}(-1)) = 0$.

Proof Denote by $t \in K$ a uniformizer at ∞ (if $t \in P_1$ as above then this t is the previous t/s for some $s \in P_1 \setminus Et$). Recall that

$$H^1(X, \mathcal{O}) = K/(B + A), \quad H^1(X, \mathcal{O}(-1)) = K/(B + tA).$$

We prove (ii). The proof of (i) is identical, replacing $<$ by $\leq$ at one place. If $K = B + tA$ then given $x, y \in B$ with $y \neq 0$ we write

$$\frac{x}{y} = b + ta$$

with $b \in B$ and $a \in A$. Then $x = yb + tay$ where $yb \in B$ and hence $tay \in B$ but since $ord_\infty(ta) \geq 1, deg(tay) < deg(y)$. In the converse direction one reverses the argument to show that if $x = by + r$ with $deg(r) < deg(y)$ one can write $r = tay$ with $a \in A$ (i.e., $ord_\infty(a) \geq 0$) so $x/y \in B + tA$. □

2.6 *Conclusion of the Proof of Theorem 42*

2.6.1 Putting Everything Together

All the ingredients are now in place. We have checked the two conditions in Theorem 45: the structure of the monoid $\coprod P_k^\times/E^\times$ and the fundamental exact sequence, leading to the description

$$P/Pt \simeq \{f \in C[T]|\, f(0) \in E\},$$

where $t \in P_1^\times$ and $C = C_t$. We have also checked that $K = B + A$, in example 44. Here $K = E(X) = Frac(P)^{\varphi=1}$,

$$B = B_e = B_F[1/t]^{\varphi=1} = P[1/t]^{\varphi=1}, \quad A = K \cap B^+_{dR,\infty}.$$

Theorem 42 follows now at once from its abstract version, Theorem 45, and Proposition 49.

2.6.2 The Field of Meromorphic Functions on $|Y|/\phi^{\mathbb{Z}}$

We want to wrap up everything by relating the field $E(X)$, derived from the schematic point of view, with the “analytically constructed” field of meromorphic functions on $|X| = |Y|/\phi^{\mathbb{Z}}$. Here we define the latter in an ad hoc fashion, as we have not defined the adic (Huber) spaces underlying $|Y|$ or $|X|$.

Recall that

$$B_F = \lim_{\leftarrow} B_I$$

(over compact intervals $I \subset (0, 1)$). The B_I are PIDs. We define

$$\mathcal{M}(Y) = \lim_{\leftarrow} Frac(B_I).$$

It can be shown ([5], Proposition 3.5.10) that $\mathcal{M}(Y) = Frac(B_F)$. Define the field of meromorphic functions on X to be

$$\mathcal{M}(X) = \mathcal{M}(Y)^{\varphi=1}.$$

Proposition 50 *We have $E(X) \simeq \mathcal{M}(X)$.*

Proof We have $E(X) = \{x/y | x, y \in P \text{ homogenous of the same degree}\} \subset Frac(P) \subset Frac(B_F) = \mathcal{M}(Y)$. Clearly $E(X)$ lands in the φ-invariant part. Suppose $f \in \mathcal{M}(Y)^{\varphi=1}$. Then by the description of the monoid $Div(Y/\phi^{\mathbb{Z}})^+ \simeq \coprod_{k=0}^{\infty} P_k^{\times}/E^{\times}$ (where $P_k = B_F^{\varphi=p^k}$) we can write

$$div(f) = div(g) - div(h),$$

where $g \in P_k^{\times}$ and $h \in P_\ell^{\times}$. It follows that

$$f = u\frac{g}{h},$$

where $u \in \mathcal{M}(Y)$ has neither poles nor zeros. From

$$\mathcal{M}(Y) = \lim_{\leftarrow} Frac(B_I)$$

and the fact that the B_I are PIDs we get that $u \in \lim_{\leftarrow} B_I^{\times} = B_F^{\times}$. Now $\varphi u = p^{\ell-k}u$ so $u \in P_{\ell-k}$ and $u^{-1} \in P_{k-\ell}$. But we have seen that $P_d = 0$ if $d < 0$. This forces $k = \ell$ and $u \in P_0^{\times} = E^{\times}$. Thus $f \in E(X)$. □

3 Vector Bundles on X_F

The classification of vector bundles on the Fargues-Fontaine curve brings into the picture the theory of p-divisible groups, their moduli spaces, and the two period morphisms, π_{GM} (the Grothendieck-Messing, or de-Rham period morphism) and π_{HT}, the Hodge-Tate period morphism.

We shall start with generalities about vector bundles on curves, but at a certain point we shall need to take a long detour into the theory of p-divisible groups. We shall therefore try, following Morrow's Bourbaki talk, to present the theory of vector bundles on X_F in a self-contained manner, "blackboxing" the necessary input from p-divisible groups. We shall then return to that theory and fill in the details as much as time will permit us.

Let us stress that for the most spectacular applications of the Fargues-Fontaine curve, to p-adic Hodge theory and local Galois representations ("weakly admissible

= admissible"), to moduli spaces of p-divisible groups ("Drinfeld tower = Lubin Tate tower"), and to local Langlands (geometrizing it via analogues of Drinfeld's Shtukas), it is the theory of vector bundles on it, and not its bare structure, that we have been studying so far, that plays the crucial role.

3.1 Harder-Narasimhan Categories

3.1.1 Vector Bundles on Curves

Let X be a compact Riemann surface, or more generally a smooth projective curve over an algebraically closed field. To every vector bundle E on X one associates

- Its rank $rk(E) \in \mathbb{N}$
- Its degree $deg(E) = deg(\bigwedge^{rk(E)} E) \in \mathbb{Z}$ (the degree of a line bundle L is the degree of the divisor D such that $L \simeq \mathcal{O}(D)$)
- Its slope $\mu(E) = deg(E)/rk(E)$.

If $0 \to E' \to E \to E'' \to 0$ is an exact sequence of vector bundles then

$$\mu(E) = \frac{rk(E')}{rk(E)}\mu(E') + \frac{rk(E'')}{rk(E)}\mu(E'')$$

is a convex combination of the slopes of E' and E''.

The vector bundle E is said to be *semi-stable* if for every $E' \subset E$ a sub-vector bundle (locally a direct summand) $\mu(E') \leq \mu(E)$. Equivalently, for every quotient bundle E'' $\mu(E) \leq \mu(E'')$. The following key theorem is due to Harder and Narasimhan (1974).

Theorem 51 *For every vector bundle E there exists a unique filtration*

$$0 = E_0 \subset E_1 \subset \cdots \subset E_m = E$$

by vector sub-bundles such that (i) E_i/E_{i-1} is semi-stable (ii) $\mu(E_i/E_{i-1})$ are strictly decreasing.

The proof is not difficult. If E is semi-stable take $m = 1$ and $E = E_1$. Otherwise let E_1 be a sub-bundle of highest slope, and if there are several such, take among them one of highest rank. It is clearly semi-stable. Apply induction to E/E_1. One only has to show that $\mu(E_2/E_1) < \mu(E_1)$. But if we had $\mu(E_2/E_1) \geq \mu(E_1)$ then $\mu(E_2) \geq \mu(E_1)$, contradicting the choice of E_1.

It turns out that there are many categories in which one can define rank and degree for which the slope satisfies the same formalism. Yves André axiomatized it, and here we take a slightly more restrictive axiomatization, that is nevertheless sufficient for our purposes.

3.1.2 A Generalization

Let $\mathcal{C}$ be a category and assume that for any object $E \in \mathcal{C}$ we can associate $rk(E) \in \mathbb{N}$ and $deg(E) \in \mathbb{Z}$ such that:

(HN1) $\mathcal{C}$ is an exact category (a full additive subcategory of an abelian category closed under extensions[7]), and rk and deg are additive on exact sequences.

(HN2) There exists an *exact faithful* functor F ("generic fiber") $F : \mathcal{C} \to \mathcal{A}$ where $\mathcal{A}$ is an abelian category, satisfying:

(i) The functor F induces a bijection between *strict subobjects* of $X \in \mathcal{C}$ and subobjects of $F(X) \in \mathcal{A}$. Here a strict subobject is a subobject X' of X sitting in an exact sequence $0 \to X' \to X \to X'' \to 0$. One should think of the inverse of this bijection as an operation of taking "schematic closure".

(ii) The rank function on $\mathcal{C}$ factors through a rank function $rk : \mathcal{A} \to \mathbb{N}$ satisfying $rk(X) = 0$ iff $X = 0$.

(iii) If $u : X' \to X$ is a morphism in $\mathcal{C}$ such that $F(u) : F(X') \to F(X)$ is an isomorphism then $deg(X') \leq deg(X)$ with an equality iff u is an isomorphism.

Example 52 Let C be a smooth complete curve over an algebraically closed field, $\mathcal{C}$ the category of vector bundles on C, the exact sequences being exact sequences in the category of sheaves. A strict subobject is a vector sub-bundle. The category $\mathcal{A}$ is the category of vector spaces over K, the function field of C, and $F(E) = E_\eta$ is the generic fiber. Note that if $E' \subset E$ is a strict subobject, then $E' = E'_\eta \cap E$ (inside E_η), giving (i). Point (ii) is clear and (iii) stems from the fact that if $E' \subset E$ and they have the same rank, then $deg(E') \leq deg(E)$ with equality iff $E' = E$. In fact, it is enough to treat the case of a line bundle.

It turns out that the Harder-Narasimhan theorem holds in $\mathcal{C}$, if one defines slope and semi-stability in the same way. To be precise, X is semi-stable if for any *strict subobject* $X' \hookrightarrow X$ we have $\mu(X') \leq \mu(X)$.

3.1.3 More Examples

(1) In the special case of *vector bundles over* $\mathbb{P}^1$ the category can be described, via Beauville-Laszlo gluing, also as the category of triples (M, N, u) where M is a free module over $\mathbb{C}[z]$, N is a free module over $\mathbb{C}[[1/z]]$, and $u : M \otimes_{\mathbb{C}[z]} \mathbb{C}((1/z)) \simeq N \otimes_{\mathbb{C}[[1/z]]} \mathbb{C}((1/z))$. Here the rank is the usual rank, and the degree can be computed as follows. Let m_i be a basis of M over $\mathbb{C}[z]$ and n_i a basis of N over $\mathbb{C}[[1/z]]$. Then $n_i = \sum a_{ij} m_j$ and

[7] The notion of an exact category can be defined intrinsically: one starts with an additive category and defines a notion of *exact structure*. The *embedding theorem* says that such a category is always a full subcategory of an abelian category closed under extensions. Some examples (such as the category of vector bundles on a curve) are better conceived as subcategories of abelian categories (of all modules on the curve), while other examples (e.g., the category of filtered objects in an abelian category) are more naturally described as an additive category with an exact structure, i.e., a class of exact sequences satisfying some axioms.

$$deg(M, N, u) = ord_\infty(det(a_{ij})).$$

The functor F of (HN2) takes (M, N, u) to the vector space $M \otimes_{\mathbb{C}[z]} \mathbb{C}((1/z))$ over $\mathbb{C}((1/z))$.

(2) Let $\mathcal{C}$ be the category of *B-pairs* introduced by Laurent Berger. Here we take B_e and B_{dR}^+ (having fixed $\infty \in X_F$) and consider triples (M, N, u) where M is a finite free module over the PID B_e, N is a finite free module over B_{dR}^+, and u is an isomorphism

$$u : M \otimes_{B_e} B_{dR} \simeq N \otimes_{B_{dR}^+} B_{dR}.$$

The rank and degree are defined as in example (1). We have seen that this category is identified with the category of vector bundles on X_F.

(3) *Filtered vector spaces.* Let L/F be a field extension. Let $\mathcal{C}$ be the category of pairs $(V, Fil^\bullet V_L)$ where V is a finite-dimensional vector space over F and $Fil^\bullet$ is a separated exhaustive decreasing filtration on V_L. Rank is the dimension. The degree is given by

$$deg(V, Fil^\bullet V_L) = \sum_i i \dim(gr^i V_L).$$

Here a strict subobject is a subspace of V whose filtration is induced by the one on V_L (not only compatible with it). The map F simply forgets the filtration and remembers only the vector space V. Verifying the axioms is elementary linear algebra.

(4) *Isocrystals* (φ-modules). Let k be a perfect field in characteristic p, $K_0 = Frac(W(k))$. Denote by φ the Frobenius of K_0. An isocrystal over k is a finite-dimensional K_0-vector space D with a φ-semi-linear bijective map $\varphi_D : D \to D$. Its rank $rk(D, \varphi_D) = \dim_{K_0} D$. Its degree is

$$deg(D, \varphi_D) = -v_{K_0}(\det(\Phi))$$

where Φ is the matrix of φ_D on the basis of D. Note that since a change in basis changes Φ to $\varphi(P)\Phi P^{-1}$, the determinant is not independent of the basis, but its valuation is. We remark that $-deg$ is also a valid degree function, and the choice of the minus sign is a matter of convention. This is because the only condition that distinguishes deg from $-deg$ is (HN2)(iii). But if $(D', \varphi'_D) \to (D, \varphi_D)$ is "generically an isomorphism", i.e., is an isomorphism as K_0-vector spaces, we must have $\varphi_D = \varphi'_D$ and the degrees are equal. With this convention, *effective* isocrystals (having a φ_D-invariant $W(k)$-lattice) have a *non-positive degree.*

In passing we remark that the fact that both deg and $-deg$ are valid degree functions, together with the Harder-Narasimhan theorem, that has the following corollary. Every isocrystal over a perfect field in characteristic p has a unique direct sum decomposition into isoclinic isocrystals, i.e., isocrystals all of whose subisocrystals have the same slope.

Example 53 Let D be spanned by $e_1, \dots, e_h$ and $\varphi_D(e_i)=e_{i+1}$ $(1\leq i<h)$ $\varphi_D(e_h) = p^{-d}e_1$. Then (D, φ_D) has rank h and degree d. If $d \leq 0$ then the $W(k)$-span of the

e_i is an invariant lattice M. The following non-isomorphic M (due to Oort) leads to an isomorphic D. Let e_i $(i \in \mathbb{Z})$ be subject to the identification $e_{i+h} = p^{-1}e_i$. Let $\varphi_D(e_i) = e_{i+d}$. Note that $\varphi_D^h e_i = p^{-d}e_i$. If $d \le 0$ we let M be the module spanned by $e_1, \ldots, e_h$. Note also that if $0 \le -d \le h$ then $pM \subset \varphi_D(M) \subset M$. Such an M (but not the previous one) will turn out to be the Dieudonné module associated with a p-divisible group of dimension $-d$ and height h.

If $\lambda = d/h \in \mathbb{Q}$ in reduced terms we denote the isocrystal (D, φ_D) by $(D_\lambda, \varphi_\lambda)$.

The following theorem is classical. For the proof see [3, 13]. We denote by $Isoc_k$ the category of isocrystals over k.

Theorem 54 (Dieudonné-Manin) *Let k be algebraically closed of characteristic p. The category of isocrystals over k is semi-simple and the $(D_\lambda, \varphi_\lambda)$ are its simple objects. The endomorphism algebra of $(D_\lambda, \varphi_\lambda)$ is the central division algebra over $\mathbb{Q}_p$ with invariant λ.*

If k is any perfect field of char. p (not necessarily algebraically closed), Dieudonné showed that $Isoc_k$ is anti-equivalent, via a functor called the Dieudonné module, to the category of p-divisible groups over k up to isogeny.

(5) *Filtered isocrystals* (filtered φ-modules). This example combines the previous two. Let K/K_0 be a totally ramified extension and consider triples $(D, \varphi_D, Fil^\bullet D_K)$ where (D, φ_D) is as in (4), $(D, Fil^\bullet)$ is as in (3). There need not be any relation between φ_D and the filtration. Setting

$$deg(D, \varphi_D, Fil^\bullet) = \sum_i i \dim(gr^i D_K) - \nu_{K_0}(\det \Phi)$$

gives the notion of a filtered isocrystal. Note that $(D, \varphi_D, Fil^\bullet)$ is *semi-stable of slope* 0 iff

$$\sum_i i \dim(gr^i D_K) = \nu_{K_0}(\det \Phi)$$

and for any strict subobject $(D', \varphi_{D'}, Fil^\bullet)$ (meaning that the filtration is induced from the one of D and Frobenius leaves D' stable) there is an inequality

$$\sum_i i \dim(gr^i D'_K) \le \nu_{K_0}(\det \Phi').$$

3.1.4 Semi-stable Objects of Slope 0

Let $\mathscr{C}$ be a Harder-Narasimhan category. We denote by $\mathscr{C}_\lambda^{ss}$ the full subcategory whose objects are semi-stable of slope λ. It is clearly an additive subcategory closed under extensions.

Proposition 55 *The category $\mathscr{C}_\lambda^{ss}$ is an abelian category.*

We do not prove the proposition. Our interest lies in $\lambda = 0$. In many examples the proposition will then be self-evident.

3.2 Classification of Vector Bundles—Statement of the Results

3.2.1 Recall of Fontaine's Rings

Fix a point ∞ on $X = X_F$ corresponding to $t \in B_F^{\varphi=p}$ (and recall that ∞ determines t up to $\mathbb{Q}_p^\times$). Recall that $B_{dR}^+ = \widehat{\mathcal{O}}_{X,\infty}$ and

$$B_e = \Gamma(X \setminus \{\infty\}, \mathcal{O}) = B_F[1/t]^{\varphi=1} = \bigcup_{i=0}^{\infty} t^{-i} P_i$$

where $P_i = B_F^{\varphi=p^i}$ (B_e is a PID). Recall that we also had $P_i = B_F^{+,\varphi=p^i}$ where B_F^+ is the closure of $B^{b,+} = W(\mathcal{O}_F)[1/p]$ in B_F, or alternatively the set of $f \in B_F$ whose Newton polygon satisfies $\mathcal{N}_f(x) \geq 0$ for all $x \in \mathbb{R}$. This was an immediate consequence of the relation

$$\mathcal{N}_f(x - i) = p\mathcal{N}_f(x)$$

which holds for any f satisfying $\varphi f = p^i f$, because this relation and the fact that $\mathcal{N}_f(x) \to \infty$ as $x \to -\infty$ force $\mathcal{N}_f(x) \geq 0$. Thus in the definition of B_e we can replace B_F by B_F^+.

3.2.2 Relation with B_{cris}

One can show that $P_i = B_{cris}^{+,\varphi=p^i}$ as well, hence $B_e = B_{cris}^{\varphi=1}$ ($B_{cris} = B_{cris}^+[1/t]$). Here, while B_F^+ is the completion of $B^{b,+} = W(\mathcal{O}_F)[1/p]$ in the family of norms $|.|_\rho$ $(0 < \rho < 1)$, $B_{cris}^+ = A_{cris}[1/p]$ where A_{cris} is the *divided power completion* in the ideal (ξ) corresponding to any point of $|Y_F|$ (any untilt) above ∞ (in the Frobenius orbit of untilts corresponding to ∞). Thus

$$B_{cris}^+ = W(\widehat{\mathcal{O}_F)[\xi^n/n!]}[1/p].$$

The relation between the two rings B_F^+ and B_{cris}^+ is

$$B_F^+ \subset B_{cris}^+, \quad B_F^+ = \bigcap_{n=0}^{\infty} \varphi^n(B_{cris}^+).$$

(We do not prove this—it boils down to computations with divided powers and the norms $|.|_\rho$.) In other words, B_F^+ is the largest subring of B_{cris}^+ on which φ is bijective.[8] It is this characterization which is responsible for

[8] In some older papers B_F^+ was called B^{rig}.

$$B_F^{+,\varphi=p^i} = B_{cris}^{+,\varphi=p^i}.$$

Indeed, the LHS is clearly contained in the RHS. But if an element $x \in B_{cris}^+$ satisfies $\varphi x = p^i x$ then

$$x = p^{-i}\varphi(x) = p^{-2i}\varphi^2(x) = \cdots$$

so $x \in \bigcap_{n=0}^{\infty} \varphi^n(B_{cris}^+) = B_F^+$. From here one gets the desired identity

$$B_F[1/t]^{\varphi=1} = \bigcup t^{-i} B_F^{\varphi=p^i} = \bigcup t^{-i} B_F^{+,\varphi=p^i} = \bigcup t^{-i} B_{cris}^{+,\varphi=p^i} = B_{cris}^{\varphi=1}.$$

Although classically B-pairs were defined by Berger using B_{cris} we shall use $B_F[1/t]$ instead. Thanks to the above computations, it gives the same modules.

3.2.3 Construction of the Vector Bundles $\mathcal{E}(D, \varphi)$

Let $k = \overline{\mathbb{F}}_p$, $L = W(k)[1/p] = \widehat{\mathbb{Q}_p^{nr}}$. Since we assumed F was algebraically closed, $W(\mathcal{O}_F)[1/p]$ and hence B_F is an L-algebra. If $(D, \varphi_D) \in Isoc_k$, define a graded $P = \oplus_{i=0}^{\infty} B_F^{\varphi=p^i}$-module

$$M(D, \varphi_D) = \oplus_{i=0}^{\infty}(D \otimes_L B_F)^{\varphi=p^i},$$

where the φ on $D \otimes_L B_F$ is $\varphi_D \otimes \varphi$. If D is the trivial isocrystal we get P itself.

Definition 56 $\mathcal{E}(D, \varphi_D)$ is the associated $\mathcal{O}_X$-module $\widetilde{M(D, \varphi_D)}$.

Proposition 57 *(i) The module $\mathcal{E}(D, \varphi_D)$ is a vector bundle.*

(ii) Its rank and degree as a vector bundle are the rank and degree of (D, φ_D) as an isocrystal.

(iii) It is associated to the (B_e, B_{dR}^+)-pair (M, N, u)

$$M = (D \otimes_L B_F[1/t])^{\varphi=1}, \quad N = D \otimes_L B_{dR}^+, \quad u = u_{can},$$

where

$$u_{can} : M \otimes_{B_e} B_{dR} \simeq N \otimes_{B_{dR}^+} B_{dR}$$

is the canonical map (which turns out to be an isomorphism).

(iv) The functor

$$\mathcal{E}(-) : Isoc_k \rightsquigarrow \mathcal{VB}_{X_F}$$

is compatible with tensor products and duals.

When $(D, \varphi_D) = (D_\lambda, \varphi_\lambda)$ $\lambda = d/h$ in reduced terms, we shall denote this vector bundle $\mathcal{O}(\lambda)$ or $\mathcal{O}(d, h)$. It has rank h, degree d, and slope λ. We shall show below that it is semi-stable.

The proof of the proposition, due to Berger (and in another form to Kedlaya), is not very difficult. The crucial point for proving (i) and (iii) is to show that the modules M, N have the same rank $h = rk(D, \varphi_D)$ over B_e and B_{dR}^+, respectively. Since k is algebraically closed, it is enough to do it with $D = D_{d,h}$, the explicit model given in example 53. To complete the proof of (ii) we have to show that the resulting vector bundle has then degree d. Finally, (iv) is more or less automatic.

We shall prove the Proposition below, after we have given an *alternative* definition of the $\mathcal{O}(d, h)$ that does not single out the point ∞ and is therefore more symmetric.

In any case, the Dieudonné-Manin theorem has the following corollary.

Corollary 58 *The vector bundle $\mathcal{E}(D, \varphi_D)$ is a direct sum of $\mathcal{O}(\lambda)$'s.*

3.2.4 The Classification Theorem

The next theorem is deep and its proof will be long.

Theorem 59 (Classification Theorem) *Every vector bundle on X_F is $\mathcal{E}(D, \varphi_D)$ for a unique isocrystal (D, φ_D). In other words, every vector bundle is of the form*

$$\mathcal{E} \simeq \bigoplus \mathcal{O}(\lambda_i)^{n_i}$$

where the slopes λ_i and their multiplicities n_i are uniquely determined.

Corollary 60 *(i) The functor $\mathcal{E}(-)$ is essentially surjective. (ii) A vector bundle is semi-stable iff it is isoclinic, i.e., of the form $\mathcal{O}(\lambda)^n$.(iii) The abelian category of semi-stable slope 0 vector bundles on X_F is equivalent to the category of finite-dimensional vector spaces over $\mathbb{Q}_p$ under $V \rightsquigarrow V \otimes \mathcal{O}_X$, $\mathcal{E} \rightsquigarrow H^0(X, \mathcal{E})$.*

We stress that $\mathcal{E}(-)$ is far from being an equivalence of categories!

3.3 *The Curves X_{F,E_h} for $h \geq 1$*

3.3.1 The Unramified Coverings X_{F,E_h}

To give a more symmetric construction of the vector bundles $\mathcal{O}(d, h)$, it is time to construct a family of unramified cyclic coverings of the Fargues-Fontaine curve. They will be denoted as $X_{F,E_h} \rightarrow X_F$, their Galois group will be cyclic of order h, and their field of definition will be E_h, the unramified extension of $\mathbb{Q}_p$ of degree h.

From an analytic (adic) point of view, not worked out in these notes, the X_{F,E_h} are very easy to construct. Just as $|X_F| = |Y_F|/\phi^{\mathbb{Z}}$, we let

$$|X_{F,E_h}| = |Y_F|/\phi^{h\mathbb{Z}}.$$

This already gives a working definition of the closed points of the would-be X_{F,E_h} and hints that $Gal(X_{F,E_h} \to X_F)$ should be cyclic with a canonical generator ϕ.

We therefore put

$$P_{E_h} = \bigoplus_{d=0}^{\infty} B_F^{\varphi^h = p^d} = \bigoplus_{d=0}^{\infty} B_F^{+,\varphi^h = p^d},$$

$$X_{F,E_h} = Proj(P_{E_h}).$$

Lemma 61 *We have* $X_{F,E_h} \simeq X_F \times_E E_h$.

Proof For any graded algebra $\oplus_{i\geq 0} R_i$ and any $h \geq 1$ we have $Proj(\oplus_{i\geq 0} R_i) \simeq Proj(\oplus_{i\geq 0} R_{hi})$. We therefore have

$$X_{F,E_h} = Proj(\bigoplus_{d=0}^{\infty} B_F^{\varphi^h = p^{dh}}).$$

The E_h-vector space $B_F^{\varphi^h = p^{dh}}$ has the semi-linear automorphism $\psi = p^{-d}\varphi$, whose invariants are $B_F^{\varphi = p^d}$. Although it is infinite dimensional, the action of ψ is locally finite (since ψ^h is the identity), so Hilbert's theorem 90 implies that

$$B_F^{\varphi^h = p^{dh}} = B_F^{\varphi = p^d} \otimes_E E_h.$$

The lemma follows. □

3.3.2 Properties of X_{F,E_h}

All the good properties of X_F hold also for X_{F,E_h}, either with the same proof, or because it is a base change of X_F. Let $y \in |Y_F|$ and let $\infty_h \in |X_{F,E_h}|$ be the $\phi^{h\mathbb{Z}}$-orbit of y, mapping to $\infty \in |X_F|$, which is its $\phi^{\mathbb{Z}}$-orbit. The points of X_{F,E_h} above ∞ are $\phi^i(\infty_h)$, $0 \leq i \leq h-1$.

Just as we constructed a $t \in B_F^{\varphi = p}$ whose divisor (in $|Y_F|$) was $\sum_{i\in\mathbb{Z}} \phi^i[y]$, we can construct a $t_h \in B_F^{\varphi^h = p}$ whose divisor is $\sum_{i\in\mathbb{Z}} \phi^{ih}[y]$. It will then be unique up to an E_h-multiple and will satisfy

$$\prod_{i=0}^{h-1} \varphi^i(t_h) = t$$

(up to a $\mathbb{Q}_p$-multiple). Recall how t was constructed. First, we found an $\varepsilon \in \mathfrak{m}_F$ such that

$$u_\varepsilon = \frac{[1+\varepsilon] - 1}{[1+\varepsilon^{1/p}] - 1}$$

was a primitive element of degree 1 in $W(\mathcal{O}_F)$ vanishing to the first order at y. Then we let

$$t = t_\varepsilon = \log([1+\varepsilon]).$$

In our case we let $q = p^h$ and consider the Lubin-Tate formal group law $F_Q(X, Y)$ over E_h attached (e.g., the choice of a model of this formal group will change things by an isomorphism) to the endomorphism

$$\{p\}_Q(X) = Q(x) = pX + X^q.$$

We let $Q_n = Q \circ \cdots \circ Q$ (n times) and define the *Q-twisted Teichmüller representative* to be

$$[\varepsilon]_Q = \lim_{n\to\infty} Q_n([\varepsilon^{1/q^n}]).$$

Here we should think of Q_n as a lifting of the n-th power of Frobenius of order q to characteristic 0. Note that if $Q = X^q$ we would get $[.]_Q = [.]$, while if $q = p$ and $Q = (1+X)^p - 1$, we would get $[\varepsilon]_Q = [1+\varepsilon] - 1$. In general, the limit exists and satisfies

$$Q([\varepsilon]_Q) = [\varepsilon^q]_Q.$$

If we put

$$u_{\varepsilon,Q} = \frac{[\varepsilon]_Q}{[\varepsilon^{1/q}]_Q}$$

then as before we get a primitive element of degree 1 of $W(\mathcal{O}_F)$, every primitive ideal of degree 1 is of the form $(u_{\varepsilon,Q})$, and ε is unique up to $\varepsilon \mapsto \{a\}_Q(\varepsilon)$, where $a \in \mathcal{O}_{E_h}^\times$. We may therefore select ε so that $u_{\varepsilon,Q}$ vanishes at a given y. Furthermore,

$$\{p\}_Q(\varepsilon) = Q(\varepsilon) = \varepsilon^q.$$

If $\log_Q(X) = X + \cdots \in XE_h[[X]]$ is the logarithm of the Lubin-Tate group and we put

$$t_h = t_{\varepsilon,Q} = \log_Q([\varepsilon]_Q)$$

then we get

$$\varphi^h(t_h) = \log_Q([\varepsilon^q]_Q) = \log_Q(\{p\}_Q([\varepsilon]_Q)) = pt_h$$

as desired. The divisor of t_h on $|Y_F|$ is $\sum_{i\in\mathbb{Z}} \phi^{ih}[y]$.

Remark 62 More generally, Fargues and Fontaine make similar constructions for any $[E' : E] < \infty$, not necessarily unramified. Lubin Tate groups over E' play a similar role.

3.4 Construction of Vector Bundles on X_F

3.4.1 Operations on Vector Bundles

Quite generally, if $\pi : Y \to X$ is a finite étale covering of complete curves (in the sense discussed in the introduction) and $\mathcal{E}$, $\mathcal{F}$ are vector bundles on X, Y, respectively, $deg(\pi) = h$, then we have the vector bundles $\pi^*\mathcal{E}$ and $\pi_*\mathcal{F}$ and they satisfy

$$rk(\pi^*\mathcal{E}) = rk(\mathcal{E}), \quad deg(\pi^*\mathcal{E}) = h \cdot deg(\mathcal{E})$$

$$rk(\pi_*\mathcal{F}) = h \cdot rk(\mathcal{F}), \quad deg(\pi_*\mathcal{F}) = deg(\mathcal{F}).$$

The first two equalities are clear: for the rank there is nothing to prove, and for the degree it is enough to consider the case of the line bundle $\mathcal{O}(D)$, where D is a divisor of degree d on X. Then $\pi^*(\mathcal{O}(D)) = \mathcal{O}(\pi^{-1}(D))$ and $\pi^{-1}(D)$ is of degree hd. The second pair of equations follows from the first if we note, say in the Galois case, that

$$\pi^*\pi_*\mathcal{F} = \oplus_{\sigma \in Gal(Y/X)} \mathcal{F}^\sigma$$

and all the $\mathcal{F}^\sigma$ have the same degree. The general case is reduced as usual to the Galois case, but we shall actually be only concerned with the cyclic cover $X_{F,E_h} \to X_F$.

3.4.2 The Vector Bundles $\mathcal{O}(d, h)$

In this subsection we do not assume that d and h are relatively prime. As before, we have on X_{F,E_h} the line bundle $\mathcal{O}(d)$, whose global sections may be identified with $B_F^{\varphi^h = p^d}$. We define the vector bundle $\mathcal{O}(d, h)$ to be $\pi_*\mathcal{O}(d)$, where $\pi : X_{F,E_h} \to X_F$ is the degree h cyclic étale covering the constructed above. The advantage of this construction, besides being symmetrical and not requiring the distinguished point ∞, is that it comes out automatically to be a vector bundle of degree d and rank h.

Claim 63 This definition agrees with the sheaf $\mathcal{E}(D, \varphi_D)$ defined before, where $D = D_{d,h}$.

Proof Starting with the identification

$$X_{F,E_h} = Proj\left(\bigoplus_{k=0}^{\infty} B_F^{\varphi^h = p^k}\right) = Proj\left(\bigoplus_{k=0}^{\infty} B_F^{\varphi^h = p^{kh}}\right),$$

the line bundle $\mathcal{O}(d)$ on X_{F,E_h} corresponds to the graded module

$$\bigoplus_{k=0}^{\infty} B_F^{\varphi^h = p^{k+d}}$$

for the graded algebra on the left, or, for the graded algebra on the right, to the graded module

$$M = \bigoplus_{k=0}^{\infty} B_F^{\varphi^h = p^{kh+d}}.$$

Since $\pi : X_{F,E_h} \to X_F$ corresponds to the graded homomorphism

$$P = \bigoplus_{k=0}^{\infty} B_F^{\varphi = p^k} \hookrightarrow \bigoplus_{k=0}^{\infty} B_F^{\varphi^h = p^{kh}},$$

the vector bundle $\mathcal{O}(d, h) = \pi_* \mathcal{O}(d)$ corresponds to the same module M, regarded as a graded module over P. On the other hand, $\mathcal{E}(D, \varphi_D)$ was associated to the graded P-module

$$M' = \bigoplus_{k=0}^{\infty} (D \otimes B_F)^{\varphi = p^k}.$$

We therefore have to identify M'_k with M_k, i.e.,

$$(D_{d,h} \otimes_E B_F)^{\varphi = p^k} \simeq B_F^{\varphi^h = p^{kh+d}}.$$

Recall that $D_{d,h}$ had a basis $e_1, \ldots, e_h$ with $\varphi(e_i) = e_{i+1}$ for $i < h$ and $\varphi(e_h) = p^{-d} e_1$. We see that $x = \sum_{i=1}^h b_i e_i$ satisfies $\varphi(x) = p^k x$ if and only if $\varphi(b_i) = p^k b_{i+1}$ for $1 \le i < h$ and $\varphi(b_h) = p^{k+d} b_1$. The coefficient $b_1 \in B_F$ determines the remaining b_i uniquely, and the only condition imposed on it is $\varphi^h(b_1) = p^{kh+d} b_1$. The homomorphism associating to x the coefficient b_1 is therefore the desired isomorphism. □

This proves (i) and (ii) of Proposition 57. Part (iii) follows from the dictionary between vector bundles and B-pairs. Part (iv) will be checked below for the $\mathcal{O}(d, h)$. Note the consequence that

$$B_F[1/t]^{\varphi^h = p^d}$$

is free of rank h over $B_e = B_F[1/t]^{\varphi=1}$. Indeed, it is free of rank 1, with t_h^d as a generator, over $B_F[1/t]^{\varphi^h=1} = B_e \otimes_E E_h$.

3.4.3 Basic Properties of the $\mathcal{O}(d, h)$

Proposition 64 *(i) If $\delta = (d, h)$ then $\mathcal{O}(d, h) = \mathcal{O}(d/h)^{\oplus \delta}$.*

(ii) Write X_n for X_{F,E_n} and X for X_F. Then $\pi_{n*}(\mathcal{O}_{X_n}(d, h)) = \mathcal{O}_X(d, nh)$ and $\pi_n^*(\mathcal{O}_X(d, h)) = \mathcal{O}_{X_n}(nd, h)$.

(iii) The vector bundle $\mathcal{O}_X(d, h)$ is semi-stable of slope d/h.

(iv) We have $\mathcal{O}_X(d_1, h_1) \otimes \mathcal{O}_X(d_2, h_2) \simeq \mathcal{O}_X(d_1 h_2 + d_2 h_1, h_1 h_2)$ and $\mathcal{O}_X(d, h)^\vee \simeq \mathcal{O}_X(-d, h)$.

(v) For $\lambda > \mu$ we have $Hom(\mathcal{O}_X(\lambda), \mathcal{O}_X(\mu)) = 0$. For $\lambda \leq \mu$ we have

$$Ext^1(\mathcal{O}_X(\lambda), \mathcal{O}_X(\mu)) = 0.$$

Proof (i) and (ii) follow from the definitions and standard facts (the projection formula). For example, if $d = d_1 h$ then

$$\begin{aligned}\mathcal{O}_X(d,h) &= \pi_{h*}(\mathcal{O}_{X_h}(d)) = \pi_{h*}(\pi_h^*\mathcal{O}_X(d_1)) = \mathcal{O}_X(d_1) \otimes \pi_{h*}\pi_h^*\mathcal{O}_{X_h} \\ &= \mathcal{O}_X(d_1) \otimes \mathcal{O}_{X_h}^h = \mathcal{O}_X(d_1)^h.\end{aligned}$$

We next show how (iii) is deduced from (i) and (ii). Since π_h^* multiplies the slope of any vector bundle by h, it is enough to show that $\pi_h^*\mathcal{O}_X(d,h)$ is semi-stable on X_h. But

$$\pi_h^*\mathcal{O}_X(d,h) = \mathcal{O}_{X_h}(dh,h) = \mathcal{O}_{X_h}(d)^{\oplus h}.$$

For any line bundle $\mathcal{O}(d)$, the vector bundle $\mathcal{O}_X(d)^{\oplus h}$ is semi-stable. In fact, on any curve an extension of semi-stable vector bundles of the same slope is again semi-stable.

For (iv) it is enough to assume, by (i), that $(h_1, h_2) = 1$. Consider then the coverings

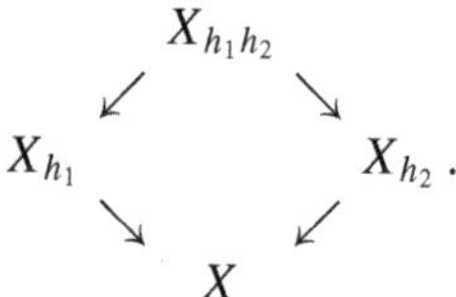

We have

$$\mathcal{O}_X(d_1,h_1) \otimes \mathcal{O}_X(d_2,h_2) = \pi_{h_1*}(\mathcal{O}_{X_{h_1}}(d_1)) \otimes \pi_{h_2*}(\mathcal{O}_{X_{h_2}}(d_2))$$

$$= \pi_{h_1h_2*}(\pi_{h_1h_2,h_1}^*\mathcal{O}_{X_{h_1}}(d_1) \otimes \pi_{h_1h_2,h_2}^*\mathcal{O}_{X_{h_2}}(d_2))$$

$$= \pi_{h_1h_2*}(\mathcal{O}_{X_{h_1h_2}}(h_2d_1 + h_1d_2)) = \mathcal{O}_X(d_1h_2 + d_2h_1, h_1h_2).$$

Here the passage from first to second line comes from the fact that for A-algebras B_1 and B_2, and modules M_i over B_i there is a canonical isomorphism

$$M_1 \otimes_A M_2 \simeq (B_2 \otimes_A M_1) \otimes_{B_2 \otimes_A B_1} (B_1 \otimes_A M_2).$$

For (v) the claim about Hom follows from the fact that $\mathcal{O}_X(\mu)$ is semi-stable of slope μ, and from the fact that if $\mathcal{E}' \to \mathcal{E}$ is an injective homomorphism of vector bundles of the same rank then $deg(\mathcal{E}') \leq deg(\mathcal{E})$. This last fact is property HN2(iii) of the abstract Harder-Narasimhan formalism, whose verification boils down to the case of line-bundles, where it becomes obvious.

For Ext note that

$$\begin{aligned} Ext^1(\mathcal{O}_X(\lambda), \mathcal{O}_X(\mu)) &\simeq H^1(X, \mathcal{O}_X(-\lambda) \otimes \mathcal{O}_X(\mu)) \\ &\simeq H^1(X, \mathcal{O}_X(\mu - \lambda))^m \end{aligned}$$

for some natural number m (see (i)). If $\mu - \lambda = d/h \geq 0$ then

$$H^1(X, \mathcal{O}_X(d/h)) = H^1(X, \pi_{h*}(\mathcal{O}_{X_h}(d))) = H^1(X_h, \mathcal{O}_{X_h}(d)) = 0$$

by what we already know about the cohomology of line bundles. □

Corollary 65 *If (D, φ_D) is a semi-stable isocrystal over $k = \bar{\mathbb{F}}_p$, then $\mathcal{E}(D, \varphi_D)$ is a semi-stable vector bundle on X_F.*

4 An Application to Galois Representations: Weakly Admissible Equals Admissible

We still have to complete the proof of the classification theorem (Theorem 59). But before we dive into it, we want to give an application to a deep theorem of p-adic Hodge theory, which for the first time brings in Galois representations.

4.1 Fontaine's Formalism of B-Admissible Galois Representations

4.1.1 Galois Representations and the General Strategy

Let $E = \mathbb{Q}_p$ (more generally, we could take E to be any finite extension of $\mathbb{Q}_p$). Let $G_E = Gal(\bar{E}/E)$. By a "Galois representation" we shall understand a finite-dimensional $\mathbb{Q}_p$-vector space V with a continuous G_E action. Let $Rep_E = Rep_E(\mathbb{Q}_p)$ be the Tannakian category of all such representations.

So far, our theory of the Fargues-Fontaine curve was functorially built from a given algebraically closed complete valued field F. If we fix an algebraic closure $\bar{E}$, its completion $\widehat{\bar{E}} = \mathbb{C}_p$ inherits a G_E-action, hence so do $F = \mathbb{C}_p^\flat$ and all the rings built from it: $W(\mathcal{O}_F)$, $B^{b,+} \subset B^b$, the completions B_I and in particular $B_F = B_{(0,1)}$, the rings B_e, $B_{dR,\infty}^+ \subset B_{dR,\infty}$ and $B_{cris,\infty}$. Here we implicitly use the fact that the G_E-action commutes with φ, and that the distinguished point ∞ (corresponding to the choice of $(\mathbb{C}_p, \iota_{can})$ as an untilt) is fixed by G_E. We shall drop the subscript ∞ from B_{dR} and B_{cris}.

Lemma 66 $B_{dR}^{G_E} = B_{cris}^{G_E} = E.$

Proof Since $B_{cris} \subset B_{dR}$ it is enough to prove that $B_{dR}^{G_E} = E$, or $B_{dR}^{+G_E} = E$. This follows from the fact that

$$Fil^i B_{dR}^+ / Fil^{i+1} B_{dR}^+ \simeq \mathbb{C}_p(i)$$

and from Tate's basic theorem that

$$H^0(G_E, \mathbb{C}_p(i)) = 0$$

if $i > 0$ and $H^0(G_E, \mathbb{C}_p) = E$. □

Fontaine's strategy for studying the category Rep_E was to associate to a Galois representation V and some "rings of periods" B, carrying a G_E action, the group

$$D_B(V) = (B \otimes_E V)^{G_E},$$

where G_E acts on the tensor product diagonally. Assuming also that $B^{G_E} = E$ (this fixed subring should be a field in general, but it need not be E, a simplifying assumption that we make here), $D_B(V)$ becomes an E-vector space, and under a fairly general assumption, which holds for B_{dR} and B_{cris}, for example, we have

$$\dim D_B(V) \leq \dim V.$$

The representation V is called "B-admissible" (or *de-Rham*, resp. *crystalline*, if $B = B_{dR}$ resp. B_{cris}) if equality holds here. The full subcategory $Rep_{E,B}$ of B-admissible Galois representations is mapped via D_B faithfully to the category of E-vector spaces and the functor D_B is exact. In general, it is impossible to reconstruct V from $D_B(V)$, even if V is B-admissible, as the functor D_B is not fully faithful.

The "game" is to endow B with "extra structure" that commutes with G_E, so is inherited by $D_B(V)$, which can now be considered as an object of a *more refined category* than just vector spaces. This category will still be considerably simpler than the category of B-admissible Galois representations. The hope is that if enough "extra structure" is imposed, (a) the functor D_B will become fully faithful, (b) its essential image will be describable in terms of the extra structure, and (c) a formula will be given to retrieve V from D, if D is in the image of D_B.

Two such examples of extra structure come to mind: B_{dR} is equipped with a filtration $Fil^\bullet$ and B_{cris} with a Frobenius φ. It turns out that each of these extra structures alone is not sufficient to answer our hopes and retrieve V, but both together do. If we use both we get an object in the category of *filtered φ-modules* (filtered isocrystals), and it can be shown rather easily that this object must be *semi-stable of slope 0*, or equivalently, *weakly admissible*. Fontaine conjectured that the category of weakly admissible filtered φ-modules, as a full subcategory of the category of all filtered φ-modules, is an *abelian tensor category*, and in fact isomorphic, under D_{cris}, to the category of crystalline representations. It should be mentioned that a B_{cris}-

admissible representation is also B_{dR}-admissible, so although the filtration comes from B_{dR}, the functor to be studied is D_{cris}.

Rather than develop the general formalism of B-admissibility, we focus from now on these two examples, and let $B = B_{dR}$ or B_{cris}. It should be mentioned that Galois representations coming from (geometric) p-adic (étale) cohomology of proper smooth varieties over $\mathbb{Q}_p$ with *good reduction* are crystalline. There is a richer ring of periods, B_{st}, that carries another piece of structure—a monodromy operator N—which enables to detect and reconstruct "semi-stable" representations, typically those coming from p-adic cohomology of proper smooth varieties with *semi-stable* reduction. We do not discuss the semi-stable case in these lectures.

Fontaine's conjecture has the non-trivial consequence that the category of weakly admissible filtered φ-modules is closed under tensor products, an analogue of the Faltings-Totaro theorem. Fontaine's conjecture was proved in 2000 by Colmez and Fontaine. A second proof was given by Berger. The Fargues-Fontaine curve, and the classification of vector bundles over it, enables us to give a short and elegant proof, which we describe in this section.

4.1.2 de-Rham and Crystalline Representations

Let V be a p-adic representation, $B = B_{dR}$ or B_{cris}, and $D = D_B(V) = (B \otimes_E V)^{G_E}$. Recall that $B_{cris} \subset B_{dR}$ and that B_{dR} is a complete discrete valuation field.

Proposition 67 *(i)* $\dim D \leq \dim V$.

(ii) If equality holds (i.e., V is B-admissible), then the natural map $\alpha_B : B \otimes_E D \to B \otimes_E V$ is an isomorphism

$$B \otimes_E D \simeq B \otimes_E V.$$

(iii) If equality holds for $D = D_{cris}$ (i.e., V is crystalline) then equality also holds for $D = D_{dR}$ (i.e., V is de Rham) and $D = D_{cris}(V) = D_{dR}(V)$.

(iv) If V is crystalline, then endowing D with a structure of a filtered φ-module (possible in view of (iii)),

$$V = V_{cris}(D) = (B_{cris} \otimes_E D)^{\varphi=1} \cap Fil^0(B_{dR} \otimes_E D).$$

Here φ acts diagonally, the filtration is the tensor product filtration, and the resulting Galois action on V comes from the G_E action on B_{cris} or B_{dR}.

Proof We start with $B = B_{dR}$, which is a field, and prove first that α_B is injective. Assume that $0 \neq \sum_{i=1}^r b_i \otimes d_i$ is an element in $\ker(\alpha_B)$ and r is minimal. Without loss of generality, $b_1 = 1$. Let $\sigma \in G_E$. Applying σ and subtracting we get

$$\sum_{i=2}^{r} (\sigma b_i - b_i) \otimes d_i \in \ker(\alpha_B),$$

so from the minimality of r all $b_i \in B_{dR}^{G_E} = E$, which implies $r = 1$, and $d_1 = 0$, a contradiction. The injectivity of α_B immediately proves (i), and (ii) is a formal consequence, counting dimensions over B_{dR}.

Now take $B = B_{cris} \subset B_{dR}$. Since $D_{cris}(V) \subset D_{dR}(V)$ (i) is clear. It is also clear that if V is crystalline, it is de Rham, and $D_{cris}(V) = D_{dR}(V)$, hence (iii). To prove (ii) for B_{cris} we need the following property of B_{cris} (incorporated in Fontaine's axioms for a "regular" ring of periods B; B_{cris} is hence regular):

- If $0 \neq b \in B_{cris}$ and the line Eb is G_E-stable, then $b \in B_{cris}^{\times}$.

Assume this property for the moment, let V be crystalline, let $v_1, \ldots, v_r$ be a basis of V and $d_1, \ldots, d_r$ a basis of $D = D_{cris}(V) = D_{dR}(V)$. Write

$$d_i = \sum b_{ij} \otimes v_j, \quad b_{ij} \in B_{cris}.$$

The injectivity of α_B implies that $b = \det(b_{ij}) \neq 0$. Since

$$d_1 \wedge \cdots \wedge d_r = b(v_1 \wedge \cdots \wedge v_r)$$

the left-hand side is G_E-invariant. As $Ev_1 \wedge \cdots \wedge v_r$ is G_E-stable, we see that Eb is G_E-stable, hence $b = \det(\alpha_B)$ is invertible and α_B is an isomorphism.

To prove the property of B_{cris} used above let

$$\breve{E} = W(\bar{\mathbb{F}}_p)[1/p] = \widehat{\mathbb{Q}_p^{nr}}.$$

We claim that if $b \in B_{cris}$ is such that $\breve{E}b$ is G_E-stable, then $\breve{E}b = \breve{E}t^i$ for some i. This will imply of course our property. Going over to B_{dR} and twisting by a suitable power of t we may assume that $b \in B_{dR}^+$, and its image $\theta(b)$ modulo $Fil^1 = \mathfrak{m}_{dR}$ is non-zero. We get an element $\theta(b) \in \mathbb{C}_p$ such that $\breve{E}\theta(b)$ is G_E-stable. A theorem of Sen [19] (answering a question of Serre) implies then that $\theta(b)$ is algebraic over $\breve{E}$. It follows that the action of $G_{\breve{E}} = I_E \subset G_E$ (the inertia subgroup) on $\theta(b)$ factors through a finite quotient. But the $\breve{E}$-line $\breve{E}b \subset B_{cris}$ is also I_E-stable and θ is injective on it. Thus the action of I_E on b factors through a finite quotient as well and b is algebraic over $\breve{E}$. It follows that $\breve{E}[b]$ is a field and b^{-1} belongs to it (in fact, it can be shown that $b \in \breve{E}$).

(iv) Using (ii), assuming V is crystalline and letting $D = D_{cris}(V)$,

$$(B_{cris} \otimes_E D)^{\varphi=1} \cap Fil^0(B_{dR} \otimes_E D) = (B_{cris} \otimes_E V)^{\varphi=1} \cap Fil^0(B_{dR} \otimes_E V)$$

$$= (B_{cris}^{\varphi=1} \otimes_E V) \cap (B_{dR}^+ \otimes_E V) = (B_{cris}^{\varphi=1} \cap B_{dR}^+) \otimes_E V = V$$

since, by the fundamental exact sequence of p-adic Hodge theory, $B_{cris}^{\varphi=1} \cap B_{dR}^+ = E$.
□

Corollary 68 *If V is one-dimensional and crystalline, then $\breve{E} \otimes_E V \simeq \breve{E}(i)$ for some i (a Tate twist). Consequently, the inertia group $I_p \subset G_E$ acts on V via a power of the cyclotomic character.*

Proof If v is a basis of V and $b \in B_{cris}$ is such that $d = b \otimes v$ is G_E-invariant, then G_E preserves the line Eb, and the proof of the Proposition shows that I_E acts on b via a power of the cyclotomic character (as it acts on t^i). □

Remark 69 If E is a finite extension of $\mathbb{Q}_p$ and V is a one-dimensional crystalline $\mathbb{Q}_p$-representation of G_E then the same stays true. However, if V is a one-dimensional crystalline E-representation then there are more possibilities: for example, I_E can act on V via a Lubin-Tate character associated with E, $\chi_{LT} : G_E \to \mathcal{O}_E^\times$. In general, p-adic Hodge theory over bases $\mathbb{Q}_p \subsetneq E$ has "cyclotomic analogues" or "Lubin-Tate analogues" of the results over $\mathbb{Q}_p$, depending on the coefficients. Similarly, there are (at least) two natural candidates for Γ in the theory of (φ, Γ)-modules: $\Gamma_{cycl} = Gal(E(\mu_{p^\infty})/E)$, or $\Gamma_{LT,\pi}$. The latter is much larger and depends on the prime π.

4.1.3 Admissible Is Weakly Admissible

We consider the functor

$$D_{cris} : Rep_{E,cris} \to \varphi ModFil_E$$

from the category of crystalline Galois representations into the category of filtered φ-modules over E. Notice that this is essentially the category of filtered isocrystals discussed before, except that we consider filtered isocrystals over $\mathbb{F}_p$ (i.e., E-vector spaces) and not over $\bar{\mathbb{F}}_p$ (i.e., $\breve{E}$-vector spaces). As we have seen in the Proposition, the functor is fully faithful. In fact, a quasi-inverse was found in part (iv) of the Proposition: it takes $D \in \varphi ModFil_E$ to

$$V_{cris}(D) = (B_{cris} \otimes_E D)^{\varphi=1} \cap Fil^0(B_{dR} \otimes_E D).$$

It remains to identify the essential image of D_{cris}.

Definition 70 A $D \in \varphi ModFil_E$ is called *admissible* if it is in the essential image of D_{cris}. It is called *weakly admissible* if it is semi-stable of slope 0.

Recall that *semi-stable of slope* 0 means that

$$t_H(D) := \sum i \dim gr^i D = \nu_E(\det(\Phi)) =: t_N(D),$$

where Φ is a matrix representing φ in some basis, and for every strict subobject $D' \subset D$ (strict = with the induced filtration) there is an inequality $\leq$ in the above equation. (The symbols t_H and t_N refer to the end points of the Hodge and Newton polygons.)

Proposition 71 *If D is admissible, then it is weakly admissible.*

Proof Assume $D = D_{cris}(V)$ and let $r = \dim V$. Then $\bigwedge^r D = D_{cris}(\bigwedge^r V)$ and so $\bigwedge^r V$ is crystalline. We have $t_H(\bigwedge^r D) = t_H(D)$ and similarly for t_N. This is tautological for t_N and an easy exercise in filtered vector spaces for t_H, which we leave to the reader. Thus to prove that $t_H(D) = t_N(D)$ we may assume that D is one-dimensional. But V is then a one-dimensional representation on which I_p acts via χ^i, where χ is the cyclotomic character. In this case both t_H and t_N can be computed directly and come out to be $-i$. *Exercise*: compute them for $\mathbb{Q}_p(i)$ and show that they are unchanged by an unramified twist. In fact, the filtration is not affected by such a twist, so t_H is clearly unchanged. As for t_N, since we assumed that $E = \mathbb{Q}_p$ for simplicity, φ is linear and is given by $\Phi \in \mathbb{Q}_p^\times$. An unramified twist of V is reflected in D in a change of Φ by $u \in \mathbb{Z}_p^\times$. Over $\mathcal{O}_{\breve{E}}$, $u = \sigma(v)/v$ where σ is the arithmetic Frobenius, so all such $D's$ become isomorphic over $\breve{E}$.

Next, we have to show that for D' a strict subobject of $D = D_{cris}(V)$, we have

$$t_H(D') \le t_N(D')$$

("The Hodge polygon lies below the Newton polygon"). For this we follow [1], Theorem 9.3.4.[9] Let $s = \dim D' \le r = \dim D$. Replacing D' and D by

$$\bigwedge^s D' \subset \bigwedge^s D$$

we may assume, without loss of generality, that $s = 1$, i.e., D' is a φ-stable line in D with the induced filtration.

Assume therefore that $D' = Ee' \subset D = D_{cris}(V)$. Applying a Tate twist to V results in a Tate twist of all the objects, so we may assume that $t_H(D') = 0$ and show that $t_N(D') \ge 0$. Let $\varphi(e') = \lambda e'$ where $t_N(D') = n = ord_p(\lambda)$. Let $v_1, \dots, v_r$ be a basis of V and write

$$e' = \sum_{i=1}^r b_i \otimes v_i \in B_{cris} \otimes_E V.$$

By the assumption that $t_H(D') = 0$ all the $b_i \in B_{dR}^+$, and one of them, say $b_1 \notin \mathfrak{m}_{dR}^+$. Since

$$\lambda e' = \varphi(e') = \sum_{i=1}^r \varphi(b_i) \otimes v_i$$

we have $\varphi(b_1) = \lambda b_1$. It is enough to prove that if $ord_p(\lambda) = n < 0$ then

$$B_{cris}^{\varphi=\lambda} \cap B_{dR}^+ = 0.$$

[9] Although Fontaine must have known the proof, the one given in "Périodes p-adiques", Proposition 5.4.2 is flawed!

Write $\lambda = up^n$ where $u \in \mathbb{Z}_p^\times$. Twisting by t^{-n} we have to show that

$$B_{cris}^{\varphi=u} \cap t^{-n} B_{dR}^+ = 0.$$

Let b belong to this group. Let $v \in \breve{E}$ be a unit such that $\varphi(v)/v = u$. Replacing b by $v^{-1}b$ we may assume that $u = 1$. But then the fundamental exact sequence implies that $B_{cris}^{\varphi=1} \cap \mathfrak{m}_{dR}^+ = 0$, contradicting $n < 0$. □

4.2 *Weak Admissibility Implies Admissibility*

4.2.1 Vector Bundles Associated with Filtered φ-Modules Over E.

If $(\breve{D}, \varphi_{\breve{D}})$ was a φ-module (over $\breve{E}$, i.e., an isocrystal over $\bar{\mathbb{F}}_p$), we associated to it the vector bundle $\mathcal{E}(\breve{D}, \varphi_{\breve{D}})$ on the Fargues-Fontaine curve X_F. In terms of (B_e, B_{dR}^+)-pairs, it was given by the pair $((B_F[1/t] \otimes_{\breve{E}} \breve{D})^{\varphi=1}, B_{dR}^+ \otimes_{\breve{E}} \breve{D})$.

Suppose $(D, \varphi_D, Fil^\bullet)$ is a filtered φ-module over $(K_0, K) = (E, E)$. We let $\mathcal{E}(D, \varphi_D, Fil^\bullet)$ be the vector bundle associated with the pair $((B_F[1/t] \otimes_E D)^{\varphi=1}, Fil^0(B_{dR} \otimes_E D))$.

Note that if $Fil^0 D = D,\ Fil^1 D = 0$, then we recover $\mathcal{E}(\breve{D}, \varphi_{\breve{D}})$, where $\breve{D} = \breve{E} \otimes_E D$. Clearly $\mathcal{E}$ is a functor from the category of filtered φ-modules to $\mathcal{V}\mathcal{B}_X$. As a particular example, consider a homomorphism

$$(D, \varphi_D, Fil_1^\bullet) \to (D, \varphi_D, Fil_2^\bullet).$$

This amounts to the identity on D and two filtrations satisfying $Fil_1^i \subset Fil_2^i$. The induced map on $(B_F[1/t] \otimes_E D)^{\varphi=1}$ is the identity, so we get that $\mathcal{E}(D, \varphi_D, Fil_1^\bullet) \to \mathcal{E}(D, \varphi_D, Fil_2^\bullet)$ is a modification at ∞, i.e., an injective homomorphism of vector bundles which is an isomorphism away from ∞. The cokernel is a skyscraper sheaf at ∞. We record even a more special case in the next Proposition.

Proposition 72 *Let $(D, \varphi_D, Fil^\bullet)$ be a filtered φ-module such that $Fil^0 D = D$. Then there is an exact sequence of sheaves on X_F*

$$0 \to \mathcal{E}(D, \varphi_D) \to \mathcal{E}(D, \varphi_D, Fil^\bullet) \to i_{\infty *}(Fil^0(B_{dR} \otimes_E D)/B_{dR}^+ \otimes_E D)) \to 0.$$

We recall that if $Fil^{i_0} D = 0$ then

$$Fil^0(B_{dR} \otimes_E D) = \sum_{i=0}^{i_0 - 1} t^{-i} B_{dR}^+ \otimes_E Fil^i D.$$

In particular, if $Fil^2 D = 0$, the last term in the exact sequence (the skyscraper sheaf) is $\mathbb{C}_p(-1) \otimes Fil^1 D$. Such a modification of $\mathcal{E}(D, \varphi_D)$, in which the skyscraper sheaf at ∞ is killed by t, hence is a $B_{dR}^+/(t) = \mathbb{C}_p$-vector space, is called *minuscule*.

By the main classification theorem (yet to be proved!)

$$\mathcal{E}(D, \varphi_D, Fil^\bullet) \simeq \mathcal{E}(D', \varphi_{D'})$$

for some φ-module $(D', \varphi_{D'})$. Which one we get is a subtle question, and depends on the relative position of the filtration w.r.t φ.

4.2.2 G_E-Equivariant Vector Bundles

There is an extra structure we can impose on the vector bundles $\mathcal{E}(D, \varphi_D, Fil^\bullet)$ for $(D, \varphi_D, Fil^\bullet) \in \varphi ModFil_E$. Recall that previously, when we constructed the vector bundles $\mathcal{O}(\lambda)$, we started with the category of isocrystals over $\bar{\mathbb{F}}_p$, i.e., the category $\varphi Mod_{\breve{E}}$. Now, even if we ignore the filtration (i.e., put the trivial filtration), we are using the category of isocrystals over $\mathbb{F}_p$, which is richer: every isocrystal over the algebraic closure has an $\mathbb{F}_p$-structure, but this structure is not unique. In addition, our F and all the resulting rings now carry a G_E-action (something that did not exist for a general F).

The Galois action allows us, as usual, to put on $\mathcal{E}(D, \varphi_D, Fil^\bullet)$ a structure of an G_E-*equivariant* vector bundle, i.e., for every $\sigma \in G_E$ an isomorphism

$$c_\sigma : \sigma^* \mathcal{E} \simeq \mathcal{E}$$

where σ^* denotes the pull-back w.r.t. the map induced by σ on X_F, such that the cocycle condition

$$c_{\sigma\tau} = c_\tau \circ \tau^*(c_\sigma)$$

holds. In terms of B-pairs, c_σ is induced by the action of σ on $((B_F[1/t] \otimes_E D)^{\varphi=1}, Fil^0(B_{dR} \otimes D))$ arising from its action on $B_F[1/t]$ and B_{dR}. The fact that this action is *semi-linear* over (B_e, B_{dR}^+), exactly means that it translates to an $\mathcal{O}_X$-*linear* isomorphism between $\sigma^*\mathcal{E}$ and $\mathcal{E}$.

The G_E-equivariant vector bundles on X_F form a category[10] $\mathcal{VB}_{X_F}^{G_E}$. We have therefore constructed a functor

$$\mathcal{E} : \varphi ModFil_E \rightsquigarrow \mathcal{VB}_{X_F}^{G_E}.$$

If $\mathcal{E}$ is a G_E-equivariant vector bundle, then $V = H^0(X_F, \mathcal{E})$ inherits a G_E-action and is therefore an (infinite dimensional in general) Galois representation over E.

---

[10] The G_E-orbit of every point of X_E other than ∞ is infinite. The quotient of X_F under G_E therefore does not exist as a scheme. If we wanted to "descend" an equivariant vector bundle we would have to do it in some stacky sense.

Similarly, $H^0(X_F \setminus \{\infty\}, \mathcal{E})$ is a semi-linear Galois representation, free over B_e of rank equal to $rk(\mathcal{E})$.

Finally, we remark that there is another "cheap" way to get Galois equivariant vector bundles, associating to $V \in Rep_E$ the equivariant vector bundle $V \otimes_E \mathcal{O}_X$, which will be semi-stable of slope 0. In fact, every semi-stable slope 0 equivariant vector bundle is of this sort and the functors

$$V \rightsquigarrow V \otimes_E \mathcal{O}_X = \mathcal{E}, \quad \mathcal{E} \rightsquigarrow H^0(X_F, \mathcal{E}) = V$$

are easily seen to define an equivalence of categories

$$Rep_E \approx \{\text{Semi-stable slope 0 equivariant vector bundles}\}.$$

The proof of this is the same as the proof of the equivalence

$$Vect_E \approx \{\text{Semi-stable slope 0 vector bundles}\}$$

(resulting from the basic classification theorem for vector bundles), enriching both categories with Galois action. For the details, see [5].

The intersection of the two families, i.e., the semi-stable, slope 0 equivariant vector bundles of the form $\mathcal{E}(D, \varphi_D, Fil^\bullet)$, will correspond to the full subcategory $Rep_{E,cris}$ of crystalline representations. But this is non-trivial, and is an equivalent formulation of the Colmez-Fontaine theorem that "admissible equals weakly admissible".

The following diagram summarizes the situation:

$$\begin{array}{ccccc} Rep_E & \ni V \mapsto \mathcal{E} = V \otimes_E \mathcal{O}_X \mapsto & H^0(X_F, \mathcal{E}) & = V \\ \cup & \| \qquad\qquad \| & \| & \| \\ Rep_{E,cris} & \ni V \mapsto \quad \mathcal{E}(D_{cris}(V)) \quad \mapsto & V_{cris}(D_{cris}(V)) & = V \end{array}.$$

4.2.3 Slopes and Semi-stability

The next theorem generalizes what we have proved before, without the filtration.

Theorem 73 *(i) The functor $\mathcal{E} : \varphi ModFil \rightsquigarrow \mathcal{VB}_{X_F}^{G_E}$ preserves rank and degree. (ii) $(D, \varphi_D, Fil^\bullet)$ is semi-stable if and only if so is $\mathcal{E}(D, \varphi_D, Fil^\bullet)$.*

Recall that

$$deg(D, \varphi_D, Fil^\bullet) = \sum_i i \dim(gr^i D) - ord_p(\det \Phi).$$

In the example given above of a minuscule modification, the graded pieces are non-zero for $i = 0, 1$ only, so this is $\dim Fil^1 D - ord_p(\det \Phi)$, while the degree of (D, φ_D) is just $-ord_p(\det \Phi)$. Thus the increase in slope due to the filtration is

$$\mu(\mathcal{E}(D, \varphi_D, Fil^\bullet)) - \mu(\mathcal{E}(D, \varphi_D)) = \frac{\dim Fil^1 D}{\dim D}.$$

Proof (i) The statement about the rank is obvious. The statement about the degree follows from the case of the trivial filtration proved already, and from the fact that if $\alpha : \mathcal{E} \hookrightarrow \mathcal{E}'$ is a modification at ∞, $\deg(\mathcal{E}') - \deg(\mathcal{E}) = length_{B_{dR}^+}(\mathrm{coker}(\alpha))$.

(ii) This follows from a more precise (and more general) statement, that if

$$0 \subset D_1 \subset \cdots \subset D_r = D$$

is the Harder-Narasimhan filtration of the filtered φ-module $D = (D, \varphi_D, Fil^\bullet)$, then

$$0 \subset \mathcal{E}(D_1) \subset \cdots \subset \mathcal{E}(D_r) = \mathcal{E}(D)$$

is the Harder-Narasimhan filtration of the vector bundle $\mathcal{E}(D)$. Now the Harder-Narasimhan filtration of a G_E-equivariant vector bundle is clearly a filtration by equivariant sub-bundles (by its uniqueness). Therefore, in conjunction with (i), it suffices to show that any equivariant vector sub-bundle $\mathcal{E}'$ of $\mathcal{E} = \mathcal{E}(D)$ is $\mathcal{E}(D')$ for a (strict) φ- sub-module $D' \subset D$.

Let $M = M_e(D) = (B_F[1/t] \otimes_E D)^{\varphi=1} = H^0(X \setminus \{\infty\}, \mathcal{E})$, a free B_e-module of rank $r = \dim D$. By the Galois equivariance of $\mathcal{E}$ it carries a G_E-action, which is of course compatible with the the Galois action on $B_F[1/t]$ and the trivial action on D. Let $M' = H^0(X \setminus \{\infty\}, \mathcal{E}')$ and $M'' = M/M'$. Since $\mathcal{E}'$ is locally a direct summand and B_e is a PID, M' and M'' are free over B_e. Put

$$D' = (B_F[1/t] \otimes_{B_e} M')^{G_E}$$

(here we use the fact that since $\mathcal{E}'$ is Galois equivariant G_E preserves M') and define D'' similarly. These are φ-modules where the action of φ comes from its action on $B_F[1/t]$.

By Galois cohomology, we have an exact sequence of φ-modules

$$0 \to D' \to D \to D''.$$

We have $\dim_E D' \leq rk_{B_e} M' = rk(\mathcal{E}')$, $\dim_E D'' \leq rk_{B_e} M'' = rk(\mathcal{E}'')$ (where $\mathcal{E}'' = \mathcal{E}/\mathcal{E}'$). The proof of this is the same as the proof that $\dim_E D_{cris}(V) \leq \dim_E V$. (It is easier to prove the stronger claim with B_{dR} replacing $B_F[1/t]$, since B_{dR} is a field.) Clearly, $rk(\mathcal{E}) = rk(\mathcal{E}') + rk(\mathcal{E}'')$. It follows, by dimension counting, that the short exact sequence is also exact on the right and that the inequalities are equalities. In particular

$$\dim_E D' = rk_{B_e} M'.$$

Now

$$B_F[1/t] \otimes_E D' \subset B_F[1/t] \otimes_{B_e} M'$$

and if $\dim_E D' = rk_{B_e} M'$ we must have an equality. The proof of this is the same as the proof of (i) and (ii) in Proposition 67. Taking φ-invariants and recalling that $B_F[1/t]^{\varphi=1} = B_e$ we get

$$(B_F[1/t] \otimes_E D')^{\varphi=1} = M' = H^0(X \setminus \{\infty\}, \mathcal{E}').$$

Similarly for M''. Finally, we put on D' and D'' the filtration induced from the one of D and verify that the completion of the stalk of $\mathcal{E}'$ at ∞ is given by $Fil^0(B_{dR} \otimes_E D')$ and similarly for $\mathcal{E}''$. This completes the proof that $\mathcal{E}' = \mathcal{E}(D')$ for a strict φ-submodule $D' \subset D$, hence, in view of (i), that the Harder-Narasimhan filtration of $\mathcal{E}$ is obtained from the one of D. □

4.2.4 The Functor $\mathcal{E}(V)$

We can now compose the functor $\mathcal{E}(-)$ with the functor D_{cris} to get a functor

$$Rep_{E,cris} \rightsquigarrow \mathcal{V}\mathcal{B}_{X_F}^{G_E},$$

that we denote also by

$$V \mapsto \mathcal{E}(V) := \mathcal{E}(D_{cris}(V)).$$

Since a crystalline representation may be reproduced from $D = D_{cris}(V)$ as

$$V = V_{cris}(D_{cris}(V)) = (B_{cris} \otimes_E D)^{\varphi=1} \cap Fil^0(B_{dR} \otimes_E D),$$

and since, as mentioned before, $(B_{cris} \otimes_E D)^{\varphi=1} = (B_F[1/t] \otimes_E D)^{\varphi=1}$, we conclude that if V is crystalline,

$$V = H^0(X_F, \mathcal{E}(V)).$$

As we have seen, if V is crystalline then $D_{cris}(V)$ is weakly admissible, i.e., semi-stable of slope 0. This implies that $\mathcal{E}(V)$ is also semi-stable of slope 0, and by the corollary of the classification theorem

$$\mathcal{E}(V) \simeq V \otimes \mathcal{O}_X.$$

4.2.5 The Main Theorem

Theorem 74 (Colmez-Fontaine) *Let $D = (D, \varphi_D, Fil^\bullet)$ be a weakly admissible filtered φ-module over E of dimension r. Then*

$$V = V_{cris}(D)$$

is an r-dimensional crystalline representation, and $D = D_{cris}(V)$, i.e., D is admissible.

Proof As we have seen, slope and semi-stability are preserved by the functor $\mathcal{E}(-)$, so $\mathcal{E}(D)$ is an equivariant semi-stable vector bundle of slope 0. Being semi-stable of slope 0, the classification theorem shows that $\mathcal{E}$ is a trivial vector bundle, so $V = V_{cris}(D) = H^0(X_F, \mathcal{E})$ is also r-dimensional and $\mathcal{E} \simeq V \otimes_E \mathcal{O}_X$ (with Galois action). This was the crucial step. □

We will show that the equality of dimensions $\dim_E V_{cris}(D) = \dim_E D$ forces

$$D_{cris}(V) = D.$$

Thus V is crystalline (because $D_{cris}(V)$ is of maximal possible dimension) and D is admissible (because it is D_{cris} of a crystalline representation).

We need a lemma, which settles the 1-dimensional case.

Lemma 75 *Let D be a 1-dimensional filtered φ-module. Then:*

i) If $t_H(D) < t_N(D)$ then $V_{cris}(D) = 0$.

(ii) If $t_H(D) = t_N(D)$ then $V_{cris}(D)$ is one-dimensional and crystalline and $D_{cris}(V_{cris}(D)) = D$. Furthermore if d is a basis of D and $v = bd$ ($b \in B_{cris}$) is a basis of V then $b \in B_{cris}^{\times}$.

(iii) If $t_H(D) > t_N(D)$ then $V_{cris}(D)$ is infinite dimensional.

Let us assume the validity of the lemma for the moment, and finish the proof of the theorem. Let C_{cris} be the field of fractions of B_{cris}. Since $C_{cris} \subset B_{dR}$, $C_{cris}^{G_E} = E$. Consider the canonical map

$$\alpha : C_{cris} \otimes_E V \to C_{cris} \otimes_E D$$

(recall $V \subset B_{cris} \otimes_E D$). Since its image is a C_{cris}-subspace stable under G_E, it is of the form $C_{cris} \otimes_E D'$ for a subspace $D' \subset D$. Since $Im(\alpha)$ is φ-stable, so is D'. Equipping D' with the filtration induced from the filtration of D, it becomes a strict sub-filtered φ-module. Let $s = \dim D' \leq r$. We have

$$V \subset V_{cris}(D') \subset V_{cris}(D) = V$$

so they are all equal. Let $v_1, \ldots, v_s$ be elements of V such that $\alpha(1 \otimes v_i)$ is a basis of $C_{cris} \otimes D'$. Let $d_1, \ldots, d_s$ be a basis of D'. Write

$$\alpha(1 \otimes v_i) = \sum b_{ij} \otimes d_j$$

with $b_{ij} \in B_{cris}$. Then $b = \det(b_{ij}) \neq 0$. Let

$$0 \neq W = \bigwedge^s V = V_{cris}(\bigwedge^s D').$$

By the assumption that D is weakly admissible, $t_H(\bigwedge^s D') = t_H(D') \le t_N(D') = t_N(\bigwedge^s D')$. Since W does not vanish, the lemma implies that $t_H(D') = t_N(D')$, that W is 1-dimensional, hence $r = s$, $D' = D$, and α is an isomorphism. Furthermore, $b \in B_{cris}^{\times}$ by part (ii) of the lemma, so α is an isomorphism also over B_{cris} :

$$\alpha : B_{cris} \otimes_E V \simeq B_{cris} \otimes_E D.$$

But this isomorphism respects the Galois action so $D = D_{cris}(V)$, showing that V is crystalline and that D is admissible.

To conclude, we prove the lemma.

Proof Let $D = Ed$, where $\varphi(d) = \lambda d$ and the filtration of d is $t_H(D)$. Tate-twisting we may assume that $t_H(D) = 0$. Then $n = t_N(D) = ord_p(\lambda)$. We have

$$V_{cris}(D) \simeq B_{cris}^{\varphi=\lambda^{-1}} \cap B_{dR}^{+}.$$

Applying an unramified twist we may assume that

$$V_{cris}(D) \simeq B_{cris}^{\varphi=p^{-n}} \cap B_{dR}^{+} = t^{-n}(B_{cris}^{\varphi=1} \cap \mathfrak{m}_{dR}^{n}).$$

The lemma is now clear. □

5 The Classification Theorem

5.1 *Preparations*

5.1.1 A Lemma on Harder-Narasimhan Filtrations in Finite étale Galois Coverings

Lemma 76 *Let $f : X \to Y$ be a finite étale Galois morphism of complete curves (in the sense discussed in these notes). Let $\mathcal{E}$ be a vector bundle on Y and*

$$0 = \mathcal{E}_0 \subset \mathcal{E}_1 \subset \cdots \subset \mathcal{E}_r = \mathcal{E}$$

its Harder-Narasimhan filtration. Then

$$0 = f^*\mathcal{E}_0 \subset f^*\mathcal{E}_1 \subset \cdots \subset f^*\mathcal{E}_r = f^*\mathcal{E}$$

is the Harder-Narasimhan filtration of $f^*\mathcal{E}$. In particular $f^*\mathcal{E}$ is semi-stable if and only if $\mathcal{E}$ is.

Proof By uniqueness, the Harder-Narasimhan filtration of $f^*\mathcal{E}$ is Galois-stable, so since f is finite étale it is pulled back from some filtration of $\mathcal{E}$. The lemma is an immediate consequence of the fact that pulling back by f multiplies the slope by $d = \deg(f)$. □

5.1.2 Equivariant Structures on a Vector Bundle

Let X be a scheme, equipped with an action of a finite group Γ. Let $(\mathcal{E}, (c_\sigma)_{\sigma\in\Gamma})$ be a Γ-equivariant vector bundle. Thus

$$c_\sigma : \sigma^*\mathcal{E} \simeq \mathcal{E}, \quad c_\tau \circ \tau^* c_\sigma = c_{\sigma\tau}.$$

Then Γ acts on the group $Aut(\mathcal{E})$ of automorphisms of $\mathcal{E}$ *on the right* via

$$f \mapsto f^\sigma = c_\sigma \circ \sigma^* f \circ c_\sigma^{-1}.$$

If (c'_σ) is another structure of an equivariant vector bundle on $\mathcal{E}$ then

$$d_\sigma = c_\sigma \circ (c'_\sigma)^{-1} \in Aut(\mathcal{E})$$

and $(d_\sigma) \in Z^1(\Gamma, Aut(\mathcal{E}))$ is a 1-cocycle, i.e.,

$$d_{\sigma\tau} = \tau(d_\sigma) \circ d_\tau.$$

Conversely, if (d_σ) satisfies this condition, $c'_\sigma = d_\sigma^{-1} \circ c_\sigma$ is another structure of an equivariant vector bundle. The following proposition is standard.

Proposition 77 *Two equivariant structures on $\mathcal{E}$ are isomorphic (i.e., there is an automorphism of $\mathcal{E}$ carrying one structure to the other) if and only if the cocycles (c_σ) and (c'_σ) differ by a coboundary (i.e., are cohomologous). The set of equivariant structures on a given $\mathcal{E}$, up to isomorphism, is either empty or a (set-theoretic) torsor for $H^1(\Gamma, Aut(\mathcal{E}))$.*

5.1.3 Pure Vector Bundles on the Fargues-Fontaine Curve

Recall the vector bundles $\mathcal{O}_X(\lambda)$ $(\lambda \in \mathbb{Q})$ that were constructed on the Fargues-Fontaine curve X. We call a vector bundle $\mathcal{E}$ *pure* if it is isomorphic to $\mathcal{O}_X(\lambda)^a$ for some a.

Corollary 78 *Let $\pi_h : X_h \to X$ be the cyclic degree h covering of the Fargues-Fontaine curve associated with the unramified field extension E_h/E of degree h. Then a vector bundle $\mathcal{E}$ on X is pure if and only if $\pi_h^*\mathcal{E}$ is pure on X_h.*

Proof We have seen the "only if" before. Assume that $\pi_h^*\mathcal{E}$ is pure. Enlarging h if necessary we may assume that

$$\pi_h^*\mathcal{E} \simeq \mathcal{O}_{X_h}(d)^a$$

where $d \in \mathbb{Z}$. Since the degree of $\pi_h^*\mathcal{E}$ must be a multiple of h, we get $h|ad$. We therefore have

$$\pi_h^*\mathcal{E} \simeq \mathcal{O}_{X_h}(d)^a \simeq \mathcal{O}_{X_h}(ad, a) = \pi_h^*\mathcal{O}_X(\frac{ad}{h}, a) =: \mathcal{F}.$$

The isomorphism classes of equivariant vector bundle structures on $\mathcal{F}$ are, according to the last Proposition, in a bijection with

$$H^1(\mathbb{Z}/h\mathbb{Z}, Aut(\mathcal{F})).$$

But $Aut(\mathcal{F}) = GL_a(E_h)$, so by Hilbert's theorem 90 the last pointed set is a point, and the equivariant structure is unique. This implies that $\mathcal{F}$ can be descended in a unique way, so $\mathcal{E} \simeq \mathcal{O}_X(\frac{ad}{h}, a)$. □

5.2 An Abstract Classification Theorem

The following criterion, leading to the conclusion that every vector bundle on the Fargues-Fontaine curve is a direct sum of $\mathcal{O}(\lambda)$'s, was inspired by the work of Kedlaya on slope filtrations.

Theorem 79 *Consider the following statement.* ***Criterion(X)****: for every vector bundle $\mathcal{E}$ on X and for every $n \geq 1$, if we have a short exact sequence*

$$0 \to \mathcal{O}_X(-\frac{1}{n}) \to \mathcal{E} \to \mathcal{O}_X(1) \to 0,$$

then $H^0(X, \mathcal{E}) \neq 0$.

Suppose we prove ***Criterion(X),*** *and that the same is true if X is replaced by its cyclic unramified covering X_h of degree h for every $h \geq 1$, i.e.,* ***Criterion(X_h)*** *also holds. Then:*

(i) The semi-stable vector bundles on X are the pure ones.

(ii) The Harder-Narasimhan filtration of X is split.

(iii) Every vector bundle on X is isomorphic to

$$\bigoplus_{i=1}^{n} \mathcal{O}_X(\lambda_i)$$

for unique rational numbers $\lambda_1 \geq \lambda_2 \geq \cdots \geq \lambda_n$.

Furthermore, if we prove ***Criterion(X_h)*** *only for vector bundles $\mathcal{E}$ of rank $\leq r$ (and every h), then (i)-(iii) hold for vector bundles of rank $\leq r$.*

Proof Observe first that (i) implies (ii) and (iii). To deduce (ii) from (i) use $Ext^1(\mathcal{O}(\lambda), \mathcal{O}(\mu)) = 0$ if $\mu \geq \lambda$. Then (iii) is a formal consequence of (i) and (ii). Furthermore, if we prove (i) for ranks $\leq r$, then the same argument gives (ii) and (iii) for ranks $\leq r$.

The proof of (i) will be by induction on the rank n of the semi-stable vector bundle $\mathcal{E}$. When $n = 1$ (i) is obvious. We assume that (i) is proved up to rank n and let $rk(\mathcal{E}) = n + 1$.

Step 1. Since $\mathcal{E}$ is semi-stable (resp. pure) if and only if $\pi_h^*\mathcal{E}$ is, we may pull back to X_h and assume that $\mu(\mathcal{E}) \in \mathbb{Z}$. Since twisting by a line bundle does not change the conclusion, we may assume now that $\mu(\mathcal{E}) = 0$. Replacing X by X_h we may therefore, without loss of generality, assume that $\mathcal{E}$ *is semi-stable of slope 0* and prove that it is trivial. This is therefore the key case, and also, incidentally, the only case of the classification theorem that was needed in the proof of "weakly admissible is admissible".

Step 2. Let $n + 1 = rk(\mathcal{E})$ and consider $\pi_n^*\mathcal{E}$. Let $\mathcal{L} \subset \pi_n^*\mathcal{E}$ be a line sub-bundle of maximal degree d and consider the short exact sequence

$$0 \to \mathcal{L} \to \pi_n^*\mathcal{E} \to \mathcal{E}' \to 0.$$

By the semi-stability of $\pi_n^*\mathcal{E}$ we have $d \leq 0 \leq \mu(\mathcal{E}')$.

Step 4. Assume $d = 0$. In this case $\mathcal{L}$ is semi-stable of slope 0. But the category of semi-stable vector bundles of slope 0 (or such objects in any Harder-Narasimhan category) is an abelian category closed under kernels, quotients, and extensions, so $\mathcal{E}'$ is also such. Applying the induction hypothesis on the rank, $\mathcal{E}'$ is trivial. Since $Ext^1(\mathcal{O}_{X_n}^n, \mathcal{O}_{X_n}) = 0$, $\pi_n^*\mathcal{E}$ is trivial, hence so is $\mathcal{E}$.

Step 5. It is impossible to have $d \leq -2$. Since $\mu(\mathcal{E}') \geq 0$, applying induction on the rank we know that there exists a $\lambda \geq 0$ with $\mathcal{O}_{X_n}(\lambda)$ a sub-bundle of $\mathcal{E}'$. Since $d \leq -2$

$$Hom(\mathcal{O}_{X_n}(d+2), \mathcal{O}_{X_n}(\lambda)) = H^0(X_n, \mathcal{O}_{X_n}(\lambda - d - 2)) \neq 0$$

so there exists a non-zero homomorphism

$$u : \mathcal{O}_{X_n}(d+2) \to \mathcal{O}_{X_n}(\lambda) \subset \mathcal{E}'.$$

Pulling back the short exact sequence which defined $\mathcal{E}'$ we get a short exact sequence

$$0 \to \mathcal{L} \to \mathcal{E}'' \to \mathcal{O}_{X_n}(d+2) \to 0$$

or, after twisting

$$0 \to \mathcal{L}(-d-1) \to \mathcal{E}''(-d-1) \to \mathcal{O}_{X_n}(1) \to 1.$$

Since $\mathcal{L}(-d-1) \simeq \mathcal{O}_{X_n}(-1)$, by our assumption the vector bundle in the middle has a global section, i.e., a non-zero homomorphism

$$\mathcal{O}_{X_n}(d+1) \to \mathcal{E}''.$$

The morphism $\mathcal{E}'' \to \pi_n^*\mathcal{E}$ is a monomorphism, so we get a non-zero homomorphism

$$v : \mathcal{O}_{X_n}(d+1) \to \pi_n^*\mathcal{E}.$$

The image of v spans a line sub-bundle whose degree is $\geq d+1$ (the inequality comes from the fact that v need not be a local direct factor, but then taking the line sub-bundle spanned by the image of v, the degree only grows). This contradicts the maximality of d.

Step 6. It is impossible to have $d = -1$. Since π_n is finite étale, π_{n*} is the left adjoint of π_n^* (on modules), if $R \to R'$ is a finite étale ring extension

$$Hom_{R'}(L', R' \otimes_R M) = Hom_R(L', M),$$

the map from the LHS to the RHS uses properties of $Tr_{R'/R}$ in finite étale extensions). Thus we have a non-zero

$$u \in Hom_{X_n}(\mathcal{L}, \pi_n^*\mathcal{E}) = Hom_X(\pi_{n*}\mathcal{L}, \mathcal{E}) = Hom_X(\mathcal{O}_X(-\frac{1}{n}), \mathcal{E}).$$

Denote by $Im(u)$ the vector sub-bundle spanned by the image of u in $\mathcal{E}$. Since $\mathcal{O}_X(-\frac{1}{n})$ is semi-stable of slope $-1/n$ and

$$\mathcal{O}_X(-\frac{1}{n})/ker(u) \to Im(u)$$

is an isomorphism in the generic fiber (i.e., up to a torsion module), we get that $\mu(Im(u)) \geq -1/n$. Since $\mathcal{E}$ was assumed to be semi-stable of slope 0 we must have

$$-\frac{1}{n} \leq \mu(Im(u)) \leq 0.$$

Since the rank of $Im(u)$ is at most n we can have only two possibilities

$$\mu(Im(u)) = 0, -\frac{1}{n}.$$

(i) If $\mu(Im(u)) = 0$ then its slope is 0, it is semi-stable as a sub-bundle of $\mathcal{E}$, and so is $\mathcal{E}/Im(u)$, so by induction they are both trivial, and so is $\mathcal{E}$ (no non-trivial extensions of trivial modules!).

(ii) If $\mu(Im(u)) = -1/n$, $rk(Im(u)) = n$ and u is a monomorphism. But since

$$u : \mathcal{O}_X(-\frac{1}{n}) \to Im(u)$$

is an isomorphism on the generic fiber and both vector bundles have degree -1, it is an isomorphism. We have

$$0 \to \mathcal{O}_X(-\frac{1}{n}) \to \mathcal{E} \to \mathcal{L}' \to 0$$

and $\mathcal{L}'$ is a line bundle of degree 1. But by our assumption this implies that there is a non-zero homomorphism $\mathcal{O}_X \to \mathcal{E}$, whose image would be a line-bundle of degree ≥ 0, pulling back to a similar line bundle in $\pi_n^*\mathcal{E}$, contradicting $d = -1$. □

5.3 *Some Constructions Related to p-Divisible Groups*

For the proof of the classification theorem on the Fargues-Fontaine curve we need to understand the relation between modification of vector bundles on X and period maps of p-divisible groups. We take now a rather long detour to review some aspects of the crystalline theory of p-divisible groups. It is due to Grothendieck, Messing, Katz, Mazur and later developments by Kottwitz, Rapoport, and Zink.

The pace will be quick, compared to the rest of the notes, and we will assume some familiarity with formal groups and p-divisible groups. For basics on p-divisible groups see [3, 7, 13, 15, 22, 23]. For the crystalline theory, see [9, 14]. For moduli and period maps, see [4, 6, 12, 17]. Much of what we do below is explained, besides the book by Fargues and Fontaine, also in the paper [18]. These lists are far from complete, of course, and notation and approach vary from one reference to another. We warn the reader that it is often a non-trivial exercise to reconcile the explicit power-series-based approach of Hazewinkel with the sheaf-theoretic point of view of Messing, or the Lubin-Tate and the Rapoport-Zink formalisms of moduli (when they both apply).

5.3.1 Adic Rings

We work over an adic ring R over $\mathbb{Z}_p$. Recall that an *adic* ring is a topological ring which is complete and separated in the I-adic topology for some ideal $I \subset R$, i.e.,

$$R \simeq \lim_{\leftarrow} R/I^n.$$

The ideal I is not part of the data, only the topology induced by it. Any ideal J satisfying $J^n \subset I$, $I^m \subset J$ for some m, n gives the same topology, and is called an ideal of definition.

Caution: If R is noetherian and I is any ideal, then the completion $\widehat{R} = \lim_{\leftarrow} R/I^n$ is an adic ring with $\widehat{I} = I\widehat{R}$ as an ideal of definition. But in an arbitrary ring R, if I is not finitely generated, funny things can happen.

Adic rings with continuous homomorphisms form a category. The category of adic rings over $\mathbb{Z}_p$ is, as usual, the relative category of continuous morphisms $\mathbb{Z}_p \to R$ in the category of adic rings. Note that if R is an adic ring over $\mathbb{Z}_p$ we may always assume that p belongs to the ideal of definition of R (in any case, a power of p will belong to any ideal of definition).

Any discrete ring in which $p^N = 0$ is trivially an adic ring over $\mathbb{Z}_p$.

In this section we write C for a complete algebraically closed valued field of characteristic 0 and residual characteristic p, such as $\mathbb{C}_p$, and $C^\flat$ for its tilt.

5.3.2 The Universal Covering of a p-Divisible Group

Let $R \in \mathrm{Adic}_{\mathbb{Z}_p}$ and let $G_{/R}$ be a p-divisible group. For $S \in \mathrm{Adic}_R$ we *re-define* the notion of points of G by

$$G(S) = \varprojlim G(S/I_S^m)$$

where I_S is any ideal of definition for S. If $G = (G_n)$ (i.e., $G_n = G[p^n]$) then

$$G(S) = \lim_{\leftarrow m} G(S/I_S^m) = \lim_{\leftarrow m} \lim_{\to n} G_n(S/I_S^m).$$

For example, if $G = \mu_{p^\infty}$ we get

$$G(\mathcal{O}_C) = \lim_{\leftarrow m} \lim_{\to n} \mu_{p^n}(\mathcal{O}_C/p^m) = 1 + \mathfrak{m}_C = \widehat{\mathbb{G}}_m(\mathcal{O}_C).$$

One cannot change the order of the limit and the colimit! This example is typical. If G is the p-divisible group associated to a p-divisible formal group law $\widehat{G}$ over R, then

$$G(S) = \widehat{G}(S).$$

On the other hand, if S is discrete (so some $I_S^m = 0$) then $G(S)$ is the usual notion of points of an ind-scheme. Note that, with our modified definition of $G(S)$, we still have

$$G(S)[p^n] = G[p^n](S).$$

If S is discrete this is so by definition, and, in general, it follows from

$$G[p^n](S) = \lim_{\leftarrow m} G[p^n](S/I_S^m).$$

We let Ab be the category of abelian groups.

Definition 80 The universal covering $\widetilde{G}$ of G is the functor $\mathrm{Adic}_R \rightsquigarrow \mathrm{Ab}$

$$\widetilde{G}(S) = \lim_{\leftarrow \times p} G(S) = \{(x_0, x_1, \ldots) \mid x_i \in G(S),\ [p]_G(x_{i+1}) = x_i\}.$$

Here are the main (easy) points about this definition.

(1) This is a presheaf of $\mathbb{Q}_p$-vector spaces. Multiplication by p^{-1} is *left shift*.
(2) If $G = \mathbb{Q}_p/\mathbb{Z}_p$ then $\widetilde{G} = \mathbb{Q}_p$. If $G = \mu_{p^\infty}$ then $\widetilde{G}(\mathcal{O}_C) = 1 + \mathfrak{m}_{C^\flat}$.
(3) If G and G' are isogenous then any isogeny between them induces an *isomorphism* on the universal coverings.

The next property is not difficult either but is of fundamental importance, so we single it out as a lemma.

Lemma 81 *(Crystalline nature of the universal covering) Let $S \in \mathrm{Adic}_R$. If I is a closed topologically nilpotent ideal of S (in particular if it is an ideal of definition) reduction modulo I induces a bijection*

$$\widetilde{G}(S) = \widetilde{G}(S/I).$$

Proof *(sketch)* Let $y = (y_0, y_1, \dots) \in \widetilde{G}(S/I)$. Let $z_i \in G(S)$ lift y_i. Defining $x_i = \lim[p^j](z_{i+j})$, the limit exists, is independent of the lifting, and defines the unique $\widetilde{G}(S) \ni x \mapsto y$. In the case of μ_{p^∞} this goes back to the computations we did when we gave the two equivalent definitions of the tilt. □

Corollary 82 *Let $T \twoheadrightarrow S \in \mathrm{Adic}_R$ be a pro-nilpotent thickening (i.e., the kernel of $T \to S$ is pro-nilpotent, or equivalently, contained in some ideal of definition of T). Let G' be a lifting of G to T. Then*

$$\widetilde{G}'(T) = \widetilde{G}'(S) = \widetilde{G}(S),$$

hence depends functorially only on G and not on the lifting G'.

Proposition 83 *Assume that R is a perfect $\mathbb{F}_p$-algebra. Let $\widehat{G}$ be a commutative formal group law in d variables, and assume that the resulting formal group functor*[11] *is p-divisible (this is equivalent to $R[X_1, \dots, X_d]$ being finite flat over $[p]^*_{\widehat{G}}(R[X_1, \dots, X_d])$). Let*

$$G = (\widehat{G}[p^n])_{n=1}^{\infty}$$

be the associated p-divisible group (such a p-divisible group is called formal *or* connected*). Then (locally on R) the functor $\widetilde{G}$ is represented by a formal scheme*

$$\widetilde{G} = Spf(R[[X_1^{1/p^\infty}, \dots, X_d^{1/p^\infty}]]).$$

Proof See [18]. The key idea is to replace the $\lim_{\times p}$ with $\lim_{\times F}$ where F is the Frobenius morphism. For this we need to consider

$$G \overset{F}{\leftarrow} G^{(p^{-1})} \overset{F}{\leftarrow} G^{(p^{-2})} \overset{F}{\leftarrow} \cdots$$

[11] For $S \in \mathrm{Adic}_R$ $\widehat{G}(S)$ is the set of d-tuples of topologically nilpotent elements of S, with addition given by $\widehat{G}$.

and this is why we need R to be perfect. □

Let H be a p-divisible group over $\mathcal{O}_C$. Let $k = \mathcal{O}_C/\mathfrak{m}_C$ (an algebraically closed field of characteristic p). Since $k = \mathcal{O}_L/p$ where $L = \breve{\mathbb{Q}}_p = W(k)[1/p]$, there is a canonical section to $\mathcal{O}_C/p \twoheadrightarrow k$, which allows us to view $\mathcal{O}_C/p$ as a k-vector space. Suppose there exists a p-divisible group $\mathbb{H}$ over k and a quasi-isogeny

$$\rho : H \times_{\mathcal{O}_C} \mathcal{O}_C/p \dashrightarrow \mathbb{H} \times_k \mathcal{O}_C/p$$

(this means that the reduction modulo p of H is *isotrivial*: is isogenous to a trivial p-divisible group, i.e., one that is extended from k by base change; by a deep theorem of Scholze and Weinstein, using the fact that C is perfectoid, this is *always* the case.)

Corollary 84 *Under the above assumptions*

$$\widetilde{H} \simeq Spf(\mathcal{O}_C[[X_1^{1/p^\infty}, \ldots, X_d^{1/p^\infty}]]).$$

Proof We can take

$$\mathbb{H} = H \times_{\mathcal{O}_C} k,$$

because ρ would anyhow specialize to a quasi-isogeny between the two, so we are allowed to replace $\mathbb{H}$ in our assumption by $H \times_{\mathcal{O}_C} k$.

By the crystalline nature of $\widetilde{H}$ we have for any $S \in \mathrm{Adic}_{\mathcal{O}_C}$, $\widetilde{H}(S) = \widetilde{H}(S/p)$. By the invariance of $\widetilde{H}$ under isogeny and our assumption the given quasi-isogeny

$$\rho_S : H \times_{\mathcal{O}_C} S/p \dashrightarrow \mathbb{H} \times_k S/p$$

induces $\widetilde{H}(S/p) \simeq \widetilde{\mathbb{H}}(S/p)$. Since k is perfect and $\mathbb{H} = H \times_{\mathcal{O}_C} k$ the corollary follows from the last Proposition. Note that to give a sequence of p-power compatible elements in S is the same as giving a similar sequence in S/p, by the usual tilting argument. □

5.3.3 The Tate Module

Let $G = (G_n)$, $G_n = G[p^n]$ be a p-divisible group over $R \in \mathrm{Adic}_{\mathbb{Z}_p}$. Define the functor $\mathrm{Adic}_R \rightsquigarrow \mathrm{Ab}$

$$T_pG(S) = \lim_{\substack{\leftarrow \\ \times p}} G[p^n](S), \quad V_pG(S) = T_pG(S) \otimes_{\mathbb{Z}_p} \mathbb{Q}_p.$$

Clearly, $T_pG(S)$ is a sub-module of $\widetilde{G}(S)$, consisting of the sequences $(x_0, x_1, \ldots)$ with $x_0 = 0$. Since the latter is a $\mathbb{Q}_p$-vector space, so is $V_pG(S)$.

Caution: If G' is a lifting of G_S to a pro-nilpotent thickening $T \twoheadrightarrow S$ then we have seen that $\widetilde{G}'(T) = \widetilde{G}(S)$ depends only on G. But the subspace $V_pG'(T)$ *very much* depends on the lifting G' and on T.

Goal: Let G be a p-divisible group over $\mathcal{O}_C$. Denote by G_0 its special fiber over k. Assume that $G \times_{\mathcal{O}_C} \mathcal{O}_C/p$ and $G_0 \times_k \mathcal{O}_C/p$ are isogenous (as mentioned above, this is the case by a theorem of Scholze and Weinstein). As we have seen, any such quasi-igoney ρ between them induces

$$\rho : \widetilde{G}(\mathcal{O}_C) = \widetilde{G}(\mathcal{O}_C/p) \simeq \widetilde{G_0}(\mathcal{O}_C/p),$$

where we emphasize that the RHS depends functorially only on G_0. We would like to construct a long exact sequence **(LOG)**

$$\begin{array}{ccccc} 0 \to V_pG(\mathcal{O}_C) \to & \widetilde{G}(\mathcal{O}_C) & \overset{\theta}{\to} Lie(G)[1/p] \to 0 \\ & ||\rho & \\ & \widetilde{G_0}(\mathcal{O}_C/p) & \end{array}$$

and later on recover it as the global sections of an exact sequence of a modification of vector bundles on the Fargues-Fontaine curve. Observe that $Lie(G)[1/p]$ is a d-dimensional C-vector space, while $V_pG(\mathcal{O}_C)$ is the usual rational Tate module, which is a $\mathbb{Q}_p$-vector space of dimension $h = ht(G)$, so the term in the middle is likely to be again one of these Banach-Colmez spaces mentioned before (and deliberately avoided in our notes).

5.3.4 Logarithms

The construction of the exact sequence we want is based on the logarithm of the p-divisible group. This is standard and we briefly describe it when G is formal (G obtained from a p-divisible formal group $\widehat{G}$ as before). By means of the connected-étale exact sequence, the case of a general G can be reduced to the case of a formal one.

Thus let $G_{/R}$ be a formal p -divisible group, and denote by $\widehat{G} = Spf(A),\ \ A = R[[X_1, \ldots, X_d]]$ the corresponding formal group so that $G_n = \widehat{G}[p^n]$.

The cotangent space of G at the origin may be identified, as usual, with the translation invariant differentials, and these are all *closed:*

$$\omega_{G/R} = \oplus_{i=1}^{d} R dX_i|_0 \simeq \{\omega \in \Omega_{A/R} |\, m^*(\omega) = \omega \otimes 1 + 1 \otimes \omega\}.$$

Assume that R is flat over $\mathbb{Z}_p$ (i.e., contains no p-torsion). The *formal Poincaré lemma,* whose proof on the level of formal power series is elementary, says that any translation invariant ω as above admits a unique primitive $\lambda_\omega \in A[1/p]$ without constant term so that $d\lambda_\omega = \omega$; furthermore, $\lambda_\omega \in Hom_{R[1/p]}(G, \hat{\mathbb{G}}_a)$.

Consider the map $\omega \mapsto \lambda_\omega$. This canonical map is a homomorphism (over $R[1/p]$)

$$(\omega \mapsto \lambda_\omega) \in Hom(\underline{\omega_{G/R}}, \underline{Hom}(G, \hat{\mathbb{G}}_a)) = Hom(G, \underline{Hom}(\omega_{G/R}, \hat{\mathbb{G}}_a)).$$

Recalling that $\underline{Lie}(G) = \underline{Hom}(\omega_{G/R}, \hat{\mathbb{G}}_a)$, we have defined the *logarithm of G*

$$\log_G \in Hom_{R[1/p]}(G, \underline{Lie}(G)).$$

Let $\theta = \log_G \circ pr_0$ where $pr_0 : \widetilde{G} \to G$ is $x \mapsto x_0$. Then it is easily verified that (LOG) is exact. If $pr_0 : \widetilde{G}(S) \to G(S)$ is surjective, θ is surjective too. This is the case for $S = \mathcal{O}_C$. We shall soon see that θ is related to the "θ" of Fontaine's rings.

5.3.5 The Universal Vectorial Extension

Let $G_{/R}$ be a p-divisible group, and assume for the moment that $p^N = 0$ in R. The sequence of fppf sheaves on the category Alg_R of R-algebras

$$0 \to G[p^n] \to G \xrightarrow{p^n} G \to 0$$

is exact. Applying $RHom(-, \mathbb{G}_a)$ we get a short exact sequence (SES)

$$0 \to Hom(G, \mathbb{G}_a)/p^n \to Hom(G[p^n], \mathbb{G}_a) \to Ext(G, \mathbb{G}_a)[p^n] \to 0.$$

Observe first that $Hom(G, \mathbb{G}_a) = 0$ since G is p-divisible but $p^N \mathbb{G}_a = 0$. Next, if $n \geq N$ then

$$Ext(G, \mathbb{G}_a) = Ext(G, \mathbb{G}_a)[p^n] = Hom(G[p^n], \mathbb{G}_a) = \{a \in A_n | \, m_G^*(a) = a \otimes 1 + 1 \otimes a\}$$

$$= Lie(G^\vee[p^n]) = Lie(G^\vee) = Hom(\omega_{G^\vee}, R).$$

Here we have used Cartier duality, and wrote $G[p^n] = Spec(A_n)$. The primitive elements of $A_n = Hom(A_n^\vee, R)$ (i.e., the elements satisfying $m_G^*(a) = a \otimes 1 + 1 \otimes a$), when regarded as linear functionals from $A_n^\vee$ to R, define ring homomorphisms

$$A_n^\vee \to R[\varepsilon]/(\varepsilon^2)$$

reducing modulo ε to the co-unit homomorphism (the dual of the structure homomorphism $R \to A_n$). This is the same as defining an $R[\varepsilon]/(\varepsilon^2)$-valued point of $G^\vee[p^n]$ reducing to the identity, and this, by definition, is a point of the Lie algebra.

Similarly, for any R-module M and the associated vector group scheme $\underline{M}$, we have

$$Ext(G, \underline{M}) \simeq Hom(\omega_{G^\vee/R}, M).$$

Taking $M = \omega_{G^\vee/R}$ and the identity homomorphism on the right-hand side yields a **"universal" extension**

$$0 \to \underline{\omega}_{G^\vee/R} \to EG \to G \to 0$$

from which *any* (fppf sheaf) extension of G by a vector-group $\underline{M}$ is gotten by a unique push-out.

This construction is functorial, *covariant* in G. Finally, passing to the limit over N gives the same thing over a p-adic base R, i.e., a base satisfying $R \simeq \lim_{\leftarrow} R/p^N R$.

5.4 *The Crystal of a p-Divisible Group*

5.4.1 Rigidified Extensions of *G*

Apply the functor $\underline{Lie}(-)$ to the universal vectorial extension. Since $\underline{Lie}$ of a vector group is the vector group itself, we get a SES of vector groups ($MG = \underline{Lie}(EG)$)

$$\boxed{0 \to \underline{\omega}_{G^\vee/R} \to MG \to \underline{Lie}(G) \to 0.}$$

In [14] Messing shows that $\forall S \in \mathrm{Alg}_R,\ MG(S)$ is a locally free module, and that $rk(MG) = ht(G)$.

Our goal is to enhance MG to a *crystal* of modules on the crystalline site of R. What we need is an interpretation whereby MG classifies *rigidified extensions* of $G^\vee$ by $\mathbb{G}_a$. This is not a mistake: *The Lie algebra of the universal vectorial extension of G classifies rigidified extensions of the* dual *p-divisible group $G^\vee$ by $\mathbb{G}_a$*. Here is the definition.

Definition 85 A **rigidification** of an (fppf) extension E of $G^\vee$ by $\mathbb{G}_a$ is a splitting (of a SES of sheaves)

$$0 \to \mathbb{G}_a \to Lie(E) \overset{\leftarrow}{\to} Lie(G^\vee) \to 0.$$

Rigidifications always exist, locally for the fppf topology. Any two rigidifications differ by a homomorphism from $Lie(G^\vee)$ to $\mathbb{G}_a$, i.e., by an element of $\omega_{G^\vee/R}$. The group of rigidified extensions $Ext^\natural(G^\vee, \mathbb{G}_a)$ therefore sits in an exact sequence of sheaves as in the top row of the following diagram:

$$\begin{array}{ccccccc} 0 \to \omega_{G^\vee/R} & \to & Ext^\natural(G^\vee, \mathbb{G}_a) & \to & Ext(G^\vee, \mathbb{G}_a) & \to & 0 \\ \| & & \| & & \| & & \\ 0 \to \omega_{G^\vee/R} & \to & MG(R) & \to & Lie(G) & \to & 0. \end{array}$$

We have already commented on the identification $Ext(G, \mathbb{G}_a) \simeq Lie(G^\vee)$ (except that we have now reversed the roles of G and $G^\vee$). It is not surprising, and in fact proved in [14], that the middle terms are canonically identified as well, in a way that makes this whole diagram commutative.

5.4.2 The Crystalline Site

The big crystalline site over $R \in \mathrm{Alg}_{\mathbb{Z}_p}$ is based on the following category. *Objects* are diagrams

$$\begin{array}{ccc} T & \overset{pd}{\twoheadrightarrow} & S \\ & & \uparrow \\ & & R \end{array}$$

where $S \in \mathrm{Alg}_R$, and T is a *nilpotent divided powers* thickening of S. If S is $\mathbb{Z}_p$-flat this means that $x \in I = \ker(T \twoheadrightarrow S) \Rightarrow x^n/n! \in I$, and $\exists N$ s.t. $(x_1^{n_1}/n_1!) \cdots (x_r^{n_r}/n_r!) = 0$ if $x_i \in I$, $\sum n_i \geq N$.

If S is not $\mathbb{Z}_p$-flat the notion of a divided power (pd) thickening is more complicated, as the divided powers $\gamma_n(x) =$ "$x^n/n!$" are part of the *structure* that one has to give on I. This structure is subject to a list of axioms that guarantee that all the good consequences that we would like to draw from divided powers (e.g., standard results on exponentials) hold.

Morphisms are maps between such diagrams that "preserve the pd structure".

To complete the definition of the crystalline site we need to say which collections of morphisms are singled out as coverings (checking that they satisfy the axioms for coverings is easy). To simplify the notation we drop the reference to the ground ring R. *Coverings* of $T \overset{pd}{\twoheadrightarrow} S$ will be, just as in the Zariski topology, collections $\{(T_i \overset{pd}{\twoheadrightarrow} S_i) \leftarrow (T \overset{pd}{\twoheadrightarrow} S)\}$ s.t. $Spec(T) = \bigcup Spec(T_i)$ is a Zariski cover, and $S_i = S \otimes_T T_i$. Note that the arrows go backwards since we consider rings, not the associated affine schemes.

Finally, the *structure sheaf* is the sheaf $\mathcal{O}(T \overset{pd}{\twoheadrightarrow} S) = \mathcal{O}(T)$.

5.4.3 The Grothendieck-Messing Crystal

The key theorem, and the reason for introducing the crystalline site, at least in the context of p-divisible groups, is the following.

Theorem 86 (Grothendieck-Messing) *If $(T \overset{pd}{\twoheadrightarrow} S)$ is as above and G'_T is a lifting of G_S to T then MG'_T depends functorially only on G. Denote it by $MG(T \twoheadrightarrow S)$.*

We shall follow Katz in sketching a proof in a favorable case, that highlights the relevance of divided powers, and at the same time indicates an important relation between MG and de Rham cohomology. First, we make two remarks.

- *Liftings G'_T of G_S to any (not necessarily pd-) thickening $T \twoheadrightarrow S$ always exist by an old (1955) theorem of Lazard, see [7]. But there are many, and the whole point* *is that MG'_T is canonically independent of the lifting if the thickening is endowed with a pd structure.*

- *$MG(T \twoheadrightarrow S)$ is a locally free coherent sheaf of rank equal to $ht(G)$, and we shall denote $MG(S \twoheadrightarrow S)$ by $MG(S) = M(G_S)$.*

Explanation *(after Katz [9]): Start with $R \in \mathrm{Adic}_{\mathbb{Z}_p}$, $\mathbb{Z}_p$-flat (so no p-torsion). Let $\mathcal{F}_{/R}$ a p-divisible formal group,*

$$\mathcal{F} = Spf(R[[X_1, \ldots, X_d]]).$$

Consider

$$H^1_{dR}(\mathcal{F}/R) = \{[\eta] |\, \eta \; closed\ 1\text{-}form,\ m^*_{\mathcal{F}}(\eta) - \eta \otimes 1 - 1 \otimes \eta \; exact\}/\{\text{exact}\; \eta\}.$$

These are the translation invariant cohomology classes. Notice that it is the cohomology class $[\eta]$, and not the form η itself, that is required to be translation invariant. However, the translation-invariant forms (which are all closed) definitely define such classes. In fact, they inject *into H^1_{dR} :*

$$\omega_{\mathcal{F}/R} = \{\eta |\, m^*_{\mathcal{F}}(\eta) = \eta \otimes 1 + 1 \otimes \eta\} \hookrightarrow H^1_{dR}(\mathcal{F}/R),$$

because R is p-adic, and $\mathcal{F}$ is p-divisible, so if η is both exact and translation invariant $\Rightarrow \eta = 0$. (Logarithms "need" $R[1/p]$.)

Proposition 87 *There is a commutative diagram with exact rows (whose terms are explained below):*

$$\begin{array}{ccccccc} 0 \rightarrow \omega_{\mathcal{F}/R} & \rightarrow & H^1_{dR}(\mathcal{F}/R) & \xrightarrow{\partial} & H^2(\mathcal{F}; \mathbb{G}_a)_s & \rightarrow 0 \\ \| & & \| & & \| & \\ 0 \rightarrow \omega_{\mathcal{F}/R} & \rightarrow & Ext^{\natural}(\mathcal{F}; \mathbb{G}_a) & \rightarrow & Ext(\mathcal{F}; \mathbb{G}_a) & \rightarrow 0 \end{array}$$

Recall that $Ext^{\natural}(\mathcal{F}; \mathbb{G}_a)$ is the group of rigidified extensions, and that the bottom row has been identified, when $G^\vee \leftrightsquigarrow \mathcal{F}$, with

$$0 \rightarrow \omega_{G^\vee/R} \rightarrow MG(R) \rightarrow Lie(G) \rightarrow 0.$$

Proof (Sketch) We first explain the map ∂ (to ease the notation only when $d = \dim \mathcal{F} = 1$). The formal group law $\mathcal{F}$ gives rise to cohomology groups based on power-series manipulations. Specifically, in degree 2 we have

$$H^2(\mathcal{F}; \mathbb{G}_a)_s = \frac{\{\Delta(X, Y) \in R[[X, Y]]^+ |\, \text{symm.}, \delta(\Delta) = 0\}}{\{\delta(f) = f(X[+]Y) - f(X) - f(Y) |\, f \in R[[X]]^+\}}$$

$$\delta(\Delta) = \Delta(Y, Z) - \Delta(X[+]Y, Z) + \Delta(X, Y[+]Z) - \Delta(X, Y)$$

$$R[[X,Y]]^{+} = XR[[X,Y]] + YR[[X,Y]].$$

The map from $H^1_{dR}(\mathcal{F}/R)$ is the following: given $[\eta]$, find a primitive $f(X) \in R[1/p][[X]]^{+}$ for η, and let $\Delta = \delta(f)$. Then $[\eta]$ is translation invariant $\Rightarrow \Delta$ is integral: we have

$$\delta(\eta) = \delta(df) = d(\delta(f)) = d\Delta.$$

But we know that $\delta(\eta)$ is exact *over* R, hence, by the uniqueness of the primitive, Δ is integral. If we add to η an exact form, say dh with $h \in R[[X]]^{+}$, then we end up adding to Δ the power series $\delta(h)$. This means that we may define ∂ by setting $\partial([\eta]) = [\Delta]$.

The identification $H^2(\mathcal{F}; \mathbb{G}_a)_s \simeq Ext(\mathcal{F}; \mathbb{G}_a)$ is standard, and that of $H^1_{dR}(\mathcal{F}/R) \simeq Ext^{\natural}(\mathcal{F}; \mathbb{G}_a)$ requires only a little more work. The key to the crystalline nature of MG is the following lemma. □

Lemma 88 *Let $\mathcal{F}', \mathcal{F}''$ be liftings of $\mathcal{F}$ to $T \overset{pd}{\twoheadrightarrow} R$. Let $\varphi : \mathcal{F}' \to \mathcal{F}''$ be a morphism* of pointed Lie varieties *reducing to the identity on R. Then (i) $\varphi^* : H^1_{dR}(\mathcal{F}''/T) \simeq H^1_{dR}(\mathcal{F}'/T)$ (preserving the invariance under the group law). (ii) φ^* is independent of φ. (iii) Similarly, if φ reduces to an* endomorphism *φ_0 of $\mathcal{F}$, φ^* is a* homomorphism *that depends only on φ_0.*

Proof ($d = 1$, see Katz for more variables). Let $\eta = df$, $f \in T[1/p][[X]]$, represent $[\eta] \in H^1_{dR}(\mathcal{F}''/T)$. Let $I = \ker(T \overset{pd}{\twoheadrightarrow} R)$, $\varphi_1, \varphi_2 \in T[[Y]]$, $\varphi_i(0) = 0$, $\varphi_1 \equiv \varphi_2$ mod I. Then by Taylor

$$\varphi_2^*(\eta) - \varphi_1^*(\eta) = d\left(\sum_{n=1}^{\infty} f^{(n)}(\varphi_1) \cdot \frac{(\varphi_2 - \varphi_1)^n}{n!}\right)$$

and $(\cdots) \in T[[Y]]$ since I has divided powers and $f^{(1)}$ is already integral. This shows (ii) $\varphi_2^*([\eta]) = \varphi_1^*([\eta])$. A similar argument proves (i) and (iii). □

We remark that in the situation of the Lemma, it is *blatantly false* that φ^* maps $\omega_{\mathcal{F}''/T}$ to $\omega_{\mathcal{F}'/T}$. The proof highlights the use of divided powers, explains the phrase: "$MG'_T(T)$ depends functorially only on G", and also the relation between crystalline and de Rham cohomology. For a proof when R is not $\mathbb{Z}_p$-flat see, Messing's thesis [14].

5.4.4 Dieudonné Modules

The Grothendieck-Messing crystal is a powerful tool. Let us explain how it captures the classical Dieudonné module of a p-divisible group over a perfect field of characteristic p. Let k be such a field, $W = W(k)$ its ring of Witt vectors, and σ the Frobenius automorphism of W.

Let $G_{/k}$ be a p-divisible group. Its Dieudonné module can be defined as

$$D(G) := M(G^\vee)(W \twoheadrightarrow k).$$

Here are its main features:

- It is a *contra*variant, free W-module of rank $h = ht(G)$.
- $M(G^\vee)(k) = D(G)/pD(G)$. If $G = \mathcal{A}[p^\infty]$ for an abelian variety $\mathcal{A}$ over k then

$$D(G)/pD(G) \simeq H^1_{dR}(\mathcal{A}/k).$$

- If $(-)^{(p)}$ denotes base-change with respect to σ, then $D(G^{(p)}) = D(G)^{(p)}$.
- Let $F_G : G \to G^{(p)}$ be the (relative) Frobenius isogeny, and let Verschiebung (of $G^\vee$) be the dual isogeny $V_{G^\vee} : G^{\vee(p)} \to G^\vee$. By functoriality of $M(-)$ we get $F : D(G)^{(p)} \to D(G)$, i.e., a σ-linear map $D(G) \to D(G)$. Similarly, $V_G : G^{(p)} \to G \rightsquigarrow \sigma^{-1}$-linear V. The relation $V_G \circ F_G = p_G$ that holds for the two isogenies implies, on the level of semi-linear algebra,

$$F \circ V = V \circ F = p.$$

- $\omega_{G/k} \simeq VD(G)^{(p^{-1})}/pD(G) \simeq D(G)^{(p^{-1})}/FD(G)$.

We conclude that $(D(G), F, V)$ is an F-crystal over k. These objects form an additive category Fcryst_k. The Manin-Dieudonné theory yields a complete classification of p-divisible groups over k in terms of their Dieudonné modules. In fact:

Theorem 89 (Dieudonné-Manin) *$D(-)$ is an anti-equivalence between* pdivgp_k *and* Fcryst_k.

For the proof, see [3, 13].

For completeness we remark that the original definition was

$$D(G) := Hom_k(G, CW),$$

where CW is the group of co-Witt vectors. The actions of F and V resulted, in this approach, from their action on CW.

5.4.5 *F*-Isocrystals

Recall that an F**-isocrystal** (N, F, V) is a finite-dim. $W[1/p]$-vector space N, equipped with a σ-linear bijective endomorphism F. We define $V = pF^{-1}$. Such an N may or may not contain a W-lattice stable under both F and V (i.e., an F-crystal), and if it contains one, its isomorphism type is not unique, in general.

The Dieudonné-Manin theorem yields an equivalence of categories between the categories of "p-divisible groups up to isogeny" and "F-isocrystals containing an invariant F-crystal".

Example 90 Let $(r, s) = 1, s > 0, \lambda = r/s$. Let $N_\lambda = \sum_{i=1}^{s} W[1/p]e_i, Fe_i = e_{i+1}$ $(i < s)$, $Fe_s = p^r e_1$. Call λ the (Frobenius) *slope of* N_λ.

Theorem 91 *Let k be algebraically closed. The category of F-isocrystals over k is semi-simple. Its simple objects are the N_λ. An F-isocrystal contains an F-crystal iff all its slopes are contained in* $[0, 1]$.

Consider in particular $\lambda = 1/h$. In this case it is not difficult to see that $N_{1/h}$ contains a *unique* underlying F-crystal (up to isomorphism). By means of the Dieudonné-Manin theorem, this implies that there is a *unique* one-dimensional p-divisible group over k of height h which is not isogenous to a product of two p-divisible groups of smaller heights. Explicitly, it can be obtained as follows. Start with a Lubin-Tate formal group over the ring of integers of $\mathbb{Q}_{p^h}$ (the unramified extension of degree h of $\mathbb{Q}_p$), see [11], reduce it modulo p, and extend scalars from $\mathbb{F}_{p^h}$ to the algebraically closed k. Honda has a different approach to the same groups, see [7].

5.4.6 Endomorphisms of F-Isocrystals

The algebra of endomorphisms (up to isogeny) of a p-divisible group over an algebraically closed field k can be calculated as the endomorphism algebra of its isocrystal. This is a pleasant exercise in semi-linear algebra whose outcome is the following.

Proposition 92 $End(N_\lambda) = D_{-\lambda}$, *the division ring over $\mathbb{Q}_p$ with invariant* $-\lambda$ mod 1.

If $N = D(G)[1/p]$ this means $qEnd(G) \simeq D_\lambda$. Note the change in invariant, from $-\lambda$ to λ. This results from the fact that $D(-)$ is a *contravariant* functor, so $End(D(G)) = End(G)^{opp}$ and not $End(G)$.

- *Exercise*: If $0 \leq r \leq s$ extend e'_i by $e'_{i+ms} = p^m e'_i$, define the F-crystal

$$M_\lambda = \sum_{i=1}^{s} We'_i,\ Fe'_i = e'_{i+r},\ Ve'_i = e'_{i+s-r}.$$

Then N_λ has a lattice isomorphic to M_λ (but there are others).

Over a non-algebraically closed perfect field k of characteristic p, the category of isocrystals is not semi-simple anymore. However, one still has a direct sum decomposition with respect to slopes.

Definition 93 Let k be perfect. Call an isocrystal N *isoclinic of slope* λ if $N \otimes_k \bar{k} \simeq N_\lambda^n$.

Proposition 94 *(Slope decomposition) Let k be perfect and N an F-isocrystal over k. Then $N = \oplus_{\lambda \in \mathbb{Q}} N(\lambda)$ where $N(\lambda)$ is isoclinic of slope λ.*

Let $\lambda_1 < \lambda_2 < \cdots < \lambda_n$ be the slopes of N. Then the **Newton polygon** $NP(N)$ is the convex polygon starting at $(0, 0)$, and having slopes λ_i with horizontal length $rk(N(\lambda_i))$. Its break points are in $\mathbb{Z}^2$.

5.5 The Grothendieck-Messing Period Map

5.5.1 The Quasi-Logarithm and a Big Diagram

Let $R, S \in Adic_{\mathbb{Z}_p}$, $\pi : S \overset{pd}{\twoheadrightarrow} R$ a divided power thickening, and $S \simeq \lim_{\leftarrow} S/(\ker \pi)^n$. Assume S to be flat over $\mathbb{Z}_p$, e.g., $(S \overset{pd}{\twoheadrightarrow} R) = (\mathcal{O}_C \twoheadrightarrow \mathcal{O}_C/p)$.

Start with a p-divisible group G_0 over R and let $G_{/S}$ lift $G_{0/R}$. Both the universal covering $\widetilde{G}$ and the Grothendieck-Messing crystal MG have a "crystalline nature", and our goal is to relate them. We do so via a big commutative diagram. The "top floor" maps to the "bottom floor" via canonical maps or via log's (to keep the diagram readable, not all the "logarithms" are shown). Its rows are exact.

$$\begin{array}{ccccccccc} & & \omega_{G^\vee/S} & \hookrightarrow & EG(S) & \cdots & \rightarrow & G(S) & \\ & & \nearrow \alpha_G & & \nearrow s_G & & \nearrow & \mid & \\ \boldsymbol{T_pG(S)} & \hookrightarrow & \boldsymbol{\widetilde{G}(S)} & \underset{pr_0}{\rightarrow} & & \boldsymbol{G(S)} & & \log_G & \\ \mid & & & & & & & \downarrow & \quad (5.1) \\ \vdots & & \omega_{G^\vee/S,\mathbb{Q}} & \hookrightarrow & MG(S)_{\mathbb{Q}} & \cdots & \rightarrow & Lie(G)_{\mathbb{Q}} & \\ \downarrow & & \nearrow & & \nearrow \text{qlog} & & \nearrow & & \\ \boldsymbol{V_pG(S)} & \hookrightarrow & \boldsymbol{\widetilde{G}(S)} & \underset{\theta}{\rightarrow} & & \boldsymbol{Lie(G)_{\mathbb{Q}}} & & & \end{array}$$

The groups appearing in the diagram have all been defined. Let us explain the various maps starting at the top.

- $s_G(x_0, x_1, \ldots) = \lim [p^n]_{EG}(\xi_n)$, if $EG(S) \ni \xi_n \mapsto x_n \in G(S)$. The convergence follows from the fact that $\bigcap_{n \geq 0} p^n \omega_{G^\vee/S} = 0$. Since $p^n(x_n) = x_0$ the right square of the top floor commutes. This defines then α_G between the two kernels.
- $\alpha_G = s_G|_{T_pG(S)}$ has the following interpretation (check directly from the definitions!):

$$x \in T_pG(S) \rightsquigarrow x_n \in G(S)[p^n] = Hom_S(G^\vee[p^n], \hat{\mathbb{G}}_m)$$

  compatible w.r.t. $G^\vee[p^n] \hookrightarrow G^\vee[p^{n+1}]$, so take its differential "$Lie(x)$" :

$$\rightsquigarrow \alpha_G(x) = Lie(x) \in Hom(Lie(G^\vee), \hat{\mathbb{G}}_a) = \omega_{G^\vee/S}.$$

- $\text{qlog}_G = \log_{EG} \circ s_G$. If $G \leftrightsquigarrow \mathcal{G}$ (a formal group), fix coordinates on $E\mathcal{G}$, let $x = (x_0, x_1, \ldots)$ and ξ_n be as above, then (check!)

$$\mathrm{qlog}_G(x) = \lim_m \lim_n \frac{1}{p^m}[p^{n+m}]_{E\mathcal{G}}(\xi_n).$$

- $\theta = \log_G \circ pr_0 = pr^{MG}_{Lie(G)} \circ \mathrm{qlog}_G$.

We emphasize that the maps s_G, qlog_G are morphisms of crystals, hence depend, like their source and target, only on $G_{0/R}$; θ, α_G depend on $G_{/S}$ and will be related to the GM (Grothendieck-Messing) / HT (Hodge-Tate) period maps, respectively.

5.5.2 Rapoport-Zink Deformation Spaces

To define the period maps we have to discuss deformation spaces of p-divisible groups. Although the first such spaces were studied by Lubin and Tate [12] and Drinfeld [4], we shall follow the more general, and somewhat different, set-up of Rapoport and Zink [17]. The latter differs from the former, both in technical details (the category of test objects is different), and in more substantial matters (allowing quasi-isogenies at the special fiber), as well as in allowing polarization, endomorphisms, and level structures. Needless to say, this is a quick survey, leading to the notions necessary for the study of vector bundles on the Fargues-Fontaine curve, and all proofs are omitted.

Let k be an algebraically closed field of char. p, $W = W(k)$, and $H_{0/k}$ a fixed p-divisible group, of height h and dimension d. Set

$$\boxed{M_0 = MH_0(W \twoheadrightarrow k) = D(H_0^\vee) \simeq W^h.}$$

As our test objects we take rings S from the category Nilp_W of W-algebras in which p is locally nilpotent, e.g., $\mathcal{O}_C/p^N$. Note that S need not be noetherian or local and if it is local noetherian, it need not be complete. With such an S we associate the set

$$\boxed{\mathcal{D}(S) = \{(G,\iota) | \, G_{/S} \ p \,\mathrm{div}\,\mathrm{gp},\ \iota : G \times_S S/p \overset{q.i.}{\dashrightarrow} H_0 \times_k S/p\}.}$$

Next, if S is a p-adic W-algebra, we let $\mathcal{D}(S) = \lim_{\leftarrow} \mathcal{D}(S/p^N)$. If S is also flat over W, ι induces, as we have seen, an isomorphism $MG(S)_{\mathbb{Q}} \simeq M_0 \otimes_W S_{\mathbb{Q}}$. Indeed, the ideal pS having canonical divided powers, we know that

$$MG(S) = MG_0(S \twoheadrightarrow S/p)$$

where G_0 is the reduction of G modulo p. The quasi-isogeny ι induces an isomorphism

$$MG_0(S \twoheadrightarrow S/p)_{\mathbb{Q}} \simeq M(H_0 \times_k S/p)(S \twoheadrightarrow S/p)_{\mathbb{Q}} = M_0 \otimes_W S_{\mathbb{Q}}.$$

5.5.3 Examples and Representability

Let us first compare with the more classical treatment in [12], followed also in [6]. In the *Lubin-Tate case* $d = 1$, and $H_{0/k}$ is the unique one-dimensional formal p-divisible group of height h (see the discussion following Theorem 91). As test objects Lubin and Tate take $R \in \mathcal{C}_k$, the category of complete local noetherian rings with residue field k. For such an R, $S = R/\mathfrak{m}_R^N \in \mathrm{Nilp}_W$. As their deformation functor, Lubin and Tate take the functor

$$\mathcal{M} : \mathcal{C}_k \rightsquigarrow \mathrm{Sets}$$

$$\mathcal{M}(R) = \{(H, \iota) |\ H \text{ formal gp}/R,\ \iota : H \times_R k \overset{q.i.}{\dashrightarrow} H_0\}.$$

Note that by the uniqueness of H_0, $H \times_R k$ and H_0 must be isomorphic (though not necessarily via ι). The endomorphism algebra of H_0 is the skew-field $D_{1/h}$ and

$$D_{1/h}^\times = \langle \Pi \rangle \cdot \mathcal{O}_{D_{1/h}}^\times$$

where Π is the Frobenius isogeny of degree p. This shows that the functor $\mathcal{M}$ breaks up, in the Lubin-Tate case, to a disjoint union indexed by $ht(\iota) \in \mathbb{Z}$ of copies of the sub-functor $\mathcal{M}^0$, in which ι is required to be an *isomorphism* between $H \times_R k$ and H_0. The paper [12] indeed concerns $\mathcal{M}^0$ and not $\mathcal{M}$, but the difference is minor.

We claim that for $S = R/p^N \in \mathrm{Nilp}_W$ we have $\mathcal{D}(S) = \mathcal{M}(S)$. This follows from rigidity of quasi-isogenies (see [10], Lemma 1.1.3)

$$qHom_{S/p}(H \times_S S/p, H_0 \times_k S/p) \simeq qHom_{R/\mathfrak{m}_R}(H \times_S S/\mathfrak{m}_R S, H_0).$$

In general, outside the Lubin-Tate case, a quasi-isogeny $G_0 \to H_0$ of height 0 need not be an isomorphism, so cannot replace $\overset{q.i.}{\dashrightarrow}$ by $\simeq$, even on $\mathcal{D}^0$ (the sub-functor of $\mathcal{D}$ where $ht(\iota) = 0$).

The deformation problems, whether in the Rapoport-Zink language, or in the one of Lubin-Tate or Drinfeld, are easy to pose. Their description can be subtle, especially if one adds PEL structure (which we have not, so far). Here are some examples.

Example 95 1) $d = 1,\ h = 2,\ H_0 = \underline{\mathbb{Q}_p/\mathbb{Z}_p} \times \mu_{p^\infty}$. Since $\underline{\mathbb{Q}_p/\mathbb{Z}_p}$ and μ_{p^∞} do not deform (the first since it is étale, the second since it is the Cartier dual of an étale group)

$$\mathcal{D}^0(S) = Ext_S(\underline{\mathbb{Q}_p/\mathbb{Z}_p}, \mu_{p^\infty}) \simeq \widehat{\mathbb{G}_m}(S)$$

("Serre-Tate canonical coordinate"). Note $\mathcal{D}^0(k)$ is a point.

2) $d = 2,\ h = 4,\ H_0 = H_{1/2} \times H_{1/2}$ where $H_{1/2}$ is the one-dimensional formal group of height 2 (the formal group of a supersingular elliptic curve). $\mathcal{D}^0(k)$ will be infinite because there are $\mathbb{P}^1(k)$'s of pairwise non-isomorphic G isogenous to H_0 (Moret-Bailly families).

3) Lubin-Tate case: H_0 the unique one-dimensional ht h *formal* p-divisible group over k. Then the main result of [12] is that $\mathcal{D}^0 = \mathcal{M}^0$ is represented by "the open unit disk in $h-1$ variables" $Spf(W[[X_1, \ldots, X_{h-1}]])$ and again $\mathcal{D}^0(k)$ is a point.

These examples are typical. The main representability theorem is the following.

Theorem 96 (Drinfeld, Rapoport-Zink) *The functor $\mathcal{D}$ is pro-representable by a formal scheme over W whose ideal of definition is locally finitely generated. Every irreducible component of its (reduced) special fiber is proper over k.*

5.5.4 The Grothendieck-Messing Period Map

To the formal scheme $\mathcal{D}$ (we use the same letter to denote the functor and the formal scheme representing it) one can associate an *analytic space*. In the old days this was a *rigid analytic space* that was called "the generic fiber of $\mathcal{D}$ in the sense of Raynaud". The suffix "in the sense of Raynaud" meant to remind the reader that formal schemes over $Spf(W)$ do not have a generic fiber, as they are simply a compatible system of schemes over W/p^nW for all $n \geq 1$. Nevertheless, Raynaud had a round-about way to attach to $\mathcal{D}$ a rigid analytic space. One of the advantages of Huber's adic spaces is that we can truly speak of generic fibers. While the intuition is clear (at least in examples 1 and 3 above), we skip in these notes the foundational aspects of adic spaces completely, and refer to the literature.

Let $\mathcal{D}_\eta^{ad}$ therefore denote the analytic space associated to $\mathcal{D}$ (over $W[1/p]$). The period map π_{GM} will be a map of *analytic spaces* (over $W[1/p]$) from $\mathcal{D}_\eta^{ad}$ to $Gr(d, M_0)_\eta^{ad}$. For simplicity we only describe it on $(C, \mathcal{O}_C)$-points.

Take $S = \mathcal{O}_C$. For $(G, \iota) \in \mathcal{D}(\mathcal{O}_C)$ we have a quotient map

$$M_0 \otimes_W C \overset{\iota^{-1}}{\simeq} MG(\mathcal{O}_C)_{\mathbb{Q}} \twoheadrightarrow Lie(G_C)$$

from our fixed $M_0 \otimes_W C \simeq C^h$ onto a d-dimensional vector space.

This defines a "period map" from the moduli space to a Grassmanian

$$\pi_{GM}(G, \iota) \in Gr(d, M_0)(C) \simeq Gr(d, h)(C).$$

Once again, the definition, at least on C-points, is straightforward and intuitively clear. The basic features of the definition are more subtle:

- *Fact:* The period map $\pi_{GM} : \mathcal{D}_\eta^{ad} \to Gr(d, M_0)_\eta^{ad}$ is an *étale analytic* map.
- *Example (Dwork): $d = 1$, $h = 2$, $H_0 = \underline{\mathbb{Q}_p/\mathbb{Z}_p} \times \mu_{p^\infty}$.* Then $\mathcal{D} = \widehat{\mathbb{G}}_m$, $\mathcal{D}_\eta^{ad}$ is the open unit disk Δ around 1, and $\pi_{GM} : \Delta \to \mathbb{A}^1 \subset \mathbb{P}^1$ is $q \mapsto \log(q)$. This is typical: (a) log is analytic étale, but (b) it is not algebraic, (c) its kernel in $\mathcal{O}_C$-points is discrete but infinite (roots of unity) and (d) $\log(1 + \mathfrak{m}_C) = C$ (it is surjective on closed points). Nevertheless (e) log is an isomorphism on W-valued points, $\log : 1 + pW \simeq pW$ $(p > 2)$, and in fact will continue to be an isomorphism

on $\mathcal{O}_K$-points as long as the absolute ramification index of K is $< p-1$. It is not an accident that this is also the estimate needed to insure that $\mathfrak{m}_K$ has divided powers. Quite generally, if $(G,\iota) \in \mathcal{D}(W)$ (*unramified*) the Grothendieck-Messing theorem allows to identify the *deformation* with its *period*:

Theorem 97 (Grothendieck-Messing) *The assignment $G \mapsto Lie(G)$ is a bijection between the liftings G of H_0 to W (up to strict isomorphism) and the liftings of $MH_0(k) \twoheadrightarrow Lie(H_0)$ to a free quotient $M_0 \twoheadrightarrow L$ over W.*

5.5.5 The Period Map in the Lubin-Tate Case

To analyze what happens to the period morphism when the ramification in K grows, let us examine once again the Lubin-Tate case.

- In the Lubin-Tate case, the map sending $(G,\iota) \in \mathcal{M}^0(W)$ to $\pi_{GM}(G,\iota) \in Gr(1, M_0)(W) \simeq \mathbb{P}^{h-1}(W)$ is one-to-one, and its image is the W-points of the residue disk R_x in $\mathbb{P}^{h-1}(W)$ reducing to $x = [MH_0(k) \twoheadrightarrow Lie(H_0)]$. *However*:
- The relation between the Lubin-Tate coordinates $(u_1, \ldots, u_{h-1}) \in \mathfrak{m}_W^{h-1}$ and the projective-space coordinates on the residue disk R_x *is* the period morphism and is highly transcendental. Look up the appendix to [6], where formulae are worked out when $h = 2$.
- If K is a finite *ramified* extension of $W[1/p]$, $\mathcal{O}_K \twoheadrightarrow k$ is in general no longer a pd thickening, so the Grothendieck-Messing theorem does not apply. We still have $\mathcal{D}^0(\mathcal{O}_K) \simeq \mathcal{M}^0(\mathcal{O}_K) \approx \mathfrak{m}_K^{h-1}$, but *a quasi-isogeny of height 0 over $\mathcal{O}_K/p$* (unlike over k!) is *not necessarily an isomorphism*, so $(G,\iota) \in \mathcal{D}^0(\mathcal{O}_K)$ only provides a map
$$M_0 \otimes_W K \simeq MG(\mathcal{O}_K)_{\mathbb{Q}} \to Lie(G_K),$$
i.e., a point of $Gr(1, M_0)(K) \simeq \mathbb{P}^{h-1}(K)$. Since it is not defined integrally, we cannot talk about its reduction.
- The resulting period map from $\mathcal{D}^0(\mathcal{O}_K)$ to $\mathbb{P}^{h-1}(K)$ is not $1:1$ in general, and its image is not confined anymore to a residue disk. In the Lubin-Tate case (but not in general), when K is replaced by C, it is even surjective, and its fibers are infinite.

The period morphism$\pi_{GM} : \mathcal{M}(\mathcal{O}_C) \to \mathbb{P}^{h-1}(C)$ was studied further in [6]. Part (i) of the following theorem is natural and expected. The action of $D^\times$ on $\mathbb{P}^{h-1}(C)$ is via the (projective) regular representation. The element Π acts (in appropriate coordinates) like

$$\begin{pmatrix} 0 & & & & p \\ 1 & 0 & & & \\ & 1 & 0 & & \\ & & \cdots & \cdots & \\ & & & 1 & 0 \end{pmatrix}.$$

Part (ii) describes the fibers of the period morphism; two deformations of H_0 to R are in the same fiber if and only if the given quasi-isogeny between their reductions modulo p lifts to R.

Point (iii) is the afore-mentioned surjectivity on closed points, special to this case, which is eventually an explicit computation (nowadays embedded in the more general question of characterizing the image by conditions of weak admissibility, see below).

In (iv) we consider, for the first time, level structure. Over the Lubin-Tate moduli space lies the Lubin-Tate tower, whose n-th layer is a formal scheme parametrizing deformations of H_0 together with a full p^n-level structure, i.e., a trivialization of the group $H(R)[p^n]$. This turns out to be a finite Galois covering with Galois group $GL_h(\mathbb{Z}/p^n\mathbb{Z})$, whose associated covering of analytic spaces ("generic fibers") is even étale (the special fibers are highly ramified though). Letting $M_{\infty,\eta}$ be the analytic space associated with the full tower (properly defined!) and $M_\eta = \mathcal{M}_\eta^{0,ad}$ the one associated with $\mathcal{M}^0$ (nothing but the $h-1$ dimensional open polydisk), we get

$$M_\eta \simeq M_{\infty,\eta}/GL_h(\mathbb{Z}_p).$$

Point (iv) is very pleasing: it re-interprets (ii) and (iii) as saying that projecting M_η further down to $\mathbb{P}_C^{h-1}$ by the period morphism is the same as dividing $M_{\infty,\eta}$ all the way by $GL_h(\mathbb{Q}_p)$. This is as close as we can get to saying that the period morphism is "like dividing by a group action"; had $GL_h(\mathbb{Z}_p)$ been normal in $GL_h(\mathbb{Q}_p)$ this would have been the case!

Theorem 98 (Gross-Hopkins) *(i) π_{GM} is $D^\times$-equivariant ($D = D_{1/h}$ the endomorphism algebra).*

(ii) $\pi_{GM}(G,\iota) = \pi_{GM}(G',\iota') \Leftrightarrow \exists f : G \overset{q.i.}{\to} G', \iota' \circ \bar{f} = \iota$.
(iii) $\pi_{GM}^0 : \mathcal{M}^0(\mathcal{O}_C) \to \mathbb{P}^{h-1}(C)$ is surjective.
(iv) $M_{\infty,\eta} \to M_\eta \to \mathbb{P}_C^{h-1}$ gives $\mathbb{P}_C^{h-1} = M_{\infty,\eta}/GL_h(\mathbb{Q}_p)$.

In general, the image of π_{GM} is restricted by the notion of "weak admissibility". Given an exact sequence (recall $M_0 = MH_0(W \twoheadrightarrow k) = D(H_0^\vee) \simeq W^h$)

$$0 \to Fil \to M_{0,C} \to M_{0,C}/Fil \to 0$$

with associated filtration $Fil^0 = M_{0,C} \supset Fil^1 = Fil \supset Fil^2 = 0$, $\underline{N} = (N, Fil) = (M_{0,C}, Fil)$ becomes a *filtered F-isocrystal*. We recall the definitions from §4.2.

- If N' is a sub-F-isocrystal let $Fil' = Fil \cap (N' \otimes_{W[1/p]} C)$, $\underline{N}' = (N', Fil')$.
- For any filtered F-isocrystal $\underline{N}$ define

$$t_{Newton}(\underline{N}) = v_p(\det(\Phi))$$

(independent of the matrix representing Φ, since this matrix is unique up to σ-conjugation),

$$t_{Hodge}(\underline{N}) = \sum i \dim gr^i_{Fil^\bullet} = \dim Fil.$$

- Call $\underline{N} = (N, Fil)$ *weakly admissible* if for any sub F-isocrystal $N' \subset N$

$$t_{Hodge}(\underline{N}') \leq t_{Newton}(\underline{N}')$$

with equality for $\underline{N}' = \underline{N}$.
- Given $H_{0/k}$, the *weakly admissible period domain* is an open subspace $\mathfrak{F}^{wa} \subset Gr(d, M_0)^{ad}_\eta$ such that $\mathfrak{F}^{wa}(C)$ consists of all d-dimensional quotients

$$M_{0,C} \to U = M_{0,C}/Fil$$

for which $\underline{N}$ is weakly admissible.

Theorem 99 *(i) The image of* $\pi_{GM} : \mathcal{D}^{ad}_\eta \to Gr(d, M_0)^{ad}_\eta$ *factors through* $\mathfrak{F}^{wa}$.
(ii) The image contains all the classical points of $\mathfrak{F}^{wa}$ *(points whose residue field is a finite extension of* $K_0 = W[1/p]$*).*

Remarks: (i) is relatively easy. (ii) is a variant of "weakly admissible filtered isocrystals are admissible". However, in contrast to what we did in §4.2, we consider p-divisible groups over C, not only over finite extensions K of $\mathbb{Q}_p$. As a result, the Galois representation "evaporates" and we cannot argue anymore directly via the functor D_{cris} as we did there. We shall later relate (ii) to the geometry of the Fargues-Fontaine curve. Finally, we mention that Hartl described the *non-classical* points in $\mathfrak{F}^a = \mathrm{Im}(\pi_{GM})$. In general, $\mathfrak{F}^a \neq \mathfrak{F}^{wa}$.

5.6 *The Hodge-Tate Decomposition and the HT Period Map*

Besides the Grothendieck-Messing period map, the Big Diagram (5.1) contains the seeds for the other period morphism, the Hodge-Tate period map. There is a beautiful duality between the two (see [5], §8.1 and the last chapter of [18]), on which we can not comment here for lack of space. However, since we already touched on it, let us at least give the definition, which goes back to Tate's fundamental paper [23].

5.6.1 The HT Exact Sequence

Recall the map $\alpha_G : T_pG(R) \to \omega_{G^\vee/R}$. Let $R = \mathcal{O}_C$ and let $-(1)$ denote Tate twist. The following theorem was the beginning of p-adic Hodge theory, 50 years ago. It holds in the generality stated here, although, strictly speaking, at the time Tate proved it only if G is defined over $\mathcal{O}_K$ for a finite extension K of $\mathbb{Q}_p$.

Theorem 100 (Tate) *(i) There is an exact sequence*

$$0 \to Lie(G_C)(1) \overset{\alpha^\vee_{G^\vee}(1)}{\to} T_pG(\mathcal{O}_C) \otimes_{\mathbb{Z}_p} C \overset{\alpha_G}{\to} \omega_{G^\vee/C} \to 0.$$

(ii) (Hodge-Tate decomposition) If G is defined over $\mathcal{O}_K$ where $K \subset C$ is a complete discrete valuation field, then the sequence splits canonically (respecting $\Gamma_K = Gal(\bar{K}/K)$ action)

$$T_pG(\mathcal{O}_C) \otimes_{\mathbb{Z}_p} C \simeq \omega_{G^\vee/C} \oplus Lie(G_C)(1).$$

Remark 101 A few remarks are in order.

(i) The map $\alpha_{G^\vee}^\vee(1)$ is constructed from $\alpha_{G^\vee}$ by duality. It sends $Lie(G_C)(1) = Hom(\omega_{G/C}, T_p\mu_{p^\infty} \otimes C)$ to

$$Hom(T_pG^\vee \otimes C, T_p\mu_{p^\infty} \otimes C) \simeq T_pG \otimes C.$$

The last equality comes from Cartier duality, and is the reason we needed not only to dualize, but also to twist by $\mathbb{Z}_p(1) = T_p\mu_{p^\infty}$.

(ii) To get (ii) from (i) invoke Tate's theorems in Galois cohomology, that $H^0(\Gamma_K, C(i)) = H^1(\Gamma_K, C(i)) = 0$ if $i \neq 0$ and both cohomology groups are 1-dimensional if $i = 0$. In the absence of a Galois action, there is no canonical splitting of (i).

(iii) Let $G = \mathscr{A}[p^\infty]$ where $\mathscr{A}$ is an abelian scheme over $\mathcal{O}_C$. Dualizing, (i) is equivalent to the existence, and degeneration, of a spectral sequence (Faltings: the Hodge-Tate spectral sequence)

$$E_{i,j}^2 - H^i(\mathscr{A}, \Omega^j_{\mathscr{A}/C})(\ j) \Rightarrow H_{et}^{i+j}(\mathscr{A}, \mathbb{Q}_p) \otimes_{\mathbb{Q}_p} C.$$

Compare with the Hodge spectral sequence that starts with $E_{i,j}^1 = H^j(\mathscr{A}, \Omega^i_{\mathscr{A}/C})$. This applies to *any* proper smooth variety over C (Faltings), and in fact to any proper and smooth rigid analytic variety, even if not algebraic, by recent work of Scholze.

(iv) The fact that the Hodge-Tate decomposition is not valid in families, only the HT *filtration*, leads to the HT *period map*, just as over $\mathbb{C}$ the fact that only the Hodge *filtration* varies holomorphically in families lies behind the classical period map to classify spaces of Hodge structures.

5.6.2 The Hodge-Tate Period Map

Let us put ourselves once again in the Lubin-Tate case. Consider the full Lubin-Tate tower and take

$$(G, \iota, \alpha_\infty) \in \mathcal{M}_\infty(\mathcal{O}_C).$$

Use $\alpha_\infty : \mathbb{Z}_p^h \simeq T_pG(\mathcal{O}_C)$ to construct the linear map $(\alpha_G \otimes 1) \circ (\alpha_\infty \otimes 1) : C^h \to \omega_{G^\vee/C}$, whose kernel is a *line* (because $G^\vee$ is $h-1$ dimensional). Mapping $(G, \iota, \alpha_\infty)$ to this line is

$$\pi_{HT} : \mathcal{M}_\infty(\mathcal{O}_C) \to \mathbb{P}^{h-1}(C).$$

Unlike π_{GM}, π_{HT} is defined only on $\mathcal{M}_\infty$. It goes *canonically* to $\mathbb{P}^{h-1}(C)$ while π_{GM} landed in $\mathbb{P}(M_0)(C) \simeq \mathbb{P}^{h-1}(C)$.

- *Fact*: π_{HT} comes from an *analytic* map $M_{\infty,\eta} \to (\mathbb{P}^{h-1})^{ad}_\eta$. In *our "basic" case* (but not always), it is also *étale.*
- For $\delta \in D^\times$, $\pi_{HT} \circ \delta = \pi_{HT}$ (obvious).
- π_{HT} intertwines the actions of $GL_h(\mathbb{Q}_p)$ on $M_{\infty,\eta}$ and $\mathbb{P}^{h-1}$ (obvious).

A global detour ($h = 2$): *modular curves at the infinite level.* Let Y_n be the (open) modular curve of full level p^n over $\mathbb{Q}_p$ and Y_∞ the scheme $\lim_\leftarrow Y_n$. A point of $Y_\infty(C)$ *is* an elliptic curve $E_{/C}$ equipped with an isomorphism $\alpha_\infty : \mathbb{Z}_p^2 \simeq T_pE$. As above, we get $\pi_{HT} : Y_\infty(C) \to \mathbb{P}^1(C)$. Let $\mathfrak{X} = \mathbb{P}^1(C) \setminus \mathbb{P}^1(\mathbb{Q}_p)$ (the Drinfeld p-adic upper half plane).

Theorem 102 *The map $\pi_{HT} : Y_\infty(C) \to \mathbb{P}^1(C)$ is surjective. We have $\pi_{HT}^{-1}(\mathbb{P}^1(\mathbb{Q}_p)) = Y_\infty(C)^{ord}$ (the pairs (E, α_∞) where E has bad, or good ordinary reduction) and $\pi_{HT}^{-1}(\mathfrak{X}) = Y_\infty(C)^{ss}$ (the pairs where E has good supersingular reduction).*

Note the anomaly: at infinite level the "fat" set $Y_\infty(C)^{ord}$ gets mapped to the "meager" $\mathbb{P}^1(\mathbb{Q}_p)$ and the meager $Y_\infty(C)^{ss}$ fills up its complement $\mathfrak{X}$.

If E has good ordinary or bad multiplicative reduction and$G = E[p^\infty]$ then T_pG^0, the Tate module of the "kernel of reduction" is a line in T_pG, and spans $\ker(\alpha_G \otimes 1)$. This proves $\pi_{HT}(E, \alpha_\infty) \in \mathbb{P}^1(\mathbb{Q}_p)$. Conversely, if E is defined over a CDVF K and $\pi_{HT}(E, \alpha_\infty) \in \mathbb{P}^1(\mathbb{Q}_p)$ then $\Gamma_K \curvearrowright T_pG$ is potentially reducible, so E is ordinary. This proves the theorem, except for the surjectivity. In general, for the Lubin-Tate tower, we have the following.

Theorem 103 *(i) The image of $\pi_{HT} : M_{\infty,\eta} \to (\mathbb{P}^{h-1})^{ad}_\eta$ is the Drinfeld p-adic symmetric domain*

$$\mathfrak{X}(C) = \mathbb{P}^{h-1}(C) \setminus \bigcup_{a \in (\mathbb{P}^{h-1})^*(\mathbb{Q}_p)} H_a.$$

(ii) π_{HT} induces $M_{\infty,\eta}/D^\times_{1/h} \simeq \mathfrak{X}(C)$ (on the level of C-points, so far).

6 Conclusion of the Classification Theorem

6.1 What We Have to Prove and a Reduction Theorem

Recall that we have to prove the following criterion, an almost formal consequence of which was the classification theorem for vector bundles (Theorem 79).

Criterion. For every vector bundle $\mathcal{E}$ on X and for every $n \geq 1$, if we have a short exact sequence

$$0 \to \mathcal{O}_X(-\frac{1}{n}) \to \mathcal{E} \to \mathcal{O}_X(1) \to 0,$$

of vector bundles, then $H^0(X,\mathcal{E})\neq 0$, and the same is true if X is replaced by X_h, its cyclic unramified covering of degree h, for every $h\geq 1$.

In what follows we shall only check things for X, but it should be understood that the same proofs go over to every X_h, with the necessary modifications.

We first replace this criterion with two similar-looking criteria, which are together equivalent to it. As will turn out at the end, and as might be implied by the formulation, there is a certain duality underlying the relation between (i) and (ii) below. Note that the exact sequences in the new criteria are not short exact sequences of vector bundles, but *modifications* of vector bundles.

Theorem 104 *([5], 5.6.29, p.233-236) Suppose: (i) Whenever there is a short exact sequence*

$$0\to\mathcal{E}\to\mathcal{O}_X(1/n)\to\mathcal{F}\to 0$$

with $\mathcal{F}$ a torsion sheaf of degree 1, then $\mathcal{E}\simeq\mathcal{O}_X^n$.

(ii) If $n\geq 1$ and there is an exact sequence

$$0\to\mathcal{O}_X^n\to\mathcal{E}\to\mathcal{F}\to 0$$

where $\mathcal{E}$ is a vector bundle and $\mathcal{F}$ is a torsion sheaf of degree 1, then for some $1\leq m\leq n$

$$\mathcal{E}\simeq\mathcal{O}_X^{n-m}\oplus\mathcal{O}_X(1/m).$$

Suppose further (i) and (ii) hold with X_h instead of X for every $h\geq 1$. Then the Criterion holds.

Exercise. *Prove (i) and (ii) assuming the classification theorem, by the arithmetic of degrees and ranks.*

Proof We skip the proof. It involves arguments on vector bundles on curves, in the style of the proof of Theorem 79. In particular, p-divisible groups and period maps do not show up (yet) in this reduction step. □

6.2 *Modification of Vector Bundles Associated with p-Divisible Groups*

6.2.1 The Modification

Let (D,φ) be an isocrystal (φ-module) over $\bar{\mathbb{F}}_p$. Thus D is a finite dimensional vector space over

$$L=\breve{\mathbb{Q}}_p=W(\bar{\mathbb{F}}_p)[1/p].$$

Fix $\infty\in X$ as before, with residue field C, so that $B_{dR}^+=\widehat{\mathcal{O}}_{X,\infty}$. Let $t=\log[\varepsilon]$ be the usual uniformizer. Giving a modification

$$0 \to \mathcal{E}' \to \mathcal{E}(D, \varphi) \to \mathcal{F} \to 0$$

with $\mathcal{F}$ coherent torsion module supported at ∞ is the same as giving a B_{dR}^{+}-lattice

$$\Lambda \subset D \otimes_L B_{dR}^{+}, \quad \mathcal{F} = i_{\infty *}(D \otimes_L B_{dR}^{+}/\Lambda).$$

Recall that the modification is called *minuscule* if $\mathcal{F}$ is killed by t, i.e., is a $C = B_{dR}^{+}/(t)$ -vector space. Giving a minuscule modification is therefore the same as giving a filtration

$$Fil D_C$$

in $D_C = D \otimes_L C = D \otimes_L B_{dR}^{+}/D \otimes_L tB_{dR}^{+}$.

Definition 105 A triple $(D, \varphi, Fil D_C)$ is *admissible* if

$$\mathcal{E}(D, \varphi, Fil D_C) = \ker(\mathcal{E}(D, \varphi) \to i_{\infty *}(D_C/Fil D_C) \to 0$$

is the trivial vector bundle.

Remark. In the proof of weakly admissible = admissible we considered vector bundles $\mathcal{E}(D, \varphi, Fil)$ where Fil was a filtration defined over E (or over a finite extension K of E, although for simplicity of the presentation we took it to be $E = \mathbb{Q}_p$). But in principle we can take the filtered φ-module to have its filtration defined over C, and now we need to consider all of these. Of course, we lose the Galois action on the Tate module of the p-divisible group, if the latter is only defined over $\mathcal{O}_C$, and we can not use Tate's theorems in Galois cohomology the way we did before. The geometry of the Fargues-Fontaine curve, in a sense, replaces these arguments.

6.2.2 The p-Divisible Group H and Its Grothendieck-Messing Period

Fix a p-divisible group $\mathbb{H}$ over $k = \bar{\mathbb{F}}_p$ (the "model p-divisible group") and a deformation H, in the sense of Rapoport and Zink, to $\mathcal{O}_C$. This means that we are *given* a quasi-isogeny

$$\rho : \mathbb{H} \otimes_k \mathcal{O}_C/p \dashrightarrow H \otimes_{\mathcal{O}_C} \mathcal{O}_C/p$$

(previously we took it in the opposite direction; we changed direction to adhere to the notation of [5]). We let

$$D = M\mathbb{H}(W \twoheadrightarrow k)[1/p]$$

be the *rational covariant* Dieudonné module of $\mathbb{H}$, so that (D, φ) is an isocrystal as above (with slope in $[0, 1]$). Here MG signifies the Grothendieck-Messing crystal, i.e., the Lie algebra of the universal vectorial extension of G, $W = W(k)$, and $W \twoheadrightarrow k$ has the standard pd structure on pW. The quasi-isogeny ρ induces, as we have seen, an isomorphism

$$MH(\mathcal{O}_C)[1/p] = MH(\mathcal{O}_C \twoheadrightarrow \mathcal{O}_C/p)[1/p] \overset{\rho}{\simeq} M\mathbb{H}(\mathcal{O}_C \twoheadrightarrow \mathcal{O}_C/p)[1/p] \overset{b.c.}{=} D_C.$$

The Hodge filtration exact sequence

$$0 \to \omega_{H^\vee/C} \to MH(\mathcal{O}_C)[1/p] \to Lie(H/C) \to 0$$

induces therefore a filtration $FilD_C \subset D_C$ in the C-points of our "model isocrystal". As the deformation (H, ρ) changes, this filtration changes too. We record the fact that if $\mathbb{H}$ is of height h and dimension d then $FilD_C$ is of dimension $h - d = \dim \mathbb{H}^\vee$. Thus

$$\pi_{GM}(H, \rho) = (D_C, FilD_C) \in \mathfrak{F} = Gr_{d,h}(C) = Gr_d(D_C).$$

This is the Grothendieck-Messing (or Hodge-de Rham) period of the point $(H, \rho) \in \mathcal{D}(\mathcal{O}_C) = \mathcal{D}^{ad}_\eta(C)$ (the $\mathcal{O}_C$-points of the Rapoport-Zink deformation space, regarded as a formal scheme, or the C-points of the associated analytic space). $\mathfrak{F}$ is the "period domain", a Grassmanian in this case (again, regarded as an analytic space).

In particular, if $d = 1$ and $\mathbb{H}$ is connected we are in the "Lubin-Tate case". In this case, as we have remarked, $\mathbb{H}$ is unique. Furthermore, $\mathfrak{F} = \mathbb{P}^{h-1}$ and $\mathcal{D}$, the Rapoport-Zink (= Lubin-Tate) deformation space, consists of $\mathbb{Z}$ copies (indexed by the height of the quasi-isogeny ρ) of $Spf(W[[X_1, \ldots, X_{h-1}]])$, the open unit polydisk of dimension $h - 1$ over W.

In this Lubin-Tate case the map

$$\pi_{GM} : \mathcal{D} \to \mathfrak{F} = \mathbb{P}^{h-1}$$

has been analyzed by Gross and Hopkins, and was shown to be *onto. This fact will play a crucial role in the proof.*

6.2.3 Admissibility of $(D, p^{-1}\varphi, FilD_C)$

In general, as we have seen, the image of π_{GM} lies in $\mathfrak{F}^{wa}$, the *weakly admissible* period domain. However, we are able now to prove that $(D, p^{-1}\varphi, FilD_C)$ is admissible in the sense defined above (which is a priori stronger than weakly admissible). The p^{-1} in front of φ is an artifact having to do with the fact that we are using now the covariant D. Passing from contravariant to covariant involves taking the Cartier dual, hence a Tate twist (otherwise the slope would end up in $[-1, 0]$ instead of $[0, 1]$).

Note that since H is defined only over C, V_pH does not carry a Galois action. Thus admissibility can not be defined now as being in the image of the functor D_{cris}, as we had before. But as we shall see, admissibility will still bring back to the picture V_pH.

Proposition 106 *The filtered φ-module $(D, p^{-1}\varphi, FilD_C)$ is admissible, and we have a natural map*

$$\mathcal{E}(D, p^{-1}\varphi, FilD_C) \simeq V_pH \otimes_E \mathcal{O}_X.$$

Proof (sketch) **Step 1.** Recall the diagram (5.1). We first remark that we have, for the universal covering,

$$\widetilde{H}(\mathcal{O}_C) = \widetilde{H}(\mathcal{O}_C/\mathfrak{a}) = \widetilde{H}_k(\mathcal{O}_C/\mathfrak{a}) \overset{\rho}{\simeq} \widetilde{\mathbb{H}}(\mathcal{O}_C/\mathfrak{a}).$$

where $\mathfrak{a}$ is an ideal of definition of $\mathcal{O}_C$ containing p, modulo which ρ induces an isomorphism

$$H_k \otimes_k \mathcal{O}_C/\mathfrak{a} \simeq H \otimes_{\mathcal{O}_C} \mathcal{O}_C/\mathfrak{a}$$

(such an $\mathfrak{a}$ can be shown to exist; in fact, its existence is equivalent to the existence of the quasi-isogeny ρ, but the further ρ is from an isomorphism, the closer $\mathfrak{a}$ would have to be to $\mathfrak{m}_C$; note that $\mathfrak{m}_C$ itself is *not* an ideal of definition).

By Proposition 83, since $\mathbb{H}$ is defined over a perfect $\mathbb{F}_p$-algebra (namely, the field k),

$$\widetilde{\mathbb{H}}(\mathcal{O}_C/\mathfrak{a}) \simeq \mathbb{H}((\mathcal{O}_C/\mathfrak{a})^\flat) = \mathbb{H}(\mathcal{O}_F).$$

We therefore get

$$V_pH(\mathcal{O}_C) \subset \widetilde{H}(\mathcal{O}_C) \simeq \mathbb{H}(\mathcal{O}_F).$$

Step 2. From the definition of the *contravariant* Dieudonné module D^* using homomorphisms from $\mathbb{H}$ to the ind-scheme of co-Witt vectors one gets the isomorphism

$$\mathbb{H}(\mathcal{O}_F) \simeq Hom_{W(k)}(D^*, B_F^+)^{\varphi=1}$$

(see [5], Corollaire 4.4.4). Since here we have been using the *rational covariant* Dieudonné module, which is the rational Dieudonné module of the Cartier dual $\mathbb{H}^\vee$, we get

$$\mathbb{H}(\mathcal{O}_F) \simeq (D \otimes_L B_F^+)^{\varphi=p} = (D \otimes_L B_F)^{\varphi=p}.$$

Under this homomorphism, the E-subspace $V_pH(\mathcal{O}_C)$ gets mapped isomorphically to

$$Fil(D \otimes_L B_F)^{\varphi=p} = \theta^{-1}(FilD_C)$$

(see [5], Proposition 4.5.14). This follows essentially from the fact that the quasi-logarithm map in the big diagram (5.1) maps V_pH to the Hodge filtration $\omega_{H^\vee/\mathcal{O}_C}$. We thus get an exact sequence

$$\begin{array}{ccccccc} 0 \to & V_pH(\mathcal{O}_C) & \to & \widetilde{H}(\mathcal{O}_C) & \overset{\log}{\to} & Lie(H/C) \to 0 \\ & \| & & \| & & \| \\ 0 \to & Fil(D \otimes_L B_F)^{\varphi=p} & \to & (D \otimes_L B_F)^{\varphi=p} & \overset{\theta_*}{\to} & D_C/FilD_C \to 0 \end{array}.$$

Step 3. This means that the induced homomorphism of vector bundles

$$V_pH(\mathcal{O}_C)\otimes_E \mathcal{O}_X \to \mathcal{E}(D, p^{-1}\varphi)$$

(expressed for example in terms of B-pairs) factors through the sub-bundle $\mathcal{E}(D, p^{-1}\varphi, FilD_C)$. The upshot is that we have used the period morphism for (H, ρ) and the "big diagram" to construct a homomorphism of vector bundles

$$u : V_pH(\mathcal{O}_C)\otimes_E \mathcal{O}_X \to \mathcal{E}(D, p^{-1}\varphi, FilD_C).$$

When we tensor over $E = B_F^{\varphi=1}$ with $B_F[1/t]$ we get an isomorphism

$$V_pH(\mathcal{O}_C)\otimes_E B_F[1/t] \simeq D\otimes_L B_F[1/t]$$

(if we used B_{cris} instead this would have recovered the old Fontaine-Messing comparison isomorphism). Thus, remembering that $H^0(X\setminus\{\infty\}, \mathcal{O}_X) = B_e = B_F[1/t]^{\varphi=1}$, u is generically an isomorphism, so it is injective. To show that u is an isomorphism all that remains is to compare degrees on both sides. On the left the degree is 0 (a trivial vector bundle). On the right it comes out to be

$$\dim \mathbb{H} - \dim H = 0$$

as well. We conclude that u is an isomorphism, and the Proposition is proved. □

6.3 Conclusion of the Proof

We sketch the main steps.

6.3.1 Proof of Point (i)

We have to show that if

$$0 \to \mathcal{E} \to \mathcal{O}_X(1/n) \to \mathcal{F} \to 0$$

is an exact sequence, and $\mathcal{F}$ is a degree 1 torsion sheaf supported at ∞, then $\mathcal{E}$ is trivial. But this $\mathcal{O}(1/n)$ is just $\mathcal{E}(D, \varphi)$ where D is the Dieudonné module of $\mathbb{H}$, the unique one-dimensional formal group of height n over k. By the theorem of Gross-Hopkins, on the surjectivity of the Grothendieck-Messing period map on C-points in the Lubin-Tate case, there exists a deformation (H, ρ) of $\mathbb{H}$ over $\mathcal{O}_C$ as above, such that $\mathcal{F} = i_{\infty,*}(D\otimes B_{dR}^+/\Lambda)$ where

$$\Lambda/(D\otimes tB_{dR}^+) = FilD_C = \ker(MH(\mathcal{O}_C)[1/p] \to Lie(H/C)).$$

This puts us in the situation considered above, where the modification is the one associated with a deformation H of $\mathbb{H}$, and the triviality of $\mathcal{E} = \mathcal{E}(D, \varphi, Fil D_C)$ follows from the last Proposition.

6.3.2 Proof of Point (ii)

The proof of (i) was the key step, and the way p-divisible groups and their period maps enter the classification of vector bundles. In a way, it was again the weak admissibility = admissibility that played a role, but we could not use the Colmez-Fontaine theorem of course, because that theorem was only about Galois representations, and we had to consider deformations of $\mathbb{H}$ over $\mathcal{O}_C$, and not only over $\mathcal{O}_K$ for a finite extension K of $E = \mathbb{Q}_p$. (Anyhow, our short proof of Colmez-Fontaine in §4.2 *relied* on the classification theorem, so the argument would have been circuitous.)

The proof of point (ii) is somewhat similar, in principle, even if the technical details are different. Recall that we start with a modification

$$0 \to \mathcal{O}_X^n \to \mathcal{E} \to \mathcal{F} \to 0 \tag{6.1}$$

where $\mathcal{F}$ is a torsion sheaf of degree 1 supported at ∞, and we must show that for some $1 \le m \le n$

$$\mathcal{E} \simeq \mathcal{O}_X^{n-m} \oplus \mathcal{O}_X(1/m).$$

After a reduction Lemma, which rules out the possibility of factoring an $\mathcal{O}_X^{n-m}$ from $\mathcal{E}$, thereby replacing n by a smaller m, one applies a duality principle, taking (derived) $\mathscr{H}om(-, \mathcal{O}_X(1/n))$ of coherent sheaves. We get a new modification of vector bundles

$$0 \to \mathcal{G}' \to \mathcal{G} \to \mathcal{H} \to 0 \tag{6.2}$$

where

$$\mathcal{G}' = \mathscr{H}om(\mathcal{E}, \mathcal{O}_X(1/n)), \;\; \mathcal{G} = \mathcal{O}_X(1/n)^n, \;\; \mathcal{H} = \mathscr{E}xt^1(\mathcal{F}, \mathcal{O}_X(1/n)).$$

Since $End(\mathcal{O}_X(1/n)) = \Delta_{1/n}$, the division algebra over $E = \mathbb{Q}_p$ of invariant $1/n$ (commonly denoted $D_{1/n}$ and sometimes $B_{1/n}$, but we have too many $D's$ and $B's$), this new modification lies in a category of modifications of sheaves *with* $\Delta_{1/n}$-*structure*, that are related to filtered φ-modules *with* $\Delta_{1/n}$-*structure,* in the same way we had before. Note that $\mathscr{H}om(\mathcal{F}, \mathcal{O}_X(1/n)) = 0$ if $\mathcal{F}$ is torsion, while $\mathcal{H} = \mathscr{E}xt^1(\mathcal{F}, \mathcal{O}_X(1/n))$ is easily computable, since $\mathcal{F}$ is a sky-scraper sheaf supported at ∞, so we are talking about extensions as B_{dR}^+-modules, and B_{dR}^+ is a DVR. It leads of course, again, to a torsion sheaf supported at ∞, and if $\mathcal{F}$ was killed by t (the modification was minuscule), so is $\mathcal{H}$.

In fact, taking $\mathscr{H}om_{\Delta_{1/n}}(-, \mathcal{O}_X(1/n))$ back recovers the original modification (6.1), so one obtains an anti-equivalence between two similar categories of modifications.

What we have to prove is that $\mathcal{G}'$ is trivial. This is analogous to what we did in point (i), and is done in the same way, essentially. What replaces the surjectivity of the period map on Lubin-Tate groups (the Gross-Hopkins theorem) is a similar argument on the period map into the Drinfeld symmetric domain, which is the period domain for the kind of formal groups with $\Delta_{1/n}$ action that show up here. In fact, in the Drinfeld case, unlike the Lubin-Tate case, the period map

$$\pi_{GM} : \mathcal{D}_{\eta}^{ad} \to \mathfrak{F}^{wa} = \mathbb{P}^{h-1} \setminus \bigcup_{a \in (\mathbb{P}^{h-1})^*(E)} H_a$$

is an isomorphism of analytic (adic, or rigid analytic) spaces.

For the details, see [5], 8.3.2.

7 Appendix: Weierstrass Division in A_{inf}

7.1 Notation

Let F be an algebraically closed perfectoid field of characteristic p. Let $\varpi \in \mathfrak{m}_F$, $u \in W(\mathcal{O}_F)^\times$ and

$$\xi = p - [\varpi]u \in W(\mathcal{O}_F),$$

primitive of degree 1 (every primitive element of degree 1 is of this form). Let

$$\theta : W(\mathcal{O}_F) \to W(\mathcal{O}_F)/(\xi) =: D$$

denote the canonical projection. The key to proving that $D = \mathcal{O}_C$ for a complete valued field C, for which $F \simeq C^\flat$, is the "Weierstrass division theorem" below. In [5], Corollaire 2.2.10, the authors prove it in an indirect way, first showing that every element of D has roots of any order in D, and then deducing it from this fact. We give a variation on this proof that is a little more direct, in the sense that it first proves a Weierstrass division theorem in $W_n(\mathcal{O}_F)$ and then uses this theorem to deduce what we want along the lines of the Fargues-Fontaine proof.

7.2 Weierstrass Division in $W_n(\mathcal{O}_F)$

We have $W_n(\mathcal{O}_F) = W(\mathcal{O}_F)/(p^{n+1})$. Fix $f \in W(\mathcal{O}_F)$.

Proposition 107 *There exists a $z \in \mathcal{O}_F$ such that $f \equiv [z] \mod (\xi, p^{n+1})$.*

Proof We work in $W_n(\mathcal{O}_F)$. Write $f = (b_0, \ldots, b_n)$ with $b_i \in \mathcal{O}_F$. Write $u = (u_0, \ldots, u_n)$ with $u_i \in \mathcal{O}_F$ and $u_0 \in \mathcal{O}_F^\times$. We need to solve

$$(b_0, \ldots, b_n) = (z, 0, \ldots, 0) + (p - [\varpi](u_0, \ldots, u_n))(x_0, \ldots, x_n).$$

Replacing ϖ by ϖu_0 we may assume that $u_0 = 1$. Reducing modulo p we see that $z = b_0 + \varpi x_0$. The equation can be written as

$$(b_0, b_1, \ldots, b_n) = (b_0 + \varpi x_0, x_0^p, \ldots, x_{n-1}^p) - (1, u_1, \ldots, u_n)(\varpi x_0, \varpi^p x_1, \ldots, \varpi^{p^n} x_n).$$

Recall that

$$(x_0, \ldots, x_n) + (y_0, \ldots, y_n) = (S_0, \ldots, S_n)$$

$$(u_0, \ldots, u_n)(x_0, \ldots, x_n) = (P_0, \ldots, P_n)$$

where the S_i and P_i are isobaric of weight p^i in the variables x_j and y_j $(0 \leq j \leq i)$. Here x_j and y_j are given weight p^j and the u_j are treated as scalars of weight 0. We therefore have to solve the n equations

$$H_i(x_0, \ldots, x_i) = x_{i-1}^p + \varpi Q_i(x_0, x_1, \ldots, x_{i-1}) - \varpi^{p^i} x_i - b_i = 0$$

$(1 \leq i \leq n)$ where the $Q_i(x_0, \ldots, x_{i-1})$ are polynomials of weight p^i (i.e., all their monomial have weight $\leq p^i$) with integral coefficients.

We claim that

$$A_n = \mathcal{O}_F[x_0, \ldots, x_n]/(H_1, \ldots, H_n)$$

is *finite* over $\mathcal{O}_F[x_n]$, and in fact generated over it as a module by $x_0^{j_0} x_1^{j_1} \ldots x_{n-1}^{j_{n-1}}$ with $0 \leq j_i \leq p-1$. Observe that H_i has weight p^i, contains a term λx_{i-1}^p with $\lambda \in \mathcal{O}_F^\times$, but also a linear term in x_i. If not for this linear term in x_i, the finiteness of A_n over $\mathcal{O}_F[x_n]$ would be trivial. That linear term complicates things a little bit.

Let M be the $\mathcal{O}_F[x_n]$-submodule of A_n spanned by $x_0^{j_0} x_1^{j_1} \ldots x_{n-1}^{j_{n-1}}$ with $0 \leq j_i \leq p-1$. For an integer w we denote by M^w and A_n^w the $\mathcal{O}_F$-submodules of M and A_n spanned by monomials $\boldsymbol{x}^{\boldsymbol{j}} = x_0^{j_0} x_1^{j_1} \ldots x_{n-1}^{j_{n-1}} x_n^{j_n}$ of weight

$$w(\boldsymbol{j}) = j_0 + pj_1 + \cdots + p^n j_n \leq w$$

(and in the case of M, $j_i \leq p-1$ for $i \leq n-1$). These are $\mathcal{O}_F$-submodules of finite rank, so it is enough to prove that

$$A_n^w = M^w + \varpi A_n^w.$$

This is clear since *modulo* ϖA_n^w we have $H_i \equiv x_{i-1}^p - b_i$, so these equations may be used to reduce the exponents of $x_0, \ldots, x_{n-1}$ in any monomial to the range $[0, p-1]$.

Thus A_n is finite over $\mathcal{O}_F[x_n]$. We claim that the map $\mathcal{O}_F[x_n] \to A_n$ is *injective*. Suppose $0 \neq h \in \mathcal{O}_F[x_n]$ were in the kernel. If $h \in \mathfrak{m}_F$ then tensoring with F we would get $F \otimes_{\mathcal{O}_F} A_n = 0$, but $F \otimes_{\mathcal{O}_F} A_n \simeq F[x_0]$ because over F the equation H_i can be used to solve for x_i in terms of the lower x_j. If h were not a scalar, we would still get that $F \otimes_{\mathcal{O}_F} A_n$ is finite over $F[x_n]/(h)$, hence finite over F, again a contradiction.

By the going-up theorem, for every homomorphism $\psi : \mathcal{O}_F[x_n] \to \mathcal{O}_F$ corresponding to a prime ideal $\ker(\psi) = \mathfrak{p} \subset \mathcal{O}_F[x_n]$, there exists a prime ideal $\mathfrak{P} \subset A_n$ lying above $\mathfrak{p}$. The integral domain $A_n/\mathfrak{P}$ is a finite extension of $\mathcal{O}_F = \mathcal{O}_F[x_n]/\mathfrak{p}$. But since F is algebraically closed we must have $A_n/\mathfrak{P} = \mathcal{O}_F$. This means that ψ extends to a homomorphism $A_n \to \mathcal{O}_F$, hence the equations H_i have a common zero in $\mathcal{O}_F$, as desired. □

7.3 Weierstrass Division in $W(\mathcal{O}_F)$

To go further we need a few lemmas.

Lemma 108 *Let R be a p-adically complete ring. If $p \neq 2$, any element of $1 + p^2R$ (if $p = 2$, any element of $1 + 8R$) has a p-th root.*

Proof Since $ord_p(k!) \leq k/(p-1)$ we find easily that

$$(1 + p^2x)^{1/p} = \sum_{k=0}^{\infty} \binom{1/p}{k} p^{2k} x^k$$

converges p-adically. If $p = 2$ we need to work modulo p^3 but the argument is the same. □

Lemma 109 *The ideal (ξ) is closed in the weak topology of $W(\mathcal{O}_F)$. In particular*

$$\bigcap_{n \geq 2} (\xi, p^n, [\varpi]^n) = (\xi).$$

Proof Suppose $f = \xi g_n + h_n$ where $h_n \in (p^n, [\varpi^n])$. Then ξg_n converges to f in the weak topology. We must show that g_n converges in the weak topology. Write $g_n = \sum_{k=0}^{\infty} [y_{n,k}] p^k$. Since convergence in the weak topology is convergence of each $\{y_{n,k}\}_n$ separately, we may assume that k_0 is the first index for which this sequence does not converge in $\mathcal{O}_F$. Subtracting the limit of $\sum_{k=0}^{k_0-1} [y_{n,k}] p^k$ and dividing by p^{k_0} we may assume, without loss of generality, that $k_0 = 0$. But then reducing ξg_n modulo p we see that $u_0 \varpi y_{n,0}$ is a Cauchy sequence, which forces $y_{n,0}$ to be Cauchy as well, hence to converge. □

Lemma 111 *Any element of D has a p-th root in D.*

Proof We must show that for any $f \in W(\mathcal{O}_F)$ there exists a $g \in W(\mathcal{O}_F)$ such that $f \equiv g^p \mod (\xi)$. By the Proposition, for every $n \geq 1$ we may write

$$f \equiv [z_n] + p^{n+2}h_n \mod (\xi).$$

For n sufficiently large $v_F(\varpi^n) > v_F(z_n)$ or else by Lemma 109 $f \equiv 0 \mod (\xi)$ and there is nothing to prove. But $p \equiv [\varpi]u \mod (\xi)$ so

$$f \equiv [z_n](1 + p^2[\varpi^n z_n^{-1}]u^n h_n) \mod (\xi).$$

The factor $[z_n]$ has a p-th root of course, and the second factor also has a p-th root by Lemma 108. □

Theorem 112 (Weierstrass division) *Let $\xi \in W(\mathcal{O}_F)$ be a primitive element of degree 1. Then for every $f \in W(\mathcal{O}_F)$ there exist $z \in \mathcal{O}_F$ and $g \in W(\mathcal{O}_F)$ such that*

$$f = [z] + \xi g.$$

In other words, $\theta(f) = \theta([z])$. Note that, contrary to Weierstrass division in $\mathbb{Z}_p[[X]]$, z and g are not unique.

Proof The ring $D = W(\mathcal{O}_F)/(\xi)$ is p-adically complete and separated and has no p-torsion. The first assertion follows from the same fact for $W(\mathcal{O}_F)$ and the fact that (ξ) is p-adically closed. The second is immediate: if $pf = (p - [\varpi]u)g$ then reducing modulo p we see that g is divisible by p, so $f \equiv 0 \mod (\xi)$.

We may therefore tilt D to form $D^\flat = \lim_{\leftarrow}(D/pD)$, the inverse limit taken with respect to Frobenius, and identify it with sequences $(\alpha^{(0)}, \alpha^{(1)}, \dots)$ of elements of D in which $\alpha^{(i+1)p} = \alpha^{(i)}$. Furthermore, Lemma 111 shows that $\alpha^{(0)}$ may be arbitrary. We have

$$D/pD = W(\mathcal{O}_F)/(\xi, p) = \mathcal{O}_F/(\varpi)$$

and its perfection $D^\flat$ is therefore canonically identified with $\mathcal{O}_F$. The map sending $z \in \mathcal{O}_F = D^\flat$ to $z^\sharp \in D$ is the map $z \mapsto \theta([z])$, as can be seen in the diagram.

$$\begin{array}{ccc} \mathcal{O}_F & \stackrel{z\mapsto[z]}{\longrightarrow} & W(\mathcal{O}_F) \\ \iota|| & & \downarrow\theta \\ D^\flat & \stackrel{\sharp}{\longrightarrow} & D \end{array}$$

Here $\iota(z) = (z^{1/p^n} \mod \varpi\mathcal{O}_F)$, and $\iota(z)^\sharp = \lim \theta([z^{1/p^n}])^{p^n} = \theta([z])$ because $\theta([z^{1/p^n}]) \in D$ is a lift of $z^{1/p^n} \mod \varpi\mathcal{O}_F \in D/pD$.

But we have just remarked that any element of D is of the form $z^\sharp$ for some z. This is what we had to show. □

Corollary 113 *D is an integral domain.*

Proof This follows from $\theta([z_1])\theta([z_2]) = \theta([z_1 z_2])$. □

References

1. O. Brinon, B. Conrad, CMI school notes on p-adic Hodge theory (preliminary notes, 2009). http://math.stanford.edu/~conrad/papers/notes.pdf
2. P. Colmez, J.-M. Fontaine, Construction des representations p-adiques semi-stables. Inv. Math. **140**, 1–43 (2000)
3. M. Demazure, *p-divisible groups*, Lecture Notes in Math, vol. 302. (Springer, New-York, 1972)
4. V. Drinfeld, Elliptic modules, Math USSR. Sbornik **23**, 561–592 (1974)
5. L. Fargues, J.-M. Fontaine, Courbes et fibrés vectoriels en théorie de Hodge p-adique. Astérisque. **406** (2018)
6. B.H. Gross, M.J. Hopkins, Equivariant vector bundles on the Lubin-Tate moduli space. Contem. Math. **158**, 23–88 (1994)
7. M. Hazewinkel, *Formal Groups and Applications* (Academic Press, New York, 1978)
8. L. Illusie, Déformations de goupes de Barsotti-Tate. Astérisque **127**, 151–198 (1985)
9. N. Katz, Crystalline cohomology, Dieudonné modules, and Jacobi sums, in *Automorphic Forms, Representation Theory and Arithmetic, Bombay Colloquium 1979* (Springer , Berlin, 1981), pp. 165–246
10. N. Katz, Serre-Tate local moduli, Surfaces Algéberiques, Springer Lecture Notes **868**, 138–202 (1981)
11. J. Lubin, J. Tate, Formal complex multiplication in local fields. Ann. Math. **81**, 380–387 (1965)
12. J. Lubin, J. Tate, Formal moduli of one parameter formal Lie groups. Bull. Soc. Math. France **94**, 49–60 (1966)
13. Y. Manin, The theory of commutative formal groups over fields of finite characteristic. Rus. Math. Surv. **18**, 1–84 (1963)
14. W. Messing, The crystals associated to Barsotti-Tate groups, Springer Lecture Notes **264** (1972)
15. M. Morrow, p-divisible groups (p-adic Hodge theory Masters lecture course, Bonn University, Winter Semester 2015/16)
16. M. Morrow, The Fargues Fontaine curve and diamonds, *Séminaire* BOURBAKI, Juin 2018, no. 1150
17. M. Rapoport, T. Zink, Period spaces for p-divisible groups, in *Annals of Mathematics Studies*, vol. 141 (Princeton University Press, Princeton, 1996)
18. P. Scholze, J. Weinstein, Moduli of p-divisible groups. Cambridge J. Math. **1**, 145–237 (2013)
19. S. Sen, Continuous cohomology and p-adic Galois representations. Invent. Math. **62**, 89–116 (1980/81)
20. J.-P. Serre, Local fields, in *Graduate Texts in Mathematics* (Springer, New York, 1979)
21. https://stacks.math.columbia.edu
22. J. Stix, A course on finite flat group schemes and p-divisible groups. https://www.uni-frankfurt.de/52288632/Stix_finflat_Grpschemes.pdf
23. J. Tate, p-divisible groups, in *Proceedings of a Conference on Local Fields* (Springer, Berlin, 1967), pp. 158–183

Simplicial Galois Deformation Functors

Yichang Cai and Jacques Tilouine

Abstract In [13], the authors showed the importance of studying simplicial generalizations of Galois deformation functors. They established a precise link between the simplicial universal deformation ring R pro-representing such a deformation problem (with local conditions) and a derived Hecke algebra. Here we focus on the algebraic part of their study which we complete in two directions. First, we introduce the notion of simplicial pseudo-characters and prove relations between the (derived) deformation functors of simplicial pseudo-characters and that of simplicial Galois representations. Secondly, we define the relative cotangent complex of a simplicial deformation functor and, in the ordinary case, we relate it to the relative complex of ordinary Galois cochains. Finally, we recall how the latter can be used to relate the fundamental group of R to the ordinary dual adjoint Selmer group, by a homomorphism already introduced in [13] and studied in greater generality in [26].

1 Introduction

Let p be an odd prime. Let K be a p-adic field, let $\mathcal{O}$ be its valuation ring, ϖ be a uniformizing parameter, and $k = \mathcal{O}/(\varpi)$ be the residue field. Let Γ be a profinite group satisfying

(Φ_p) the p-Frattini quotient $\Gamma/\Gamma^p(\Gamma,\Gamma)$ is finite.

For instance, Γ could be $\mathrm{Gal}(F_S/F)$, the Galois group of the maximal S-ramified extension of a number field F with S finite. Let G be a split connected reductive group scheme over $\mathcal{O}$. Let $\overline{\rho}\colon \Gamma \to G(k)$ be a continuous Galois representation. Assume it

This research was supported in part by the International Centre for Theoretical Sciences (ICTS) during a visit for participating in the program Perfectoid spaces (Code: ICTS/perfectoid2019/09). The authors are partially supported by the ANR grant CoLoSS ANR-19-PRC. One of the authors (J. Tilouine) is partially supported by the NSF grant: DMS 1464106.

Y. Cai (✉) · J. Tilouine
LAGA, UMR 7539, Institut Galilée, Université de Paris 13, USPN, 99 av. J.-B. Clément, 93430 Villetaneuse, France
e-mail: yccai1993@126.com

© The Author(s), under exclusive license to Springer Nature Singapore Pte Ltd. 2022
D. Banerjee et al. (eds.), *Perfectoid Spaces*, Infosys Science Foundation Series in Mathematical Sciences https://doi.org/10.1007/978-981-16-7121-0_8

is absolutely G-irreducible, which means its image is not contained in $P(k)$ for any proper parabolic subgroup P of G. The goal of this paper is to present and develop some aspects of the fundamental work [13] and the subsequent papers [26] and [2], by putting emphasis on the algebraic notion of simplicial deformation over simplicial Artin local $\mathcal{O}$-algebras of $\overline{\rho}$.

In the papers mentioned above, it is assumed that the given residual Galois representation is automorphic: $\overline{\rho} = \overline{\rho}_\pi$ for a cohomological cuspidal automorphic representation on the dual group of G over a number field F, then the (classical and simplicial) deformation problems considered impose certain local deformation conditions satisfied by $\overline{\rho}$ at primes above p and at ramification primes for π. The fundamental insight of [13] is to relate the corresponding universal simplicial deformation ring to a derived version of the Hecke algebra acting on the graded cohomology of a locally symmetric space. Actually, the main result [13, Theorem 14.1] (slightly generalized in [2]) is that after localization at the non-Eisenstein maximal ideal $\mathfrak{m}$ of the Hecke algebra corresponding to $\overline{\rho}$, the integral graded cohomology in which π occurs is free over the graded homotopy ring of the universal simplicial deformation ring (and the degree zero part of this ring is isomorphic to the top degree integral Hecke algebra). This is therefore a result of automorphic nature.

Here, on the other hand, we want to focus on the purely algebraic machinery of simplicial deformations and pseudo-deformations and their (co)tangent complex for a general profinite group Γ satisfying (Φ_p).

In [16, Sect. 11], V. Lafforgue introduced the notion of a pseudo-character for a split connected reductive group G. He proved that this notion coincides with that of G-conjugacy classes of G-valued Galois representations over an algebraically closed field E. The main ingredient of his proof is a criterion of semisimplicity for elements in $G(E)^n$ in terms of closed conjugacy class; it is due to Richardson in characteristic zero. It has been generalized to the case of an algebraically closed field of arbitrary characteristic by [5] replacing semisimplicity by G-complete reducibility (see also [23] and [4, Theorem 3.4]). Note that absolute G-irreducibility implies G-complete reducibility.

Using this (and a variant for Artin rings), Boeckle-Khare-Harris-Thorne [4, Theorem 4.10] proved a generalization of Carayol's result for any split reductive group G: any pseudo-deformation over G of an absolutely G-irreducible representation $\overline{\rho}$ is a G-deformation.

In Sect. 3.2.2, we reformulate the theory of [4, Sect. 4] in the language of simplicial deformation. Our main results are Theorem 3.16 and Theorem 3.20. In Sect. 3.3, we propose a generalization of this theory for derived deformations. Unfortunately, the result in this context is only partial, but still instructive.

In Sect. 4, after recalling the definition of the tangent and cotangent complexes and its calculation for a Galois deformation functor, we introduce a relative version of the cotangent complex. In order to relate the cotangent complex of the universal simplicial ring $\mathcal{R}$ pro-representing a deformation functor to a Selmer group, we shall take $\Gamma = G_{F,S}$ for a number field F and for S equal to the set of places above p and ∞, and we shall deal with the simplest sort of local conditions, namely, unramified outside p and ordinary at each place above p. We show that the cotangent complex

$L_{\mathcal{R}/\mathcal{O}} \otimes_{\mathcal{R}} T$ is related to the ordinary Galois cochain complex. Note that here the base T is arbitrary, whereas in [13] and [2] it was mostly the case $T = k$.

Finally, in Sect. 5, we recall how this is used to define a homomorphism, first constructed in [13, Lemma 15.1] and generalized and studied in [26], which relates the fundamental group of the simplicial ordinary universal deformation ring and the ordinary dual adjoint Selmer group.

This work started during the conference on p-adic automorphic forms and Perfectoids held in Bangalore in September 2019. The authors greatly appreciated the excellent working atmosphere during their stay.

2 Classical and Simplicial Galois Deformation Functors

2.1 *Classical Deformations*

Let Γ be a profinite group which satisfies (Φ_p). When necessary, we view Γ as projective limit of finite groups Γ_i. Let $\mathbf{Art}_{\mathcal{O}}$ be the category of Artinian local $\mathcal{O}$-algebras with residue field k. Recall that the framed deformation functor $\mathcal{D}^{\square}: \mathbf{Art}_{\mathcal{O}} \to \mathbf{Sets}$ of $\overline{\rho}$ is defined by associating $A \in \mathbf{Art}_{\mathcal{O}}$ to the set of continuous liftings $\rho\colon \Gamma \to G(A)$ which make the following diagram commute:

$$\begin{array}{ccc} \Gamma & \xrightarrow{\rho} & G(A) \\ & \searrow^{\bar{\rho}} & \downarrow \\ & & G(k). \end{array} \tag{1}$$

Let Z be the center of G over $\mathcal{O}$. We assume throughout it is a smooth group scheme over $\mathcal{O}$. Let $\widehat{G}(A) = \mathrm{Ker}(G(A) \to G(k))$, resp., $\widehat{Z}(A) = \mathrm{Ker}(Z(A) \to Z(k))$. Let $\mathfrak{g} = \mathrm{Lie}(G/\mathcal{O})$, resp., $\mathfrak{z} = \mathrm{Lie}(Z/\mathcal{O})$ be the $\mathcal{O}$-Lie algebra of G, resp., Z, and let $\mathfrak{g}_k = \mathfrak{g} \otimes_{\mathcal{O}} k$, resp., $\mathfrak{z}_k = \mathfrak{z} \otimes_{\mathcal{O}} k$. The universal deformation functor $\mathcal{D} = \mathrm{Def}_{\overline{\rho}}\colon \mathbf{Art}_{\mathcal{O}} \to \mathbf{Sets}$ is defined by associating $A \in \mathbf{Art}_{\mathcal{O}}$ to the set of $\widehat{G}(A)$-conjugacy classes of $\mathcal{D}^{\square}(A)$. As an application of Schlessinger's criterion (see [21, Theorem 2.11]), the functor $\mathcal{D}^{\square}$ is pro-representable, and when $\bar{\rho}$ satisfies $H^0(\Gamma, \mathfrak{g}_k) = \mathfrak{z}_k$, the functor $\mathcal{D}$ is pro-representable (see [25, Theorem 3.3]).

We shall consider (nearly) ordinary deformations. In this case, we always suppose $\Gamma = G_{F,S}$, where F is a number field and $S = S_p \cup S_\infty$ is the set of places above p and ∞. Note that Γ is profinite and satisfies (Φ_p). For any $v \in S_p$, let $\Gamma_v = \mathrm{Gal}(\overline{F}_v/F_v)$. Let $B = TN \subset G$ be a Borel subgroup scheme (T is a maximal split torus and N is the unipotent radical of B); all these groups are defined over $\mathcal{O}$. Let Φ be the root system associated to (G, T) and Φ^+ the subset of positive roots associated to (G, B, T). Assume that for any place $v \in S_p$, we have

(Ord_v) there exists $\overline{g}_v \in G(k)$ such that $\overline{\rho}|_{\Gamma_v}$ takes values in $\overline{g}_v^{-1} \cdot B(k) \cdot \overline{g}_v$.

Let $\overline{\chi}_v\colon \Gamma_v \to T(k)$ be the reduction modulo $N(k)$ of $\overline{g}_v \cdot \overline{\rho}|_{\Gamma_v} \cdot \overline{g}_v^{-1}$. Let $\omega\colon \Gamma_v \to k^\times$ be the mod. p cyclotomic character. We shall need the following conditions for $v \in S_p$:

(Reg_v) for any $\alpha \in \Phi^+$, $\alpha \circ \overline{\chi}_v \neq 1$, and

(Reg_v^*) for any $\alpha \in \Phi^+$, $\alpha \circ \overline{\chi}_v \neq \omega$.

We can define the subfunctor $\mathcal{D}^{\square,\mathrm{n.o}} \subset \mathcal{D}^{\square}$ of nearly ordinary liftings by the condition that $\rho \in \mathcal{D}^{\square,\mathrm{n.o}}$ if and only if for any place $v \in S_p$ there exists $g_v \in G(A)$ which lifts $\bar{g}_v$ such that $\rho|_{\Gamma_v}$ takes values in $g_v^{-1} \cdot B(A) \cdot g_v$. Note that this implies that the homomorphism $\chi_{\rho,v}\colon \Gamma_v \to T(A)$ given by $g_v \cdot \rho|_{\Gamma_v} \cdot g_v^{-1}$ lifts $\overline{\chi}_v$.

We define the subfunctor $\mathcal{D}^{\mathrm{n.o}} \subset \mathcal{D}$ of nearly ordinary deformations by $\mathcal{D}^{\mathrm{n.o}}(A) = \mathcal{D}^{\square,\mathrm{n.o}}(A)/\widehat{G}(A)$.

Recall [25, Proposition 6.2]:

Proposition 2.1 Assume that $\mathrm{H}^0(\Gamma, \mathfrak{g}_k) = \mathfrak{z}_k$ and that (Ord_v) and (Reg_v) hold for all places $v \in S_p$. Then $\mathcal{D}^{\mathrm{n.o}}$ is pro-representable, say by the complete noetherian local $\mathcal{O}$-algebra $R^{\mathrm{n.o}}$.

Note that the condition (Reg_v^*) will occur later in the study of the cotangent complex in terms of the (nearly) ordinary Selmer complex.

Remark 2.2 As noted in [25, Chapter 8], under the assumption (Reg_v) $(\forall v \in S_p)$, the morphism of functors $\mathcal{D}^{\mathrm{n.o}} \to \prod_{v\in S_p} \mathrm{Def}_{\overline{\chi}_v}$ given by $[\rho] \mapsto (\chi_{\rho,v})_{v\in S_p}$ provides a structure of Λ-algebra on $R^{\mathrm{n.o}}$ for an Iwasawa algebra Λ called the Hida-Iwasawa algebra.

Remark 2.3 A lifting $\rho\colon \Gamma \to G(A)$ of $\bar{\rho}$ is called ordinary of weight μ if for any $v \in S_p$, after conjugation by g_v, the cocharacter $\rho|_{I_v}\colon I_v \to T(A) = B(A)/N(A)$ is given (via the Artin reciprocity map rec_v) by $\mu \circ rec_v^{-1}\colon I_v \to \mathcal{O}_{F_v}^\times \to T(A)$.

If we assume that $\bar{\rho}$ admits a lifting $\rho_0\colon \Gamma \to G(\mathcal{O})$ which is ordinary of weight μ, we can also consider the weight μ ordinary deformation problem, defined as the subfunctor $\mathcal{D}^{\mathrm{n.o},\mu} \subset \mathcal{D}^{\mathrm{n.o}}$ where we impose the extra condition to $[\rho]$ that for any $v \in S_p$, after conjugation by some g_v, $\rho|_{I_v}\colon I_v \to T(A) = B(A)/N(A)$ is given (via the Artin reciprocity map rec_v) by $\mu \circ rec_v^{-1}\colon I_v \to \mathcal{O}_{F_v}^\times \to T(\mathcal{O}) \to T(A)$. This problem is pro-representable as well, say by $R_\mu^{\mathrm{n.o}}$. The difference is that $R^{\mathrm{n.o}}$ has a natural structure of algebra over an Iwasawa algebra, while, if ρ_0 is automorphic, $R_\mu^{\mathrm{n.o}}$ is often proven to be a finite $\mathcal{O}$-algebra (see [29] or [9], for instance).

These functors have natural simplicial interpretations.

2.2 *Simplicial Reformulation of Classical Deformations*

In this section, we'll try to introduce the basic notions of simplicial homotopy theory and proceed at the same time to give a simplicial definition of the deformation functor of $\bar{\rho}$.

Recall that a groupoid is a category such that all homomorphisms between two objects are isomorphisms. Let **Gpd** be the category of small groupoids. We have a functor $\mathbf{Gp} \to \mathbf{Gpd}$ from the category **Gp** of groups to **Gpd** sending a group G to the groupoid with one object $\bullet$ and such that $\mathsf{End}(\bullet) = G$.

A model category is a category with three classes of morphisms called weak equivalences, cofibrations, and fibrations, satisfying five axioms, see [15, Definition 7.1.3]. The category of groups is not a model category. But it is known (see [24, Theorem 6.7]) that the category of groupoids **Gpd** is a model category, where a morphism $f : G \to H$ is

(1) a weak equivalence if it is an equivalence of categories;
(2) a cofibration if it is injective on objects; and
(3) a fibration if for all $a \in G, b \in H$ and $h : f(a) \to b$ there exists $g : a \to a'$ such that $f(a') = b$ and $f(g) = h$.

If $\mathcal{C}$ is a model category, its homotopy category $\mathrm{Ho}(\mathcal{C})$ is the localization of $\mathcal{C}$ at weak equivalences. It comes with a functor $\mathcal{C} \to \mathrm{Ho}(\mathcal{C})$ universal for the property of sending weak equivalences to isomorphisms.

In **Gpd**, the empty groupoid is the initial object and the unit groupoid consisting in a unique object with a unique isomorphism is the final object. In a model category, a fibration, resp., cofibration, over the final object, resp., from the initial object, is called a fibrant, resp., cofibrant object. Note that every object of **Gpd** is both cofibrant and fibrant, and the homotopy category Ho(**Gpd**) is the quotient category of **Gpd** modding out natural isomorphisms. If we regard a group G as a one point groupoid, the functor $\mathbf{Gp} \to \mathrm{Ho}(\mathbf{Gpd})$ so obtained has the effect of modding out conjugations, so, for any finite group Γ_i, we have

$$\mathsf{Hom}_{\mathbf{Gp}}(\Gamma_i, G(A))/G^{\mathrm{ad}}(A) \cong \mathsf{Hom}_{\mathrm{Ho}(\mathbf{Gpd})}(\Gamma_i, G(A)). \tag{2}$$

To construct the deformation functor, we first need to recall the construction of the classifying simplicial set BG associated to a groupoid G.

Let $\mathbf{\Delta}$ be the category whose objects are sets $[n] = \{0, \dots, n\}$ and morphisms are non-decreasing maps. It is called the cosimplicial indexing category (see [15, Definition 15.1.8]). Given a category $\mathcal{C}$, the category $s\mathcal{C}$ of simplicial objects of $\mathcal{C}$ is the category of contravariant functors $F : \mathbf{\Delta} \to \mathcal{C}$. In particular, s**Sets** is the category of simplicial sets. For any $n \geq 0$, let $\mathbf{\Delta}[n]$ be the simplicial set

$$[k] \mapsto \mathsf{Hom}_{\mathbf{\Delta}}([k], [n]).$$

Note that the category s**Sets** admits enriched homomorphisms: if X, Y are two simplicial sets, there is a natural simplicial set $\mathsf{sHom}(X, Y)$ whose degree zero term is $\mathsf{Hom}_{s\mathbf{Sets}}(X, Y)$. Actually,

$$\mathsf{sHom}(X, Y)_n = \mathsf{Hom}_{s\mathbf{Sets}}(X \times \mathbf{\Delta}[n], Y).$$

For $X \in s\mathbf{Sets}$, the morphism $(d_1, d_0)\colon X_1 \to X_0 \times X_0$ generates an equivalence relation $\widetilde{X}_1$. The zeroth homotopy set $\pi_0 X$ is defined as the quotient set $X_0/\widetilde{X}_1$. Let X be fibrant and let $x \in X_0$; one can define for $i \geq 1$, the i-th homotopy set $\pi_i(X, x)$ as the quotient of the set of pointed morphisms $\mathsf{Hom}_{s\mathbf{Sets}_*}(\mathbf{\Delta}[n], X)$ (morphisms sending the boundary $\partial\mathbf{\Delta}[n]$ to x) by the homotopy relation (see [27, Sect. 8.3]). Then $\pi_i(X, x)$ is naturally a group which is Abelian when $i \geq 2$ (see [11, Theorem I.7.2]).

For $X \in s\mathbf{Sets}$, let $\mathbf{\Delta}X$ be the category whose objects are pairs (n, σ) where $n \geq 0$ and $\sigma\colon \mathbf{\Delta}[n] \to X$ is a morphism of simplicial sets, and morphisms $(n, \sigma) \to (m, \tau)$ are given by a non-decreasing map $\varphi\colon [n] \to [m]$ such that $\sigma = \tau \circ \varphi$. The category $\mathbf{\Delta}X$ is called the category of simplices of X (see [15, Definition 15.1.16]).

The following lemma is well known:

Lemma 2.4 *Suppose $\mathcal{C}$ is a category admitting colimits; let $F\colon \mathbf{\Delta} \to \mathcal{C}$ be a covariant functor. Let $F_*\colon \mathcal{C} \to s\mathbf{Sets}$ be the functor which sends $A \in \mathcal{C}$ to the simplicial set $X = (X_n)_{n\geq 0}$ given by $X_n = \mathsf{Hom}_{\mathcal{C}}(F([n]), A)$ at n-th simplicial degree, and let $F^*\colon s\mathbf{Sets} \to \mathcal{C}$ be the functor which sends $X \in s\mathbf{Sets}$ to $\varinjlim_{(n,\sigma)\in\mathbf{\Delta}X} F(\sigma)$. Then F^* is left adjoint to F_*.*

Proof It's clear that F_* is well defined, and F^* is well defined since every simplicial set morphism $f\colon X \to Y$ induces a functor $\mathbf{\Delta}X \to \mathbf{\Delta}Y$. For $X \in s\mathbf{Sets}$ and $A \in \mathcal{C}$, we have

$$\begin{aligned}
\mathsf{Hom}_{\mathcal{C}}(F^*(X), A) &\cong \varprojlim_{(\Delta[n]\to X)\in(\mathbf{\Delta}X)^{\mathrm{op}}} \mathsf{Hom}_{\mathcal{C}}(F([n]), A) \\
&\cong \varprojlim_{(\Delta[n]\to X)\in(\mathbf{\Delta}X)^{\mathrm{op}}} \mathsf{Hom}_{s\mathbf{Sets}}(\Delta[n], F_*(A)) \\
&\cong \mathsf{Hom}_{s\mathbf{Sets}}(\varinjlim_{(\Delta[n]\to X)\in\mathbf{\Delta}X} \Delta[n], F_*(A)) \\
&\cong \mathsf{Hom}_{s\mathbf{Sets}}(X, F_*(A)),
\end{aligned}$$

where the last equation follows from [15, Proposition 15.1.20]. So F^* is left adjoint to F_*. □

Example 2.5 (1) Let $\mathbf{\Delta} \to \mathbf{Cat}$ be the functor defined by regarding $[n]$ as a posetal category: its objects are $0, 1, \ldots n$ and $\mathsf{Hom}_{[n]}(k, \ell)$ has at most one element, and is non-empty if and only if $k \leq \ell$. We write $P\colon s\mathbf{Sets} \to \mathbf{Cat}$ and $B\colon \mathbf{Cat} \to s\mathbf{Sets}$ for the associate left adjoint functor and right adjoint functor, respectively. The functor B is called the nerve functor. The simplicial set $B\mathcal{C} = (X_n)$ is defined by sets $X_n \subset \mathrm{Ob}(\mathcal{C})^{[n]}$ of $(n+1)$-tuples $(C_0, \ldots, C_n)$ of objects of $\mathcal{C}$ with morphisms $C_k \to C_\ell$ when $k \leq \ell$, which are compatible when n varies; it is a fibrant simplicial set if and only if $\mathcal{C} \in \mathbf{Gpd}$ (see [11, Lemma I.3.5]). In a word, for $B\mathcal{C}$ to be fibrant, it must have the extension property with respect to inclusions of horns in $\Delta[n]$ ($\forall n \geq 1$). For $n = 2$, it amounts to saying that all homomorphisms in $\mathcal{C}$ are invertible; for $n > 2$, the extension condition is automatic (details in the reference above). For $\mathcal{C} \in \mathbf{Cat}$, we have $PB\mathcal{C} \cong \mathcal{C}$, so $\mathsf{Hom}_{\mathbf{Cat}}(\mathcal{C}, \mathcal{D}) \cong \mathsf{Hom}_{s\mathbf{Sets}}(B\mathcal{C}, B\mathcal{D})$ ($\forall \mathcal{C}, \mathcal{D} \in \mathbf{Cat}$). Note that $B(\mathcal{C} \times [1]) \cong B\mathcal{C} \times \Delta[1]$ (product is the degreewise product); in consequence, when $\mathcal{C} \in \mathbf{Cat}$ and $\mathcal{D} \in \mathbf{Gpd}$, two functors $f, g \colon \mathcal{C} \to \mathcal{D}$ are naturally isomorphic if and only if Bf and Bg are homotopic.

(2) As a corollary of (1), we have $\mathsf{Hom}_{\mathbf{Gpd}}(GPX, H) \cong \mathsf{Hom}_{s\mathbf{Sets}}(X, BH)$ for $X \in s\mathbf{Sets}$ and $H \in \mathbf{Gpd}$, where GPX is the free groupoid associated to PX. We remark that GPX and $\pi_1|X|$ (the fundamental groupoid of the geometric realization) are isomorphic in Ho($\mathbf{Gpd}$) (see [11, Theorem III.1.1]).

Recall that a functor between two model categories is called right Quillen if it preserves fibrations and trivial fibrations (i.e., fibrations which are weak equivalences).

Lemma 2.6 *The nerve functor $B \colon \mathbf{Gpd} \to s\mathbf{Sets}$ is fully faithful and takes fibrant values (Kan-valued). Moreover, it is right Quillen.*

Proof For the first statement, we know by Example 2.2 that: $\mathsf{Hom}_{\mathbf{Cat}}(\mathcal{C}, \mathcal{D}) \cong \mathsf{Hom}_{s\mathbf{Sets}}(B\mathcal{C}, B\mathcal{D})$ ($\forall \mathcal{C}, \mathcal{D} \in \mathbf{Cat}$, hence the fully faithfulness. Moreover $B\mathcal{C}$ is fibrant for a groupoid $\mathcal{C}$.

For the second statement, note that B obviously preserves weak equivalences; moreover, by definition, $Bf \colon BG \to BH$ is a fibration if and only if it has the right lifting property with respect to inclusions of horns in $\Delta[n]$, $\forall n \geq 1$ (see [11, page 10]). For $n = 1$ this means exactly that f is a fibration, while for $n \geq 2$ it's automatic (see the proof of [11, Lemma I.3.5]). □

Let $A \in \mathbf{Art}_{\mathcal{O}}$. Consider the group $G(A)$ of A-points of our reductive group scheme G. Passing to homotopy categories, we get the isomorphism

$$\begin{aligned} \mathsf{Hom}_{\mathrm{Ho}(\mathbf{Gpd})}(\Gamma_i, G(A)) &\cong \mathsf{Hom}_{\mathrm{Ho}(s\mathbf{Sets})}(B\Gamma_i, BG(A)) \\ &\cong \pi_0\, \mathsf{sHom}_{s\mathbf{Sets}}(B\Gamma_i, BG(A)). \end{aligned}$$

Let $X = (B\Gamma_i)_i$ be the pro-simplicial set associated to the profinite group Γ. We define

$$\mathsf{Hom}_{s\mathbf{Sets}}(X, -) = \varinjlim_i \mathsf{Hom}_{s\mathbf{Sets}}(B\Gamma_i, -).$$

Then the Galois representation $\bar{\rho} \colon \Gamma \to G(k)$ gives rise to an element of $\mathsf{Hom}_{s\mathbf{Sets}}(X, BG(k))$, which we also denote by $\bar{\rho}$. In order to take into account the deformations of $\bar{\rho}$, we introduce the overcategory $\mathcal{M} = s\mathbf{Sets}/_{BG(k)}$ of pairs (Y, π) where Y is a simplicial set and $\pi \colon Y \to BG(k)$ is a morphism of simplicial sets. The category $\mathcal{M}$ has a natural simplicial model category structure: the cofibrations, fibrations, weak equivalences, and tensor products are those of $s\mathbf{Sets}$ (see [11, Lemma II.2.4] for the only non-trivial part of the statement). When we consider $X \in \mathcal{M}$, we specify the morphism $\bar{\rho} \colon X \to BG(k)$; similarly, when we consider $BG(A) \in \mathcal{M}$ for $A \in \mathbf{Art}_{\mathcal{O}}$, we specify the natural projection $BG(A) \to BG(k)$. For $X, Y \in \mathcal{M}$, we can define an object of $\mathcal{M}$ of enriched homomorphisms $\mathsf{sHom}_{\mathcal{M}}(X, Y)$ for which

$\mathsf{sHom}_{\mathcal{M}}(X, Y)_n$ consists in the morphisms $X \times \mathbf{\Delta}[n] \to Y$ compatible to the projections to $BG(k)$. Since $BG(A) \to BG(k)$ is a fibration, $BG(A) \in \mathcal{M}$ is fibrant. Similar to the discussion of the preceding paragraph, we have

$$\mathcal{D}(A) \cong \mathsf{Hom}_{\mathrm{Ho}(\mathcal{M})}(X, BG(A)) \cong \pi_0\, \mathsf{sHom}_{\mathcal{M}}(X, BG(A)) \tag{3}$$

for $A \in \mathbf{Art}_{\mathcal{O}}$. Note that $\mathsf{sHom}_{\mathcal{M}}(X, BG(A))$ is the fiber over $\bar{\rho}$ of the fibration map

$$\mathsf{sHom}_{s\mathbf{Sets}}(X, BG(A)) \to \mathsf{sHom}_{s\mathbf{Sets}}(X, BG(k)),$$

so it actually calculates the homotopy fiber (see [15, Theorem 13.1.13 and Proposition 13.4.6]).

When $\Gamma = G_{F,S}$, $S = S_p \cup S_\infty$ and $\bar{\rho}$ satisfies (Ord_v) for $v \in S_p$, we reformulate the definition of the nearly ordinary deformation subfunctor $\mathcal{D}^{\mathrm{n.o}} \subset \mathcal{D}$ as follows. For each $v \in S_p$, we form $\Gamma_v = \varprojlim_i \Gamma_{i,v}$ where $\Gamma_v \to \Gamma$ induces morphisms $\Gamma_{i,v} \to \Gamma_i$ of finite groups. Let $X_v = (B\Gamma_{i,v})_i$ be the pro-simplicial set associated. For the fixed Borel subgroup B of G, we have a natural cofibration $BB(A) \subset BG(A)$. Recall that $\bar{g}_v \cdot \bar{\rho}|_{\Gamma_v} \cdot \bar{g}_v^{-1}$ takes values in $B(k)$. Let $\mathcal{D}_v(A)$ be π_0 of the fiber over $\bar{\rho}|_{\Gamma_v}$ of the fibration map

$$\mathsf{sHom}_{s\mathbf{Sets}}(X_v, BG(A)) \to \mathsf{sHom}_{s\mathbf{Sets}}(X_v, BG(k)),$$

and let $\mathcal{D}_v^{\mathrm{n.o}}(A)$ be π_0 of the fiber over $\bar{g}_v \cdot \bar{\rho}|_{\Gamma_v} \cdot \bar{g}_v^{-1}$ of the fibration map

$$\mathsf{sHom}_{s\mathbf{Sets}}(X_v, BB(A)) \to \mathsf{sHom}_{s\mathbf{Sets}}(X_v, BB(k)).$$

Then there is a natural functorial inclusion i_v of $\mathcal{D}_v^{\mathrm{n.o}}(A)$ into $\mathcal{D}_v(A)$. Let $\mathcal{D}_{\mathrm{loc}}(A) = \prod_{v \in S_p} \mathcal{D}_v(A)$ and $\mathcal{D}_{\mathrm{loc}}^{\mathrm{n.o}}(A) = \prod_{v \in S_p} \mathcal{D}_v^{\mathrm{n.o}}(A)$. There is a natural functorial map $\mathcal{D}(A) \to \mathcal{D}_{\mathrm{loc}}(A)$, resp., $\mathcal{D}_{\mathrm{loc}}^{\mathrm{n.o}}(A) \to \mathcal{D}_{\mathrm{loc}}(A)$, induced by $\rho \mapsto (\rho|_{\Gamma_v})_{v \in S_p}$, resp., by $\prod_{v \in S_p} i_v$.

We define $\mathcal{D}^{\mathrm{n.o}}(A)$ as the fiber product

$$\mathcal{D}^{\mathrm{n.o}}(A) = \mathcal{D}(A) \times_{\mathcal{D}_{\mathrm{loc}}(A)} \mathcal{D}_{\mathrm{loc}}^{\mathrm{n.o}}(A).$$

Lemma 2.7 *Suppose (Reg_v) holds for each place $v \in S_p$. Then the functor $\mathcal{D}^{\mathrm{n.o}}$ is isomorphic to the classical nearly ordinary deformation functor.*

Proof It follows easily from what precedes. See [2] or [26]. □

2.3 Simplicial Reformulation of Classical Framed Deformations

Let $\mathbf{Gpd}_*$ and $s\mathbf{Sets}_*$ be the model categories of based groupoids and based simplicial sets (in other words, under categories $_*\backslash\mathbf{Gpd}$ and $_*\backslash s\mathbf{Sets}$), respectively. Then we have

$$\mathsf{Hom}_{\mathbf{Gp}}(\Gamma_i, G(A)) \cong \mathsf{Hom}_{\mathrm{Ho}(\mathbf{Gpd}_*)}(\Gamma_i, G(A)). \tag{4}$$

Let $\mathcal{M}_*$ be the over and under category $_*\backslash s\mathbf{Sets}/_{BG(k)}$. Note that X and $BG(A)$ for $A \in \mathbf{Alg}_{\mathcal{O}}$ are naturally objects of $\mathcal{M}_*$. Proceeding as the unframed case, we see that

$$\mathcal{D}^{\square}(A) \cong \mathsf{Hom}_{\mathrm{Ho}(\mathcal{M}_*)}(X, BG(A)) \cong \pi_0\, \mathsf{sHom}_{\mathcal{M}_*}(X, BG(A)). \tag{5}$$

We remark that $\mathsf{sHom}_{\mathcal{M}_*}(X, BG(A))$ is weakly equivalent to $\mathrm{hofib}_*(\mathsf{sHom}_{\mathcal{M}}(X, BG(A)) \to \mathsf{sHom}_{\mathcal{M}}(*, BG(A)))$, since $\mathsf{sHom}_{\mathcal{M}}(X, BG(A)) \to \mathsf{sHom}_{\mathcal{M}}(*, BG(A))$ is a fibration.

2.4 Derived Deformation Functors

We have defined the functor $\mathsf{sHom}_{\mathcal{M}}(X, BG(-))$ from $\mathbf{Art}_{\mathcal{O}}$ to $s\mathbf{Sets}$. Our next goal is to extend this functor to simplicial Artinian $\mathcal{O}$-algebras over k, which we define below.

Let $s\mathbf{CR}$ be the category of simplicial commutative rings (these are simplicial sets which are rings in all degrees and for which all face and degeneracy maps are ring homomorphisms). A usual commutative ring A can be regarded as an element of $s\mathbf{CR}$, which consists of A on each simplicial degree with identity face and degeneracy maps. In this way, we regard $\mathcal{O}$ and k as objects of $s\mathbf{CR}$. With the natural reduction map $\mathcal{O} \to k$, the over and under category $_{\mathcal{O}}\backslash s\mathbf{CR}/_k$ has a simplicial model category structure, such that the cofibrations, fibrations, and weak equivalences are those of $s\mathbf{CR}$, and the tensor product of $A \in {_{\mathcal{O}}\backslash s\mathbf{CR}/_k}$ and $K \in s\mathbf{Sets}$ is the pushout of $\mathcal{O} \leftarrow \mathcal{O} \otimes K \to A \otimes K$. Note that degreewise surjective morphisms $A \to B$ are fibrations.

Since $s\mathbf{CR}$ is cofibrantly generated, any $A \in {_{\mathcal{O}}\backslash s\mathbf{CR}}$ admits a functorial cofibrant replacement $c(A)$:

$$\mathcal{O} \hookrightarrow c(A) \overset{\sim}{\twoheadrightarrow} A.$$

Concretely, for any $n \geq 0$ the $\mathcal{O}$-algebra $c(A)_n$ is a suitable polynomial $\mathcal{O}$-algebra mapping surjectively onto A_n. The key property of the cofibrant replacement is that

-$c(A)$ is a cofibrant object and

-$c(A) \to A$ is a trivial fibration (i.e., a fibration which is a weak equivalence).

Note that the functor $B \mapsto \mathsf{sHom}(c(A), B)$ commutes to weak equivalence (this is called homotopy invariance), while it is not necessarily the case of the functor $B \mapsto \mathsf{sHom}(A, B)$.

For $A \in {}_{\mathcal{O}}\backslash s\mathbf{CR}$, for any $i \geq 0$, $\pi_i A$ is a commutative group and $\bigoplus_i \pi_i A$ is naturally a graded $\mathcal{O}$-algebra, hence a $\pi_0 A$-algebra (see [10, Lemma 8.3.2]).

Definition 2.8 The simplicial Artinian $\mathcal{O}$-algebras over k, which we denote by ${}_{\mathcal{O}}\backslash s\mathbf{Art}/_k$, is the full subcategory of ${}_{\mathcal{O}}\backslash s\mathbf{CR}/_k$ consisting of objects $A \in {}_{\mathcal{O}}\backslash s\mathbf{CR}/_k$ such that:

(1) $\pi_0 A$ is Artinian local in the usual sense.
(2) $\pi_* A = \oplus_{i \geq 0} \pi_i A$ is finitely generated as a module over $\pi_0 A$.

Note that ${}_{\mathcal{O}}\backslash s\mathbf{Art}/_k$ is not a model category, and cofibrations, fibrations, and weak equivalences in ${}_{\mathcal{O}}\backslash s\mathbf{Art}/_k$ are used to indicate those in ${}_{\mathcal{O}}\backslash s\mathbf{CR}/_k$. Nevertheless, ${}_{\mathcal{O}}\backslash s\mathbf{Art}/_k$ is closed under weak equivalences since the definition only involves homotopy groups. We also remark that every $A \in {}_{\mathcal{O}}\backslash s\mathbf{Art}/_k$ is fibrant since $A \to k$ is degreewise surjective.

We define $\mathcal{O}_{N_\bullet G} \in \mathbf{Alg}_{\mathcal{O}}^{\mathbf{\Delta}}$ (i.e., a functor $\mathbf{\Delta} \to \mathbf{Alg}_{\mathcal{O}}$, also called a cosimplicial object in $\mathbf{Alg}_{\mathcal{O}}$) as follows: in codegree p we have $\mathcal{O}_{N_p G} = \mathcal{O}_G^{\otimes p}$, and the coface and codegeneracy maps are induced from the comultiplication and the coidentity of the Hopf algebra $\mathcal{O}_G$, respectively. Then for $A \in \mathbf{Alg}_{\mathcal{O}}$, the nerve $BG(A)$ is nothing but $\mathsf{Hom}_{\mathbf{Alg}_{\mathcal{O}}}(\mathcal{O}_{N_\bullet G}, A)$, with face and degeneracy maps induced by the coface and codegeneracy maps in $\mathcal{O}_{N_\bullet G}$. When $A \in {}_{\mathcal{O}}\backslash s\mathbf{CR}$, the naïve analogy is the diagonal of the bisimplicial set $([p], [q]) \mapsto \mathsf{Hom}_{\mathbf{Alg}_{\mathcal{O}}}(\mathcal{O}_{N_p G}, A_q)$ (recall that the diagonal of a bisimplicial set is a simplicial set model for its geometric realization). However, we need to make some modifications using cofibrant replacements to ensure the homotopy invariance.

Definition 2.9 (1) For $A \in {}_{\mathcal{O}}\backslash s\mathbf{CR}$, we define $\mathrm{Bi}(A)$ to be the bisimplicial set

$$([p], [q]) \mapsto \mathsf{Hom}_{{}_{\mathcal{O}}\backslash s\mathbf{CR}}(c(\mathcal{O}_{N_p G}), A^{\Delta[q]}),$$

with face and degeneracy maps induced by the coface and codegeneracy maps in $\mathcal{O}_{N_\bullet G}$ and the face and degeneracy maps in $A^{\Delta[\bullet]}$.

(2) The diagonal of $\mathrm{Bi}(A)$, which is denoted by $\operatorname{diag}\mathrm{Bi}(A)$, is the simplicial set induced from the diagonal embedding $\mathbf{\Delta}^{\mathrm{op}} \to \mathbf{\Delta}^{\mathrm{op}} \times \mathbf{\Delta}^{\mathrm{op}} \xrightarrow{\mathrm{Bi}(A)} \mathbf{Sets}$.

When A is an $\mathcal{O}$-algebra regarded as a constant object in ${}_{\mathcal{O}}\backslash s\mathbf{CR}$, we have

$$\mathrm{Bi}(A)_{p,q} = \mathsf{Hom}_{{}_{\mathcal{O}}\backslash s\mathbf{CR}}(c(\mathcal{O}_{N_p G}), A^{\Delta[q]}) \cong \mathsf{Hom}_{\mathbf{Alg}_{\mathcal{O}}}(\mathcal{O}_{N_p G}, A),$$

where the latter isomorphism is because the constant embedding functor is right adjoint to $\pi_0 \colon {}_{\mathcal{O}}\backslash s\mathbf{CR} \to \mathbf{Alg}_{\mathcal{O}}$. Hence, $\mathrm{Bi}(A)$ is just a disjoint union of copies of $BG(A)$ in index q. In particular, for $A \in {}_{\mathcal{O}}\backslash s\mathbf{Art}/_k$ there is a natural map $\mathrm{Bi}(A)_{\bullet,q} \to BG(k)$ for each $q \geq 0$, so we may regard $\mathrm{Bi}(A) \in \mathcal{M}^{\mathbf{\Delta}^{\mathrm{op}}}$ via the association $[q] \mapsto$

$\mathrm{Bi}(A)_{\bullet,q}$ (recall that $\mathcal{M}$ is the overcategory $s\mathbf{Sets}/_{BG(k)}$), and $\operatorname{diag}\mathrm{Bi}(A)$ is an object of $\mathcal{M}$). Recall that any morphism $X \to Y$ in $s\mathbf{Sets}$ admits a functorial factorization

$$X \overset{\sim}{\hookrightarrow} \widetilde{X} \twoheadrightarrow Y$$

into a trivial cofibration and a fibration.

Definition 2.10 For $A \in {}_{\mathcal{O}}\backslash s\mathbf{Art}/_k$, the simplicial set $\mathcal{B}G(A)$ is defined by the functorial trivial cofibration-fibration factorization $\operatorname{diag}\mathrm{Bi}(A) \overset{\sim}{\hookrightarrow} \mathcal{B}G(A) \twoheadrightarrow BG(k)$.

It's clear that $\mathcal{B}G\colon {}_{\mathcal{O}}\backslash s\mathbf{Art}/_k \to \mathcal{M}$ defines a functor. If $A \in \mathbf{Art}_{\mathcal{O}}$ is regarded as a constant simplicial ring, then $\operatorname{diag}\mathrm{Bi}(A) = BG(A) \twoheadrightarrow BG(k)$ is a fibration, so $BG(A)$ is a strong deformation retract of $\mathcal{B}G(A)$ in $\mathcal{M}$ (see [15, Definition 7.6.10]). In particular, these two are indistinguishable in our applications.

Remark 2.11 Our $\mathcal{B}G(A)$ is weakly equivalent to the simplicial set $\mathrm{Ex}^\infty \operatorname{diag}\mathrm{Bi}(A)$ which is the definition chosen in [13, Definition 5.1]. There is a slight difference: we want to emphasize the fibration $\mathcal{B}G(A) \twoheadrightarrow BG(k)$, so that it's more convenient to handle the homotopy pullbacks.

As mentioned above, the reason for taking cofibrant replacements is

Lemma 2.12 *If $A \to B$ is a weak equivalence, then so is $\mathcal{B}G(A) \to \mathcal{B}G(B)$.*

Proof If $A \to B$ is a weak equivalence, then $\mathsf{sHom}_{\mathcal{O}\backslash s\mathbf{CR}}(c(\mathcal{O}_{N_pG}), A) \to \mathsf{sHom}_{\mathcal{O}\backslash s\mathbf{CR}}(c(\mathcal{O}_{N_pG}), B)$ is a weak equivalence for each $p \geq 0$, so is $\operatorname{diag}\mathrm{Bi}(A) \to \operatorname{diag}\mathrm{Bi}(B)$ (see [15, Theorem 15.11.11]), and so is $\mathcal{B}G(A) \to \mathcal{B}G(B)$. □

Definition 2.13 (1) The derived universal deformation functor $s\mathcal{D}\colon {}_{\mathcal{O}}\backslash s\mathbf{Art}/_k \to s\mathbf{Sets}$ is defined by

$$s\mathcal{D}(A) = \mathsf{sHom}_{\mathcal{M}}(X, \mathcal{B}G(A)).$$

(2) The derived universal framed deformation functor $s\mathcal{D}^\square\colon {}_{\mathcal{O}}\backslash s\mathbf{Art}/_k \to s\mathbf{Sets}$ is defined by

$$s\mathcal{D}^\square(A) = \mathrm{hofib}_*(s\mathcal{D}(A) \to \mathsf{sHom}_{\mathcal{M}}(*, \mathcal{B}G(A))).$$

Remark 2.14 In [13, Definition 5.4], the derived universal deformation functor is defined by

$$s\mathcal{D}(A) = \mathrm{hofib}_{\bar{\rho}}(\mathsf{sHom}_{s\mathbf{Sets}}(X, \mathrm{Ex}^\infty \operatorname{diag}\mathrm{Bi}(A)) \to \mathsf{sHom}_{s\mathbf{Sets}}(X, BG(k))).$$

Since $\mathrm{Ex}^\infty \operatorname{diag}\mathrm{Bi}(A))$ and $\mathcal{B}G(A)$ are weakly equivalent fibrant simplicial sets, $\mathsf{sHom}_{s\mathbf{Sets}}(X, \mathrm{Ex}^\infty \operatorname{diag}\mathrm{Bi}(A))$ is weakly equivalent to $\mathsf{sHom}_{s\mathbf{Sets}}(X, \mathcal{B}G(A))$. But $\mathsf{sHom}_{s\mathbf{Sets}}(X, \mathcal{B}G(A)) \to \mathsf{sHom}_{s\mathbf{Sets}}(X, BG(k))$ is a fibration, so $\mathsf{sHom}_{\mathcal{M}}(X, \mathcal{B}G(A))$ is weakly equivalent to the homotopy fiber.

When $\Gamma = G_{F,S}$, $S = S_p \cup S_\infty$ and $\bar{\rho}$ satisfies (Ord_v) for $v \in S_p$, we can define for each $v \in S_p$ a functor $s\mathcal{D}_v \colon \mathcal{O}\backslash s\mathbf{Art}/_k \to s\mathbf{Sets}$ as $A \mapsto \mathsf{sHom}_{s\mathbf{Sets}/_{BG(k)}}(X_v, \mathcal{B}G(A))$, and a functor $s\mathcal{D}_v^{\text{n.o}} \colon \mathcal{O}\backslash s\mathbf{Art}/_k \to s\mathbf{Sets}$ as $A \mapsto \mathsf{sHom}_{s\mathbf{Sets}/_{BB(k)}}(X_v, \mathcal{B}B(A))$. Let $s\mathcal{D}_{\text{loc}} = \prod_{v \in S_p} s\mathcal{D}_v$ and let $s\mathcal{D}_{\text{loc}}^{\text{n.o}} = \prod_{v \in S_p} s\mathcal{D}_v^{\text{n.o}}$. Define $s\mathcal{D}^{\text{n.o}}$ as the homotopy fiber product

$$s\mathcal{D}^{\text{n.o}} = s\mathcal{D} \times^h_{s\mathcal{D}_{\text{loc}}} s\mathcal{D}_{\text{loc}}^{\text{n.o}}.$$

Definition 2.15 Let $\mathcal{F} \colon \mathcal{O}\backslash s\mathbf{Art}/_k \to s\mathbf{Sets}$ be a functor. We say $\mathcal{F}$ is formally cohesive if it satisfies the following conditions:

(1) $\mathcal{F}$ is homotopy invariant (*i.e.*, preserves weak equivalences).
(2) Suppose that

$$\begin{array}{ccc} A & \longrightarrow & B \\ \downarrow & & \downarrow \\ C & \longrightarrow & D \end{array} \tag{6}$$

is a homotopy pullback square with at least one of $B \to D$ and $C \to D$ degree-wise surjective, then

$$\begin{array}{ccc} \mathcal{F}(A) & \longrightarrow & \mathcal{F}(B) \\ \downarrow & & \downarrow \\ \mathcal{F}(C) & \longrightarrow & \mathcal{F}(D) \end{array} \tag{7}$$

is a homotopy pullback square.
(3) $\mathcal{F}(k)$ is contractible.

We summarize our preceding discussions:

Proposition 2.16 The functors $s\mathcal{D}$, $s\mathcal{D}^{\square}$, $s\mathcal{D}_v^{?}$ (here $? = \emptyset$ or n. o) and $s\mathcal{D}^{\text{n.o}}$ are all formally cohesive.

Proof We first verify three conditions in the above definition for $s\mathcal{D}$:

(1) If $A \to B$ is a weak equivalence, then $\mathcal{B}G(A) \to \mathcal{B}G(B)$ is a weak equivalence between fibrant objects in $\mathcal{M}$, so $\mathsf{sHom}_{\mathcal{M}}(X, \mathcal{B}G(A)) \to \mathsf{sHom}_{\mathcal{M}}(X, \mathcal{B}G(B))$ is also a weak equivalence.
(2) First we show that

$$\begin{array}{ccc} \mathcal{B}G(A) & \longrightarrow & \mathcal{B}G(B) \\ \downarrow & & \downarrow \\ \mathcal{B}G(C) & \longrightarrow & \mathcal{B}G(D) \end{array} \tag{8}$$

is a homotopy pullback square in $\mathcal{M}$. Note that regarding the above diagram as a diagram in $s\mathbf{Sets}$ doesn't affect the homotopy pullback nature. By [13, Lemma 4.31], it suffices to check:

(1) the functor $\Omega\mathcal{B}G\colon {}_{\mathcal{O}}\backslash s\mathbf{Art}/_k \to s\mathbf{Sets}$ preserves homotopy pullbacks and
(2) $\pi_1\mathcal{B}G(C) \to \pi_1\mathcal{B}G(D)$ is surjective whenever $C \to D$ is degreewise surjective.

Part (a) follows from [13, Lemma 5.2], and part (b) follows from [13, Corollary 5.3].
Since small filtered colimits of simplicial sets preserve homotopy pullbacks, we may suppose the pro-object X lies in $\mathcal{M}$. Then $\mathsf{sHom}_{\mathcal{M}}(X, -)\colon \mathcal{M} \to s\mathbf{Sets}$ is a right Quillen functor, hence its right derived functor commutes with homotopy pullbacks in the homotopy categories. But we are dealing with fibrant objects, so in the homotopy category $\mathsf{sHom}_{\mathcal{M}}(X, -)$ is isomorphic to its right derived functor. The conclusion follows:

(3) It's clear that $s\mathcal{D}(k)$ is contractible.

The same argument applies for $A \to \mathsf{sHom}_{\mathcal{M}}(*, \mathcal{B}G(A))$. So $s\mathcal{D}^{\square}$ is formally cohesive because it is the homotopy pullback of formally cohesive functors.

In the nearly ordinary case, we may replace X by X_v and replace G by B and the same argument applies. Hence $s\mathcal{D}_v^{?}$ ($? = \emptyset$ or n. o) is formally cohesive. Since $s\mathcal{D}^{\text{n.o}}$ is the homotopy limits of formally cohesive functors, it is also formally cohesive.

□

2.4.1 Modifying the Center

None of these functors cannot be pro-representable unless G is of adjoint type. If G has a non-trivial center Z, we need a variant $s\mathcal{D}_Z$, resp., $s\mathcal{D}_Z^{\text{n.o}}$, of the functor $s\mathcal{D}$, resp., of $s\mathcal{D}^{\text{n.o}}$, in order to allow pro-representability. For this modification, we follow [13, Section 5.4]. For a classical ring $A \in \mathbf{Art}$, we have a short exact sequence

$$1 \to Z(A) \to G(A) \to PG(A) \to 1.$$

It yields a fibration sequence $BG(A) \to BPG(A) \to B^2Z(A)$. Indeed, given a simplicial group H and a simplicial sets X with a left H-action, we can form the bar construction $N_\bullet(*, H, X)$ at each simplicial degree (see [10, Example 3.2.4]), which gives the bisimplicial set $([p], [q]) \mapsto H_p^q \times X_p =: N_q(*, H_p, X_p)$. Consider the action $Z(A) \times G(A) \to G(A)$, and the corresponding simplicial action $N_pZ(A) \times N_pG(A) \to N_pG(A)$ (note that $N_\bullet Z(A)$ is a simplicial group because $Z(A)$ is Abelian). We identify for each $p \geq 0$,

$$BG(A)_p = N_p(*, *, N_pG(A)),$$

$$BPG(A)_p = N_p(*, N_pZ(A), N_pG(A)),$$

and we put

$$B^2Z(A)_p = N_p(*, N_pZ(A), *)$$

(with diagonal face and degeneracy maps). The desired fibration is given by the canonical morphisms of simplicial sets which in degree p are

$$N_p(*, *, N_pG(A)) \to N_p(*, N_pZ(A), N_pG(A)) \to N_p(*, N_pZ(A), *).$$

Let us generalize this to $A \in {}_{\mathcal{O}}\backslash s\mathbf{Art}/_k$. For this, we note first that $BPG(A)$ can also be defined as the functorial fibrant replacement of diag(N) where N is the trisimplicial set associated to $(p, q, r) \mapsto N_q(*, N_pZ(A_r), N_p(G(A_r))$ (replacing $\mathcal{O}_{NpG(A_r)}$ by its functorial cofibrant replacement as above).

Then, we define $B^2Z(A)$ as the functorial fibrant replacement of diag(N') where N' is the trisimplicial set associated to $(p, q, r) \mapsto N_q(*, N_pZ(A_r), *)$ (replacing $\mathcal{O}_{NpG(A_r)}$ by its functorial cofibrant replacement as above). The obvious system of maps $N_q(*, N_pZ(A_r), N_pG(A_r)) \to N_q(*, N_pZ(A_r), *)$ gives the desired map

$$\mathcal{B}PG(A) \to B^2Z(A).$$

The functor $s\mathcal{D}_Z\colon {}_{\mathcal{O}}\backslash s\mathbf{Art}/_k \to s\mathbf{Sets}$ is defined by the homotopy pullback square (here for simplicity we use $\mathcal{M}$, but the base maps are those induced from $\mathcal{B}G(k) \to \mathcal{B}PG(k) \to B^2Z(k)$)

$$\begin{array}{ccc} s\mathcal{D}_Z(A) & \longrightarrow & \mathsf{sHom}_{\mathcal{M}}(*, B^2Z(A)) \\ \downarrow & & \downarrow \\ \mathsf{sHom}_{\mathcal{M}}(X, \mathcal{B}PG(A)) & \longrightarrow & \mathsf{Hom}_{\mathcal{M}}(X, B^2Z(A)). \end{array}$$

Then $s\mathcal{D}_Z$ is formally cohesive because it is the homotopy pullback of formally cohesive functors. Observe that $s\mathcal{D}_Z$ and $s\mathcal{D}$ coincide when Z is trivial.

Remark 2.17 (1) We'll see later that $s\mathcal{D}_Z$ is pro-representable, under the assumption $H^0(\Gamma, \mathfrak{g}_k) = \mathfrak{z}_k$.

(2) In the nearly ordinary case, one defines similarly $s\mathcal{D}_{\mathrm{loc},Z} = \prod_{v\in S_p} s\mathcal{D}_{v,Z}$ and $s\mathcal{D}^{\mathrm{n.o}}_{\mathrm{loc},Z} = \prod_{v\in S_p} s\mathcal{D}^{\mathrm{n.o}}_{v,Z}$. Note that the construction for $s\mathcal{D}_Z$ is functorial in X and G, we can form the homotopy pullback

$$s\mathcal{D}^{\mathrm{n.o}}_Z = s\mathcal{D}_Z \times^h_{s\mathcal{D}_{\mathrm{loc},Z}} s\mathcal{D}^{\mathrm{n.o}}_{\mathrm{loc},Z}.$$

All these functors are formally cohesive. We'll see later that $s\mathcal{D}^{\mathrm{n.o}}_Z$ is pro-representable, under the assumption $H^0(\Gamma, \mathfrak{g}_k) = \mathfrak{z}_k$.

Proposition 2.18 When A is homotopy discrete, we have $\pi_0 s\mathcal{D}_Z(A) \cong \mathcal{D}(\pi_0 A)$ and $\pi_0 s\mathcal{D}^?_{v,Z}(A) \cong \mathcal{D}^?_v(\pi_0 A)$ (here $? = \emptyset$ or n. o). If in addition (Reg_v) holds for each $v \in S_p$, then $\pi_0 s\mathcal{D}^{\mathrm{n.o}}_Z(A) \cong \mathcal{D}^{\mathrm{n.o}}(\pi_0 A)$.

Proof We may suppose $A \in \mathbf{Art}_{\mathcal{O}}$ by the formal cohesiveness.

From the definition of $s\mathcal{D}_Z$, it follows that we have a natural fibration sequence

$$s\mathcal{D}(A) \to s\mathcal{D}_Z(A) \to \mathsf{sHom}_{\mathcal{M}}(*, B^2 Z(A)).$$

Since $\pi_i\, \mathsf{sHom}_{\mathcal{M}}(*, B^2 Z(A))$ vanishes for $i \neq 2$, we have $\pi_0 s\mathcal{D}_Z(A) = \pi_0 s\mathcal{D}(A)$. By Equation 3 of Sect. 2.2, we have $\pi_0 s\mathcal{D}(A) = \mathcal{D}(A)$, hence also $\pi_0 s\mathcal{D}_Z(A) = \mathcal{D}(A)$.

By applying the same argument with X replaced by X_v and G replaced by B when necessary, we obtain $\pi_0 s\mathcal{D}^?_{v,Z}(A) \cong \mathcal{D}^?_v(A)$ ($? = \emptyset$ or n. o).

We have the exact sequence

$$\begin{aligned} \pi_1 s\mathcal{D}_Z(A) \oplus (\bigoplus_{v\in S_p} \pi_1 s\mathcal{D}^{\mathrm{n.o}}_{v,Z}(A)) &\to \bigoplus_{v\in S_p} \pi_1 s\mathcal{D}_{v,Z}(A) \\ \to \pi_0 s\mathcal{D}^{\mathrm{n.o}}_Z(A) \to \pi_0 s\mathcal{D}_Z(A) \oplus (\bigoplus_{v\in S_p} \pi_0 s\mathcal{D}^{\mathrm{n.o}}_{v,Z}(A)) &\to \bigoplus_{v\in S_p} \pi_0 s\mathcal{D}_{v,Z}(A). \end{aligned}$$

We will see later (Lemma 4.20) that $s\mathcal{D}_v(A)$ is weakly equivalent to $\mathrm{holim}_{\Delta_X}\mathrm{hofib}_*(BG(A) \to BG(k))$, and (by Lemma 4.22) $\pi_1 s\mathcal{D}_v(A) \cong H^0(\Gamma_v, \widehat{G}(A))$. Similarly $\pi_1 s\mathcal{D}^{\mathrm{n.o}}_v(A) \cong H^0(\Gamma_v, \widehat{B}(A))$.

By the assumption (Reg_v) and Artinian induction, the map $\pi_1 s\mathcal{D}^{\mathrm{n.o}}_v(A) \to \pi_1 s\mathcal{D}_v(A)$ is an isomorphism, and so is $\pi_1 s\mathcal{D}^{\mathrm{n.o}}_{v,Z}(A) \to \pi_1 s\mathcal{D}_{v,Z}(A)$. We deduce that $\pi_0 s\mathcal{D}^{\mathrm{n.o}}_Z(A)$ is the kernel of $\mathcal{D}(A) \oplus (\bigoplus_{v\in S_p} \mathcal{D}^{\mathrm{n.o}}_v(A)) \to \bigoplus_{v\in S_p} \mathcal{D}_v(A)$, which is isomorphic to $\mathcal{D}^{\mathrm{n.o}}(A)$ by Lemma 2.7. □

3 Pseudo-Deformation Functors

3.1 Classical Pseudo-Characters and Functors on FFS

Recall the notion of a (classical) G-pseudo-character due to V. Lafforgue (see [16, Définition-Proposition 11.3] and [4, Definition 4.1]):

Definition 3.1 Let A be an $\mathcal{O}$-algebra. A G-pseudo-character Θ on Γ over A is a collection of $\mathcal{O}$-algebra morphisms $\Theta_n \colon \mathcal{O}^{\mathrm{ad}G}_{N_n G} \to \mathrm{Map}(\Gamma^n, A)$ for each $n \geq 1$, satisfying the following conditions:

(1) For each $n, m \geq 1$ and for each map $\zeta \colon \{1, \dots, n\} \to \{1, \dots, m\}$, $f \in \mathcal{O}^{\mathrm{ad}G}_{N_m G}$, and $\gamma_1, \dots, \gamma_m \in \Gamma$, we have

$$\Theta_m(f^\zeta)(\gamma_1, \dots, \gamma_m) = \Theta_n(f)(\gamma_{\zeta(1)}, \dots, \gamma_{\zeta(n)}),$$

where $f^\zeta(g_1, \dots, g_m) = f(g_{\zeta(1)}, \dots, g_{\zeta(n)})$.

(2) For each $n \geq 1$, for each $\gamma_1, \ldots, \gamma_{n+1} \in \Gamma$, and for each $f \in \mathcal{O}^{\mathrm{ad}G}_{N_n G}$, we have

$$\Theta_{n+1}(\hat{f})(\gamma_1, \ldots, \gamma_{n+1}) = \Theta_n(f)(\gamma_1, \ldots, \gamma_{n-1}, \gamma_n \gamma_{n+1}),$$

where $\hat{f}(g_1, \ldots, g_{n+1}) = f(g_1, \ldots, g_{n-1}, g_n g_{n+1})$.

We denote by $\mathrm{PsCh}(A)$ the set of pseudo-characters over A.

We want to give a simplicial reformulation of this notion. As a first step, following [28], let us consider **FS** the category of finite sets and **FFS** be the category of finite free semigroups. For any finite set X, let M_X be the finite free semigroup generated by X; we have $\Gamma^X = \mathsf{Hom}_{\mathbf{semGp}}(M_X, \Gamma)$ and $G^X = \mathsf{Hom}_{\mathbf{semGp}}(M_X, G)$. For a semigroup $M \in \mathbf{FFS}$, note that $\mathsf{Hom}_{\mathbf{semGp}}(M_X, G)$ is a group scheme, so we can define a covariant functor $\mathbf{FFS} \to \mathbf{Alg}_{\mathcal{O}}$, $M \mapsto \mathcal{O}_{\mathsf{Hom}_{\mathbf{semGp}}(M,G)}$. We can also define the covariant functor $M \mapsto \mathrm{Map}(\mathsf{Hom}_{\mathbf{semGp}}(M, \Gamma), A)$. These functors on **FFS** extend canonically those defined on the category **FS** by $X \mapsto \mathcal{O}_{G^X}$ and $X \mapsto \mathrm{Map}(\Gamma^X, A)$. Moreover, the natural transformation

$$\mathcal{O}^{\mathrm{ad}G}_{G^X} \to \mathrm{Map}(\Gamma^X, A)$$

extends uniquely to a natural transformation of functors on **FFS**. Actually, there are several useful functors on **FFS**; by the canonical extension from **FS** to **FFS** mentioned above, it is enough to define them on the objects $[n]$, as in [28, Example 2.4 and Example 2.5]:

(1) The association $[n] \mapsto \Gamma^n$ defines an object $\Gamma^\bullet \in \mathbf{Sets}^{\mathbf{FFS}^{\mathrm{op}}}$.
(2) For $A \in \mathbf{Alg}_{\mathcal{O}}$, the association $[n] \mapsto \mathrm{Map}(\Gamma^n, A)$ defines an object $\mathrm{Map}(\Gamma^\bullet, A) \in \mathbf{Alg}_{\mathcal{O}}^{\mathbf{FFS}}$.
(3) The association $[n] \mapsto \mathcal{O}^{\mathrm{ad}G}_{N_n G}$ defines an object $\mathcal{O}^{\mathrm{ad}G}_{N_\bullet G} \in \mathbf{Alg}_{\mathcal{O}}^{\mathbf{FFS}}$.
(4) Let $G^n /\!\!/ G = \mathrm{Spec}(\mathcal{O}^{\mathrm{ad}G}_{N_n G})$. Then for $A \in \mathbf{Alg}_{\mathcal{O}}$, the association $[n] \mapsto (G^n /\!\!/ G)(A)$ defines an object $(G^\bullet /\!\!/ G)(A) \in \mathbf{Sets}^{\mathbf{FFS}^{\mathrm{op}}}$.

As noted in [28, Theorem 2.12], one sees that a G-pseudo-character Θ of Γ over A is exactly a natural transformation from $\mathcal{O}^{\mathrm{ad}G}_{N_\bullet G}$ to $\mathrm{Map}(\Gamma^\bullet, A)$ (we call these natural transformations $\mathbf{Alg}_{\mathcal{O}}^{\mathbf{FFS}}$-morphisms).

Lemma 3.2 *For $A \in \mathbf{Alg}_{\mathcal{O}}$, there is a bijection between* $\mathrm{PsCh}(A)$ *and* $\mathsf{Hom}_{\mathbf{Sets}^{\mathbf{FFS}^{\mathrm{op}}}}(\Gamma^\bullet, (G^\bullet /\!\!/ G)(A))$.

Proof It suffices to note that there is a bijection between $\mathbf{Sets}^{\mathbf{FFS}^{\mathrm{op}}}$-morphisms $\Gamma^\bullet \to (G^\bullet /\!\!/ G)(A)$ and $\mathbf{Alg}_{\mathcal{O}}^{\mathbf{FFS}}$-morphisms $\mathcal{O}^{\mathrm{ad}G}_{N_\bullet G} \to \mathrm{Map}(\Gamma^\bullet, A)$.

For an algebraically closed field A and a (continuous) homomorphism $\rho \colon \Gamma \to G(A)$, we say that ρ is G-completely reducible if any parabolic subgroup containing $\rho(\Gamma)$ has a Levi subgroup containing $\rho(\Gamma)$. Recall the following results in [4, Sect. 4]:

Theorem 3.3 *(1)* [4, Theorem 4.5] *Suppose that $A \in \mathbf{Alg}_{\mathcal{O}}$ is an algebraically closed field. Then we have a bijection between the following two sets:*

(a) The set of $G(A)$-conjugacy classes of G-completely reducible group homomorphisms $\rho\colon \Gamma \to G(A)$.
(b) The set of pseudo-characters over A.

(2) [4, Theorem 4.10] *Fix an absolutely G-completely reducible representation $\bar{\rho}\colon \Gamma \to G(k)$, and suppose further that the centralizer of $\bar{\rho}$ in G_k^{ad} is scheme-theoretically trivial. Let $\bar{\Theta}$ be the pseudo-character, which regarded as an element of $\mathrm{Hom}_{\mathbf{Sets}^{\mathbf{FFS}^{\mathrm{op}}}}(\Gamma^\bullet, (G^\bullet /\!\!/ G)(k))$ is induced from $(\gamma_1, \ldots, \gamma_n) \mapsto (\bar{\rho}(\gamma_1), \ldots, \bar{\rho}(\gamma_n))$. Let $A \in \mathbf{Art}_{\mathcal{O}}$. Then we have a bijection between the following two sets:*

(a) The set of $\widehat{G}(A)$-conjugacy classes of group homomorphisms $\rho\colon \Gamma \to G(A)$ which lift $\bar{\rho}$.
(b) The set of pseudo-characters over A which reduce to $\bar{\Theta}$ modulo $\mathfrak{m}_A$.

Note that there are similarities between $\mathbf{Sets}^{\mathbf{FFS}^{\mathrm{op}}}$ and $\mathbf{Sets}^{\mathbf{\Delta}^{\mathrm{op}}} = s\mathbf{Sets}$. In the following, we shall prove similar results with $\mathbf{Sets}^{\mathbf{FFS}^{\mathrm{op}}}$ replaced by $s\mathbf{Sets}$.

3.2 Classical Pseudo-Characters and Simplicial Objects

Recall that on $\mathcal{O}_{N_\bullet G}$ there are natural coface and codegeneracy maps, and we can regard $\mathcal{O}_{N_\bullet G}$ as an object in $\mathbf{Alg}_{\mathcal{O}}^{\mathbf{\Delta}}$ (*i.e.*, a cosimplicial $\mathcal{O}$-algebra). The adjoint action of G on $G^\bullet$ induces an action of G on $\mathcal{O}_{N_\bullet G}$, which obviously commutes with the coface and codegeneracy maps. In consequence, $\mathcal{O}_{N_\bullet G}^{\mathrm{ad}G}$ is well defined in $\mathbf{Alg}_{\mathcal{O}}^{\mathbf{\Delta}}$.

Definition 3.4 We define the functor $\bar{B}G\colon \mathbf{Alg}_{\mathcal{O}} \to s\mathbf{Sets}$ by associating $A \in \mathbf{Alg}_{\mathcal{O}}$ to $\mathrm{Hom}_{\mathbf{Alg}_{\mathcal{O}}}(\mathcal{O}_{N_\bullet G}^{\mathrm{ad}G}, A)$ with face and degeneracy maps induced from the coface and codegeneracy maps in $\mathcal{O}_{N_\bullet G}^{\mathrm{ad}G}$.

Note that the inclusion $\mathcal{O}_{N_\bullet G}^{\mathrm{ad}G} \to \mathcal{O}_{N_\bullet G}$ gives a natural transformation $BG \to \bar{B}G$.

3.2.1 Algebraically Closed Field

Let $A \in \mathbf{Alg}_{\mathcal{O}}$ be an algebraically closed field. We would like to characterize the elements of $\mathrm{Hom}_{s\mathbf{Sets}}(B\Gamma, \bar{B}G(A))$. They correspond to the quasi-homomorphisms, which we define below.

Definition 3.5 Let Γ and G be two groups. We say a map $\rho\colon \Gamma \to G$ is a quasi-homomorphism if there exists a map $\phi\colon \Gamma \to G$ such that $\rho(x)^{-1}\rho(xy) = \phi(x)\rho(y)\,\phi(x)^{-1}$ for any $x, y \in \Gamma$.

Obviously a group homomorphism is a quasi-homomorphism. Note that every quasi-homomorphism preserves the identity, and the set of quasi-homomorphisms is closed under G-conjugations.

Remark 3.6 A quasi-homomorphism can fail to be a group homomorphism. We can construct a quasi-homomorphism as follows: let $\sigma\colon \Gamma \to G$ be a group homomorphism, let $\phi\colon \Gamma \to Z(\sigma(\Gamma))$ be a group homomorphism and let $g \in G$, then $\rho(x) = g^{-1}\sigma(x)\phi(x)g\phi(x)^{-1}$ is a quasi-homomorphism. Such ρ is not necessarily a group homomorphism, an example could be the following: take $G = H \times H$, $\sigma\colon \Gamma \to H \times \{\mathbf{e}\}$ and $\phi\colon \Gamma \to \{\mathbf{e}\} \times H$, and choose g such that $g \notin Z(\phi(\Gamma))$.

Lemma 3.7 *Let ρ be a quasi-homomorphism and let ϕ as above. Then the map ϕ induces a group homomorphism $\Gamma \to G/Z(\rho(\Gamma))$ which doesn't depend on the choice of ϕ.*

Proof For $x, y, z \in \Gamma$, we have

$$\begin{aligned}
\phi(xy)\rho(z)\phi(xy)^{-1} &= \rho(xy)^{-1}\rho(xyz)\\
&= (\phi(x)\rho(y)\phi(x)^{-1})^{-1}(\phi(x)\rho(yz)\phi(x)^{-1})\\
&= \phi(x)\rho(y)^{-1}\rho(yz)\phi(x)^{-1}\\
&= \phi(x)\phi(y)\rho(z)\phi(y)^{-1}\phi(x)^{-1}.
\end{aligned}$$

Hence $\phi(xy)^{-1}\phi(x)\phi(y) \in Z(\rho(\Gamma))$ for any $x, y \in \Gamma$, and ϕ induces a group homomorphism $\Gamma \to G/Z(\rho(\Gamma))$. For any other choice ϕ_1 such that $\rho(x)^{-1}\rho(xy) = \phi_1(x)\rho(y)\phi_1(x)^{-1}$, we see $\phi_1^{-1}(x)\phi(x) \in Z(\rho(\Gamma))$, and the conclusion follows. □

Lemma 3.8 *Suppose that $A \in \mathbf{Alg}_{\mathcal{O}}$ is an algebraically closed field. Let $f \in \mathsf{Hom}_{s\mathbf{Sets}}(B\Gamma, \bar{B}G(A))$. Then we can associate a quasi-homomorphism $\rho\colon \Gamma \to G(A)$ to f such that f sends $(\gamma_1, \dots, \gamma_n) \in B\Gamma_n$ to the class in $\bar{B}G(A)_n$ represented by $(\rho(\prod_{j=1}^{i-1}\gamma_j)^{-1}\rho(\prod_{j=1}^{i}\gamma_j))_{i=1,\dots,n}$.*

Proof For each $n \geq 1$ and $\underline{\gamma} = (\gamma_1, \dots, \gamma_n) \in \Gamma^n$, we choose a representative $T(\underline{\gamma}) = (g_1, \dots, g_n) \in G(A)^n$ of $f(\underline{\gamma})$ with closed orbit, note that any other representative with closed orbit is conjugated to $(g_1, \dots, g_n)$. Let $H(\underline{\gamma})$ be the Zariski closure of the subgroup of $G(A)$ generated by the entries of $T(\underline{\gamma})$. Let $n(\underline{\gamma})$ be the dimension of a parabolic $P \subseteq G_A$ minimal among those containing $H(\underline{\gamma})$, we see $n(\underline{\gamma})$ is independent of the choice of P. Let $N = \sup_{n\geq 1, \underline{\gamma}\in\Gamma^n} n(\underline{\gamma})$. We fix a choice of $\underline{\delta} = (\delta_1, \dots, \delta_n)$ satisfying the following conditions:

(1) $n(\underline{\delta}) = N$.
(2) For any $\underline{\delta}' \in \Gamma^{n'}$ satisfying (1), we have $\dim Z_{G_A}(H(\underline{\delta})) \leq \dim Z_{G_A}(H(\underline{\delta}'))$.
(3) For any $\underline{\delta}' \in \Gamma^{n'}$ satisfying (1) and (2), we have $\#\pi_0(Z_{G_A}(H(\underline{\delta}))) \leq \#\pi_0(Z_{G_A}(H(\underline{\delta}')))$.

Write $T(\underline{\delta}) = (h_1, \dots, h_n)$. As in the proof of [4, Theorem 4.5], we have the following facts:

(1) For any $(\gamma_1, \ldots, \gamma_m) \in \Gamma^m$, there exists a unique tuple $(g_1, \ldots, g_m) \in G(A)^m$ such that $(h_1, \ldots, h_n, g_1, \ldots, g_m)$ is conjugated to $T(\delta_1, \ldots, \delta_n, \gamma_1, \ldots, \gamma_m)$.
(2) Let $(h_1, \ldots, h_n, g_1, \ldots, g_m)$ be as above. Any finite subset of the group generated by $(h_1, \ldots, h_n, g_1, \ldots, g_m)$ which contains $(h_1, \ldots, h_n)$ has a closed orbit.

We define $\rho(\gamma)$ to be the unique element such that $(h_1, \ldots, h_n, \rho(\gamma))$ is conjugated to $T(\delta_1, \ldots, \delta_n, \gamma)$.

Suppose for $\gamma_1, \ldots, \gamma_m \in \Gamma$, the unique tuple conjugated to $T(\delta_1, \ldots, \delta_n, \gamma_1, \ldots, \gamma_m)$ is $(h_1, \ldots, h_n, g_1, \ldots, g_m)$. Consider the following diagram, where the horizontal arrows are compositions of face maps:

$$\begin{array}{ccc} (\delta_1, \ldots, \delta_n, \gamma_1, \ldots, \gamma_m) & \longrightarrow & (h_1, \ldots, h_n, g_1, \ldots, g_m) \\ \downarrow & & \downarrow \\ (\delta_1, \ldots, \delta_n, \prod_{j=1}^{i} \gamma_j) & \longrightarrow & (h_1, \ldots, h_n, \prod_{j=1}^{i} g_j). \end{array}$$

Since $(h_1, \ldots, h_n, \prod_{j=1}^{i} g_j)$ has a closed orbit and is a pre-image of $f(\delta_1, \ldots, \delta_n, \prod_{j=1}^{i} \gamma_j)$, we have $\prod_{j=1}^{i} g_j = \rho(\prod_{j=1}^{i} \gamma_j)$, and $g_i = \rho(\prod_{j=1}^{i-1} \gamma_j)^{-1}\rho(\prod_{j=1}^{i} \gamma_j)$ $(\forall i = 1, \ldots, m)$.

Let $x, y \in \Gamma$. Then the element in $G(A)^{2n+2}$ associated to $(\delta_1, \ldots, \delta_n, x, \delta_1, \ldots, \delta_n, y)$ is

$$(h_1, \ldots, h_n, \rho(x), \rho(x)^{-1}\rho(x\delta_1), \ldots, \rho(x\prod_{j=1}^{n-1}\delta_j)^{-1}\rho(x\prod_{j=1}^{n}\delta_j), \rho(x\prod_{j=1}^{n}\delta_j)^{-1}\rho(x\prod_{j=1}^{n}\delta_j \cdot y)),$$

and the element in $G(A)^{2n+1}$ associated to $(\delta_1, \ldots, \delta_n, \delta_1, \ldots, \delta_n, y)$ is

$$(h_1, \ldots, h_n, \rho(\delta_1), \ldots, \rho(\prod_{j=1}^{n-1}\delta_j)^{-1}\rho(\prod_{j=1}^{n}\delta_j), \rho(\prod_{j=1}^{n}\delta_j)^{-1}\rho(\prod_{j=1}^{n}\delta_j \cdot y)).$$

We see both $(\rho(x\prod_{j=1}^{i-1}\delta_j)^{-1}\rho(x\prod_{j=1}^{i}\delta_j))_{i=1,\ldots,n}$ and $(\rho(\prod_{j=1}^{i-1}\delta_j)^{-1}$ $\rho(\prod_{j=1}^{i}\delta_j))_{i=1,\ldots,n}$ have a closed orbit and are pre-images of $f(\delta_1, \ldots, \delta_n)$, so they are conjugated by some $\phi(x) \in G(A)$. Since $Z_{G_A}(H(\underline{\delta}))$ is minimal by the defining property, $\phi(x)$ must conjugate $\rho(\prod_{j=1}^{n}\delta_j)^{-1}\rho(\prod_{j=1}^{n}\delta_j \cdot y)$ to $\rho(x\prod_{j=1}^{n}\delta_j)^{-1}$ $\rho(x\prod_{j=1}^{n}\delta_j \cdot y)$. We deduce that $\forall x, y \in \Gamma$, $\rho(x)^{-1}\rho(xy) = \phi(x)\rho(y)\phi(x)^{-1}$, and ρ is a quasi-homomorphism. It's obvious that for any $(\gamma_1, \ldots, \gamma_n) \in \Gamma^n$, $(\rho(\prod_{j=1}^{i-1}\gamma_j)^{-1}\rho(\prod_{j=1}^{i}\gamma_j))_{i=1,\ldots,n}$ is a pre-image of $f(\gamma_1, \ldots, \gamma_n)$. □

3.2.2 Artinian Coefficients

Let $\bar{\rho}\colon \Gamma \to G(k)$ be an absolutely G-completely reducible representation, and suppose that $H^0(\Gamma, \mathfrak{g}) = \mathfrak{z}$. We write $\bar{f} \in \mathsf{Hom}_{s\mathbf{Sets}}(B\Gamma, \bar{B}G(k))$ for the map induced from $(\gamma_1, \ldots, \gamma_n) \mapsto (\bar{\rho}(\gamma_1), \ldots, \bar{\rho}(\gamma_n))$.

Definition 3.9 For $A \in \mathbf{Art}_{\mathcal{O}}$, the set $\mathrm{aDef}_{\bar{f}}(A)$ is the fiber over $\bar{f}$ of the map

$$\mathsf{Hom}_{s\mathbf{Sets}}(B\Gamma, \bar{B}G(A)) \to \mathsf{Hom}_{s\mathbf{Sets}}(B\Gamma, \bar{B}G(k)).$$

Definition 3.10 Let $A \in \mathbf{Art}_{\mathcal{O}}$. We say a map $\rho\colon \Gamma \to G(A)$ is a quasi-lift of $\bar{\rho}$ if $\rho \mod \mathfrak{m}_A = \bar{\rho}$ and ρ is a quasi-homomorphism.

Remark 3.11 In general, a quasi-lift may not be a group homomorphism. Let $0 \to I \to A_1 \twoheadrightarrow A_0$ be an infinitesimal extension in $\mathbf{Art}_{\mathcal{O}}$. Let $\rho_0 : \Gamma \to G(A_0)$ be a group homomorphism, let $\sigma : G(A_0) \to G(A_1)$ be a set-theoretic section of $G(A_1) \to G(A_0)$ and let $\tilde{\rho} = \sigma \circ \rho_0$. Let's construct a quasi-lift $\rho_1 = \exp(X_\alpha)\tilde{\rho}$ where $X : \Gamma \to \mathfrak{g} \otimes_k I$ is a cochain to be determined.

For $\alpha, \beta \in \Gamma$, there exists $c_{\alpha,\beta} \in \mathfrak{g} \otimes_k I$ such that $\tilde{\rho}(\alpha)\tilde{\rho}(\beta) = \exp(c_{\alpha,\beta})\tilde{\rho}(\alpha\beta)$ since $\rho_0 : \Gamma \to G(A_0)$ is a group homomorphism. It's easy to check that $c \in Z^2(\Gamma, \mathfrak{g} \otimes_k I)$. Let $\phi(\alpha) = \exp(Y_\alpha)$ where $Y : \Gamma \to \mathfrak{g} \otimes_k I$ is a group homomorphism also to be determined. We require $\rho_1(\alpha\beta) = \rho_1(\alpha)\phi(\alpha)\rho_1(\beta)\phi(\alpha)^{-1}$ for all $\alpha, \beta \in \Gamma$. Note that $\rho_1(\alpha\beta) = \exp(X_{\alpha\beta})\tilde{\rho}(\alpha\beta)$ and

$$\begin{aligned}
\rho_1(\alpha)\phi(\alpha)\rho_1(\beta)\phi(\alpha)^{-1} &= \exp(X_\alpha)\tilde{\rho}(\alpha)\exp(Y_\alpha)\exp(X_\beta)\tilde{\rho}(\beta)\exp(Y_\alpha)^{-1} \\
&= \exp(X_\alpha)\tilde{\rho}(\alpha)\exp(X_\beta + Y_\alpha - \mathrm{Ad}\,\tilde{\rho}(\beta)Y_\alpha)\tilde{\rho}(\beta) \\
&= \exp(X_\alpha + \mathrm{Ad}\,\tilde{\rho}(\alpha)X_\beta)\exp(\mathrm{Ad}\,\tilde{\rho}(\alpha)(1 - \mathrm{Ad}\,\tilde{\rho}(\beta))Y_\alpha)\tilde{\rho}(\alpha)\tilde{\rho}(\beta) \\
&= \exp(X_\alpha + \mathrm{Ad}\,\tilde{\rho}(\alpha)X_\beta)\exp(\mathrm{Ad}\,\tilde{\rho}(\alpha)(1 - \mathrm{Ad}\,\tilde{\rho}(\beta))Y_\alpha)\exp(c_{\alpha,\beta})\tilde{\rho}(\alpha\beta),
\end{aligned}$$

so we need to find a group homomorphism $Y : \Gamma \to \mathfrak{g} \otimes_k I$ such that $\mathrm{Ad}\,\tilde{\rho}(\alpha)(1 - \mathrm{Ad}\,\tilde{\rho}(\beta))Y_\alpha) + c_{\alpha,\beta}$ is a coboundary. In particular, in the case $H^2(\Gamma, \mathfrak{g}) = 0$, we can take an arbitrary group homomorphism $Y : \Gamma \to \mathfrak{g}$. Note that ρ_1 is a group homomorphism if and only if $\phi(\alpha) = \exp(Y_\alpha) \in Z(A)$ for any $\alpha \in \Gamma$.

Lemma 3.12 *Let $A \in \mathbf{Art}_{\mathcal{O}}$ and let $\rho\colon \Gamma \to G(A)$ be a quasi-lift of $\bar{\rho}$. Then $Z(\rho(\Gamma)) = Z(A)$.*

Proof See [25, Lemma 3.1] (note that the condition that ρ is a group homomorphism is not used in the proof).

Corollary 3.13 *Let $A \in \mathbf{Art}_{\mathcal{O}}$ and let $\rho\colon \Gamma \to G(A)$ be a quasi-lift of $\bar{\rho}$. Then ρ induces a uniquely determined group homomorphism $\phi\colon \Gamma \to \mathrm{Ker}(G^{\mathrm{ad}}(A) \to G^{\mathrm{ad}}(k))$ such that $\rho(x)^{-1}\rho(xy) = \phi(x)\rho(y)\phi(x)^{-1}$ for any $x, y \in \Gamma$.*

Proof By combining the above lemma with Lemma 3.7, we see $\phi\colon \Gamma \to G^{\mathrm{ad}}(A)$ is uniquely determined. Since $\bar{\rho}$ is a group homomorphism, $\phi \mod \mathfrak{m}_A$ commutes with $\bar{\rho}(\Gamma)$, and hence $\phi \mod \mathfrak{m}_A$ is trivial. □

Now we can characterize $\mathrm{aDef}_{\bar{f}}(A)$ in terms of quasi-lifts. The following proposition owing to [4] plays a crucial role (see also its use in the proof of [4, Theorem 4.10]):

Proposition 3.14 Suppose that X is an integral affine smooth $\mathcal{O}$-scheme on which G acts. Let $\underline{x} = (x_1, \ldots, x_n) \in X(k)$ be a point with $G_k \cdot x$ closed, and $Z_{G_k}(\underline{x})$ scheme-theoretically trivial. We write $X^{\wedge,\underline{x}}$ for the functor $\mathbf{Art}_{\mathcal{O}} \to \mathbf{Sets}$ which sends A to the set of pre-images of $\underline{x}$ under $X(A) \to X(k)$, and write $G^{\wedge}$ for the functor $\mathbf{Art}_{\mathcal{O}} \to \mathbf{Sets}$ which sends A to $\mathrm{Ker}(G(A) \to G(k))$. Then

1. The $G^{\wedge}$-action on $X^{\wedge,\underline{x}}$ is free on A-points for any $A \in \mathbf{Art}_{\mathcal{O}}$.
2. Let $X /\!\!/ G = \mathrm{Spec}\, \mathcal{O}[X]^G$, let $\pi\colon X \to X /\!\!/ G$ be the natural map, and let $(X /\!\!/ G)^{\wedge,\pi(\underline{x})}$ be the functor $\mathbf{Art}_{\mathcal{O}} \to \mathbf{Sets}$ which sends A to the set of pre-images of $\pi(\underline{x})$ under $(X /\!\!/ G)(A) \to (X /\!\!/ G)(k)$. Then $\pi\colon X \to X /\!\!/ G$ induces an isomorphism $X^{\wedge,\underline{x}}/G \cong (X /\!\!/ G)^{\wedge,\pi(\underline{x})}$.

Proof See [4, Proposition 3.13]. □

Corollary 3.15 *If $(\gamma_1, \ldots, \gamma_m)$ is a tuple in Γ^m such that $(\bar{\rho}(\gamma_1), \ldots, \bar{\rho}(\gamma_m))$ has a closed orbit and a scheme-theoretically trivial centralizer in G_k^{ad}, then $(\bar{\rho}(\gamma_1), \ldots, \bar{\rho}(\gamma_m))$ has a lift $(g_1, \ldots, g_m) \in G(A)^m$ which is a pre-image of $f(\gamma_1, \ldots, \gamma_m) \in \bar{B}G(A)_m$, and any other choice is conjugated to this one by a unique element of $G^{\mathrm{ad}}(A)$.*

Theorem 3.16 *Let $A \in \mathbf{Art}_{\mathcal{O}}$. Then $\mathrm{aDef}_{\bar{f}}(A)$ is isomorphic to the set of $\widehat{G}(A)$-conjugacy classes of quasi-lifts of $\bar{\rho}$.*

Proof Given a quasi-lift $\rho\colon \Gamma \to G(A)$, then the association $(\gamma_1, \ldots, \gamma_m) \mapsto (\rho(\prod_{j=1}^{i-1} \gamma_j)^{-1} \rho(\prod_{j=1}^{i} \gamma_j))_{i=1,\ldots,m}$ defines an element of $\mathrm{aDef}_{\bar{f}}(A)$.

In the following, we will construct a quasi-lift from a given $f \in \mathrm{aDef}_{\bar{f}}(A)$.

Let $n \geq 1$ be sufficiently large and choose $\delta_1, \ldots, \delta_n \in \Gamma$ such that $(\bar{h}_1 = \bar{\rho}(\delta_1), \ldots, \bar{h}_n = \bar{\rho}(\delta_n))$ is a system of generators of $\bar{\rho}(\Gamma)$, then the tuple $(\bar{h}_1, \ldots, \bar{h}_n)$ has a scheme-theoretically trivial centralizer in G_k^{ad}. By [5, Corollary 3.7], the absolutely G-completely reducibility implies that the tuple $(\bar{h}_1, \ldots, \bar{h}_n)$ has a closed orbit. By the above corollary, we can choose a lift $(h_1, \ldots, h_n) \in G(A)^n$ of $(\bar{h}_1, \ldots, \bar{h}_n)$ which is at the same time a pre-image of $f(\delta_1, \ldots, \delta_n)$.

For any $\gamma \in \Gamma$, the tuple $(\bar{h}_1, \ldots, \bar{h}_n, \bar{\rho}(\gamma))$ obviously has a closed orbit and a trivial centralizer in G_k^{ad}, so we can choose a tuple in $G(A)^{n+1}$ which lifts $(\bar{h}_1, \ldots, \bar{h}_n, \bar{\rho}(\gamma))$ and is a pre-image of $f(\delta_1, \ldots, \delta_n, \gamma)$. For this tuple, the first n elements are conjugated to $(h_1, \ldots, h_n)$ by a unique element of $G^{\mathrm{ad}}(A)$, so there is a unique $g \in G(A)$ such that the tuple is conjugated to $(h_1, \ldots, h_n, g)$. We define $\rho(\gamma)$ to be this g. It follows immediately that $\rho \mod \mathfrak{m}_A = \bar{\rho}$.

Now suppose $\gamma_1, \ldots, \gamma_m \in \Gamma$. As above, let $(g_1, \ldots, g_m)$ be the unique tuple such that $(h_1, \ldots, h_n, g_1, \ldots, g_m)$ lifts $(\bar{h}_1, \ldots, \bar{h}_n, \bar{\rho}(\gamma_1), \ldots, \bar{\rho}(\gamma_m))$ and is a pre-image of $f(\delta_1, \ldots, \delta_n, \gamma_1, \ldots, \gamma_m)$, consider the following diagram, where the horizontal arrows are compositions of face maps:

$$\begin{array}{ccc}(\delta_1,\ldots,\delta_n,\gamma_1,\ldots,\gamma_m) & \longrightarrow & (h_1,\ldots,h_n,g_1,\ldots,g_m)\\ \downarrow & & \downarrow\\ (\delta_1,\ldots,\delta_n,\prod_{j=1}^i\gamma_j) & \longrightarrow & (h_1,\ldots,h_n,\prod_{j=1}^i g_j).\end{array}$$

Then $(h_1,\ldots,h_n,\prod_{j=1}^i g_j)$ lifts $(\bar h_1,\ldots,\bar h_n,\bar\rho(\prod_{j=1}^i\gamma_j))$ and is a pre-image of $f(\delta_1,\ldots,\delta_n,\prod_{j=1}^i\gamma_j)$. Hence $\prod_{j=1}^i g_j=\rho(\prod_{j=1}^i\gamma_j)$, and $g_i=\rho(\prod_{j=1}^{i-1}\gamma_j)^{-1}\rho(\prod_{j=1}^i\gamma_j)$ $(\forall i=1,\ldots,m)$.

Let $x,y\in\Gamma$. Then the element in $G(A)^{2n+2}$ associated to $(\delta_1,\ldots,\delta_n,x,\delta_1,\ldots,\delta_n,y)$ is

$$(h_1,\ldots,h_n,\rho(x),\rho(x)^{-1}\rho(x\delta_1),\ldots,\rho(x\prod_{j=1}^{n-1}\delta_j)^{-1}\rho(x\prod_{j=1}^{n}\delta_j),\rho(x\prod_{j=1}^{n}\delta_j)^{-1}\rho(x\prod_{j=1}^{n}\delta_j\cdot y)),$$

and the element in $G(A)^{2n+1}$ associated to $(\delta_1,\ldots,\delta_n,\delta_1,\ldots,\delta_n,y)$ is

$$(h_1,\ldots,h_n,\rho(\delta_1),\ldots,\rho(\prod_{j=1}^{n-1}\delta_j)^{-1}\rho(\prod_{j=1}^{n}\delta_j),\rho(\prod_{j=1}^{n}\delta_j)^{-1}\rho(\prod_{j=1}^{n}\delta_j\cdot y)).$$

We see both $(\rho(x\prod_{j=1}^{i-1}\delta_j)^{-1}\rho(x\prod_{j=1}^{i}\delta_j))_{i=1,\ldots,n}$ and $(\rho(\prod_{j=1}^{i-1}\delta_j)^{-1}\rho(\prod_{j=1}^{i}\delta_j))_{i=1,\ldots,n}$ are lifts of $(\bar h_1,\ldots,\bar h_n)$ and pre-images of $f(\delta_1,\ldots,\delta_n)$, so they are conjugated by some $\phi(x)\in G(A)$. We can even suppose $\phi(x)\in\mathrm{Ker}(G(A)\to G(k))$ because the centralizer of $(\bar h_1,\ldots,\bar h_n)$ is Z. Since $\phi(x)$ is uniquely determined modulo $Z(A)$, it must conjugate $\rho(\prod_{j=1}^n\delta_j)^{-1}\rho(\prod_{j=1}^n\delta_j\cdot y)$ to $\rho(x\prod_{j=1}^n\delta_j)^{-1}\rho(x\prod_{j=1}^n\delta_j\cdot y)$. We deduce that $\forall x,y\in\Gamma$, $\rho(x)^{-1}\rho(xy)=\phi(x)\rho(y)\phi(x)^{-1}$, and ρ is a quasi-lift.

For the ρ constructed as above, we can recover f from the formula $(\gamma_1,\ldots,\gamma_m)\mapsto(\rho(\prod_{j=1}^{i-1}\gamma_j)^{-1}\rho(\prod_{j=1}^{i}\gamma_j))_{i=1,\ldots,m}$.

So it remains to prove that if ρ_1 and ρ_2 have the same image in $\mathrm{aDef}_{\bar f}(A)$, then they are equal modulo $\mathrm{Ker}(G(A)\to G(k))$-conjugation. Since $(\rho_1(\prod_{j=1}^{i-1}\delta_j)^{-1}\rho_1(\prod_{j=1}^{i}\delta_j))_{i=1,\ldots,n}$ and $(\rho_2(\prod_{j=1}^{i-1}\delta_j)^{-1}\rho_2(\prod_{j=1}^{i}\delta_j))_{i=1,\ldots,n}$ are both lifts of $(\bar h_1,\ldots,\bar h_n)$ and pre-images of $f(\delta_1,\ldots,\delta_n)$, they are conjugated by some $g\in G(A)$, and we may choose $g\in\mathrm{Ker}(G(A)\to G(k))$ because the centralizer of $(\bar h_1,\ldots,\bar h_n)$ is Z. After conjugation by g, we may suppose $(\rho_1(\prod_{j=1}^{i-1}\delta_j)^{-1}\rho_1(\prod_{j=1}^{i}\delta_j))_{i=1,\ldots,n}=(\rho_2(\prod_{j=1}^{i-1}\delta_j)^{-1}\rho_2(\prod_{j=1}^{i}\delta_j))_{i=1,\ldots,n}=(h'_1,\ldots,h'_n)$. Then for $\gamma\in\Gamma$, $\rho_k(\prod_{j=1}^n\delta_j)^{-1}\rho_k(\prod_{j=1}^n\delta_j\cdot\gamma)$ $(k=1,2)$ is uniquely determined by the condition: $(h'_1,\ldots,h'_n,\rho_k(\prod_{j=1}^n\delta_j)^{-1}\rho_k(\prod_{j=1}^n\delta_j\cdot\gamma))$ lifts $(\bar h_1,\ldots,\bar h_n,\bar\rho(\gamma))$ and is a pre-image of $f(\delta_1,\ldots,\delta_n,\gamma)$. In consequence, we have $\rho_1=\rho_2$.

For $A \in \mathbf{Art}_{\mathcal{O}}$, let $\mathrm{aDef}_{\bar{f},c}(A)$ be the subset of $\mathrm{aDef}_{\bar{f}}(A)$ consisting of $f: B\Gamma \to \bar{B}G(A)$ which factorizes through some finite quotient of Γ. In fact, we have $\mathrm{aDef}_{\bar{f},c}(A) = \mathsf{Hom}_{s\mathbf{Sets}_{/\bar{B}G(k)}}(X, \bar{B}G(A))$ (recall that X is the pro-simplicial set $(B\Gamma_i)_i$). The following corollary is obvious:

Corollary 3.17 *Let $A \in \mathbf{Art}_{\mathcal{O}}$. Then $\mathrm{aDef}_{\bar{f},c}(A)$ is isomorphic to the set of $\widehat{G}$-conjugacy classes of continuous quasi-lifts of $\bar{\rho}$.*

As a by-product of the proof of Theorem 3.16, we also have

Corollary 3.18 *For $A \in \mathbf{Art}_{\mathcal{O}}$, the set $\mathrm{aDef}_{\bar{f}}(A)$ (resp., $\mathrm{aDef}_{\bar{f},c}(A)$) is isomorphic to $\mathsf{Hom}_{\mathcal{M}}(B\Gamma, BG(A)/G^{\wedge}(A))$ (resp., $\mathsf{Hom}_{\mathcal{M}}(X, BG(A)/G^{\wedge}(A))$).*

But unfortunately, the simplicial set $BG(A)/G^{\wedge}(A)$ isn't generally fibrant.

We attempt to compare the difference between $\mathrm{aDef}_{\bar{f},c}(A)$ and $\mathcal{D}(A)$. Motivated by the front-to-back duality in [27, 8.2.10], we make the following definition. Let the reflection action r act on $B\Gamma$ and $\bar{B}G(A)$ as follows:

(1) r acts on $B\Gamma_n \cong \Gamma \times \cdots \times \Gamma$ by $r(\gamma_1, \ldots, \gamma_n) = (\gamma_n, \ldots, \gamma_1)$.
(2) r acts on $\mathcal{O}_{N_n G}$ by $r(f)(g_1, \ldots, g_n) = f(g_n, \ldots, g_1)$. We see that r preserves $\mathcal{O}_{N_n G}^{\mathrm{ad}G}$, hence r acts on $\bar{B}G(A)_n$.

Definition 3.19 For $A \in \mathbf{Art}_{\mathcal{O}}$, we define $\mathrm{bDef}_{\bar{f}}(A)$ (resp., $\mathrm{bDef}_{\bar{f},c}(A)$) to be the subset of $\mathrm{aDef}_{\bar{f}}(A)$ (resp., $\mathrm{aDef}_{\bar{f},c}(A)$) consisting of $f: B\Gamma \to \bar{B}G(A)$ which commutes with r.

Theorem 3.20 *Let $A \in \mathbf{Art}_{\mathcal{O}}$. Suppose the characteristic of k is not 2. Then $\mathrm{bDef}_{\bar{f}}(A)$ is in bijection with the set of group homomorphisms $\Gamma \to G(A)$ which lift $\bar{\rho}$, and $\mathrm{bDef}_{\bar{f},c}(A)$ is in bijection with $\mathcal{D}(A)$.*

Proof Let $f \in \mathrm{bDef}_{\bar{f}}(A)$. It suffices to prove that the quasi-lift ρ obtained in Theorem 3.16 is a group homomorphism. We choose the tuple $(\delta_1, \ldots, \delta_n)$ such that $\delta_i = \delta_{n+1-i}$ and $\prod_{j=1}^{n} \delta_j = e$. Write ρ for the quasi-lift constructed from this tuple as in Theorem 3.16, note that the choice of $(\delta_1, \ldots, \delta_n)$ only affects ρ by some conjugation. Let $\phi\colon \Gamma \to G(A)/Z(A)$ be the group homomorphism such that $\rho(xy) = \rho(x)\phi(x)\rho(y)\phi(x)^{-1}$ for any $x, y \in \Gamma$. Note that $\phi(x) \mod \mathfrak{m}_A = 1$ because $\bar{\rho}$ is a group homomorphism.

Since f commutes with r, we have

(1) $\rho(x) = \rho(x^{-1})^{-1}, \forall x \in \Gamma$.
(2) $\rho(x)^{-1}\rho(xy) = \rho(yx)\rho(x)^{-1}, \forall x, y \in \Gamma$.

By substituting (1) into $\rho(xy) = \rho(x)\phi(x)\rho(y)\phi(x)^{-1}$, we get $\rho(y^{-1}x^{-1})^{-1} = \rho(x^{-1})^{-1}\phi(x)\rho(y^{-1})^{-1}\phi(x)^{-1}$, then consider $(x, y) \mapsto (x^{-1}, y^{-1})$ and take the inverse, we get $\rho(yx) = \phi(x)^{-1}\rho(y)\phi(x)\rho(x)$. Now (2) implies $\rho(xy)\rho(x) = \rho(x)\rho(yx)$, which in turn gives

$$\rho(x)\phi(x)\rho(y)\phi(x)^{-1}\rho(x) = \rho(x)\phi(x)^{-1}\rho(y)\phi(x)\rho(x).$$

So $\phi(x)^2$ commutes with $\rho(\Gamma)$ for any $x \in \Gamma$, and $\phi^2 = 1$. Since the characteristic of k is not 2 and $\phi(x) \mod \mathfrak{m}_A = 1 \in G(k)/Z(k)$, we deduce $\phi = 1$ and ρ is a group homomorphism. $\square$

3.3 Derived Deformations of Pseudo-Characters

The functor $\mathrm{aDef}_{\bar{f},c} = \mathsf{Hom}_{s\mathbf{Sets}_{/\bar{B}G(k)}}(X, \bar{B}G(-))$ is analogous to the functor $\mathcal{D}^{\square} = \mathsf{Hom}_{s\mathbf{Sets}_{/BG(k)}}(X, BG(-))$, so it's natural to consider the function complex $\mathsf{sHom}_{s\mathbf{Sets}_{/\bar{B}G(k)}}(X, \bar{B}G(-))$ and then to extend the domain of definition to ${}_{\mathcal{O}}\backslash s\mathbf{Art}/_k$, as constructing the functor $s\mathcal{D}\colon {}_{\mathcal{O}}\backslash s\mathbf{Art}/_k \to s\mathbf{Sets}$.

Definition 3.21 For $A \in {}_{\mathcal{O}}\backslash s\mathbf{Art}/_k$, we define $\bar{\mathcal{B}}G(A)$ to be the Ex^{∞} of the diagonal of the bisimplicial set

$$([p],[q]) \mapsto \mathsf{Hom}_{\mathcal{O}\backslash s\mathbf{CR}}(c(\mathcal{O}^{\mathrm{ad}G}_{N_pG}), A^{\Delta[q]}),$$

and define $sa\mathcal{D}(A) = \mathrm{hofib}_{\bar{f}}(\mathsf{Hom}_{s\mathbf{Sets}}(X, \bar{\mathcal{B}}G(A)) \to \mathsf{Hom}_{s\mathbf{Sets}}(X, \bar{\mathcal{B}}G(k)))$.

If $A \in \mathbf{Art}_{\mathcal{O}}$, then the bisimplicial set $([p],[q]) \mapsto \mathsf{Hom}_{\mathcal{O}\backslash s\mathbf{CR}}(c(\mathcal{O}^{\mathrm{ad}G}_{N_pG}), A^{\Delta[q]})$ doesn't depend on the index q, and each of its lines is isomorphic to $\mathrm{Ex}^{\infty}\bar{B}G(A)$. Hence $\bar{f}$ can be regarded as an element of $\mathsf{Hom}_{s\mathbf{Sets}}(X, \bar{\mathcal{B}}G(k))$. As the derived deformation functors $s\mathcal{D}$, we see that $sa\mathcal{D}\colon {}_{\mathcal{O}}\backslash s\mathbf{Art}/_k \to s\mathbf{Sets}$ is homotopy invariant.

Note that the inclusion $\mathcal{O}^{\mathrm{ad}G}_{N_\bullet G} \hookrightarrow \mathcal{O}_{N_\bullet G}$ induces a natural transformation $s\mathcal{D} \to sa\mathcal{D}$.

We would like to understand $\pi_0 sa\mathcal{D}(A)$. Let's first analyze the case $A \in \mathbf{Art}_{\mathcal{O}}$. For simplicity, we don't take the Ex^{∞} here. Since $BG(A) \to BG(k)$ is a fibration, $\mathsf{sHom}_{s\mathbf{Sets}_{/\bar{B}G(k)}}(X, \bar{B}G(A))$ is a good model for $s\mathcal{D}(A)$. However, if $\bar{B}G(A) \to \bar{B}G(k)$ is a not fibration, then $\mathsf{sHom}_{s\mathbf{Sets}_{/\bar{B}G(k)}}(X, \bar{B}G(A))$ is not weakly equivalent to $sa\mathcal{D}(A)$.

We have the commutative diagram

$$\begin{array}{ccc}
\mathsf{sHom}_{s\mathbf{Sets}_{/\bar{B}G(k)}}(X, BG(A))_0 & \longrightarrow & \mathsf{sHom}_{s\mathbf{Sets}_{/\bar{B}G(k)}}(X, \bar{B}G(A))_0 \\
\downarrow & & \downarrow \\
\pi_0\, \mathsf{sHom}_{s\mathbf{Sets}_{/BG(k)}}(X, BG(A)) & \longrightarrow & \pi_0\, \mathsf{sHom}_{s\mathbf{Sets}_{/\bar{B}G(k)}}(X, \bar{B}G(A)).
\end{array}$$

Note that $\pi_0 sa\mathcal{D}(A)$ is the coequalizer of $sa\mathcal{D}(A)_1 \rightrightarrows sa\mathcal{D}(A)_0 = \mathrm{aDef}_{\bar{f},c}(A)$ by definition.

Proposition 3.22 The above diagram is naturally isomorphic to

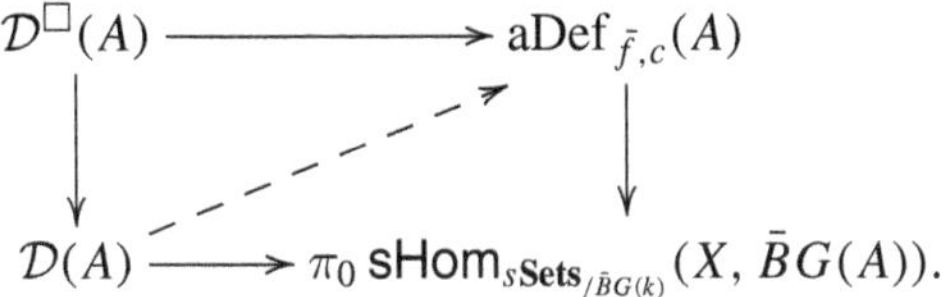

And there is a dotted arrow which makes the diagram commutative, whose image is $\mathrm{bDef}_{\bar{f},c}(A) \subseteq \mathrm{aDef}_{\bar{f},c}(A)$.

Proof We have $\mathsf{sHom}_{s\mathbf{Sets}_{/BG(k)}}(X, BG(A))_0 = \mathsf{Hom}_{\mathcal{M}}(X, BG(A))$, which is exactly $\mathcal{D}^{\square}(A)$, since $B\colon \mathbf{Gpd} \to s\mathbf{Sets}$ is fully faithful. The other isomorphisms follow by definition.

The dotted arrow signifies the inclusion of usual deformations into pseudo-deformations, whose image is $\mathrm{bDef}_{\bar{f},c}(A)$ by Theorem 3.20. □

Remark 3.23 Note however that the functor $sa\mathcal{D}\colon {}_{\mathcal{O}}\backslash s\mathbf{Art}/_k \to s\mathbf{Sets}$ remains quite mysterious. It may be asked whether there is a more adequate derived deformation functor for pseudo-characters.

4 (Co)tangent Complexes and Pro-Representability

4.1 Dold-Kan Correspondence

Let's briefly review the Dold-Kan correspondence. Let R be a commutative ring. Our goal here is to recall an equivalence (of model categories) between the category of simplicial R-modules $s\mathbf{Mod}_R$ and the category of chain complexes of R-modules concentrated on non-negative degrees $\mathbf{Ch}_{\geq 0}(R)$. Recall the model category structures on $s\mathbf{Mod}_R$ and $\mathbf{Ch}_{\geq 0}(R)$:

(1) For $s\mathbf{Mod}_R$, the fibrations and weak equivalences are linear morphisms which are in $s\mathbf{Sets}$, and the cofibrations are linear morphisms satisfying a lifting property (see [15, Proposition 7.2.3]).
(2) For $\mathbf{Ch}_{\geq 0}(R)$, the cofibrations, fibrations, and weak equivalences are linear morphisms satisfying the following:
    (a) $f\colon C_\bullet \to D_\bullet$ is a cofibration if $C_n \to D_n$ is injective with projective cokernel for $n \geq 0$.
    (b) $f\colon C_\bullet \to D_\bullet$ is a fibration if $C_n \to D_n$ is surjective for $n \geq 1$.
    (c) $f\colon C_\bullet \to D_\bullet$ is a weak equivalence if the morphism $H_* f$ induced on homology is an isomorphism.

We write $M \in s\mathbf{Mod}_R$ for the simplicial R-module with M_n on n-th simplicial degree. Let $N(M)$ be the chain complexes of R-modules such that $N(M)_n = \bigcap_{i=0}^{n-1} \mathrm{Ker}(d_i) \subseteq M_n$ with differential maps

$$(-1)^n d_n\colon \bigcap_{i=0}^{n-1} \mathrm{Ker}(d_i) \subseteq M_n \to \bigcap_{i=0}^{n-2} \mathrm{Ker}(d_i) \subseteq M_{n-1}.$$

Obviously $M \mapsto N(M)$ is functorial. We call $N(M) \in \mathbf{Ch}_{\geq 0}(R)$ the normalized complex of M.

The Dold-Kan functor DK: $\mathbf{Ch}_{\geq 0}(R) \to s\mathbf{Mod}_R$ is the quasi-inverse of N. Explicitly, for a chain of R-modules $C_\bullet = (C_0 \leftarrow C_1 \leftarrow C_2 \leftarrow \ldots)$, we define $\mathrm{DK}(C_\bullet) \in s\mathbf{Mod}_R$ as follows:

(1) $\mathrm{DK}(C_\bullet)_n = \bigoplus_{[n] \twoheadrightarrow [k]} C_k$.

(2) For $\theta\colon [m] \to [n]$, we define the corresponding $\mathrm{DK}(C_\bullet)_n \to \mathrm{DK}(C_\bullet)_m$ on each component of $\mathrm{DK}(C_\bullet)_n$ indexed by $[n] \overset{\sigma}{\twoheadrightarrow} [k]$ as follows: suppose $[m] \overset{t}{\twoheadrightarrow} [s] \overset{d}{\hookrightarrow} [k]$ is the epi-monic factorization of the composition $[m] \overset{\theta}{\to} [n] \overset{\sigma}{\twoheadrightarrow} [k]$, then the map on component $[n] \overset{\sigma}{\twoheadrightarrow} [k]$ is

$$C_k \overset{d^*}{\to} C_s \hookrightarrow \bigoplus_{[m] \twoheadrightarrow [r]} C_r.$$

Theorem 4.1 *(1) (Dold-Kan) The functors* DK *and* N *are quasi-inverse and hence form an equivalence of categories. Moreover, two morphisms* $f, g \in \mathrm{Hom}_{s\mathbf{Mod}_R}(M, N)$ *are simplicially homotopic if and only if* $N(f)$ *and* $N(g)$ *are chain homotopic.*

(2) The functors DK *and* N *preserve the model category structures of* $\mathbf{Ch}_{\geq 0}(R)$ *and* $s\mathbf{Mod}_R$ *defined above.*

Proof See [27, Theorem 8.4.1] and [11, Lemma 2.11]. Note that (1) is valid for any Abelian category instead of $s\mathbf{Mod}_R$.

Remark 4.2 Let $\mathbf{Ch}(R)$ be the category of complexes $(C_i)_{i \in \mathbb{Z}}$ of R-modules and $\mathbf{Ch}_{\geq 0}(R)$ the subcategory of complexes for which $C_i = 0$ for $i < 0$. The category $\mathbf{Ch}_{\geq 0}(R)$ is naturally enriched over simplicial R-modules, and we have

$$\mathrm{sHom}_{\mathbf{Ch}_{\geq 0}(R)}(C_\bullet, D_\bullet) \cong \mathrm{sHom}_{s\mathbf{Mod}_R}(\mathrm{DK}(C_\bullet), \mathrm{DK}(D_\bullet)).$$

Given $C_\bullet, D_\bullet \in \mathbf{Ch}_{\geq 0}(R)$. Let $[C_\bullet, D_\bullet] \in \mathbf{Ch}(R)$ be the mapping complex, more precisely, $[C_\bullet, D_\bullet]_n = \prod_m \mathrm{Hom}_R(C_m, D_{m+n})$ and the differential maps are natural ones. Let $\tau_{\geq 0}$ be the functor which sends a chain complex $X_\bullet$ to the truncated complex

$$0 \leftarrow \mathrm{Ker}(X_0 \to X_{-1}) \leftarrow X_1 \leftarrow \ldots$$

Then there is a weak equivalence

$$\mathsf{sHom}_{\mathbf{Ch}_{\geq 0}(R)}(C_\bullet, D_\bullet) \simeq \mathrm{DK}(\tau_{\geq 0}[C_\bullet, D_\bullet])$$

(see [18, Remark 11.1]).

It's clear that $\pi_i \,\mathsf{sHom}_{\mathbf{Ch}_{\geq 0}(R)}(C_\bullet, D_\bullet)$ is isomorphic to the chain homotopy classes of maps from $C_\bullet$ to $D_{\bullet+n}$.

4.2 (Co)tangent Complexes of Simplicial Commutative Rings

We recall Quillen's cotangent and tangent complexes of simplicial commutative rings.

Let A be a commutative ring. For R an A-algebra, let $\Omega_{R/A}$ be the module of differentials with the canonical R-derivation $d\colon R \to \Omega_{R/A}$. Let $\mathrm{Der}_A(R, -)$ be the covariant functor which sends an R-module M to the R-module

$$\mathrm{Der}_A(R, M) = \{D\colon R \to M \mid D \text{ is } A\text{-linear and } D(xy) = xD(y) + yD(x),\ \forall x, y \in R\}.$$

It's well known that $\mathsf{Hom}_R(\Omega_{R/A}, -)$ is naturally isomorphic to $\mathrm{Der}_A(R, -)$ via $\phi \mapsto \phi \circ d$.

Let T be an A-algebra, and let ${}_A\backslash\mathbf{CR}/_T$ be the category of commutative rings R over T and under A. Then for any T-module M and any $R \in {}_A\backslash\mathbf{CR}/_T$, we have natural isomorphisms

$$\mathsf{Hom}_T(\Omega_{R/A} \otimes_R T, M) \cong \mathrm{Der}_A(R, M) \cong \mathsf{Hom}_{{}_A\backslash\mathbf{CR}/_T}(R, T \oplus M),$$

where $T \oplus M$ is the T-algebra with square-zero ideal M. So the functor $R \mapsto \Omega_{R/A} \otimes_R T$ is left adjoint to the functor $M \mapsto T \oplus M$.

The above isomorphisms have level-wise extensions to simplicial categories (see [11] Lemma II.2.9 and Example II.2.10). For $R \in {}_A\backslash s\mathbf{CR}$, we can form $\Omega_{R/A} \otimes_R T \in s\mathbf{Mod}_T$.

We have

$$\mathsf{sHom}_{s\mathbf{Mod}_T}(\Omega_{R/A} \otimes_R T, M) \cong \mathsf{sHom}_{{}_A\backslash s\mathbf{CR}/_T}(R, T \oplus M).$$

The functor $M \mapsto T \oplus M$ from $s\mathbf{Mod}_T$ to ${}_A\backslash s\mathbf{CR}/_T$ preserves fibrations and weak equivalences (we may see this via the Dold-Kan correspondence), so the left adjoint functor $R \mapsto \Omega_{R/A} \otimes_R T$ is left Quillen and it admits a total left derived functor. We introduce the cotangent complex $L_{R/A}$ in the following definition, so that the total left derived functor has the form $R \mapsto L_{R/A} \underline{\otimes}_R T$. Note that given two simplicial modules M, N over a simplicial ring S, one can form (degreewise) a tensor product, denoted $M \underline{\otimes}_S N$, which is a simplicial S-module.

Definition 4.3 For $R \in {}_A\backslash s\mathbf{CR}$, we define $L_{R/A} = \Omega_{c(R)/A}\underline{\otimes}_{c(R)} R \in s\mathbf{Mod}_R$, where $c(R)$ is the middle object of some cofibration-trivial fibration factorization $A \hookrightarrow c(R) \overset{\sim}{\twoheadrightarrow} R$, and we call $L_{R/A}$ the cotangent complex of R.

Note that it is an abuse of language, as it should be called cotangent simplicial R-module, because for R simplicial, $L_{R/A} \in s\mathbf{Mod}_R$ but there is no notion of complexes of R-modules.

By construction, $L_{R/A}\underline{\otimes}_R T$ is cofibrant as it's the image of the cofibrant object $c(R)$ under a total left derived functor, and it is fibrant in $s\mathbf{Mod}_R$ (all objects are fibrant there). Note also that the weak equivalence class of $L_{R/A}\underline{\otimes}_R T$ is independent of the choice of $c(R)$. It follows from these two observations that $L_{R/A}$ is determined up to homotopy equivalence (by the Whitehead theorem [15, Theorem 7.5.10]). Using the Dold-Kan equivalence, we can form the normalized complex (determined up to homotopy equivalence)

$$N(L_{R/A}\underline{\otimes}_R T) \in \mathbf{Ch}_{\geq 0}(T).$$

From now on, we keep the functor N understood and simply write

$$L_{R/A} \otimes_R T \in \mathbf{Ch}_{\geq 0}(T).$$

Recall that for $M, N \in \mathbf{Ch}(T)$, the internal Hom $[M, N] \in \mathbf{Ch}(T)$ is defined as

$$[M, N]_n = \prod_m \mathsf{Hom}_T(M_m, N_{m+n}).$$

Note that if $M \in \mathbf{Ch}_{\geq 0}(T)$, then $[M, T] \in \mathbf{Ch}_{\leq 0}(T)$. For $C \in \mathbf{Ch}_{\leq 0}(T)$, we write $C^i = C_{-i}$ for $i \geq 0$; we thus identify $\mathbf{Ch}_{\leq 0}(T) = \mathbf{Ch}^{\geq 0}(T)$.

For $R \in {}_A\backslash s\mathbf{CR}/_T$ and $C_\bullet \in \mathbf{Ch}_{\geq 0}(T)$, we have (by Remark 4.2)

$$\begin{aligned} \mathsf{sHom}_{{}_A\backslash s\mathbf{CR}/_T}(c(R), T \oplus \mathrm{DK}(C_\bullet)) &\cong \mathsf{sHom}_{s\mathbf{Mod}_T}(L_{R/A}\underline{\otimes}_R T, \mathrm{DK}(C_\bullet)) \\ &\simeq \mathrm{DK}(\tau_{\geq 0}[L_{R/A} \otimes_R T, C_\bullet]). \end{aligned}$$

Definition 4.4 The T-tangent complex $\mathfrak{t}R_T$ of $R \to T$ is the internal hom complex

$$[L_{R/A} \otimes_R T, T] \in \mathbf{Ch}^{\geq 0}(T).$$

Note that $\mathfrak{t}R_T$ is well defined up to chain homotopy equivalence since it is the case for $L_{R/A} \otimes_R T$.

4.3 *Tangent Complexes of Formally Cohesive Functors and Lurie's Criterion*

In [13, Sect. 4], the authors define the tangent complexes of formally cohesive functors. To summarize, we have the following proposition:

Proposition 4.5 Let $\mathcal{F}\colon {}_{\mathcal{O}}\backslash s\mathbf{Art}/_k \to s\mathbf{Sets}$ be a formally cohesive functor. Then there exists $L_{\mathcal{F}} \in \mathbf{Ch}(k)$ such that $\mathcal{F}(k \oplus \mathrm{DK}(C_\bullet))$ is weakly equivalent to $\mathrm{DK}(\tau_{\geq 0}[L_{\mathcal{F}}, C_\bullet])$ for every $C_\bullet \in \mathbf{Ch}_{\geq 0}(k)$ with $H_*(C_\bullet)$ finite over k. □

Proof See [13, Lemma 4.25].

Definition 4.6 Let $\mathcal{F}\colon {}_{\mathcal{O}}\backslash s\mathbf{Art}/_k \to s\mathbf{Sets}$ be a formally cohesive functor.

(1) We call $L_{\mathcal{F}}$ the cotangent complex of $\mathcal{F}$.
(2) The tangent complex $\mathfrak{t}\mathcal{F}$ of $\mathcal{F}$ is the chain complex defined by the internal hom complex $[L_{\mathcal{F}}, k]$.

Note that $L_{\mathcal{F}}$ and $\mathfrak{t}\mathcal{F}$ are uniquely determined up to quasi-isomorphism. We shall use $\mathfrak{t}^i\mathcal{F}$ to abbreviate the homology groups $H_{-i}\mathfrak{t}\mathcal{F}$.

Remark 4.7 If $R \in {}_{\mathcal{O}}\backslash s\mathbf{CR}/_k$ is cofibrant, then the functor $\mathcal{F}_R = \mathsf{sHom}_{{}_{\mathcal{O}}\backslash s\mathbf{CR}/_k}(R, -)\colon {}_{\mathcal{O}}\backslash s\mathbf{Art}/_k \to s\mathbf{Sets}$ is formally cohesive. Since $\mathrm{DK}(\tau_{\geq 0}[L_{\mathcal{F}_R}, k[n]]) \simeq \mathsf{sHom}_{{}_{\mathcal{O}}\backslash s\mathbf{CR}/_k}(R, k \oplus k[n]) \simeq \mathrm{DK}(\tau_{\geq 0}[L_{R/\mathcal{O}} \otimes_R k, k[n]])$, the cotangent complexes $L_{\mathcal{F}_R}$ and $L_{R/\mathcal{O}} \otimes_R k$ are quasi-isomorphic.

Definition 4.8 We say a functor $\mathcal{F}\colon {}_{\mathcal{O}}\backslash s\mathbf{Art}/_k \to s\mathbf{Sets}$ is pro-representable, if there exists a projective system $R = (R_n)_{n\in\mathbb{N}}$ with each $R_n \in {}_{\mathcal{O}}\backslash s\mathbf{Art}/_k$ cofibrant, such that $\mathcal{F}$ is weakly equivalent to $\varinjlim_n \mathsf{sHom}_{{}_{\mathcal{O}}\backslash s\mathbf{Art}/_k}(R_n, -)$. In this case, we say $R = (R_n)_{n\in\mathbb{N}}$ is a representing ring for $\mathcal{F}$. We shall write $\mathsf{sHom}_{{}_{\mathcal{O}}\backslash s\mathbf{Art}/_k}(R, -)$ for $\varinjlim_n \mathsf{sHom}_{{}_{\mathcal{O}}\backslash s\mathbf{Art}/_k}(R_n, -)$.

Remark 4.9 The pro-representability defined above is called sequential pro-representability in [13], but we will only deal with this case.

Theorem 4.10 (Lurie's criterion) *Let $\mathcal{F}$ be a formally cohesive functor. If $\dim_k \mathfrak{t}^i\mathcal{F}$ is finite for every $i \in \mathbb{Z}$, and $\mathfrak{t}^i\mathcal{F} = 0$ for every $i < 0$, then $\mathcal{F}$ is (sequentially) pro-representable.*

Proof See [17, Corollary 6.2.14] and [13, Theorem 4.33]. □

The following lemma illustrates the conservativity of the tangent complex functor:

Lemma 4.11 *Suppose $\mathcal{F}_1, \mathcal{F}_2\colon {}_{\mathcal{O}}\backslash s\mathbf{Art}/_k \to s\mathbf{Sets}$ are formally cohesive functors. Then a natural transformation $\mathcal{F}_1 \to \mathcal{F}_2$ is a weak equivalence if and only if it induces isomorphisms $\mathfrak{t}^i\mathcal{F}_1 \to \mathfrak{t}^i\mathcal{F}_2$ for all i.*

Proof If the natural transformation induces isomorphisms $\mathfrak{t}^i\mathcal{F}_1 \to \mathfrak{t}^i\mathcal{F}_2$, then $\mathcal{F}_1(k \oplus k[n]) \to \mathcal{F}_2(k \oplus k[n])$ is a weak equivalence. So by simplicial Artinian induction [13, Sect. 4], it induces a weak equivalence $\mathcal{F}_1(A) \to \mathcal{F}_2(A)$ for $A \in {}_{\mathcal{O}}\backslash s\mathbf{Art}/_k$. □

4.4 Pro-Representability of Derived Deformation Functors

In the following, we suppose $p > 2$, and $\Gamma = G_{F,S}$ for $S = S_p \cup S_\infty$. Suppose further that $\bar{\rho}$ satisfies (Ord_v) and (Reg_v) for $v \in S_p$, and $H^0(\Gamma, \mathfrak{g}_k) = \mathfrak{z}_k$. Recall that we've introduced derived deformation functors $s\mathcal{D}$ and $s\mathcal{D}^{\mathrm{n.o}}$, as well as the modifying-center variants $s\mathcal{D}_Z$ and $s\mathcal{D}_Z^{\mathrm{n.o}}$. These functors are all formally cohesive. Their tangent complexes are related to the Galois cohomology groups $H^i_*(\Gamma, \mathfrak{g}_k)$ of adjoint representations, where $* = \emptyset$ or n. o.

4.4.1 Galois Cohomology

We briefly review the Galois cohomology theory. To define the nearly ordinary cohomology, we fix the standard Levi decomposition $B = TN$ of the standard Borel of G; it induces a decomposition of Lie algebras over k: $\mathfrak{b}_k = \mathfrak{t}_k \oplus \mathfrak{n}_k$. Recall the definition of the Greenberg-Wiles nearly ordinary Selmer group

$$\widetilde{H}^1_{\mathrm{n.o}}(\Gamma, \mathfrak{g}_k) = \mathrm{Ker}\left(H^1(\Gamma, \mathfrak{g}_k) \to \prod_{v \in S_p} \frac{H^1(\Gamma_v, \mathfrak{g}_k)}{L_v} \right),$$

where $L_v = \mathrm{im}(H^1(\Gamma_v, \mathfrak{b}_k) \to H^1(\Gamma_v, \mathfrak{g}_k))$.

For $v \in S_p$, let $\tilde{L}_v \subseteq Z^1(\Gamma_v, \mathfrak{g}_k)$ be the pre-image of L_v. Let $C^\bullet_{\mathrm{n.o}}(\Gamma, \mathfrak{g}_k)$ be the mapping cone of the natural cochain morphism

$$\begin{array}{ccccccccc}
0 & \longrightarrow & C^0(\Gamma, \mathfrak{g}_k) & \longrightarrow & C^1(\Gamma, \mathfrak{g}_k) & \longrightarrow & C^2(\Gamma, \mathfrak{g}_k) & \longrightarrow & \cdots \\
 & & \downarrow & & \downarrow & & \downarrow & & \\
0 & \longrightarrow & 0 & \longrightarrow & \bigoplus_{v \in S_p} C^1(\Gamma_v, \mathfrak{g}_k)/\tilde{L}_v & \longrightarrow & \bigoplus_{v \in S_p} C^2(\Gamma_v, \mathfrak{g}_k) & \longrightarrow & \cdots
\end{array}$$

Then we define the nearly ordinary cohomology groups $H^*_{\mathrm{n.o}}(\Gamma, \mathfrak{g}_k)$ as the cohomology of the complex $C^\bullet_{\mathrm{n.o}}(\Gamma, \mathfrak{g}_k)$. They fit into the exact sequence (★):

$$\begin{aligned}
0 \to & H^0_{\mathrm{n.o}}(\Gamma, \mathfrak{g}_k) \to H^0(\Gamma, \mathfrak{g}_k) \to 0 \\
\to & H^1_{\mathrm{n.o}}(\Gamma, \mathfrak{g}_k) \to H^1(\Gamma, \mathfrak{g}_k) \to \bigoplus_{v \in S_p} H^1(\Gamma_v, \mathfrak{g}_k)/L_v \\
\to & H^2_{\mathrm{n.o}}(\Gamma, \mathfrak{g}_k) \to H^2(\Gamma, \mathfrak{g}_k) \to \bigoplus_{v \in S_p} H^1(\Gamma_v, \mathfrak{g}_k) \\
\to & H^3_{\mathrm{n.o}}(\Gamma, \mathfrak{g}_k) \to 0.
\end{aligned}$$

In particular, $\widetilde{H}^1_{\mathrm{n.o}}(\Gamma, \mathfrak{g}_k) = H^1_{\mathrm{n.o}}(\Gamma, \mathfrak{g}_k)$.

Definition 4.12 For a finite $\mathcal{O}[\Gamma]$-module M, we write $M^\vee = \mathrm{Hom}_{\mathcal{O}}(M, K/\mathcal{O})$ and $M^* = \mathrm{Hom}_{\mathcal{O}}(M, K/\mathcal{O}(1))$. In particular, if M is a k-vector space, $M^\vee = \mathrm{Hom}_k(M, k)$ and $M^* = \mathrm{Hom}_k(M, k(1))$.

Recall the local Tate duality $H^1(\Gamma_v, \mathfrak{g}_k) \times H^1(\Gamma_v, \mathfrak{g}_k^*) \to k$. Let $L_v^\perp \subseteq H^1(\Gamma_v, \mathfrak{g}_k^*)$ be the dual of L_v. We define similarly the cohomology groups $H^*_{\mathrm{n.o},\perp}(\Gamma, \mathfrak{g}_k^*)$. In particular,

$$\bigoplus_{v \in S_p} L_v \to H^1(\Gamma, \mathfrak{g}_k^*)^\vee \to H^1_{\mathrm{n.o},\perp}(\Gamma, \mathfrak{g}_k^*)^\vee \to 0$$

is exact. By fitting this into the Poitou-Tate exact sequence (see [20, Theorem I.4.10]), we obtain the exact sequence (★★):

$$\begin{aligned} & H^1(\Gamma, \mathfrak{g}_k) \to \bigoplus_{v \in S_p} H^1(\Gamma_v, \mathfrak{g}_k)/L_v \\ \to\ & H^1_{\mathrm{n.o},\perp}(\Gamma, \mathfrak{g}_k^*)^\vee \to H^2(\Gamma, \mathfrak{g}_k) \to \bigoplus_{v \in S_p} H^2(\Gamma_v, \mathfrak{g}_k) \\ \to\ & H^0(\Gamma, \mathfrak{g}_k^*)^\vee \to 0. \end{aligned}$$

We deduce the Poitou-Tate duality:

Theorem 4.13 *For each $i \in \{0, 1, 2, 3\}$, there is a perfect pairing*

$$H^i_{\mathrm{n.o},\perp}(\Gamma, \mathfrak{g}_k^*) \times H^{3-i}_{\mathrm{n.o}}(\Gamma, \mathfrak{g}_k) \to k.$$

Proof For $i \in \{0, 1\}$, it suffices to compare the exact sequences (★) and (★★). The cases $i \in \{2, 3\}$ follow by duality. □

4.4.2 Tangent Complex

Lemma 4.14 *(1) We have $\mathfrak{t}^i s\mathcal{D} \cong H^{i+1}(\Gamma, \mathfrak{g}_k)$ for all $i \in \mathbb{Z}$. On the other hand, $\mathfrak{t}^i s\mathcal{D}_Z \cong \mathfrak{t}^i s\mathcal{D}$ when $i \neq -1$, and $\mathfrak{t}^{-1} s\mathcal{D}_Z = 0$.*
(2) Let $v \in S_p$. Then we have $\mathfrak{t}^i s\mathcal{D}_v \cong H^{i+1}(\Gamma_v, \mathfrak{g}_k)$ for all $i \in \mathbb{Z}$. On the other hand, $\mathfrak{t}^i s\mathcal{D}_{v,Z} \cong \mathfrak{t}^i s\mathcal{D}_v$ when $i \neq -1$, and $\mathfrak{t}^{-1} s\mathcal{D}_{v,Z} \cong H^0(\Gamma_v, \mathfrak{g}_k)/\mathfrak{z}_k$.
(3) Let $v \in S_p$. Then we have $\mathfrak{t}^i s\mathcal{D}_v^{\mathrm{n.o}} \cong H^{i+1}(\Gamma_v, \mathfrak{b}_k)$ for all $i \in \mathbb{Z}$. On the other hand, $\mathfrak{t}^i s\mathcal{D}_{v,Z}^{\mathrm{n.o}} \cong \mathfrak{t}^i s\mathcal{D}_v^{\mathrm{n.o}}$ when $i \neq -1$, and $\mathfrak{t}^{-1} s\mathcal{D}_{v,Z}^{\mathrm{n.o}} \cong H^0(\Gamma_v, \mathfrak{b}_k)/\mathfrak{z}_k$. Moreover, $\mathfrak{t}^1 s\mathcal{D}_v^{\mathrm{n.o}} = 0$ if (Reg_v^*) holds.

Proof Note that $\mathfrak{t}^{j-i}\mathcal{F} \cong \pi_i \mathcal{F}(k \oplus k[j])$ for any formally cohesive functor $\mathcal{F}$ and any $i, j \geq 0$. Later in Sect. 4.5 we shall give a slightly generalized version of the lemma. See also [13, Sect. 7.3].

In particular, by Lurie's criterion (Theorem 4.10), this lemma together with the finiteness of the cohomology groups implies

Corollary 4.15 *The center-modified functor $s\mathcal{D}_Z$ is pro-representable.*

Now we treat the nearly ordinary case $s\mathcal{D}_Z^{\text{n.o}}$. Let's recall that $s\mathcal{D}_{\text{loc},Z} = \prod_{v\in S_p} s\mathcal{D}_{v,Z}$, $s\mathcal{D}^{\text{n.o}}_{\text{loc},Z} = \prod_{v\in S_p} s\mathcal{D}^{\text{n.o}}_{v,Z}$, and $s\mathcal{D}^{\text{n.o}}_Z = s\mathcal{D}_Z \times^h_{s\mathcal{D}_{\text{loc},Z}} s\mathcal{D}^{\text{n.o}}_{\text{loc},Z}$. Recall that $\bar\rho$ satisfies (Ord_v) and (Reg_v) for $v \in S_p$, so $s\mathcal{D}^{\text{n.o}}_Z$ is indeed the derived generalization of $\mathcal{D}^{\text{n.o}}$, i.e., $\pi_0 s\mathcal{D}^{\text{n.o}}_Z(A) \cong \mathcal{D}^{\text{n.o}}(\pi_0 A)$ for homotopy discrete $A \in {}_{\mathcal{O}}\backslash s\mathbf{Art}/_k$ (see Proposition 2.18).

Lemma 4.16 *Suppose furthermore (Reg_v^*) for $v \in S_p$. Then $\mathfrak{t}^i s\mathcal{D}^{\text{n.o}}_Z \cong H^{i+1}_{\text{n.o}}(\Gamma, \mathfrak{g})$ when $i \geq 0$, and $\mathfrak{t}^i s\mathcal{D}^{\text{n.o}}_Z = 0$ when $i < 0$.*

Proof We have the Mayer-Vietoris exact sequence (see [13, Lemma 4.30 (iv)] and [27, Sect. 1.5])

$$\mathfrak{t}^i s\mathcal{D}^{\text{n.o}}_Z \to \mathfrak{t}^i s\mathcal{D}_Z \oplus \mathfrak{t}^i s\mathcal{D}^{\text{n.o}}_{\text{loc},Z} \to \mathfrak{t}^i s\mathcal{D}_{\text{loc},Z} \xrightarrow{[1]} \dots .$$

By Lemma 4.14, we obtain an exact sequence

$$\begin{aligned} 0 &\to \mathfrak{t}^{-1} s\mathcal{D}^{\text{n.o}}_Z \to \bigoplus_{v\in S_p} H^0(\Gamma_v, \mathfrak{b}_k)/\mathfrak{z}_k \to \bigoplus_{v\in S_p} H^0(\Gamma_v, \mathfrak{g}_k)/\mathfrak{z}_k \\ &\to \mathfrak{t}^0 s\mathcal{D}^{\text{n.o}}_Z \to H^1(\Gamma, \mathfrak{g}_k) \oplus \Big(\bigoplus_{v\in S_p} H^1(\Gamma_v, \mathfrak{b}_k)\Big) \to \bigoplus_{v\in S_p} H^1(\Gamma_v, \mathfrak{g}_k) \\ &\to \mathfrak{t}^1 s\mathcal{D}^{\text{n.o}}_Z \to H^2(\Gamma, \mathfrak{g}_k) \to \bigoplus_{v\in S_p} H^2(\Gamma_v, \mathfrak{b}_k) \\ &\to \mathfrak{t}^2 s\mathcal{D}^{\text{n.o}}_Z \to 0. \end{aligned}$$

By assumption (Reg_v), the map $H^0(\Gamma_v, \mathfrak{b}_k)/\mathfrak{z}_k \to H^0(\Gamma_v, \mathfrak{g}_k)/\mathfrak{z}_k$ is an isomorphism. The conclusion follows from comparing the above exact sequence with (★). □

In particular $\mathfrak{t}^{-1} s\mathcal{D}^{\text{n.o}}_Z = 0$ (note that for this we don't need (Reg_v^*)). By Lurie's criterion (Theorem 4.10) and the finiteness of the cohomology groups, we have the following corollary:

Corollary 4.17 *The functor $s\mathcal{D}^{\text{n.o}}_Z$ is pro-representable.*

Let $R^{s,\text{n.o}}$ be a representing (pro-)simplicial ring. Since $\pi_0 s\mathcal{D}^{\text{n.o}}_Z(A) \cong \mathcal{D}^{\text{n.o}}(A)$ for $A \in \mathbf{Art}_{\mathcal{O}}$, the ring $\pi_0 R^{s,\text{n.o}}$ represents the classical nearly ordinary deformation functor $\mathcal{D}^{\text{n.o}}$.

4.5 Relative Derived Deformations and Relative Tangent Complexes

Let $T \in \mathbf{Art}_{\mathcal{O}}$ and let $\rho_T : \Gamma \to G(T)$ be a nearly ordinary representation. For $v \in S_p$, we write $\rho_{T,v}$ for the restriction of ρ_T on Γ_v and we suppose the image of

$\rho_{T,v}$ lies in $B(T)$ (more precisely, we should say the image of some conjugation of $\rho_{T,v}$ lies in $B(T)$, but there is no crucial difference). Let X and X_v be the pro-simplicial sets associated to the profinite groups Γ and Γ_v. We identify ρ_T as a map of (pro-)simplicial sets $X \to BG(T) \to \mathcal{B}G(T)$ (here $BG(T)$ is the classical classifying space of the finite group $G(T)$ and $\mathcal{B}G(T)$ is a fibrant replacement, see Definition 2.10) and identify $\rho_{T,v}$ as $X_v \to BB(T) \to \mathcal{B}B(T) \to \mathcal{B}G(T)$.

Let's consider the derived deformation functors over ρ_T.

Definition 4.18 (1) Let $s\mathcal{D}_{\rho_T} : \mathcal{O}\backslash s\mathbf{Art}/_T \to s\mathbf{Sets}$ be the functor

$$A \mapsto \mathrm{hofib}_{\rho_T}(\mathsf{sHom}_{s\mathbf{Sets}}(X, \mathcal{B}G(A)) \to \mathsf{sHom}_{s\mathbf{Sets}}(X, \mathcal{B}G(T))).$$

(2) For $v \in S_p$, let $s\mathcal{D}_{\rho_{T,v}} : \mathcal{O}\backslash s\mathbf{Art}/_T \to s\mathbf{Sets}$ be the functor

$$A \mapsto \mathrm{hofib}_{\rho_{T,v}}(\mathsf{sHom}_{s\mathbf{Sets}}(X_v, \mathcal{B}G(A)) \to \mathsf{sHom}_{s\mathbf{Sets}}(X_v, \mathcal{B}G(T))).$$

(3) For $v \in S_p$, let $s\mathcal{D}^{\mathrm{n.o}}_{\rho_{T,v}} : \mathcal{O}\backslash s\mathbf{Art}/_T \to s\mathbf{Sets}$ be the functor

$$A \mapsto \mathrm{hofib}_{\rho_{T,v}}(\mathsf{sHom}_{s\mathbf{Sets}}(X_v, \mathcal{B}B(A)) \to \mathsf{sHom}_{s\mathbf{Sets}}(X_v, \mathcal{B}B(T))).$$

Our goal is to prove the following proposition (see also [13, Example 4.38 and Lemma 5.10]):

Proposition 4.19 Let M be a finite module over an arbitrary Artin ring T. Then for $i, j \geq 0$ we have

$$\pi_i s\mathcal{D}_{\rho_T}(T \oplus M[j]) \cong H^{1+j-i}(\Gamma, \mathfrak{g}_T \otimes_T M).$$

Note that $\mathsf{sHom}_{s\mathbf{Sets}}(X, -)$ is defined by the filtered colimit $\varinjlim_i \mathsf{sHom}_{s\mathbf{Sets}}(B\Gamma_i, -)$, which commutes with homotopy pullbacks. So it suffices to prove the proposition with Γ replaced by Γ_i and X replaced by $B\Gamma_i$. To simplify the notations, we suppose Γ is a finite group during the proof.

Lemma 4.20 *Let $A \in \mathcal{O}\backslash s\mathbf{Art}/_T$. Then $s\mathcal{D}_{\rho_T}(A)$ is weakly equivalent to*

$$\mathrm{holim}_{\mathbf{\Delta}X}\mathrm{hofib}_*(\mathcal{B}G(A) \to \mathcal{B}G(T)).$$

Proof By [15, Proposition 18.9.2], X is weakly equivalent to $\mathrm{hocolim}_{(\mathbf{\Delta}X)^{\mathrm{op}}}*$ (i.e., the homotopy colimit of the single-point simplicial set indexed by $(\mathbf{\Delta}X)^{\mathrm{op}}$). Hence (see [15, Theorem 18.1.10])

$$\mathsf{sHom}_{s\mathbf{Sets}}(X, \mathcal{B}G(A)) \simeq \mathrm{holim}_{\mathbf{\Delta}X}\, \mathsf{sHom}_{s\mathbf{Sets}}(*, \mathcal{B}G(A)) \simeq \mathrm{holim}_{\mathbf{\Delta}X}\mathcal{B}G(A),$$

and

$$\mathsf{sHom}_{s\mathbf{Sets}}(X, \mathcal{B}G(T)) \simeq \mathrm{holim}_{\mathbf{\Delta}X}\, \mathsf{sHom}_{s\mathbf{Sets}}(*, \mathcal{B}G(T)) \simeq \mathrm{holim}_{\mathbf{\Delta}X}\mathcal{B}G(T).$$

Note that ρ_T, as the single-point simplicial subset of $\mathsf{sHom}_{s\mathbf{Sets}}(X, \mathcal{B}G(T))$, is identified with $\mathrm{holim}_{\mathbf{\Delta}X} * \to \mathrm{holim}_{\mathbf{\Delta}X}\mathcal{B}G(T)$. Since homotopy limits commute with homotopy pullbacks, we conclude that

$$s\mathcal{D}_{\rho_T}(A) \simeq \mathrm{holim}_{\mathbf{\Delta}X}\mathrm{hofib}_*(\mathcal{B}G(A) \to \mathcal{B}G(T)).$$

□

Let's first analyze $\mathrm{hofib}_*(\mathcal{B}G(T \oplus M[j]) \to \mathcal{B}G(T))$.

Lemma 4.21 *The homotopy groups of* $\mathrm{hofib}_*(\mathcal{B}G(T \oplus M[j]) \to \mathcal{B}G(T))$ *are trivial except at degree* $j+1$, *where it is* $\mathfrak{g}_T \otimes_T M$.

Proof Note that $A \mapsto \mathrm{hofib}_*(\mathcal{B}G(A) \to \mathcal{B}G(T))$ preserves weak equivalences and homotopy pullbacks.

Since $T \oplus M[j] \to T$ is j-connected, the map $\mathcal{B}G(T \oplus M[j]) \to \mathcal{B}G(T)$ is $(j+1)$-connected (see [13, Corollary 5.3]), and the homotopy groups of the homotopy fiber vanish up to degree j. Since the functor $A \mapsto \mathrm{hofib}_*(\mathcal{B}G(A) \to \mathcal{B}G(T))$ maps the homotopy pullback square

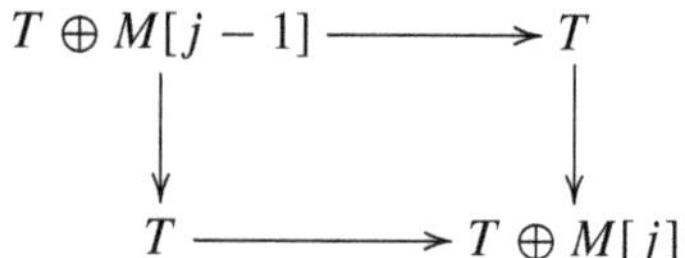

to a homotopy pullback square, we get

$$\pi_{j+k}\mathrm{hofib}_*(\mathcal{B}G(T \oplus M[j]) \to \mathcal{B}G(T)) \cong \pi_{j+k-1}\mathrm{hofib}_*(\mathcal{B}G(T \oplus M[j-1]) \to \mathcal{B}G(T))$$

for any $k \geq 0$. Consequently

$$\pi_{j+k}\mathrm{hofib}_*(\mathcal{B}G(T \oplus M[j]) \to \mathcal{B}G(T)) \cong \pi_k\mathrm{hofib}_*(\mathcal{B}G(T \oplus M[0]) \to \mathcal{B}G(T)),$$

and $\mathrm{hofib}_*(\mathcal{B}G(T \oplus M[j]) \to \mathcal{B}G(T))$ has homotopy groups concentrated on degree $j+1$, where it is $\mathfrak{g}_T \otimes_T M$. □

Let Y be the $\mathbf{\Delta}X$-diagram in $s\mathbf{Sets}$ (i.e., functor $\mathbf{\Delta}X \to s\mathbf{Sets}$) which takes the value $\mathrm{hofib}_*(\mathcal{B}G(A) \to \mathcal{B}G(T))$. Then Y is a local system (see [13, Definition 4.34], it's called the cohomological coefficient system in [12, Page 28]) on X. There is hence a $\pi_1(X, *)$-action on the homotopy group $\mathfrak{g}_T \otimes_T M$. By unwinding the constructions, we see this is the conjugacy action of ρ_T on $\mathfrak{g}_T \otimes_T M$.

It suffices to calculate $\mathrm{holim}\, Y$. Under the Dold-Kan correspondence, we may identify $\mathrm{hofib}_*(\mathcal{B}G(A) \to \mathcal{B}G(T))$ with the chain complex with homology $\mathfrak{g}_T \otimes_T M$ concentrated on degree $j+1$. But in fact it's more convenient to regard $\mathrm{hofib}_*(\mathcal{B}G(A) \to \mathcal{B}G(T))$ as a cochain complex with cohomology $\mathfrak{g}_T \otimes_T M$ concentrated on degree $-(j+1)$, because the homotopy limit of cochain complexes is drastically simple (see [8, Sect. 19.8]). By shifting degrees, it suffices to suppose that the cohomology is concentrated on degree 0.

Lemma 4.22 *Let N be a $T[\Gamma]$-module, and we regard N as a cochain complex concentrated on degree* 0. *Let Y be the $\mathbf{\Delta}X$-diagram in $\mathbf{Ch}^{\geq 0}(T)$ (i.e., functor $\mathbf{\Delta}X \to \mathbf{Ch}^{\geq 0}(T)$) which takes the value N. Then* $\operatorname{holim} Y \simeq C^\bullet(\Gamma, N)$. *Here $C^\bullet(\Gamma, N)$ is the cochain which computes the usual group cohomology.*

Proof By [15, Lemma 18.9.1], $\operatorname{holim} Y$ is naturally isomorphic to the homotopy limit of the cosimplicial object Z in $\mathbf{Ch}^{\geq 0}(T)$ whose codegree $[n]$ term is $\prod_{\sigma \in X_n} Y_\sigma = \prod_{\sigma \in X_n} N$. We have to explain the coface maps of Z. For this purpose, we describe $Z = (Z^n)_n$ as follows:

The $T[\Gamma]$-module N defines a functor D from the one-object groupoid $\bullet$ with $\mathsf{End}(\bullet) = \Gamma$ to $\mathbf{Ch}^{\geq 0}(T)$, such that $D(\bullet) = N$, and $D(\Gamma)$ acts on N by the Γ-action. Then Z^n is $\prod\limits_{i_0 \to \cdots \to i_n} D(i_n)$ (all i_k's are equal to the object $\bullet$ here, but keeping the difference helps to clarify the process). Let d_k be the k-th face map from Γ^{n+1} to Γ^n, in other words, d_k maps $(i_0 \to \cdots \to i_{n+1})$ to $(j_0 \to \cdots \to j_n)$ by "covering up" i_k. Then the corresponding $D(j_n) \to D(i_{n+1})$ is the identity map if $k \neq n+1$ and is $D(i_n \to i_{n+1})$ if $k = n+1$.

By [8, Proposition 19.10], $\operatorname{holim} Z$ is quasi-isomorphic to the total complex of the alternating double complex defined by Z. Since each Z^n is concentrated on degree 0, the total complex is simply

$$\cdots \to \prod_{\Gamma^n} N \to \prod_{\Gamma^{n+1}} N \to \ldots$$

and the alternating sum $\prod\limits_{\Gamma^n} N \to \prod\limits_{\Gamma^{n+1}} N$ is exactly the one which occurs in computing group cohomology. We conclude that $\operatorname{holim} Y \simeq \operatorname{holim} Z \simeq C^\bullet(\Gamma, N)$. □

Now we can prove Proposition 4.19:

Proof From the above discussions, $s\mathcal{D}_{\rho_T}(T \oplus M[j])$ corresponds to $\tau^{\leq 0} C^{\bullet + j + 1}(\Gamma, \mathfrak{g}_T \otimes_T M)$ under the Dold-Kan correspondence (with $\mathbf{Ch}_{\geq 0}(T)$ replaced by $\mathbf{Ch}^{\leq 0}(T)$). Hence $\pi_i s\mathcal{D}_{\rho_T}(T \oplus M[j]) \cong H^{1+j-i}(\Gamma, \mathfrak{g}_T \otimes_T M)$. □

We can define the modifying-center version $s\mathcal{D}_{\rho_T, Z}$ as in Sect. 2.4.1. Note the fibration sequence (see [13, (5.7)]) $\operatorname{hofib}(\mathsf{sHom}_{s\mathbf{Sets}}(X, \mathcal{B}Z(A)) \to \mathsf{sHom}_{s\mathbf{Sets}}(X, BZ(T))) \to s\mathcal{D}_{\rho_T}(A) \to s\mathcal{D}_{\rho_T, Z}(A)$. From this, we deduce that $\pi_i s\mathcal{D}_{\rho_T, Z}(T \oplus M[j]) \cong \pi_i s\mathcal{D}_{\rho_T}(T \oplus M[j])$ when $i \neq j+1$, and $\pi_{i+1} s\mathcal{D}_{\rho_T, Z}(T \oplus M[i]) = 0$.

For each $v \in S_p$, there is also a modifying-center version $s\mathcal{D}_{\rho_{T,v}, Z}$, resp., $s\mathcal{D}^{\mathrm{n.o}}_{\rho_{T,v}, Z}$ of $s\mathcal{D}_{\rho_{T,v}}$, resp., $s\mathcal{D}^{\mathrm{n.o}}_{\rho_{T,v}}$. Similar to the global situation, we have

$$\pi_i s\mathcal{D}_{\rho_{T,v}, Z}(T \oplus M[j]) \cong \begin{cases} H^{1+j-i}(\Gamma_v, \mathfrak{g}_T \otimes_T M) \text{ when } i \neq j+1; \\ H^0(\Gamma_v, \mathfrak{g}_T \otimes_T M)/(\mathfrak{z}_T \otimes_T M) \text{ when } i = j+1. \end{cases}$$

And

$$\pi_i s\mathcal{D}^{\mathrm{n.o}}_{\rho_{T,v}, Z}(T \oplus M[j]) \cong \begin{cases} H^{1+j-i}(\Gamma_v, \mathfrak{b}_T \otimes_T M) \text{ when } i \neq j+1; \\ H^0(\Gamma_v, \mathfrak{b}_T \otimes_T M)/(\mathfrak{z}_T \otimes_T M) \text{ when } i = j+1. \end{cases}$$

The global nearly ordinary derived deformation functor over ρ_T is defined as follows:

$$s\mathcal{D}^{\text{n.o}}_{\rho_T,Z} = s\mathcal{D}_{\rho_T,Z} \times^h_{\prod_{v\in S_p} s\mathcal{D}_{\rho_{T,v},Z}} \prod_{v\in S_p} s\mathcal{D}^{\text{n.o}}_{\rho_{T,v},Z}.$$

Then $\pi_i s\mathcal{D}^{?}_{\rho_T,Z}(T\oplus M[j])$ ($? = \text{n.o}$ or $\emptyset$) depends only on $j-i$. We denote $\mathfrak{t}^{j-i}_{T,M} s\mathcal{D}^{?}_{\rho_T,Z} = \pi_i s\mathcal{D}^{?}_{\rho_T,Z}(T\oplus M[j])$.

Proposition 4.23 Suppose (Reg_v) and (Reg_v^*). Let $j\geq i\geq 0$ and let M be a finitely generated (classical) T-module. Then $\pi_i s\mathcal{D}^{\text{n.o}}_{\rho_T,Z}(T\oplus M[j]) \cong H^{1+j-i}_{\text{n.o}}(\Gamma,\mathfrak{g}_T\otimes_T M)$.

Proof By preceding discussions, we have the exact sequence

$$\begin{aligned}
0 &\to \mathfrak{t}^{-1}_{T,M} s\mathcal{D}^{\text{n.o}}_{\rho_T,Z} \to \bigoplus_{v\in S_p} H^0(\Gamma_v,\mathfrak{b}_T\otimes_T M)/(\mathfrak{z}_T\otimes_T M) \to \bigoplus_{v\in S_p} H^0(\Gamma_v,\mathfrak{g}_T\otimes_T M)/(\mathfrak{z}_T\otimes_T M)\\
&\to \mathfrak{t}^{0}_{T,M} s\mathcal{D}^{\text{n.o}}_{\rho_T,Z} \to H^1(\Gamma,\mathfrak{g}_T\otimes_T M)\oplus(\bigoplus_{v\in S_p} H^1(\Gamma_v,\mathfrak{b}_T\otimes_T M)) \to \bigoplus_{v\in S_p} H^1(\Gamma_v,\mathfrak{g}_T\otimes_T M)\\
&\to \mathfrak{t}^{1}_{T,M} s\mathcal{D}^{\text{n.o}}_{\rho_T,Z} \to H^2(\Gamma,\mathfrak{g}_T\otimes_T M) \to \bigoplus_{v\in S_p} H^2(\Gamma_v,\mathfrak{b}_T\otimes_T M)\\
&\to \mathfrak{t}^{2}_{T,M} s\mathcal{D}^{\text{n.o}}_{\rho_T,Z} \to 0.
\end{aligned}$$

Note that we have used $H^2(\Gamma_v,\mathfrak{b}_T\otimes_T M)=0$ for $v\in S_p$. To see this, it suffices to show $H^2(\Gamma_v,\mathfrak{b}_k)=0$ by Artinian induction. By local Tate duality, it suffices to prove $H^0(\Gamma_v,\mathfrak{b}_k^*)=0$. But we have a Galois-equivariant isomorphism $\mathfrak{b}_k^*\cong\mathfrak{g}_k/\mathfrak{n}_k(1)$, so the result follows from the assumption (Reg_v^*).

Under the condition (Reg_v), the map $H^0(\Gamma_v,\mathfrak{b}_T\otimes_T M)\to H^0(\Gamma_v,\mathfrak{g}_T\otimes_T M)$ is an isomorphism. Let $L_{v,T,M}=\mathrm{im}(H^1(\Gamma_v,\mathfrak{b}_T\otimes_T M)\to H^1(\Gamma_v,\mathfrak{g}_T\otimes_T M))$, then we have the following exact sequence similar to (★):

$$\begin{aligned}
0 &\to H^0_{\text{n.o}}(\Gamma,\mathfrak{g}_T\otimes_T M)\to H^0(\Gamma,\mathfrak{g}_T\otimes_T M)\to 0\\
&\to H^1_{\text{n.o}}(\Gamma,\mathfrak{g}_T\otimes_T M)\to H^1(\Gamma,\mathfrak{g}_T\otimes_T M)\to \bigoplus_{v\in S_p} H^1(\Gamma_v,\mathfrak{g}_T\otimes_T M)/L_{v,T,M}\\
&\to H^2_{\text{n.o}}(\Gamma,\mathfrak{g}_T\otimes_T M)\to H^2(\Gamma,\mathfrak{g}_T\otimes_T M)\to \bigoplus_{v\in S_p} H^1(\Gamma_v,\mathfrak{g}_T\otimes_T M)\\
&\to H^3_{\text{n.o}}(\Gamma,\mathfrak{g}_T\otimes_T M)\to 0.
\end{aligned}$$

By comparing the two exact sequences above, we get $\mathfrak{t}^{i}_{T,M} s\mathcal{D}^{\text{n.o}}_{\rho_T,Z}\cong H^{i+1}_{\text{n.o}}(\Gamma,\mathfrak{g}_T\otimes_T M)$. □

Recall that we have a pro-simplicial ring $R^{s,\text{n.o}}$ which represents $s\mathcal{D}^{\text{n.o}}_Z$. Then ρ_T defines a map

$$R^{s,\text{n.o}}\to\pi_0 R^{s,\text{n.o}}\to T.$$

With this specified map, we regard $R^{s,\mathrm{n.o}} \in \text{pro}-{}_{\mathcal{O}}\backslash s\mathbf{Art}/_T$, and it's easy to see that $R^{s,\mathrm{n.o}}$ represents $s\mathcal{D}^{\mathrm{n.o}}_{\rho_T,Z}$. Write $R^{s,\mathrm{n.o}} = (R_k)$ for a projective system (R_k) in ${}_{\mathcal{O}}\backslash s\mathbf{Art}/_T$. Then

$$\begin{aligned}\pi_i s\mathcal{D}^{\mathrm{n.o}}_Z(T \oplus M[j]) &\cong \pi_i \varinjlim_k \mathsf{sHom}_{\mathcal{O}\backslash s\mathbf{Art}/_T}(R_k, T \oplus M[j]) \\ &\cong \pi_i \varinjlim_k \mathrm{DK}(\tau_{\geq 0}[L_{R_k} \otimes_{R_k} T, M[j]]) \\ &\cong H_i \varinjlim_k [L_{R_k} \otimes_{R_k} T, M[j]].\end{aligned}$$

Let's define $[L_{R/\mathcal{O}} \otimes_R T, M] = \varinjlim_k [L_{R_k/\mathcal{O}} \otimes_{R_k} T, M]$ for $R = (R_k) \in \text{pro}-{}_{\mathcal{O}}\backslash s\mathbf{Art}/_T$. Then $[L_{R^{s,\mathrm{n.o}}/\mathcal{O}} \otimes_{R^{s,\mathrm{n.o}}} T, M]$, when regarded as a cochain complex, has the same cohomology groups as the complex $\tau^{\geq 0} C^{\bullet+1}_{\mathrm{n.o}}(\Gamma, \mathfrak{g}_T \otimes_T M)$. We thus obtain the following corollary:

Corollary 4.24 *For every finite T-module M, there is a quasi-isomorphism*

$$[L_{R^{s,\mathrm{n.o}}/\mathcal{O}} \otimes_{R^{s,\mathrm{n.o}}} T, M] \simeq \tau^{\geq 0} C^{\bullet+1}_{\mathrm{n.o}}(\Gamma, \mathfrak{g}_T \otimes_T M).$$

Remark 4.25 Recall that there is a natural transformation $\mathcal{D}^{\mathrm{n.o}} \to \prod_{v \in S_p} \mathrm{Def}_{\overline{\chi}_v}$ (see Remark 2.2). We can construct the derived analogue $s\mathcal{D}^{\mathrm{n.o}}_{\rho_T,Z} \to \prod_{v \in S_p} s\,\mathrm{Def}_{\overline{\chi}_v,T}$, so it's natural to ask if $R^{s,\mathrm{n.o}}$ is a Λ-simplicial ring. Indeed, this is the case when $\mathcal{O}^{*,p-\mathrm{ab}}_{F_v}$ has no torsion for every $v \in S_p$. In general, let $\mathcal{O}^{*,p-\mathrm{ab}}_{F_v} = \Delta_v \times W_v$ be a decomposition of $\mathcal{O}^{*,p-\mathrm{ab}}_{F_v}$ into a finite group Δ_v and a pro-p group W_v, and let $\varphi_v \colon \Delta_v \to \mathcal{O}$ be a fixed character, we can modify $s\mathcal{D}^{\mathrm{n.o}}_{\rho_T,Z}$ by taking into account $\varphi = (\varphi_v)_{v \in S_p}$, the resulting $s\mathcal{D}^{\mathrm{n.o},\varphi}_{\rho_T,Z}$ is then pro-represented by a simplicial Artinian Λ-algebra $R^{s,\mathrm{n.o},\varphi}$, and we can relate $[L_{R^{s,\mathrm{n.o},\varphi}/\Lambda} \otimes_{R^{s,\mathrm{n.o},\varphi}} T, M]$ to the ordinary cochain complex as Corollary 4.24. See [2] for details.

Comments: Let $\rho_T : \Gamma \to G(T)$ be an ordinary representation of weight μ, which satisfies (Reg_v) for all $v \in S_p$. This means that the cocharacter given by $\rho_T|_{I_v} : I_v \to B(T)/N(T)$ is given (via Artin reciprocity) by $\mu \circ rec_v^{-1} : I_v \to \mathcal{O}_v^\times \to \Theta(T)$ (here $\Theta = B/N$ is the standard maximal split torus of B). In this whole section, if ρ_T is ordinary of weight μ, we could consider instead of the functor $s\mathcal{D}^{\mathrm{n.o}}_{\rho_T}$ the subfunctor $s\mathcal{D}^{\mathrm{n.o},\mu}_{\rho_T}$ of ordinary deformations of fixed weight μ. This means we impose as local condition at $v \in S_p$ that

$$s\mathcal{D}^{\mathrm{n.o},\mu}_{\rho_{T,v}}(A) = \mathrm{hofib}_{\mu \circ rec_v^{-1}}\left(s\mathcal{D}^{\mathrm{n.o}}_{\rho_{T,v}}(A) \to \mathsf{sHom}(BI_v, B\Theta(A))\right).$$

Then, $s\mathcal{D}^{\mathrm{n.o},\mu}_{\rho_T}$ is pro-representable by a simplicial pro-Artinian ring $R^{s,\mathrm{n.o}}_\mu$ and we have an analogue of Proposition 4.23:

Proposition 4.26 Suppose (Reg_v) and (Reg_v^*). Let $j \geq i \geq 0$ and let M be a finitely generated (classical) T-module. Then $\pi_i s\mathcal{D}^{\mathrm{n.o},\mu}_{\rho_T,Z}(T \oplus M[j]) \cong H^{1+j-i}_{\mathrm{n.o},str}(\Gamma, \mathfrak{g}_T \otimes_T M)$.

Here $H^{1+j-i}_{\mathrm{n.o},str}(\Gamma, \mathfrak{g}_T \otimes_T M)$ is the cohomology of the subcomplex $C^\bullet_{\mathrm{n.o},\mu}(\Gamma, \mathfrak{g}_T \otimes_T M)$ defined as in Sect. 4.4.1, replacing $(L_v, \widetilde{L}_v)$ by $(L'_v, \widetilde{L}'_v)$ where L'_v is the image in $H^1(\Gamma_v, \mathfrak{g}_T \otimes M)$ of the kernel of $H^1(\Gamma_v, \mathfrak{b}_T \otimes M) \to H^1(I_v, (\mathfrak{b}_T/\mathfrak{n}_T) \otimes M)$, and $\widetilde{L}'_v$ is the inverse image of L'_v in $Z^1(\Gamma_v, \mathfrak{g}_T \otimes M)$.

The proof is identical to Proposition 4.23. As a corollary, we get

Corollary 4.27 *For every finite T-module M, there is a quasi-isomorphism*

$$[L_{R^{s,\mathrm{n.o}}_\mu/\mathcal{O}} \otimes_{R^{s,\mathrm{n.o}}_\mu} T, M] \simeq \tau^{\geq 0} C^{\bullet+1}_{\mathrm{n.o},\mu}(\Gamma, \mathfrak{g}_T \otimes_T M).$$

In the next section, we shall use these objects with a fixed weight μ.

5 Application to the Galatius-Venkatesh Homomorphism

Let $\Gamma = \mathrm{Gal}(F_S/F)$ for $S = S_p \cup S_\infty$. Let $\bar{\rho}\colon \Gamma \to G(k)$ be an ordinary representation of weight μ, which satisfies (Reg_v) for all $v \in S_p$. Let T be a finite local $\mathcal{O}$-algebra and $\rho_T : \Gamma \to G(T)$ be an ordinary lifting of weight μ of $\bar{\rho}$. Let M be a T-module which is of $\mathcal{O}$-cofinite type, that is, whose Pontryagin dual $\mathrm{Hom}_{\mathcal{O}}(M, K/\mathcal{O})$ is finitely generated over $\mathcal{O}$. We use the notations of Definition 4.12. Recall that if $\bar{\rho}\colon \Gamma \to G(k)$ is ordinary automorphic, it is proven under certain assumptions (see [6, Th.5.11] and [26, Lemma 11]) that $H^1_{\mathrm{n.o}}(\Gamma, \mathfrak{g}_T \otimes_T M)$ is finite and $H^1_{\mathrm{n.o}^\perp}(\Gamma, \mathfrak{g}^*_T \otimes_T M)^\vee$ is of $\mathcal{O}$-cofinite type. Let $T_n = T/(\varpi^n)$; it is a finite algebra over $\mathcal{O}_n = \mathcal{O}/(\varpi^n)$. Let $\mathcal{R} = R^{s,\mathrm{n.o}}_\mu$, which pro-represents simplicial ordinary deformations of weight μ. We consider the simplicial ring homomorphism

$$\phi_n : \mathcal{R} \to T_n$$

given by the universal property for the deformation $\rho_n = \rho_{\mathrm{m}} \pmod{(\varpi^n)}$. Let $T_n = T/(\varpi^n)$; it is a finite algebra over $\mathcal{O}_n = \mathcal{O}/(\varpi^n)$. We consider the simplicial ring homomorphism

$$\phi_n : \mathcal{R} \to T_n$$

given by the universal property for the deformation $\rho_n = \rho_{\mathrm{m}} \pmod{(\varpi^n)}$. Let M_n be a finite T_n-module. Consider the simplicial ring $\Theta_n = T_n \oplus M_n[1]$ concentrated in degrees 0 and 1 up to homotopy. It is endowed with a simplicial ring homomorphism $\mathrm{pr}_n : \Theta_n \to T_n$ given by the first projection. Let $L_n(\mathcal{R})$ be the set of homotopy equivalence classes of simplicial ring homomorphisms $\Phi\colon \mathcal{R} \to \Theta_n$ such that $\mathrm{pr}_n \circ \Phi = \phi_n$. By Proposition 4.26, there is a canonical bijection

$$L_n(\mathcal{R}) \cong H^2_{\mathrm{n.o},str}(\Gamma, \mathfrak{g}_{T_n} \otimes_{T_n} M_n).$$

Moreover, as noticed in [13, Lemma 15.1], there is a natural map

$$\pi(n, \mathcal{R})\colon L_n(\mathcal{R}) \to \mathsf{Hom}_T(\pi_1(\mathcal{R}), M_n)$$

defined as follows. Let $[\Phi]$ be the homotopy class of $\Phi \in \mathsf{Hom}_T(\mathcal{R}, \Theta_n)$; then $\pi(\mathcal{R})(\Phi)$ is the homomorphism which sends the homotopy class $[\gamma]$ of a loop γ to $\Phi \circ \gamma \in \mathsf{Hom}_{s\mathbf{Sets}}(\Delta[1], M_n[1]) = M_n$. Recall a loop γ is a morphism of $s\mathbf{Sets}$

$$\gamma\colon \Delta[1] \to \Theta_n$$

from the simplicial interval $\Delta[1]$ to the simplicial set Θ_n which sends the boundary $\partial\Delta[1]$ to 0. For $G = \mathsf{GL}_N$ and F a CM field (assuming Calegari-Geraghty assumptions), it is proven in [26] that

Proposition 5.1 For any $n \geq 1$, the map $\pi(n, \mathcal{R})$ is surjective.

Then, we choose $M_n = \mathsf{Hom}(T, \varpi^{-n}\mathcal{O}/\mathcal{O})$; we take the Pontryagin dual $\pi(n, \mathcal{R})^\vee$ and apply Poitou-Tate duality

$$H^2_{\mathrm{n.o},str}(\Gamma, \mathfrak{g}_{T_n} \otimes_{T_n} M_n) \cong H^1_{\mathrm{n.o},str}(\Gamma, (\mathfrak{g}_{T_n} \otimes_{T_n} M_n)^*).$$

We obtain a T-linear homomorphism called the mod. ϖ^n Galatius-Venkatesh homomorphism:

$$GV_n\colon \mathsf{Hom}_T(\pi_1(\mathcal{R}), M_n)^\vee \hookrightarrow H^1_{\mathrm{n.o},str}(\Gamma, (\mathfrak{g}_{T_n} \otimes_{T_n} M_n)^*).$$

The left-hand side is $\pi_1(\mathcal{R}) \otimes \varpi^{-n}/\mathcal{O}$ and the right-hand side is $\mathrm{Sel}_{\mathrm{n.o},str}(\mathrm{Ad}(\rho_n)(1))$. Taking inductive limit on both sides, we obtain

Proposition 5.2 There is a canonical T-linear injection

$$GV_T\colon \pi_1(R^{s,\mathrm{n.o}}) \otimes_{\mathcal{O}} K/\mathcal{O} \hookrightarrow \mathrm{Sel}(\mathrm{Ad}(\rho_T)^\vee(1)).$$

For $G = \mathsf{GL}_N$, F CM, and under Calegary-Geraghty assumptions, and for T the non-Eisenstein localization of the Hecke algebra acting faithfully on the Betti cohomology, it follows from [6, Theorem 5.11] that the left-hand side is ϖ-divisible of corank $rk(T)$ and it is proven in [26, Lemma 11] that the right-hand side has corank $rk(T)$. For any $\mathcal{O}$-finitely generated ordinary Γ-module M such that the Selmer group $H^1_{\mathrm{n.o},str}(\Gamma, M \otimes \mathbb{Q}/\mathbb{Z})$ is $\mathcal{O}$-cofinitely generated, we define its Tate-Shafarevich module as

$$\text{Ш}(M) = H^1_{\mathrm{n.o},str}(\Gamma, M \otimes \mathbb{Q}/\mathbb{Z})/H^1_{\mathrm{n.o},str}(\Gamma, M \otimes \mathbb{Q}/\mathbb{Z})_{\varpi-div}.$$

It is the torsion quotient of $\mathrm{H}^1_{\mathrm{n.o},str}(\Gamma, M \otimes \mathbb{Q}/\mathbb{Z})$. For any $\mathcal{O}$-algebra homomorphism $\lambda\colon T \to \mathcal{O}$, let $\rho_\lambda = \rho_T \otimes_\lambda \mathcal{O}$. For $M = \mathrm{Ad}(\rho_\lambda)^\vee(1))$, one shows in [26, Lemma 11], using Poitou-Tate duality, that $\text{Ш}(\mathrm{Ad}(\rho_\lambda)^\vee(1)))$ is Pontryagin dual to $\mathrm{Sel}_{\mathrm{n.o},str}(\mathrm{Ad}(\rho_\lambda))$.

It follows from [26, Lemma 11] that the cokernel of GV_λ can be identified to the Tate-Shafarevich group $\text{III}(\text{Ad}(\rho_\lambda)^\vee(1))$ in the sense of Bloch-Kato. So that

$$\text{Coker } GV_\lambda \cong \text{Sel}_{\text{n.o},str}(\text{Ad}(\rho_\lambda))^\vee.$$

References

1. P. Allen, F. Calegari, A. Caraiani, T. Gee, D. Helm, B. Le Hung, J. Newton, P. Scholze, R. Taylor, and J. Thorne, Potential automorphy over CM fields. Arxiv:1812.09999v1 [math.NT]
2. Y. Cai, Derived minimal deformation rings for small groups, preprint
3. L. Clozel, M. Harris, R. Taylor, *Automorphy for some* ℓ-adic *lifts of automorphic modulo* ℓ *Galois representations* (Publ. Math. Inst, Hautes Études Sci, 2008)
4. G. Boeckle, M. Harris, C. Khare, J. Thorne, $\widehat{G}$-local systems on smooth projective curves are potentially automorphic. Preprint at arXiv:1609.03491 (2016) to appear in Acta Mathematica
5. M. Bate, B. Martin, G. Rohrle, A geometric approach to complete reducibility. Invent. Math. **161**(1), 177–218 (2005)
6. F. Calegari, D. Geraghty, Modularity lifting beyond the Taylor-Wiles method. Invent. Math. **211**, 297–433 (2018)
7. A.M. Cegarra, J. Remedios, The relationship between the diagonal and the bar constructions on a bisimplicial set. Topology and its Applications **153**(1), 21–51 (2005)
8. D. Dugger, A primer on homotopy colimits. http://math.uoregon.edu/~ddugger/hocolim.pdf (2008)
9. D. Geraghty, Modularity lifting theorems for ordinary Galois representations. Math. Ann. **373**(3–4), 1341–1427 (2019)
10. W. D. Gillam, *Simplicial methods in algebra and algebraic geometry*, Course Notes, Brown Univ. 2013, http://www.math.boun.edu.tr/instructors/wdgillam/simplicialalgebra.pdf
11. P. Goerss, R. Jardine, *Simplicial homotopy theory* (Springer Science & Business Media, New York, 2009)
12. S.I. Gelfand, Y.I. Manin, *Methods ofhomological algebra* (Springer Science & Business Media, New York, 2013)
13. S. Galatius, A. Venkatesh, Derived Galois deformation rings. Preprint at ArXiv:1608.07236v2 [math NT] 2018
14. M. Harris, K.-W. Lan, R. Taylor, and J. Thorne, On the rigid cohomology of certain Shimura varieties. Res. Math. Sci. 3, 308 (2016)
15. P. S. Hirschhorn, *Model Categories and Their Localizations*, Mathematical surveys and monographs, Volume 99, 2003
16. V. Lafforgue, Chtoucas pour les groupes réductifs et paramétrisation de Langlands globale. J. Am. Math. Soc. **31**(3), 719–891 (2018)
17. J. Lurie, *Derived Algebraic Geometry* (ProQuest LLC, Ann Arbor, MI, 2004)
18. J. Lurie, *Derived algebraic geometry I: stable* ∞ *-categories*. arXiv preprint math/0608228 (2009)
19. H. Matsumura, *Commutative algebra* (Benjamin/Cummings Publishing Company, Redwood City, CA, 1980)
20. J.S. Milne, *Arithmetic Duality Theorems* (BookSurge, LLC, Charleston, SC, 2006)
21. M. Schlessinger, Functors of Artin Rings. Trans. Am. Math. Soc. **130**(2), 208–222 (1968)
22. P. Scholze, On torsion in the cohomology of locally symmetric varieties. Ann. Math. **182**(3), 945–1066 (2015)
23. J.-P. Serre, Complète réductibilité, *Séminaire Bourbaki Vol. 2003-2004*, Exp. No. 932, viii,, Astérisque 299, 195-217, 2005

24. N. P. Strickland, $K(N)$-local duality for finite groups and groupoids. arXiv preprint math/0011109 (2000)
25. J. Tilouine, *Deformations of Galois representations and Hecke algebras* (Narosa Publ. House, Delhi, 1996)
26. J. Tilouine, E. Urban, On the cohomology of $\mathsf{GL}(N)$ and adjoint Selmer groups, preprint
27. C. Weibel, *An Introduction to Homological Algebra*, Cambridge Studies in Adv. Math. 38 (Cambridge University Press, Cambridge, 1994)
28. M. Weidner, Pseudocharacters of classical groups, arXiv preprint arXiv:1809.03644 (2018)
29. A. Wiles, Modular Elliptic Curves and Fermat's Last Theorem. Ann. Math. **142**, 443–551 (1995)

GPSR Compliance
The European Union's (EU) General Product Safety Regulation (GPSR) is a set of rules that requires consumer products to be safe and our obligations to ensure this.

If you have any concerns about our products, you can contact us on

ProductSafety@springernature.com

In case Publisher is established outside the EU, the EU authorized representative is:

Springer Nature Customer Service Center GmbH
Europaplatz 3
69115 Heidelberg, Germany

www.ingramcontent.com/pod-product-compliance
Ingram Content Group UK Ltd.
Pitfield, Milton Keynes, MK11 3LW, UK
UKHW021833270726
14058UKWH00001B/118
9789811671234